T0094024

# Extremal Finite Set Theory

## Discrete Mathematics and Its Applications

Series Editors
**Miklós Bóna**
**Donald L. Kreher**
**Douglas West**
**Patrice Ossona de Mendez**

**Representation Theory of Symmetric Groups**
*Pierre-Loïc Méliot*

**Advanced Number Theory with Applications**
*Richard A. Mollin*

**A Multidisciplinary Introduction to Information Security**
*Stig F. Mjølsnes*

**Combinatorics of Compositions and Words**
*Silvia Heubach and Toufik Mansour*

**Handbook of Linear Algebra, Second Edition**
*Leslie Hogben*

**Combinatorics, Second Edition**
*Nicholas A. Loehr*

**Handbook of Discrete and Computational Geometry, Third Edition**
*C. D. Tóth, Jacob E. Goodman and Joseph O'Rourke*

**Handbook of Discrete and Combinatorial Mathematics, Second Edition**
*Kenneth H. Rosen*

**Crossing Numbers of Graphs**
*Marcus Schaefer*

**Graph Searching Games and Probabilistic Methods**
*Anthony Bonato and Paweł Prałat*

**Handbook of Geometric Constraint Systems Principles**
*Meera Sitharam, Audrey St. John, and Jessica Sidman,*

**Additive Combinatorics**
*Béla Bajnok*

**Algorithmics of Nonuniformity:**
Tools and Paradigms
*Micha Hofri and Hosam Mahmoud*

**Extremal Finite Set Theory**
*Dániel Gerbner and Balázs Patkós*

https://www.crcpress.com/Discrete-Mathematics-and-Its-Applications/book-series/CHDISMTHAPP?page=1&order=dtitle&size=12&view=list&status=published,forthcoming

# Extremal Finite Set Theory

Dániel Gerbner
Balázs Patkós

CRC Press is an imprint of the
Taylor & Francis Group, an informa business

CRC Press
Taylor & Francis Group
6000 Broken Sound Parkway NW, Suite 300
Boca Raton, FL 33487-2742



Printed on acid-free paper
Version Date: 20180821

International Standard Book Number-13: 978-1-1381-9784-8 (Hardback)



**Library of Congress Cataloging-in-Publication Data**

Catalog record is available from the Library of Congress

**Visit the Taylor & Francis Web site at**
**http://www.taylorandfrancis.com**

**and the CRC Press Web site at**
**http://www.crcpress.com**

# Contents

# Acknowledgement

We thank Gyula O.H. Katona, our former PhD supervisor, for introducing us to the theory of extremal set families.

During the writing of this book, both authors were supported by the János Bolyai Research Fellowship of the Hungarian Academy of Sciences and the National Research, Development and Innovation Office – NKFIH under the grant K116769.

# Preface

The preface of an academic book should contain information about its topic and its intended audience. Starting with the former, let us say that extremal combinatorics is a huge topic that can be described more easily than that of our book: it is concerned with finding the largest, smallest, or otherwise optimal combinatorial structures with a given property. Here we deal with structures that consist of subsets of a finite set. Most typically, we focus on maximizing the cardinality of a family of sets satisfying some properties.

It is equally important to describe what is *not* in this book. We only mention (or use) results concerning 2-uniform families, i.e. ordinary graphs (luckily, many good books and survey papers are available on extremal graph theory). Many of the results we state in the book have analogous versions in more structured creatures than sets (e.g. permutations, multisets, vector spaces, block designs, etc) or infinite versions - we hardly even mention them. We do not deal with colorings (this book contains hardly any Ramsey-type or anti-Ramsey-type results). We usually focus on the cardinality of a family of sets, and often ignore results about something else, like the minimum degree (co-degree). One particular topic that would really fit in this book, but is completely avoided due to space constraints, is coding theory.

When writing the book we had the following two goals in our minds:

- Presenting as many proof techniques as possible. Chapter 1 is completely devoted to this aim via presenting several proofs of classical results of extremal set theory (apart from the very recent method of Croot, Lev, and Pach) but later chapters also contain brief or more detailed introductions to more novel methods like flag algebras, the container method, etc.

- Giving a detailed and almost complete list of recent developments, results (mostly without proof, but mentioning the proof techniques applied) in the areas covered.

As a consequence, the intended audience consists of

- Graduate students who are eager to to be introduced to the theory of extremal set systems. Note however, that the exercises are not those of a proper textbook. Some of them are rather easy, they check whether the reader understood the definitions, some of them state lemmas needed for theorems with longer proofs, some others are "easier" results from research papers.

- University professors giving courses in (extremal) combinatorics. The above warning about the exercises applies here as well, but we hope that our book could help to design the syllabus of graduate courses in different areas of extremal set theory.

- Researchers interested in recent developments in the field. There has been progress in many of the subtopics we cover since the publication of other books dealing with set systems or extremal combinatorics. In some cases even the latest survey articles are not very new. We are convinced that senior academic people can profit in their research by reading or leafing through some or all of the chapters.

Another way to describe the topic of this book is that we wanted to gather results and proofs that could have been presented at the Extremal Sets Systems Seminar of the Alfréd Rényi Institute of Mathematics. (Some of the theorems in the book were indeed presented there.) This might not be very informative for some of our readers, but we wanted to mention it as both authors gave their first talk at this seminar about fifteen years ago.

We also have Chapter 9 that glances at a different topic. We chose the topic and the results mentioned to illustrate how extremal finite set theory results can be applied in other areas of mathematics.

The topic of Chapter 9 runs under several names in the literature. It is often referred to as combinatorial group testing or pool designs and many others, as most applications of this area are in biology with its own terminology. The reason for which we decided to go with the name combinatorial search theory is again personal: the Combinatorial Search Seminar (formerly known as Combinatorial Search and Communicational Complexity Seminar) is the *other* seminar at the Rényi Institute that the two authors attend (apart from the Combinatorics Seminar, probably the longest running of its sort worldwide, started by Vera Sós and András Hajnal).

**A very important final note**: We cannot deny that we put more emphasis on topics we know and like. The number of our own results in this book is in no way proportional to their importance.

# *Notation and Definitions*

Here below we gather the basic notions (mostly with definitions, some without) that we will use in the book. We use standard notation.

**Sets.** We denote by $[n]$ the set $\{1,2,\dots,n\}$ of the first $n$ positive integers and for integers $i<j$ let $[i,j]=\{i,i+1,\dots,j\}$. For a set $S$ we use the notation $2^S$ for its power set $\{T : T \subseteq S\}$ and $\binom{S}{k}$ for the family of its $k$-element subsets ($k$-subsets for short) $\{T \subseteq S : |T| = k\}$. The latter will be referred to as the $k$th *(full) level of* $2^S$. We will use the notation $\binom{X}{\le k} := \cup_{i=0}^{k}\binom{X}{i}$ and $\binom{X}{\ge k} := \cup_{i=k}^{|X|}\binom{X}{i}$. The *symmetric difference* $(F \setminus G) \cup (G \setminus F)$ of two sets $F$ and $G$ will be denoted by $F \triangle G$. If a set $F$ is a subset of $[n]$, then its complement $[n] \setminus F$ is denoted by $\overline{F}$. To denote the fact that $A$ is a subset of $B$ we use $A \subseteq B$ and $A \subsetneq B$ with the latter meaning $A$ is a proper subset of $B$. The Cartesian product of $r$ sets $A_1,\dots,A_r$ is $A_1 \times A_2 \times \cdots \times A_r = \{(a_1,a_2,\dots,a_r) : a_i \in A_i \text{ for every } 1 \le i \le r\}$.

**Families of sets.** We will mostly use the terminology *families* for collections of sets, although *set systems* is also widely used in the literature. If all sets in a family $\mathcal{F}$ are of the same size, we say $\mathcal{F}$ is *uniform*, if this size is $k$, then we say $\mathcal{F}$ is $k$-uniform. Sometimes (mostly in the chapters on Turán type problems, and saturation problems) we will use the word hypergaph for $k$-uniform families.

The *shadow* of a $k$-uniform family $\mathcal{F} \subseteq \binom{X}{k}$ is $\Delta(\mathcal{F}) = \{G : |G| = k-1, \exists F \in \mathcal{F}\ G \subset F\}$, and its *up-shadow* or *shade* is $\nabla(\mathcal{F}) = \{G : |G| = k+1, \exists F \in F\ F \subset G \subseteq X\}$. For not necessarily uniform families $\mathcal{F} \subseteq 2^X$ we will use $\Delta_m(\mathcal{F}) = \{G : |G| = m, \exists F \in \mathcal{F}\ G \subset F\}$ and $\nabla_m(\mathcal{F}) = \{G : |G| = m, \exists F \in F\ F \subset G \subseteq X\}$.

**Graphs.** A graph $G$ is a pair $(V(G), E(G))$ with $V(G)$ its vertex set and $E(G) \subseteq \binom{V(G)}{2}$ its edge set. $G$ is bipartite if there exists a partition $V = A \cup B$ such that every edge of $G$ contains one vertex from $A$ and one from $B$. In this case, we will write $G = (A, B, E)$. If $U \subseteq V$, the induced subgraph of $G$ on $U$ has vertex set $U$ and edge set $\{e = \{u,v\} \in E(G) : u, v \in U\}$. It is denoted by $G[U]$. In particular, if $V(G)$ is a family of sets, we will write $G[\mathcal{F}]$ for the subgraph induced on the subfamily $\mathcal{F}$ of $V$. The number of edges in $G$ is denoted by $e(G)$. Similarly the number of edges in $G[U]$ or between two disjoint sets $X, Y \subseteq V(G)$ is denoted by $e(U)$ and $e(X,Y)$, respectively.

A *path of length* $l$ in a graph $G$ between vertices $v_0$ and $v_l$ is a sequence

$v_0, e_1, v_1, e_2 \ldots, e_l, v_l$ such that $v_i$ is a vertex for all $i = 0, 1, \ldots, l$ and $e_{i+1} = \{v_i, v_{i+1}\}$ is an edge in $G$ for all $i = 0, 1, \ldots, l-1$. We say that a graph $G$ is *connected* if for any two vertices $u, v \in V(G)$ there exists a path between $u$ and $v$ in $G$. The *(graph) distance* $d_G(u_1, u_2)$ of two vertices $u_1, u_2$ in $G$ is the length of the shortest path between $u_1$ and $u_2$. The *ball* $B_r(v)$ *of radius* $r$ *at vertex* $v$ is the set of all vertices at distance at most $r$ from $v$. The *open neighborhood* $N(v)$ *of a vertex* $v$ is the set of vertices that are adjacent to $v$. The closed neighborhood $N[v]$ of $v$ is $N(v) \cup \{v\}$. For a set $U \subseteq V(G)$ we define its open and closed neighborhood by $N(U) = \cup_{u \in U} N(u)$ and $N[U] = \cup_{u \in U} N[u]$. The degree $d_G(v)$ of a vertex $v \in G$ is $|N(v)|$. If all vertices of a graph $G$ have the same degree, then we say that $G$ is *regular.* A set of vertices is *independent* in $G$ if there is no edge between them, and a set of edges is *independent* in $G$ if they do not share any vertices. A *matching* in $G$ is a set of independent edges, and a *perfect matching* is a matching covering all the vertices of $G$.

We will use the following specific graphs: $P_l$ is a path on $l$ vertices (thus it has $l-1$ edges). If we add an edge between the endpoints of a path, we obtain the *cycle* $C_l$ on $l$ vertices. $K_n$ is the *complete graph* on $n$ vertices, having all the possible $\binom{n}{2}$ edges, while $K_{s,t}$ is the *complete bipartite graph* with one part having size $s$ and the other size $t$, that has all the possible $st$ edges between the two parts. A graph that does not contain any cycle is a *forest*, and a connected forest is a *tree*, usually denoted by $T$. A vertex of degree 1 is called a *leaf*, and $S_r$ is the *star* with $r$ leaves, i.e. $S_r = K_{1,r}$. $M_n$ is the *matching* on $n$ vertices, i.e. the graph consisting of $n/2$ independent edges.

**Binomial coefficients**. The binomial coefficient $\binom{n}{k} = \frac{n!}{k!(n-k)!}$ is the size of $\binom{X}{k}$ for an $n$-set $X$. For any $n$, we have $\binom{n}{k} \leq \binom{n}{k+1}$ if and only if $k < n/2$, with equality if $n$ is odd and $k = \lfloor n/2 \rfloor$. For any real $x$ and positive integer $k$ we define $\binom{x}{k} = \prod_{i=0}^{k-1} \frac{x-i}{k-i}$. It is easy to see that for every positive real $y$ and positive integer $k$ there exists exactly one $x$ with $\binom{x}{k} = y$. We will often use the well-known bounds $(\frac{n}{k})^k \leq \binom{n}{k} \leq (\frac{en}{k})^k$ and the fact that $\binom{x}{k}$ is a convex function.

**Posets**. A partially ordered set (poset) is a pair $(P, \leq)$ where $P$ is a set and $\leq$ is a binary relation on $P$ satisfying (i) $p \leq p$ for all $p \in P$, (ii) $p \leq q, q \leq p$ imply $p = q$, (iii) $p \leq q$, $q \leq r$ imply $p \leq r$. If the relation is clear from context, we will denote the poset by $P$. For any poset $(P, \leq_P)$, the *opposite poset* $(P', \leq_{P'})$ is defined on the same set of elements with $p \leq_{P'} q$ if and only if $q \leq_P p$. Two elements $p, q \in P$ are *comparable* if $p \leq q$ or $q \leq p$ holds, and *incomparable* otherwise. A set $C \subseteq P$ of pairwise comparable elements is called a *chain*, its cardinality is called its *length.* A chain of length $k$ is also called a $k$-chain and it is denoted by $P_k$. A set $S \subseteq P$ of pairwise incomparable elements is called an *antichain*.

An element $p \in P$ is *minimal (maximal)* if there does not exist $q \neq p$ in $P$ with $q \leq p$ ($p \leq q$). Most often we will consider the *Boolean lattice/hypercube* $Q_n$. Its set of elements is $2^S$ and $T_1 \leq T_2$ if and only if $T_1 \subseteq T_2$ holds. Therefore, a chain of length $k$ (a $k$-chain) in $Q_n$ is a family $\mathcal{C} = \{C_1, C_2, \ldots, C_k\}$ with $C_i \subsetneq C_{i+1}$. A family $\mathcal{F} \subseteq 2^{[n]}$ can be considered as a subposet of the hypercube. We will say that a set $F \in \mathcal{F}$ is minimal/maximal if it is so in this partial ordering, i.e. no proper subset/superset of $F$ is contained in $\mathcal{F}$.

We say that a poset is *totally ordered* if any two elements are comparable. The *initial segment of size s* of a totally ordered set $P$ is the set of the $s$ smallest elements of $P$. A poset is *ranked* if all maximal chains in it have the same length.

**Permutations**. A permutation of a set $X$ is a bijective function from $X$ to itself. We will denote permutations by Greek letters $\sigma, \pi$, etc. The set of all permutations of $X$ is denoted by $S_X$. If $X$ is finite, the number of permutations in $S_X$ is $|X|!$. For any two elements $\sigma, \pi \in S_X$, the composition $\sigma \circ \pi$ also belongs to $S_X$. For a set $X = \{x_1, x_2, \ldots, x_n\}$, let $\alpha$ denote the permutation with $\alpha(x_1) = x_2, \alpha(x_2) = x_3, \ldots, \alpha(x_{n-1}) = x_n, \alpha(x_n) = x_1$. Two permutations $\sigma, \pi \in S_x$ are equivalent if $\sigma = \alpha^k \circ \pi$ for some integer $k$. This defines an equivalence relation on $S_X$ with each equivalence class containing $|X|$ permutations. The equivalence classes are the *cyclic permutations* of $X$. The number of cyclic permutations of $X$ is $(|X| - 1)!$.

**Linear algebra**. We will assume that the reader is familiar with the notions of linear independence, vector space dimension, matrix, positive definite and semidefinite matrix. A reader interested in combinatorics but not feeling comfortable with linear algebra might consult Sections 2.1-2.3 of the excellent book [28] by Babai and Frankl. If $v_1, v_2, \ldots, v_k$ are vectors in a vector space $V$, then $\langle v_1, v_2, \ldots, v_k \rangle$ denotes the subpace of $V$ spanned by these vectors. The scalar product of two vectors $\underline{u} = (u_1, u_2, \ldots, u_n), \underline{v} = (v_1 v_2, \ldots, v_n)$ is $\underline{u} \cdot \underline{v} = \sum_{i=1}^n u_i v_i$. $\mathbb{F}_q$ denotes the finite field of $q$ elements, and $\mathbb{F}_q^n$ the vector space of dimension $n$ over $\mathbb{F}_q$. The multiplicative group of nonzero elements of $\mathbb{F}_q$ is denoted by $\mathbb{F}_q^{\times}$. $\mathbb{F}^n[x]$ denotes the vector space of polynomials of $n$ variables over $\mathbb{F}$. A matrix or vector with only 0 and 1 entries will be called *binary matrix* or *binary vector*.

There are several vectors that can be naturally associated with sets. The *characteristic vector* of a subset $F \subseteq [n]$ is the binary vector $v_F = (v_1, v_2, \ldots, v_n)$ with $v_i = 1$ if and only if $i \in F$.

**Functions** (or mappings). If $A$ is a subset of the domain of a function $f$, then $f(A)$ will denote the set $\{f(a) : a \in A\}$. The inverse of a function $f$ will be denoted by $f^{-1}$. To compare the order of magnitude of two functions $f(n)$ and $g(n)$ we will write $f(n) = O(g(n))$ if there exists a positive constant $C$ such that $\left|\frac{f(n)}{g(n)}\right| \leq C$ for all $n \in \mathbb{N}$. Similarly, $f(n) = \Omega(g(n))$ means that there exists a positive constant $C$ such that $\left|\frac{f(n)}{g(n)}\right| \geq C$ for all $n \in \mathbb{N}$. If the constant $C$ in the above definitions depends on some other parameter $k$, then we write $f(n) = O_k(g(n))$ and $f(n) = \Omega_k(g(n))$. If both $f(n) = O(g(n))$ and $f(n) = \Omega(g(n))$ hold, then we write $f(n) = \Theta(n)$. Finally, $f(n) = o(g(n))$, or equivalently $f(n) \ll g(n)$ denotes that $\lim_{n\to\infty}\left|\frac{f(n)}{g(n)}\right| = 0$. Similarly $f(n) = \omega(g(n))$, or equivalently $f(n) \gg g(n)$ denotes that $\lim_{n\to\infty}\left|\frac{f(n)}{g(n)}\right| = \infty$. We say that a real-valued function $f$ is *convex* on an interval $[x_1, x_2]$ if it lies above its tangents. We will use Jensen's inequality, which states that in this case $f(tx_1 + (1-t)x_2) \leq tf(x_1) + (1-t)f(x_2)$.

Throughout the book, log and ln stand for logarithms of base 2 and $e$, respectively.

**Probability**. The probabilities of an event $\mathcal{E}$ and the expected value of a random variable $X$ are denoted by $\mathbb{P}(\mathcal{E})$ and $\mathbb{E}(X)$, respectively. Its standard deviation is $\sigma(X) = \sqrt{\mathbb{E}(X - \mathbb{E}(x))^2} = \sqrt{\mathbb{E}(X^2) - \mathbb{E}(X)^2}$. We say that a sequence $\mathcal{E}_n$ of events holds *with high probability (w.h.p.)* if $\mathbb{P}(\mathcal{E}_n) \to 1$ as $n$ tends to infinity. Standard inequalities concerning probabilities of events will be introduced at the beginning of Chapter 4. For a detailed introduction to applications of probability theory in combinatorics we recommend the book by Alon and Spencer [19]. The binary entropy function $-x\log x - (1-x)\log(1-x)$ is denoted by $h(x)$.

# 1

# *Basics*

CONTENTS



As its title suggests, this chapter contains some of the very first results of extremal finite set theory (Sperner's theorem and the Erdős-Ko-Rado theorem), and tries to present some of the major techniques used in the field (shifting, the permutation method, the polynomial method). To this end, in many cases we will include several proofs of the same theorem.

## 1.1 Sperner's theorem, LYM-inequality, Bollobás inequality

In this section we will mostly consider families that form an *antichain* in the hypercube (they are also called *Sperner families*). Recall that it means there are no two different members of the family such that one of them contains the other. We will be interested in the maximum size that an antichain $\mathcal{F} \subseteq 2^{[n]}$ can have. An easy and natural way to create antichains is to collect sets of the same size, i.e. levels. Among full levels of $2^{[n]}$ the largest is (are) the middle one(s). Sperner proved [520] that this is best possible. His result is the first theorem in extremal finite set theory. We will give two proofs of his theorem, then consider a generalization by Erdős and a related notion by Bollobás.

**Theorem 1 (Sperner [520])** *If $\mathcal{F} \subseteq 2^{[n]}$ is an antichain, then we have $|\mathcal{F}| \le \binom{n}{\lfloor n/2 \rfloor}$. Moreover $|\mathcal{F}| = \binom{n}{\lfloor n/2 \rfloor}$ holds if and only if $\mathcal{F} = \binom{[n]}{\lfloor n/2 \rfloor}$ or $\mathcal{F} = \binom{[n]}{\lceil n/2 \rceil}$.*

**First proof of Theorem 1.** We start with the following simple lemma.

**Lemma 2** *Let $G(A,B,E)$ be a connected bipartite graph such that for every vertex $a \in A$ and $b \in B$ the inequality $d_G(a) \geq d_G(b)$ holds. Then for any subset $A' \subseteq A$ the size of its neighborhood $N(A')$ is at least the size of $A'$. Moreover if $|N(A')| = |A'|$ holds, then $A' = A$, $N(A') = B$ and $G$ is regular.*

**Proof of Lemma.** The number of edges between $A'$ and $N(A')$ is exactly $\sum_{a\in A'} d_G(a)$ and is at most $\sum_{b\in N(A')} d_G(b)$. By the assumption on the degrees, the number of summands in the latter sum should be at least the number of summands in the former sum; thus we obtain $|N(A')| \geq |A'|$. If the number of summands is the same in both sums, then we must have $d_G(a) = d_G(b)$ for any $a \in A', b \in N(A')$ and all edges incident to $N(A')$ must be incident to $A'$. By the connectivity of $G$ it follows that $A' = A$ and $N(A') = B$ hold. □

Let $G_{n,k,k+1}$ be the bipartite graph with parts $\binom{[n]}{k}$ and $\binom{[n]}{k+1}$ in which two sets $S \in \binom{[n]}{k}$ and $T \in \binom{[n]}{k+1}$ are joined by an edge if and only if $S \subset T$. We want to apply Lemma 2 to $G_{n,k,k+1}$. It is easy to see that it is connected. The degree of a set $S \in \binom{[n]}{k}$ is $n-k$, while the degree of $T \in \binom{[n]}{k+1}$ is $k+1$, therefore as long as $k \leq \lfloor n/2 \rfloor$ holds, $\binom{[n]}{k}$ can play the role of $A$, and as soon as $k \geq \lceil n/2 \rceil$ holds, $\binom{[n]}{k+1}$ can play the role of $A$. Moreover, if $k < \lfloor n/2 \rfloor$ or $k \geq \lceil n/2 \rceil$, then $G_{n,k,k+1}$ is not regular. Note that if $\mathcal{F} \subseteq \binom{[n]}{k}$, then $N(\mathcal{F}) = \nabla(\mathcal{F})$, while if $\mathcal{F} \subseteq \binom{[n]}{k+1}$, then $N(\mathcal{F}) = \Delta(\mathcal{F})$.

To prove Theorem 1 let $\mathcal{F} \subseteq 2^{[n]}$ be an antichain. First we prove that if $\mathcal{F}$ is of maximum size, then it contains sets only of size $\lfloor n/2 \rfloor$ and $\lceil n/2 \rceil$. Suppose not and, say, there exists a set of size larger than $\lceil n/2 \rceil$. Then let $m$ be the largest set size in $\mathcal{F}$ and let us consider the graph $G_{n,m-1,m}$. Applying Lemma 2 to $\mathcal{F}_m = \{F \in \mathcal{F} : |F| = m\}$ we obtain that $|\mathcal{F} \setminus \mathcal{F}_m \cup \Delta(\mathcal{F}_m)| > |\mathcal{F}|$ holds. To finish the proof we need to show that $\mathcal{F}' = \mathcal{F} \setminus \mathcal{F}_m \cup \Delta(\mathcal{F}_m)$ is an antichain. Sets of $\Delta(\mathcal{F}_m)$ have largest size in $\mathcal{F}'$, therefore they cannot be contained in other sets of $\mathcal{F}'$. No set $F' \in \Delta(\mathcal{F}_m)$ can contain any other set $F$ from $\mathcal{F}' \cap \mathcal{F}$ as there exists $F'' \in \mathcal{F}_m$ with $F' \subset F''$, thus $F \subset F''$ would follow, and that contradicts the antichain property of $\mathcal{F}$.

We showed that $\mathcal{F} \subseteq \binom{[n]}{\lfloor n/2 \rfloor} \cup \binom{[n]}{\lceil n/2 \rceil}$, which proves the theorem if $n$ is even. If $n$ is odd, then suppose $\mathcal{F}$ contains sets of size both $\lceil n/2 \rceil$ and $\lfloor n/2 \rfloor$. Applying the moreover part of Lemma 2 to $\mathcal{F}_{\lceil n/2 \rceil}$, we again obtain a larger antichain $\mathcal{F} \setminus \mathcal{F}_{\lceil n/2 \rceil} \cup \Delta(\mathcal{F}_{\lceil n/2 \rceil})$ unless $\mathcal{F}_{\lceil n/2 \rceil} = \binom{[n]}{\lceil n/2 \rceil}$. This shows that if $\mathcal{F}$ is a maximum size antichain that contains a set of size $\lceil n/2 \rceil$, then $\mathcal{F} = \binom{[n]}{\lceil n/2 \rceil}$. Similarly, if $\mathcal{F}$ is a maximum size antichain that contains a set of size $\lfloor n/2 \rfloor$, then $\mathcal{F} = \binom{[n]}{\lfloor n/2 \rfloor}$. ■

We give a second proof of Theorem 1 based on the inequality proved independently by Lubell, Yamamoto, and Meshalkin. As Bollobás obtained an even more general inequality that we will prove in Theorem 6, it is sometimes referred to as YBLM-inequality (Miklós Ybl was a famous Hungarian architect in the nineteenth century, who designed, among others, the State

Opera House and St. Stephen's Basilica in Budapest), but we will use the more common acronym.

**Lemma 3 (LYM-inequality, [397, 415, 554])** *If $\mathcal{F} \subseteq 2^{[n]}$ is an antichain, then the following inequality holds:*

$$\sum_{F \in \mathcal{F}} \frac{1}{\binom{n}{|F|}} \leq 1.$$

*Moreover, the above sum equals 1 if and only if $\mathcal{F}$ is a full level.*

Before the proof let us introduce a further definition. A chain $\mathcal{C} \subseteq 2^{[n]}$ is called a *maximal chain* if it is of length $n+1$, i.e. it contains a set of size $i$ for every $0 \leq i \leq n$.

**Proof.** Let $\mathcal{F} \subseteq 2^{[n]}$ be an antichain and let us consider the pairs $(F, \mathcal{C})$ such that $\mathcal{C}$ is a maximal chain in $[n]$ and $F \in \mathcal{F} \cap \mathcal{C}$. There are exactly $|F|!(n-|F|)!$ maximal chains containing $F$, therefore the number of such pairs is $\sum_{F \in \mathcal{F}} |F|!(n-|F|)!$. On the other hand, by the antichain property of $\mathcal{F}$, every maximal chain contains at most one set from $\mathcal{F}$ and thus the number of pairs is at most $n!$. We obtained

$$\sum_{F \in \mathcal{F}} |F|!(n-|F|)! \leq n! \tag{1.1}$$

and dividing by $n!$ yields the LYM-inequality.

Let us now prove the moreover part of the lemma. Clearly, if $\mathcal{F}$ is a full level $\binom{[n]}{k}$ for some $k$, then $\sum_{F \in \mathcal{F}} \frac{1}{\binom{n}{|F|}} = \binom{n}{k} \cdot \frac{1}{\binom{n}{k}} = 1$ holds. If $\mathcal{F}$ is not a full level, then there exist two sets $F \in \mathcal{F}, G \notin \mathcal{F}$ with $|F| = |G| = |F \cap G| + 1$. We construct maximal chains that do not contain any set from $\mathcal{F}$ and thus (1.1) cannot hold with equality. Consider any maximal chain $\mathcal{C}$ that contains $F \cap G$, $G$, and $F \cup G$. Any set $C$ of $\mathcal{C}$ with $C \subseteq F \cap G \subset F$ cannot be in $\mathcal{F}$ by the antichain property of $\mathcal{F}$, $G$ is not in $\mathcal{F}$ by definition, finally all sets $C \in \mathcal{C}$ with $C \supseteq F \cup G \supset F$ are not in $\mathcal{F}$ by the antichain property. By the choice of $F$ and $G$ there are no other sets in $\mathcal{C}$ and thus $\mathcal{F} \cap \mathcal{C} = \emptyset$ holds. ■

The function $\sum_{F \in \mathcal{F}} \frac{1}{\binom{n}{|F|}}$ in the LYM-inequality is called the *Lubell function* of $\mathcal{F}$ and it is often used (for several applications see Chapter 7 on forbidden subposet problems). It is denoted by $\lambda(\mathcal{F}, n)$ and we omit $n$ if it is clear from the context. Having Lemma 3 in hand, the second proof of Theorem 1 is immediate.

**Second proof of Theorem 1.** Let $\mathcal{F} \subseteq 2^{[n]}$ be an antichain. Then we have $\sum_{F \in \mathcal{F}} \frac{1}{\binom{n}{\lfloor n/2 \rfloor}} \leq \sum_{F \in \mathcal{F}} \frac{1}{\binom{n}{|F|}} \leq 1$ by Lemma 3. Therefore $|\mathcal{F}|$, that equals the number of summands, is at most $\binom{n}{\lfloor n/2 \rfloor}$. The moreover part of the theorem follows from the moreover part of Lemma 3. ■

Let us illustrate the strength of the LYM-inequality by the following generalization of Theorem 1. We say that a family $\mathcal{F}$ of sets is *$k$-Sperner* if all chains in $\mathcal{F}$ have length at most $k$. We define $\Sigma(n,k)$ to be the sum of the $k$ largest binomial coefficients of order $n$, i.e. $\Sigma(n,k) = \sum_{i=1}^{k} \binom{n}{\lfloor \frac{n-k}{2} \rfloor + i}$. Let $\Sigma^*(n,k)$ be the collection of families consisting of the corresponding full levels, i.e. if $n+k$ is odd, then $\Sigma^*(n,k)$ contains one family $\cup_{i=1}^{k}\binom{[n]}{\lfloor \frac{n-k}{2} \rfloor + i}$, while if $n+k$ is even, then $\Sigma^*(n,k)$ contains two families of the same size $\cup_{i=0}^{k-1}\binom{[n]}{\frac{n-k}{2}+i}$ and $\cup_{i=1}^{k}\binom{[n]}{\frac{n-k}{2}+i}$.

**Theorem 4 (Erdős, [149])** *If $\mathcal{F} \subseteq 2^{[n]}$ is a $k$-Sperner family, then $|\mathcal{F}| \le \Sigma(n,k)$ holds. Moreover, if $|\mathcal{F}| = \Sigma(n,k)$, then $\mathcal{F} \in \Sigma^*(n,k)$.*

**Proof.** We start with the following simple observation.

**Lemma 5** *A family $\mathcal{F}$ of sets is $k$-Sperner if and only if it is the union of $k$ antichains.*

**Proof of Lemma.** If $\mathcal{F}$ is the union of $k$ antichains, then any chain in $\mathcal{F}$ has length at most $k$ as any chain can contain at most one set from each antichain. Conversely, if $\mathcal{F}$ is $k$-Sperner, we define the required $k$ antichains recursively. Let $\mathcal{F}_1$ denote the family of all minimal sets in $\mathcal{F}$ and if $\mathcal{F}_j$ is defined for all $1 \le j < i$, then let $\mathcal{F}_i$ denote the family of all minimal sets in $\mathcal{F} \setminus \cup_{j=1}^{i-1}\mathcal{F}_j$. The $\mathcal{F}_i$'s are antichains by definition and for every $F \in \mathcal{F}_i$, there exists an $F' \in \mathcal{F}_{i-1}$ with $F' \subset F$. Hence, the existence of a set in $\mathcal{F}_{k+1}$ would imply the existence of a $(k+1)$-chain in $\mathcal{F}$. The partition we obtained is often referred to as the *canonical partition* of $\mathcal{F}$. □

Let $\mathcal{F} \subseteq 2^{[n]}$ be a $k$-Sperner family. By Lemma 5, $\mathcal{F}$ is the union of $k$ antichains $\mathcal{F}_1, \mathcal{F}_2, \ldots, \mathcal{F}_k$. Adding up the LYM-inequalities for all $\mathcal{F}_i$'s we obtain

$$\sum_{F \in \mathcal{F}} \frac{1}{\binom{n}{|F|}} \le k.$$

This immediately yields $|\mathcal{F}| \le \Sigma(n,k)$. If $|\mathcal{F}| = \Sigma(n,k)$, then all these LYM-inequalities must hold with equality. Therefore, by the moreover part of Lemma 3, all $\mathcal{F}_i$'s are full levels and thus $\mathcal{F} \in \Sigma^*(n,k)$. ■

We finish this section by stating and proving an inequality of Bollobás that generalizes the LYM-inequality. We say that a family of pairs of sets $(A_1,B_1),(A_2,B_2),\ldots,(A_t,B_t)$ is an *intersecting set pair system* (ISPS) if $A_i \cap B_j \neq \emptyset$ holds if and only if $i \neq j$.

**Theorem 6 (Bollobás, [62])** *If $(A_1,B_1),(A_2,B_2),\ldots,(A_t,B_t)$ is an ISPS,* *then the following inequality holds:*

$$\sum_{i=1}^{t} \frac{1}{\binom{|A_i|+|B_i|}{|A_i|}} \leq 1.$$

*In particular, if all $A_i$'s have size at most $k$ and all $B_i$'s have size at most $l$, then $t \leq \binom{k+l}{k}$ holds.*

**Proof.** Let $(A_1, B_1), (A_2, B_2), \ldots, (A_t, B_t)$ be an ISPS and let us define $M = \cup_{i=1}^{t}(A_i \cup B_i)$ and $m = |M|$. Consider all pairs $(i, \sigma)$ such that $\sigma$ is a permutation of $M$ and all elements of $A_i$ "come before" the elements of $B_i$, i.e. $\sigma^{-1}(a) < \sigma^{-1}(b)$ holds for all $a \in A_i, b \in B_i$. We count the number of such pairs in two ways.

If we fix an index $i$, then the number of permutations $\sigma$ of $M$ such that $(i, \sigma)$ has the above property is exactly $|A_i|! \cdot |B_i|!(m - |A_i| - |B_i|)!\binom{m}{|A_i|+|B_i|}$. Indeed, one determines the order of the elements inside $A_i$, $B_i$ and $M \setminus (A_i \cup B_i)$ independently from each other, and then determines where the elements of $A_i \cup B_i$ are placed in $\sigma$.

On the other hand for every permutation $\sigma$ of $M$ there is at most one pair $(i, \sigma)$ with the above property. Indeed, if the elements of $A_i$ come before the elements of $B_i$ in $\sigma$, then for any $j \neq i$ an element $b \in B_j \cap A_i$ comes before an element $a \in A_j \cap B_i$. As $(A_1, B_1), (A_2, B_2), \ldots, (A_t, B_t)$ is an ISPS, the elements $a$ and $b$ exist. We obtained

$$m! \geq \sum_{i=1}^{t} |A_i|! \cdot |B_i|!(m - |A_i| - |B_i|)!\binom{m}{|A_i| + |B_i|} = \sum_{i=1}^{t} \frac{|A_i|! \cdot |B_i|!}{(|A_i| + |B_i|)!} m!.$$

Dividing by $m!$ yields the statement of the theorem. ■

As we mentioned earlier, Theorem 6 generalizes the LYM-inequality as if $\mathcal{F} = \{F_1, F_2, \ldots, F_m\} \subseteq 2^{[n]}$ is an antichain, then the pairs $A_i = F_i, B_i = \overline{F_i}$ form an ISPS and $\frac{1}{\binom{|A_i|+|B_i|}{|A_i|}} = \frac{1}{\binom{n}{|F_i|}}$.

**Exercise 7** *Let $\mathcal{F} \subseteq 2^{[n]}$ be a family such that for any $F, F' \in \mathcal{F}$ we have $|F \setminus F'| \geq l$. Prove that the following inequality holds:*

$$\sum_{F \in \mathcal{F}} \left( \frac{\binom{|F|}{|F|-l}}{\binom{n}{|F|-l}} + \frac{\binom{n-|F|}{l}}{\binom{n}{|F|+l}} \right) \leq 2.$$

## 1.2 The Erdős-Ko-Rado theorem - several proofs

The main notion of this section is the following: a family $\mathcal{F}$ is *intersecting* if for any $F, F' \in \mathcal{F}$ the intersection $F \cap F'$ is non-empty. We will be interested in the maximum possible size that an intersecting family can have. If there are no restrictions on the sizes of sets in $\mathcal{F}$, then we have the following proposition.

**Proposition 8** *If $\mathcal{F} \subseteq 2^{[n]}$ is intersecting, then $|\mathcal{F}| \le 2^{n-1}$.*

**Proof.** Clearly, an intersecting family cannot contain both $F$ and $\overline{F}$ for any subset $F$ of $[n]$. ■

Note that the bound is sharp as shown by the family $\{F \subseteq [n] : 1 \in F\}$, but there are many other intersecting families of that size (see Exercise 15 at the end of this section). This situation changes if we consider $k$-uniform families. Again, if all the sets contain the element 1, the family is intersecting (Such families are called *trivially intersecting families*.) The largest trivially intersecting families are the *star* with *center* $x$ $\{F \in \binom{[n]}{k} : x \in F\}$ for some fixed $x \in [n]$. They give constructions of size $\binom{n-1}{k-1}$. The celebrated theorem of Erdős, Ko, and Rado states that this is best possible if $2k \le n$ holds (the additional condition is necessary, as if $2k > n$, then any pair of $k$-subsets of $[n]$ intersects). They proved their result in 1938, but they thought that it might not be interesting enough, so they published it only in 1961. They could not have been more wrong about the importance of their theorem: the next chapter is completely devoted to theorems that deal with families satisfying properties defined by intersection conditions.

**Theorem 9 (Erdős, Ko, Rado, [163])** *If $\mathcal{F} \subseteq \binom{[n]}{k}$ is an intersecting family, then $|\mathcal{F}| \le \binom{n-1}{k-1}$ holds provided $2k \le n$. Moreover, if $2k < n$, then $|\mathcal{F}| = \binom{n-1}{k-1}$ if and only if $\mathcal{F}$ is a star.*

The first proof we give is more or less the one given in the original paper by Erdős, Ko, and Rado. It uses the very important technique of *shifting* (or compression). Some of the many shifting operations will be defined in this book. For a more detailed introduction see the not very recent survey of Frankl [193].

**First proof of Theorem 9.** We proceed by induction on $k$, the base case $k = 1$ being trivial. For $k \ge 2$, let $\mathcal{F} \subseteq \binom{[n]}{k}$ be an intersecting family. For $x, y \in [n]$, let us define

$$S_{x,y}(F) = \begin{cases} F \setminus \{y\} \cup \{x\} & \text{if } y \in F, x \notin F \text{ and } F \setminus \{y\} \cup \{x\} \notin \mathcal{F} \\ F & \text{otherwise} \end{cases} \tag{1.2}$$

and write $S_{x,y}(\mathcal{F}) = \{S_{x,y}(F) : F \in \mathcal{F}\}$.

**Lemma 10** *Let $\mathcal{F} \subseteq \binom{[n]}{k}$ be an intersecting family and $x, y \in [n]$. Then $S_{x,y}(\mathcal{F}) \subseteq \binom{[n]}{k}$ is intersecting with $|\mathcal{F}| = |S_{x,y}(\mathcal{F})|$.*

**Proof of Lemma.** The statements $S_{x,y}(\mathcal{F}) \subseteq \binom{[n]}{k}$ and $|\mathcal{F}| = |S_{x,y}(\mathcal{F})|$ are clear by definition. To prove the intersecting property of $S_{x,y}(\mathcal{F})$ let us call a set $G \in S_{x,y}(\mathcal{F})$ *new* if $G \notin \mathcal{F}$ and *old* if $G \in \mathcal{F}$. Two old sets intersect by the intersecting property of $\mathcal{F}$ and two new sets intersect as by definition they both contain $x$. Finally, suppose $F$ is an old, and $G$ is a new set of $S_{x,y}(\mathcal{F})$. By definition $x \in G$, so if $x \in F$, then $F$ and $G$ intersect. Suppose $x \notin F$ and consider $F' := G \setminus \{x\} \cup \{y\}$. As $G$ is a new set of $S_{x,y}(\mathcal{F})$ we have $F' \in \mathcal{F}$ and therefore $F \cap F' \neq \emptyset$. If there exists $z \in F \cap F'$ with $z \neq y$, then $z \in F \cap G$ and we are done. If $F \cap F' = \{y\}$, then consider $F'' = F \setminus \{y\} \cup \{x\}$. As $F$ is an old set of $S_{x,y}(\mathcal{F})$, we must have $F'' \in \mathcal{F}$, but then $F' \cap F'' = \emptyset$ contradicting the intersecting property of $\mathcal{F}$. Therefore $F \cap F' \neq \{y\}$ holds. This finishes the proof of the lemma. □

Let us define the weight of a family $\mathcal{G}$ to be $w(\mathcal{G}) = \sum_{G \in \mathcal{G}} \sum_{i \in G} i$. Observe that if $x < y$ and $S_{x,y}(\mathcal{F}) \neq \mathcal{F}$, then $w(S_{x,y}(\mathcal{F})) < w(\mathcal{F})$ holds. Therefore, as the weight is a non-negative integer, after applying a finite number of such shifting operations to $\mathcal{F}$, we obtain a family $\mathcal{F}'$ with the following property: $S_{x,y}(\mathcal{F}') = \mathcal{F}'$ for all $1 \leq x < y \leq n$. We call such a family *left-shifted*. By Lemma 10, to obtain the bound of the theorem it is enough to prove that a left-shifted intersecting family $\mathcal{F}$ has size at most $\binom{n-1}{k-1}$.

**Lemma 11** *If $\mathcal{F} \subseteq \binom{[n]}{k}$ is a left-shifted intersecting family, then for any $F_1, F_2 \in \mathcal{F}$ we have $F_1 \cap F_2 \cap [2k-1] \neq \emptyset$.*

**Proof of Lemma.** Suppose not and let $F_1, F_2 \in \mathcal{F}$ be such that $F_1 \cap F_2 \cap [2k-1] = \emptyset$ holds and $|F_1 \cap F_2|$ is minimal. Let us choose $y \in F_1 \cap F_2$ and $x \in [2k-1] \setminus (F_1 \cup F_2)$. As $\mathcal{F}$ is intersecting, $y$ exists. By the assumption $F_1 \cap F_2 \cap [2k-1] = \emptyset$, we have $y \geq 2k$, and thus $|F_1 \cap [2k-1]|, |F_2 \cap [2k-1]| \leq k-1$, therefore $x$ exists. By definition, we have $x < y$. As $\mathcal{F}$ is left-shifted, we have $F_1' := F_1 \setminus \{y\} \cup \{x\} \in \mathcal{F}$, but then $F_1' \cap F_2 \cap [2k-1] = \emptyset$ and $|F_1' \cap F_2| < |F_1 \cap F_2|$, contradicting the choice of $F_1$ and $F_2$. □

We are now ready to prove the bound of the theorem. For $i = 1, 2, \ldots, k$ let us define $\mathcal{F}_i = \{F \in \mathcal{F} : |F \cap [2k]| = i\}$ and $\mathcal{G}_i = \{F \cap [2k] : F \in \mathcal{F}_i\}$. By Lemma 11, $\mathcal{G}_i$ is intersecting for all $i$ and $\mathcal{F} = \cup_{i=1}^{k} \mathcal{F}_i$. By the inductive hypothesis, we obtain $|\mathcal{G}_i| \leq \binom{2k-1}{i-1}$ for all $i \leq k-1$. By definition, $\mathcal{G}_k = \mathcal{F}_k$ holds, and for every $k$-subset $S$ of $[2k]$ at most one of $S$ and $[2k] \setminus S$ belongs to $\mathcal{G}_k$, therefore we have $|\mathcal{G}_k| \leq \frac{1}{2}\binom{2k}{k} = \binom{2k-1}{k-1}$. Every set $G \in \mathcal{G}_i$ can be extended to a $k$-set of $\mathcal{F}_i$ in at most $\binom{n-2k}{k-i}$ ways. We obtain

$$|\mathcal{F}| = \sum_{i=1}^{k} |\mathcal{F}_i| \leq \sum_{i=1}^{k} |\mathcal{G}_i| \binom{n-2k}{k-i} \leq \sum_{i=1}^{k} \binom{2k-1}{i-1} \binom{n-2k}{k-i} = \binom{n-1}{k-1}.$$

We still have to prove that if $2k < n$ holds, then the only intersecting families with size $\binom{n-1}{k-1}$ are stars. First we show this for left-shifted intersecting families, where obviously the center of the star should be 1. If $k = 2$, then there are only two types of maximal intersecting families: the star and the triangle $\{\{1,2\}, \{1,3\}, \{2,3\}\}$. If $k \geq 3$, then in the inductive argument above, all $\mathcal{F}_i$'s are stars with center 1 and all possible extensions of these sets to $k$-sets must be present in $\mathcal{F}$, i.e. $\mathcal{F}' := \{F' \in \binom{[n]}{k} : 1 \in F', |F' \cap [2k-1]| \leq k-1\} \subseteq \mathcal{F}$ holds. But for any $k$-set $S$ with $1 \notin S$ there exists a set $F' \in \mathcal{F}'$ with $S \cap F' = \emptyset$, therefore all sets in $\mathcal{F}$ must contain 1.

For general $\mathcal{F}$, let us start applying shifting operations $S_{x,y}$ with $x < y$. If we obtain a left-shifted family $\mathcal{H}$ that is not a star, then we are done by the above. If not, then at some point we obtain a family $\mathcal{G}$ such that $S_{x,y}(\mathcal{G})$ is a star. Therefore $G \cap \{x, y\} \neq \emptyset$ for all $G \in \mathcal{G}$. Then we apply shifting operations $S_{x',y'}$ with $x' < y'$ and $x', y' \in [n] \setminus \{x, y\}$ until we obtain a family $\mathcal{H}$ with $S_{x',y'}(\mathcal{H}) = \mathcal{H}$ for all $x', y' \in [n] \setminus \{x, y\}$ with $x' < y'$. Clearly, we still have $H \cap \{x, y\} \neq \emptyset$ for all $H \in \mathcal{H}$. The next lemma is the equivalent of Lemma 11.

**Lemma 12** *Let $A$ denote the first $2k-2$ elements of $[n] \setminus \{x, y\}$ and let $B = A \cup \{x, y\}$. Then $H \cap H' \cap B \neq \emptyset$ for all $H, H' \in \mathcal{H}$.*

**Proof of Lemma.** Suppose not and let $H, H'$ be members of $\mathcal{H}$ with $H \cap H' \cap B = \emptyset$ and $|H \cap H'|$ minimal. This means that $H$ and $H'$ meet $\{x, y\}$ in different elements and as $\mathcal{H}$ is intersecting, there exists $y' \in (H \cap H') \setminus B$. Also there exists $x' \in A \setminus (H \cup H')$ as $H \setminus \{x, y\}$ and $H' \setminus \{x, y\}$ are $(k-1)$-sets both containing $y' \notin A$. Then $H \setminus \{y'\} \cup \{x'\} \in \mathcal{H}$ by $S_{x',y'}(\mathcal{H}) = \mathcal{H}$, which contradicts the choice of $H$ and $H'$. □

Now the exact same calculation and reasoning can be carried out as in the left-shifted case. So $\mathcal{H}$ to have size $\binom{n-1}{k-1}$ we must have that for all $i \leq k-1$ there exists $z_i$ with $z_i \in H$ for all $H \in \mathcal{H}_i = \{H \in \mathcal{H} : |H \cap B| = i\}$ and $|\mathcal{H}_i| = \binom{k-1}{i-1}\binom{n-2k}{k-i}$. Every $H \in \mathcal{H}_k$ must contain $z_i$ for all $i \leq k-1$ as otherwise we would find a $H' \in \mathcal{H}_i$ disjoint from $H$. So if $z_i \neq z_j$, then $|\mathcal{H}_k| \leq \binom{2k-2}{k-2}$ and thus $|\mathcal{H}| < \binom{n-1}{k-1}$ or $z_i$ is the same for all $i$ and thus $\mathcal{H}$ is a star contradicting our assumption. ■

The next proof we present is due to G.O.H. Katona [337]. This proof is the first application of the *cycle method*. The idea is very similar to that in the proof of the LYM-inequality where one considers maximal chains. One addresses the original problem on a simpler structure $\mathcal{S}$ which is symmetric in the sense that if two sets $S_1, S_2$ have the same size, then the number of copies of $\mathcal{S}$ containing $S_1$ equals the number of copies of $\mathcal{S}$ containing $S_2$. And when this simpler problem is solved, one tries to reduce the original problem to the simpler one. The way to do that is to count the pairs $(\pi, F)$, where $\pi$ is a permutation, $F$ is a member of the original family and $\pi(F) \in \mathcal{S}$. For an $F$ it is usually easy to count the number of permutations that form a pair with it, while for every $\pi$ the solution on the simpler structure gives an upper bound.

In this general form, the method is called the *permutation method* and the cycle method is the following special case: if $\sigma$ is a cyclic permutation of $[n]$, then a set $S$ is an *interval* of $\sigma$ if it is a set of consecutive elements. More precisely either $S = \emptyset$ or $S = [n]$ or $S = \{\sigma(i), \sigma(i+1), \ldots, \sigma(i+j)\}$ for some integers $1 \leq i, j \leq n-1$, where addition is considered modulo $n$. An interval of size $k$ is said to be a *$k$-interval.* The element $\sigma(i)$ is the *left* *endpoint* of the interval and $\sigma(i+j)$ is the *right endpoint* of the interval. For a fixed permutation $\sigma$, there are $n(n-1)+2$ intervals of $\sigma$. For a fixed $k$-set $S \subseteq [n]$ with $1 < k < n$ there are $|F|!(n-|F|)!$ cyclic permutations $\sigma$ of which $S$ is an interval. Note that $\emptyset$ and $[n]$ are intervals of all the $(n-1)!$ cyclic permutations. Thus in the future calculations they need a separate treatment, but we usually omit that, as it is trivial.

**Second proof of Theorem 9.** We start with the solution of the problem on the cycle.

**Lemma 13** *Let $\sigma$ be a cyclic permutation of $[n]$ and let $G_1, G_2, \ldots, G_r$ be $k$-intervals of $\sigma$ that form an intersecting family $\mathcal{G}$. Then $r \leq k$ holds provided $2k \leq n$. Furthermore, if $2k < n$ and $r = k$, then the $G_i$'s are all the $k$-intervals that contain a fixed element of $[n]$.*

**Proof of Lemma.** Without loss of generality we may assume that $G_1 = \{\sigma(1), \sigma(2), \ldots, \sigma(k)\}$. Then as the $G_i$'s all have the same size and they form an intersecting family, all $G_i$'s have either left endpoint $\sigma(j)$ for some $2 \leq j \leq k$ or right endpoint $\sigma(j)$ for some $1 \leq j \leq k-1$. Furthermore, because of the intersecting property and $2k \leq n$, the $k$-interval with right endpoint $\sigma(j)$ and the $k$-interval with left endpoint $\sigma(j+1)$ cannot be both among the $G_i$'s. This proves $r \leq k$.

Let us assume now $2k < n$ and let $\sigma(j)$ be the right endpoint of some $G_i$ with $1 \leq j \leq k-1$. Then $k$-intervals with left endpoint $\sigma(j+1)$ and $\sigma(j+2)$ cannot be in $\mathcal{G}$. Therefore, as $|\mathcal{G}| = k$, the $k$-interval with right endpoint $\sigma(j+1)$ must belong to $\mathcal{G}$. Similarly, if for some $2 \leq j \leq k$ the left endpoint of some $G_i$ is $\sigma(j)$, then the $k$-interval with left endpoint $\sigma(j-1)$ must belong to $\mathcal{G}$. This means that if $|\mathcal{G}| = k$ and $\sigma(1), \sigma(k)$ do not belong to all members of $\mathcal{G}$, then there is exactly one $j$ $(2 \leq j \leq k-1)$ for which $\sigma(j)$ is both a left and a right endpoint and then $\sigma(j)$ is an element of all the intervals in $\mathcal{G}$. □

Let us consider the number of pairs $(F, \sigma)$, where $\sigma$ is a cyclic permutation of $[n]$ and $F \in \mathcal{F}$ is an interval of $\sigma$. By Lemma 13, for fixed $\sigma$ the number of such cycles is at most $k$. On the other hand, every set $F$ in $\mathcal{F}$ is an interval of $k!(n-k)!$ cyclic permutations. We obtained

$$|\mathcal{F}|k!(n-k)! = \sum_{F \in \mathcal{F}} k!(n-k)! \leq k(n-1)!.$$

Dividing by $k!(n-k)!$ yields $|\mathcal{F}| \leq \binom{n-1}{k-1}$.

If $2k < n$ and $|\mathcal{F}| = \binom{n-1}{k-1}$, then by Lemma 13 for every cyclic permutation $\sigma$ there exists $x_\sigma \in [n]$ such that all intervals of $\sigma$ containing $x_\sigma$ belong to $\mathcal{F}$. Suppose toward a contradiction that $\mathcal{F}$ is not a star. Let $\sigma$ be an arbitrary cyclic permutation and let $x_\sigma$ be the element that is contained in $k$ intervals of $\sigma$ all in $\mathcal{F}$. Let $F_1 \in \mathcal{F}$ be the interval that has right endpoint $x_\sigma$ in $\sigma$ and let $F_2 \in \mathcal{F}$ be the interval that has left endpoint $x_\sigma$ in $\sigma$. Finally let $F \in \mathcal{F}$ be a set with $x_\sigma \notin F$ (if no such $F$ exists, then $\mathcal{F}$ is a star with center $x_\sigma$). Writing

$A = F \cap F_1, B = F \cap F_2$, and $C = F \setminus (A \cup B)$ let $\pi$ be a cyclic permutation such that for some $1 \le j_1 < j_2 \le j_3 < j_4 < j_5$ we have $(F_1 \setminus \{x_\sigma\}) \setminus F = \{\pi(1), \ldots, \pi(j_1)\}$, $A = \{\pi(j_1+1), \ldots, \pi(j_2)\}$, $C = \{\pi(j_2+1), \ldots, \pi(j_3)\}$, $B = \{\pi(j_3+1), \ldots, \pi(j_4)\}$, and $(F_2 \setminus \{x_\sigma\}) \setminus F = \{\pi(j_4+1), \ldots, \pi(j_5)\}$. In words, $F$ is an interval of $\pi$ such that elements of $A$ are at one end of the interval, elements of $B$ are at the other end, and elements of $C$ (if any) are in the middle. Furthermore, all remaining elements of $F_1$ except $x_\sigma$ are placed next to $A$ and all remaining elements of $F_2$ except $x_\sigma$ are placed next to $B$. Note that we have not decided about the position of $x_\sigma$ in $\pi$. We distinguish two cases:

CASE I: $C \neq \emptyset$.

Then let $\pi(j_5+1) = x_\sigma$ and therefore $F_2$ is an interval of $\pi$. We obtain that $x_\pi \in B$. As $C$ is non-empty, the size of $(C \cup F_2) \setminus \{x_\sigma\}$ is at least $k$. Therefore any of its $k$-subintervals containing $x_\pi$ should belong to $\mathcal{F}$. But by definition, it (at least one of them) is disjoint from $F_1$. This contradiction finishes the proof in Case I.

CASE II: $C = \emptyset$ and thus $F \subset (F_1 \cup F_2) \setminus \{x_\sigma\}$ holds and $(F_1 \cup F_2) \setminus \{x_\sigma\}$ has size $2k-2$.

Then let $\pi(2k) = x_\sigma$. As $F$ is an interval of $\pi$, we have $x_\pi \in F$. If $x_\pi \in F \cap F_1 = A$, then $(F_1 \cup \pi(n)) \setminus \{x_\sigma\} = \{\pi(n), \pi(1), \ldots, \pi(k-1)\}$ is a $k$-interval of $\pi$ that contains $x_\pi$, therefore it must belong to $\mathcal{F}$. But as $n > 2k$, we have $\pi(n) \neq x_\sigma$ and therefore $(F_1 \cup \pi(n)) \setminus \{x_\sigma\}$ is disjoint from $F_2 \in \mathcal{F}$, a contradiction to the intersecting property of $\mathcal{F}$. If $x_\pi \in F \cap F_2$, then $(F_2 \cup \pi(2k-1)) \setminus \{x_\sigma\} = \{\pi(k), \pi(k+1), \ldots, \pi(2k-1)\}$ is a $k$-interval of $\pi$ containing $x_\pi$, therefore it belongs to $\mathcal{F}$. But it is disjoint from $F_1 \in \mathcal{F}$ contradicting the intersecting property of $\mathcal{F}$. This finishes the proof of Case II. ■

We present a third proof of the Erdős-Ko-Rado theorem that uses the polynomial method. The main idea of this method is to assign polynomials $p_F(x)$ to every set $F \in \mathcal{F}$ and show that these polynomials are linearly independent in the appropriate vector space $V$. If this is so, then $|\mathcal{F}| \le \dim(V)$ follows. Let us start with a general lemma giving necessary conditions for polynomials to be independent.

**Lemma 14** *Let $p_1(x), p_2(x), \ldots, p_m(x) \in \mathbb{F}^n[x]$ be polynomials and $v_1, v_2, \ldots, v_m \in \mathbb{F}^n$ be vectors such that $p_i(v_i) \neq 0$ and $p_i(v_j) = 0$ holds for all $1 \le j < i \le m$. Then the polynomials are linearly independent.*

**Proof.** Suppose that $\sum_{i=1}^m c_i p_i(x) = 0$. As $p_i(v_1) = 0$ for all $1 < i$ we obtain $c_1 p_1(v_1) = 0$ and therefore $c_1 = 0$ holds. We proceed by induction on $j$. If $c_h = 0$ holds for all $h < j$, then using this and $p_i(v_j) = 0$ for all $i > j$, we obtain $c_j p_j(v_j) = 0$ and therefore $c_j = 0$. ■

Recall that the characteristic vector $v_F$ of a set $F \subseteq [n]$ is the binary vector of length $n$ with $i$th entry 1 if and only if $i \in F$. By definition, we have $v_F \cdot v_G = |F \cap G|$ for any pair of sets $F, G \subseteq [n]$. Therefore if $\mathcal{F} \subseteq \binom{[n]}{k}$ is an intersecting family, then the polynomials $p_F(x) = \prod_{i=1}^{k-1}(x \cdot v_F - i)$ and the characteristic vectors $v_F$ satisfy the conditions of Lemma 14 (independently of the order of sets of $\mathcal{F}$). What is the dimension of the smallest vector space that contains these polynomials? As characteristic vectors have only 0 and 1 entries, we can replace all powers $x_j^h$ by $x_j$ in the $p_F$'s and the polynomials $p'_F$ obtained will still satisfy the conditions of Lemma 14. The vector space generated by these polynomials is a subspace of $V = \langle\{x_{i_1} \cdot x_{i_2} \cdot \ldots \cdot x_{i_h}, 1 \le i_1 < i_2 < \ldots < i_h, 0 \le h \le k-1\}\rangle$ and clearly, we have $\dim(V) = \sum_{i=0}^{k-1}\binom{n}{i}$ holds. This yields the weak bound $|\mathcal{F}| \le \sum_{i=0}^{k-1}\binom{n}{i}$. The next proof that improves this and attains the bound of Theorem 9 is due to Füredi, Hwang, and Weichsel [238] and uses an idea that first appeared in a paper of Blokhuis [55].

**Third proof of Theorem 9.** Our plan is to add $\sum_{i=0}^{k-1}\binom{n}{i} - \binom{n-1}{k-1} = 2\sum_{j=0}^{k-2}\binom{n-1}{j}$ many sets to $\mathcal{F}$ and define corresponding polynomials and vectors such that the conditions of Lemma 14 are still satisfied. More precisely, we will also change some of the polynomials corresponding to sets in $\mathcal{F}$, and replace the characteristic vectors by other vectors. Let $a \in [n]$ be arbitrary and let us define $\mathcal{A} = \mathcal{F}_0 \cup \mathcal{H} \cup \mathcal{F}_1 \cup \mathcal{G}$ where $\mathcal{F}_0 = \{F \in \mathcal{F} : a \notin F\}$, $\mathcal{H} = \{H \subset [n] : a \notin H, 0 \le |H| \le k-2\}$, $\mathcal{F}_1 = \{F \in \mathcal{F} : a \in F\}$, $\mathcal{G} = \{G \subset [n] : a \in G : 1 \le |G| \le k-1\}$. Clearly, $|\mathcal{H}| = |\mathcal{G}| = \sum_{j=0}^{k-2}\binom{n-1}{j}$ holds. We define the following polynomials:

- For $F \in \mathcal{F}_0$ let $p_F(x) = \prod_{i=1}^{k-1}(v_F \cdot x - i)$, therefore for a set $S \subseteq [n]$ we have $p_F(v_S) = 0$ if and only if $1 \le |(F \cap S| \le k-1$. As opposed to what we had before, let $u_F = v_{\overline{F}\setminus\{a\}}$.

- For $H \in \mathcal{H}$ we define $p_H(x) = (\mathbf{1} \cdot x - (n-k-1))\prod_{h \in H} x_h$, where $\mathbf{1}$ denotes the vector of length $n$ with all 1 entries. For a set $S \subseteq [n]$ we have $p_H(v_S) = 0$ if and only if $|S| = n-k-1$ or $H \not\subseteq S$. We let $u_H = v_H$.

- For $F \in \mathcal{F}_1$ we consider $p_F(x) = \prod_{i=0}^{k-2}(v_{F\setminus\{a\}} \cdot x - i)$; therefore for a set $S \subseteq [n]$ we have $p_F(v_S) = 0$ if and only if $0 \le |(F \setminus \{a\}) \cap S| \le k-2$. Let $u_F = v_{F\setminus\{a\}}$.

- For $G \in \mathcal{G}$ we define $p_G(x) = \prod_{g \in G} x_g$. For a set $S \subseteq [n]$ we have $p_G(v_S) = 0$ if and only if $G \not\subset S$. We let $u_G = v_G$.

Again, as all vectors are binary, we can change every power $x_i^s$ to $x_i$ in these polynomials, so they are contained in the vector space of polynomials with $n$ real variables that have degree at most 1 in each variable. Therefore $|\mathcal{F}| + |\mathcal{H}| + |\mathcal{G}| = |\mathcal{A}| \le \sum_{i=0}^{k-1}\binom{n}{i}$ and thus $|\mathcal{F}| \le \binom{n-1}{k-1}$ provided these polynomials are independent. To check that the conditions of Lemma 14 hold, we still need to define an order of these sets and polynomials. First we enumerate the polynomials $p_F$ belonging to sets in $\mathcal{F}_0$ in arbitrary order, then the polynomials $p_H$ belonging to sets in $\mathcal{H}$ such that if $|H| < |H'|$ then $H$ comes before $H'$. Then come polynomials $p_F$ belonging to sets in $\mathcal{F}_1$ in arbitrary order, finally polynomials $p_G$ belonging to sets in $\mathcal{G}$ such that if $|G| < |G'|$ then $G$ comes before $G'$. We verify that the conditions of Lemma 14 hold by a simple case analysis.

- For $F, F' \in F_0$ we have $p_F(u_F) \neq 0$ as $F \cap \overline{F} = \emptyset$ and $p_F(u'_F) = 0$ as $1 \leq |F' \cap (\overline{F} \setminus \{a\})| \leq k-1$ if $F \neq F'$. Indeed, $1 \leq |F' \cap \overline{F}| \leq k-1$ as $\mathcal{F}$ is intersecting, and also we have $a \notin F'$. For $H \in \mathcal{H}$ we have $p_H(u_F) = 0$ as $|\overline{F} \setminus \{a\}| = n-k-1$. For $F_1 \in \mathcal{F}_1$ we have $p_{F_1}(u_F) = 0$ as $|(F_1 \setminus \{a\}) \cap (\overline{F} \setminus \{a\})| \leq k-2$. Indeed, $|F_1 \cap \overline{F}| \leq k-1$ as $\mathcal{F}$ is intersecting, and $a$ is contained in both. For $G \in \mathcal{G}$ we have $p_G(u_F) = 0$ as $G \not\subset \overline{F} \setminus \{a\}$ since $a \in G$ for all $G \in \mathcal{G}$.
- For $H, H' \in \mathcal{H}$ with $|H| \leq |H'|$ we have $p_H(u_H) \neq 0$ as $H \subseteq H$ and $|H| \leq k-2 < n-k-1$ (this is the only time we use the assumption $2k \leq n$). We also have $p_{H'}(u_H) = 0$ as $H' \not\subset H$. For $F \in \mathcal{F}_1$ we have $p_F(u_H) = 0$ as $|H| \leq k-2$ and for $G \in \mathcal{G}$ we have $p_G(u_H) = 0$ as $G \not\subset H$ since $a \in G, a \notin H$.
- Trivially, for $F, F' \in \mathcal{F}_1$ we have $p_F(u_F) \neq 0$ and $p_{F'}(u_F) = 0$. For $G \in \mathcal{G}$ we have $p_G(u_F) = 0$ as $G \not\subset F \setminus \{a\}$ since $a \in G$ for all $G \in \mathcal{G}$.
- For $G, G' \in \mathcal{H}$ with $|G| \leq |G'|$ we have $p_G(u_G) \neq 0$ as $G \subseteq G$, and $p_{G'}(u_G) = 0$ as $G' \not\subset G$.

■

**Exercise 15** *Show that any maximal intersecting family $\mathcal{F} \subseteq 2^{[n]}$ of sets has size $2^{n-1}$.*

## 1.3 Intersecting Sperner families

This short section is devoted to antichains that also possess the intersecting property. They are called *intersecting Sperner families*. They are interesting in their own right, but we would also like to give more examples of easy applications of the cycle method. First we prove the result of Milner that determines the maximum possible size of an intersecting Sperner family $\mathcal{F} \subseteq 2^{[n]}$. Obviously, if $n$ is odd, then $\binom{[n]}{\lceil n/2 \rceil}$ is intersecting Sperner and by Theorem 1 it has maximum size even among all antichains. Therefore the important part of the following theorem is when $n$ is even. The proof we present here is due to Katona [340].

**Theorem 16 (Milner, [419])** *If $\mathcal{F} \subseteq 2^{[n]}$ is an intersecting Sperner family, then $|\mathcal{F}| \leq \binom{n}{\lceil \frac{n+1}{2} \rceil}$.*

**Proof.** First we prove the following lemma.

**Lemma 17** *Let $\sigma$ be a cyclic permutation of $[n]$ and let $G_1, G_2, \ldots, G_r$ be intervals of $\sigma$ that form an intersecting Sperner family. Then the following inequality holds:*

$$\sum_{i=1}^{r} \binom{n}{|G_i|} \leq n \binom{n}{\lceil \frac{n+1}{2} \rceil}.$$

**Proof of Lemma.** By being an antichain, we see that all the $G_i$'s have distinct left endpoints and therefore $r \leq n$ holds. This finishes the proof if $n$ is odd (as in that case $\binom{n}{\lceil \frac{n+1}{2} \rceil}$ is the largest binomial coefficient). If $n$ is even, we distinguish two cases.

CASE I: $r = n$.

We can assume that the left endpoint of $G_i$ is $\sigma(i)$. Then by the Sperner property, we must have $|G_i| \leq |G_{i+1}|$. Consequently, we obtain $|G_1| \leq |G_2| \leq \ldots |G_n| \leq |G_1|$ and therefore all $G_i$'s must have the same size. By the intersecting property this size is at least $\lceil \frac{n+1}{2} \rceil$.

CASE II: $r < n$.

By the intersecting property and Lemma 13, at most $n/2$ intervals have size $n/2$. Therefore we have

$$\sum_{i=1}^{r} \binom{n}{|G_i|} \leq \frac{n}{2}\binom{n}{\frac{n}{2}} + \left(\frac{n}{2} - 1\right)\binom{n}{\frac{n}{2}+1} = n\binom{n}{\frac{n}{2}+1} = n\binom{n}{\lceil \frac{n+1}{2} \rceil}.$$

□

Let $\mathcal{F} \subseteq 2^{[n]}$ be an intersecting Sperner family and let us consider the sum

$$\sum_{\sigma, F} \binom{n}{|F|},$$

where the summation is over all cyclic permutations $\sigma$ of $[n]$ and sets $F \in \mathcal{F}$ that are intervals of $\sigma$. For a fixed set $F$, the number of cyclic permutations of which $F$ is an interval is $|F|!(n-|F|)!$, therefore the above sum equals $|\mathcal{F}| \cdot n!$. On the other hand, by Lemma 17, the sum is at most $(n-1)! \cdot n\binom{n}{\lceil \frac{n+1}{2} \rceil}$. This yields $|\mathcal{F}| \cdot n! \leq (n-1)! \cdot n\binom{n}{\lceil \frac{n+1}{2} \rceil}$ and the theorem follows. ■

**Theorem 18 (Bollobás, [64])** *If $\mathcal{F} \subseteq 2^{[n]}$ is an intersecting Sperner family such that all sets of $\mathcal{F}$ have size at most $n/2$, then the following inequality holds:*

$$\sum_{F \in \mathcal{F}} \frac{1}{\binom{n-1}{|F|-1}} \leq 1.$$

Note that this inequality is a strengthening of Theorem 9 as any family containing sets of the same size is an antichain.

**Proof.** We start with a generalization of Lemma 13.

**Lemma 19** *Let $\sigma$ be a cyclic permutation of $[n]$ and let $G_1, G_2, \dots, G_r$ be intervals of $\sigma$ of size at most $n/2$ that form an intersecting Sperner family. Then $r \leq \min\{|G_i| : 1 \leq i \leq r\}$ holds.*

**Proof of Lemma.** By symmetry, it is enough to show that $r \leq |G_1| =: j$. If $G_1 = \{\sigma(i), \sigma(i+1), \dots, \sigma(i+j-1)\}$, then by the intersecting Sperner property, all $G_k$'s must have one of their endpoints in $G_1$ and $\sigma(i)$ cannot be a left endpoint and $\sigma(i+j-1)$ cannot be a right endpoint. This would give $2j-1$ possible other intervals in the family. Notice that as all $G_k$'s have size at most $n/2$, if $\sigma(i+h)$ is a right endpoint, then $\sigma(i+h+1)$ cannot be a left endpoint and vice versa. This leaves at most $j-1$ possible other $G_k$'s. □

Let us consider the sum

$$\sum_{\sigma,F} \frac{1}{|F|},$$

where the summation is over all cyclic permutations $\sigma$ of $[n]$ and sets $F \in \mathcal{F}$ that are intervals of $\sigma$. For fixed $\sigma$, by Lemma 19, we have that if $F_1, F_2, \dots, F_r$ are the intervals of $\sigma$ belonging to $\mathcal{F}$, then $\sum_{i=1}^{r} \frac{1}{|F_i|} \leq 1$. For fixed $F \in \mathcal{F}$ the number of cyclic permutations of which $F$ is an interval is $|F|!(n-|F|)!$, therefore $\sum_{\sigma} \frac{1}{|F|} = (|F|-1)!(n-|F|)!$. We obtained

$$\sum_{F \in \mathcal{F}} (|F|-1)!(n-|F|)! \leq (n-1)!,$$

and dividing by $(n-1)!$ finishes the proof. ■

**Theorem 20 (Greene, Katona, Kleitman, [273])** *If $\mathcal{F} \subseteq 2^{[n]}$ is an intersecting Sperner family, then the following inequality holds:*

$$\sum_{F \in \mathcal{F}, |F| \leq n/2} \frac{1}{\binom{n}{|F|-1}} + \sum_{F \in \mathcal{F}, |F| > n/2} \frac{1}{\binom{n}{|F|}} \leq 1.$$

**Proof.** The core of the proof is again an inequality for intersecting Sperner families of intervals equipped with an appropriate weight function.

**Lemma 21** *Let $\sigma$ be a cyclic permutation of $[n]$ and let $G_1, G_2, \dots, G_r$, $G'_1, G'_2, \dots, G'_s$ be intervals of $\sigma$ that form an intersecting Sperner family such that $|G_i| \leq n/2$ for all $1 \leq i \leq r$ and $|G'_j| > n/2$ for all $1 \leq j \leq s$. Then the following inequality holds:*

$$\sum_{i=1}^{r} \frac{n-|G_i|+1}{|G_i|} + \sum_{j=1}^{s} 1 \leq n.$$

**Proof.** We distinguish two cases.

Case I: $r = 0$.

Then the first sum of the left hand side of the inequality is empty and all we have to prove is $s \leq n$. This follows from the Sperner property as the left endpoints of the intervals must be distinct.

CASE II: $r > 0$.

We may assume that $k = |G_1|$ is the smallest size among all $G_i$'s and that $G_1 = \{\sigma(1), \sigma(2), \ldots, \sigma(k)\}$. Note that the weight $\frac{n-w+1}{w}$ is monotone decreasing in $w$, therefore all weights are at most $\frac{n-k+1}{k}$. Because of the intersecting Sperner property every $G_i$ and $G'_j$ has either left endpoint $\sigma(u)$ for some $u = 2, 3, \ldots, k$ or right endpoint $\sigma(u')$ for some $u' = 1, 2, \ldots, k-1$. Therefore apart from $G_1$ the possible remaining sets can be partitioned into at most $k-1$ pairs (not all pairs and not both sets of a pair are necessarily present): the $(u-1)$st such pair consists of the set with left endpoint $\sigma(u)$ and the set with right endpoint $\sigma(u-1)$. Observe that because of the intersecting property every pair contains at most one set with size smaller than $n/2$. Therefore the sum of all weights is at most

$$\frac{n-k+1}{k} + (k-1)\Big(\frac{n-k+1}{k} + 1\Big) = n.$$

□

Having Lemma 21 in hand we prove the theorem by considering the sum

$$\sum_{\sigma, F, |F| \leq n/2}^{r} \frac{n - |F| + 1}{|F|} + \sum_{\sigma, F, |F| > n/2} 1,$$

where the summation in both sums is over all cyclic permutations $\sigma$ and all sets $F$ of $\mathcal{F}$ that are intervals of $\sigma$. By Lemma 21, for fixed $\sigma$ the sum is at most $n$. Recall that for fixed $F$ the number of cyclic permutations of which $F$ is an interval is $|F|!(n-|F|)!$. We obtain

$$\sum_{F \in \mathcal{F}, |F| \leq n/2}^{r} \frac{n - |F| + 1}{|F|} |F|!(n-|F|)! + \sum_{F \in \mathcal{F} |F| > n/2} |F|!(n-|F|)! \leq$$
$$\leq (n-1)! \cdot n = n!.$$

Note that $\frac{n-|F|+1}{|F|}|F|!(n-|F|)! = (|F|-1)!(n-|F|+1)!$, therefore dividing by $n!$ finishes the proof of the theorem. ■

**Exercise 22 (Scott [501])** *Prove Theorem 16 for even $n$ in the following three steps:*

- *By an argument similar to that in the first proof of Theorem 1, show that a maximum size intersecting Sperner family $\mathcal{F} \subseteq \binom{[n]}{\frac{n}{2}} \cup \binom{[n]}{\frac{n}{2}+1}$.*
- *Observe that $\nabla(\mathcal{F} \cap \binom{[n]}{\frac{n}{2}})$ is disjoint from $\mathcal{F} \cap \binom{[n]}{\frac{n}{2}+1}$.*
- *Using the cycle method show $|\mathcal{F} \cap \binom{[n]}{\frac{n}{2}}| \leq |\nabla(\mathcal{F} \cap \binom{[n]}{\frac{n}{2}})|$ holds.*

## 1.4 Isoperimetric inequalities: the Kruskal-Katona theorem and Harper's theorem

The original geometric isoperimetric problem dates back to antiquity and states that if $R$ is a region in the plane of area 1 and the boundary of $R$ is a simple curve $C$, then the length of $C$ is minimized when $C$ is a cycle. There are mainly two discrete versions of this problem. For a graph $G$ and a subset $U$ of the vertex $V(G)$, the *vertex boundary of* $U$ is $\partial U = \{v \in V(G) \setminus U : \exists u \in U$ such that $u$ is adjacent to $v\}$ and the *exterior of* $U$ is $ext(U) = V(G) \setminus (U \cup \partial U)$. For fixed integer $m$ and a graph $G$ the *vertex isoperimetric problem* is to minimize $|\partial U|$ over all subsets $U \subseteq V(G)$ with $|U| = m$. The *edge isoperimetric problem* is to minimize the number of edges between $U$ and $\partial U$ over all subsets $U \subseteq V(G)$ with $|U| = m$. Note that the edge isoperimetric problem is equivalent for values $m$ and $|V(G)| - m$. Discrete isoperimetric appear frequently in the literature, we only consider some examples for graphs the vertex set of which are associated with set families.

Our first example is the bipartite graph $G = G_{n,k-1,k}$ that we defined in the first proof of Theorem 1. Let us remind the reader that $G$ is bipartite with parts $\binom{[n]}{k}$ and $\binom{[n]}{k-1}$ and the $(k-1)$-set $F$ is adjacent to the $k$-set $F'$ if and only if $F \subset F'$. Note that if $\mathcal{F} \subseteq \binom{[n]}{k}$, then $\partial\mathcal{F} = N(\mathcal{F}) = \Delta(\mathcal{F})$ holds.

The *lexicographic ordering* of sets is defined by $A$ being smaller than $B$ if and only if $\min A \setminus B < \min B \setminus A$ holds (i.e. the first difference matters). However, the following theorem uses an ordering where the last different element matters. Let us define the *colex ordering* $\prec_k$ on $\binom{\mathbb{Z}^+}{k}$ by setting $A \prec_k B$ if and only if $\max A \setminus B < \max B \setminus A$ holds. The smallest element of this ordering is the set $[k]$ and for every $n \geq k$ the family $\binom{[n]}{k}$ is an initial segment of $(\prec_k, \binom{\mathbb{Z}^+}{k})$. For general $m$, we denote its initial segment of size $m$ by $\mathcal{I}^k_m$.

**Theorem 23 (Kruskal, Katona, [335, 385])** *Let $\mathcal{F}$ be a $k$-uniform family of size $m$. Then the inequality $|\Delta(\mathcal{F})| \geq |\Delta(\mathcal{I}^k_m)|$ holds.*

The Kruskal-Katona theorem is not a real vertex isoperimetric inequality as it states a sharp lower bound on the size of $\partial U$ only for some sets $U \subseteq V(G)$: those that do not contain vertices from the class corresponding to sets of size $k-1$.

**Proof.** The proof we present here is due to Bollobás and Leader [70]. It uses the following generalization of the shifting operation introduced in the first proof of Theorem 9. For disjoint sets $X, Y$ with $|X| = |Y|$ we define

$$S_{X,Y}(F) = \begin{cases} F \setminus Y \cup X & \text{if } Y \subseteq F, X \cap F = \emptyset \text{ and } F \setminus Y \cup X \notin \mathcal{F} \\ F & \text{otherwise,} \end{cases} \tag{1.3}$$

and write $S_{X,Y}(\mathcal{F}) = \{S_{X,Y}(F) : F \in \mathcal{F}\}$. We say that $\mathcal{F}$ is $(X, Y)$*-shifted* if $S_{X,Y}(\mathcal{F}) = \mathcal{F}$, and $\mathcal{F}$ is *s-shifted* if it is $(X, Y)$-shifted for all disjoint pairs $X, Y$ with $|X| = |Y| = r$, $X \prec_r Y$ and $r \leq s$. Finally, we say that $\mathcal{F}$ is *shifted* if it is shifted for all $s \leq k$. Note that by definition, every family $\mathcal{F}$ is $(\emptyset, \emptyset)$-shifted and thus 0-shifted. Also, being 1-shifted and left-shifted are equivalent.

First we prove a lemma that characterizes initial segments of the colex ordering with the above compression operation.

**Lemma 24** *A family $\mathcal{F} \subseteq \binom{\mathbb{Z}^+}{k}$ is an initial segment of the colex ordering if and only if $\mathcal{F}$ is shifted.*

**Proof of Lemma.** Suppose first that $\mathcal{F}$ is shifted. If $\mathcal{F}$ is not an initial segment, then there exist $F \in \mathcal{F}, G \notin \mathcal{F}$ with $G \prec_k F$. Writing $X = G \setminus F, Y = F \setminus G$ and $r = |F \cap G|$, we have $X \cap Y = \emptyset$, $X \prec_r Y$ and $(F \setminus Y) \cup X = G$. As $\mathcal{F}$ is shifted, we must have $G \in \mathcal{F}$, a contradiction.

Conversely, if $\mathcal{F}$ is an initial segment of the colex ordering, then for any $F, X, Y$ with $Y \subseteq F, X \cap F = \emptyset, X \prec_r Y$ we have $(F \setminus Y) \cup X \prec_k F$ and thus $(F \setminus Y) \cup X \in \mathcal{F}$. This shows $S_{X,Y}(\mathcal{F}) = \mathcal{F}$ and $\mathcal{F}$ is shifted. □

Our strategy to prove the theorem is to apply shifting operations $S_{X_i,Y_i}$ to $\mathcal{F}$ until we obtain a shifted family $\mathcal{F}'$ and show that the size of the shadow never increased when applying a shifting operation. Unfortunately, it would not be true for an arbitrary sequence of operations; however, the following lemma holds.

**Lemma 25** *Let $\mathcal{F}$ be a family of $k$-subsets of $\mathbb{Z}^+$ and $X, Y$ be disjoint sets with $|X| = |Y|$ and suppose that for every $x \in X$ there exists $y \in Y$ such that $\mathcal{F}$ is $(X \setminus \{x\}, Y \setminus \{y\})$-shifted. Then $\Delta(S_{X,Y}(\mathcal{F})) \leq \Delta(\mathcal{F})$ holds.*

**Proof of Lemma.** We define an injective mapping from $\Delta(S_{X,Y}(\mathcal{F})) \setminus \Delta(\mathcal{F})$ to $\Delta(\mathcal{F}) \setminus \Delta(S_{X,Y}(\mathcal{F}))$. If $A$ is in $\Delta(S_{X,Y}(\mathcal{F})) \setminus \Delta(\mathcal{F})$, then there exists $x \in [n]$ such that $A \cup \{x\} \in S_{X,Y}(\mathcal{F}) \setminus \mathcal{F}$. Therefore $X \subseteq A \cup \{x\}$, $Y \cap (A \cup \{x\}) = \emptyset$ and $S^{-1}_{X,Y}(A \cup \{x\}) = A \cup \{x\} \setminus X \cup Y \in \mathcal{F}$. If $x \in X$ held, then by the condition of the lemma there would exist $y \in Y$ for which $S_{X \setminus \{x\}, Y \setminus \{y\}}(A \cup \{x\} \setminus X \cup Y) = A \cup \{y\}$ would be in $\mathcal{F}$ and thus $A$ would belong to $\Delta(\mathcal{F})$, a contradiction. Therefore we have $x \notin X$ and so $A \cup Y \setminus X \subsetneq A \cup \{x\} \setminus X \cup Y$ and $A \cup Y \setminus X \in \Delta(\mathcal{F})$.

We show that $A \cup Y \setminus X \notin \Delta(S_{X,Y}(\mathcal{F}))$. Suppose towards a contradiction that there exists $u$ such that $(A \cup Y \setminus X) \cup \{u\} \in S_{X,Y}(\mathcal{F})$ holds. As $Y \subset (A \cup Y \setminus X) \cup \{u\}$ we have $(A \cup Y \setminus X) \cup \{u\} \in \mathcal{F}$. If $u \in X$, then by the condition of the lemma, there exists a $v \in Y$ such that $S_{X \setminus \{u\}, Y \setminus \{v\}}(\mathcal{F}) = \mathcal{F}$. It follows that $(A \cup Y \setminus X) \cup \{u\} \setminus (Y \setminus \{v\}) \cup (X \setminus \{u\}) = A \cup \{v\}$ belongs to $\mathcal{F}$, in particular $A \in \Delta(\mathcal{F})$, a contradiction. If $u \notin X$, then $S_{X,Y}((A \cup Y \setminus X) \cup \{u\}) = A \cup \{u\} \in S_{X,Y}(\mathcal{F})$. Recall that we indirectly assumed $(A \cup Y \setminus X) \cup \{u\} \in S_{X,Y}(\mathcal{F})$ holds also. But then both these sets must belong to $\mathcal{F}$, too, which contradicts $A \notin \Delta(\mathcal{F})$.

This means that the images of the mapping $i : A \mapsto A \cup Y \setminus X$ are indeed in $\Delta(\mathcal{F}) \setminus \Delta(S_{X,Y}(\mathcal{F}))$. Moreover, $i$ is injective as for all $A \in \Delta(S_{X,Y}(\mathcal{F})) \setminus \Delta(\mathcal{F})$ we showed that $X \subset A$ and it is obvious that $Y \cap A = \emptyset$. □

All that is left to prove is that starting with any family $\mathcal{F} \subseteq \binom{\mathbb{Z}^+}{k}$ using only shifting operations allowed in Lemma 25 one can arrive at a shifted family $\mathcal{F}'$. The next claim finishes the proof of the theorem.

**Claim 26** *For every $1 \leq s \leq k$ and $\mathcal{F} \subseteq \binom{\mathbb{Z}^+}{k}$, using only shifting operations $S_{X,Y}$ with $X \prec_r Y$, and $r \leq s$ that satisfy the conditions of Lemma 25, one can turn $\mathcal{F}$ into an $s$-shifted family $\mathcal{F}'$.*

**Proof of Claim.** We proceed by induction on $s$. As every family is 0-shifted, all shifting operations $S_{X,Y}$ are allowed when $X$ and $Y$ are singletons; thus, as seen in the first proof of Theorem 9, one can achieve a left-shifted family. Assume that for some $s \geq 2$ the statement of the claim is proved for $s-1$ and let $\mathcal{F} \subseteq \binom{\mathbb{Z}^+}{k}$. Suppose there is an integer $r$ with $2 \leq r \leq s$, and a disjoint pair of $r$-sets $X, Y$ with $X \prec_r Y$ and $S_{X,Y}(\mathcal{F}) \neq \mathcal{F}$. By induction, we can achieve an $(r-1)$-shifted family $\mathcal{F}^*$ using only allowed operations. Now we show that $S_{X,Y}$ is an allowed operation, i.e. for every $x \in X$ there exists $y \in Y$ such that $\mathcal{F}^*$ is $(X \setminus \{x\}, Y \setminus \{y\})$-shifted. Take any $y \in Y$ with $y \neq \max Y$, then we have $X \setminus \{x\} \prec_{r-1} Y \setminus \{y\}$, and by the $(r-1)$-shifted property of $\mathcal{F}^*$ we have $S_{X\setminus\{x\},Y\setminus\{y\}}(\mathcal{F}^*) = \mathcal{F}^*$.

Finally, this process ends, as whenever a shifting operation $S_{X,Y}$ with $X \prec_r Y$ changes a family $\mathcal{F}$, then $\sum_{G \in S_{X,Y}(\mathcal{F})} |G| < \sum_{F \in \mathcal{F}} |F|$ holds. □■

It is easy to see that for every pair of positive integers $m$ and $k$ there exists a sequence $a_k > a_{k-1} > \cdots > a_l > 0$ with $a_i \geq i$ such that $m = \sum_{i=l}^{k} \binom{a_i}{i}$ holds. If so, then again it is easy to see that $\mathcal{I}_m^k$ is the union of $\binom{[a_k]}{k}$, $\{S \cup \{a_k+1\} : S \in \binom{[a_{k-1}]}{k-1}\}$, $\{S \cup \{a_k+1, a_{k-1}+1\} : S \in \binom{[a_{k-2}]}{k-2}\}$, etc. It follows that $|\Delta(\mathcal{I}_m^k)| = \sum_{i=l}^{k} \binom{a_i}{i-1}$. Unfortunately, it is very inconvenient to calculate using this expression, therefore in most applications of the shadow theorem, one uses the following version of Theorem 23 due to Lovász.

**Theorem 27 (Lovász, [392])** *Let $\mathcal{F}$ be a $k$-uniform family with $|\mathcal{F}| = \binom{x}{k}$ for some real number $x \geq k$. Then $|\Delta(\mathcal{F})| \geq \binom{x}{k-1}$ holds. Moreover if $\Delta(\mathcal{F}) = \binom{x}{k-1}$, then $x$ is an integer and $\mathcal{F} = \binom{X}{k}$ for a set $X$ of size $x$.*

We present a proof by Keevash [347].

**Proof.** We start with stating an equivalent form of the theorem. Let $\mathcal{K}_{r+1}^r$ be the family consisting of all the $r+1$ $r$-sets on an underlying set of size $r+1$. For an $r$-uniform family $\mathcal{G}$ we denote by $\mathcal{K}_{r+1}^r(\mathcal{G})$ the set of copies of $\mathcal{K}_{r+1}^r$ in $\mathcal{G}$ and by $\mathcal{K}_{r+1}^r(v) = \mathcal{K}_{r+1}^r(\mathcal{G}, v)$ the set of copies of $K_{r+1}^r$ in $\mathcal{G}$ containing $v$.

**Lemma 28** *If $\mathcal{G}$ is a $(k-1)$-uniform family with $|\mathcal{G}| = \binom{x}{k-1}$, then*

$|\mathcal{K}_k^{k-1}(\mathcal{G})| \le \binom{x}{k}$ *and equality holds if and only if* $x$ *is an integer and* $\mathcal{G} = \binom{X}{k-1}$ *for some set* $X$ *of size* $x$.

Before proving the lemma we show that it is indeed equivalent to the theorem. We prove first that the theorem implies the lemma. If $\mathcal{G}$ satisfies the conditions of the lemma, then let us define $\mathcal{F} = \mathcal{K}_k^{k-1}(\mathcal{G})$. If $|\mathcal{F}| > \binom{x}{k}$ held, then by the theorem we would have $\binom{x}{k-1} < |\Delta(\mathcal{F})| \le |\mathcal{G}|$, a contradiction. Assume now the lemma holds and let $\mathcal{F}$ satisfy the conditions of the theorem. Suppose $|\Delta(\mathcal{F})| = \binom{y}{k-1}$ with $y < x$ and apply the lemma to $\mathcal{G} = \Delta(\mathcal{F})$. We obtain $|\mathcal{F}| \le |\mathcal{K}_k^{k-1}(\mathcal{G})| \le \binom{y}{k} < \binom{x}{k}$, a contradiction.

The equivalence of the cases of equality follows similarly.

**Proof of Lemma.** We proceed by induction on $k$ with the case $k-1 = 1$ being trivial. We will need the following definition. If $\mathcal{G}$ is a $k$-uniform family and $v$ is an element of the underlying set, then the *link* $\mathcal{L}(v)$ is a $(k-1)$-uniform family with $S \in \mathcal{L}(v)$ if and only if $S \cup \{v\} \in \mathcal{G}$.

**Claim 29** *(i)* $|\mathcal{K}_k^{k-1}(v)| \le |\mathcal{G}| - d_{\mathcal{G}}(v)$,

*(ii)*$|\mathcal{K}_k^{k-1}(v)| \le |K_{k-1}^{k-2}(\mathcal{L}(v))|$,

*(iii)* $|\mathcal{K}_k^{k-1}(v)| \le (\frac{x}{k-1} - 1)d_{\mathcal{G}}(v)$ *for every vertex* $v$ *and equality holds only if* $d_{\mathcal{G}}(v) = \binom{x-1}{k-2}$.

**Proof of Claim.** Observe that for a $k$-set $S$ not containing $v$, $S \cup \{v\}$ spans a copy of $\mathcal{K}_k^{k-1}$ in $\mathcal{G}$ if and only if $S \in \mathcal{G}$ and $S$ spans a copy of $\mathcal{K}_{k-1}^{k-2}$ in $\mathcal{L}(v)$. The first condition implies **(i)**, the second implies **(ii)**.

To see **(iii)**, suppose first that $d_{\mathcal{G}}(v) \ge \binom{x-1}{k-2}$. Then by **(i)**, we have $|\mathcal{K}_k^{k-1}(v)| \le \binom{x}{k-1} - d_{\mathcal{G}}(v) \le (\frac{x}{k-1} - 1)d_{\mathcal{G}}(v)$. If $d_{\mathcal{G}}(v) \le \binom{x-1}{k-2}$, then let $x_v \le x$ be the real number with $d_{\mathcal{G}}(v) = \binom{x_v-1}{k-2}$. Using induction and **(ii)**, we obtain $|\mathcal{K}_k^{k-1}(v)| \le |\mathcal{K}_{k-1}^{k-2}(\mathcal{L}(v))| \le \binom{x_v-1}{k-1} = (\frac{x_v}{k-1} - 1)d_{\mathcal{G}}(v) \le (\frac{x}{k-1} - 1)d_{\mathcal{G}}(v)$. In both cases, $d_{\mathcal{G}}(v) = \binom{x-1}{k-2}$ is necessary for all inequalities to hold with equality. □

Using **(iii)** of Claim 29 we have

$$k|\mathcal{K}_k^{k-1}(\mathcal{G})| = \sum_v |\mathcal{K}_k^{k-1}(v)| \le \sum_v (\frac{x}{k-1} - 1)d_{\mathcal{G}}(v)$$
$$= (\frac{x}{k-1} - 1)|\mathcal{G}|(k-1) = k\binom{x}{k}.$$

This finishes the proof of the inductive step for the inequality, and equality holds if all the degrees are equal to $\binom{x-1}{k-2}$. But then writing $n = |\cup_{G \in \mathcal{G}} G|$ we have $n\binom{x-1}{k-2} = \sum_v d_{\mathcal{G}}(v) = (k-1)|\mathcal{G}| = (k-1)\binom{x}{k-1} = x\binom{x-1}{k-2}$. Therefore $x$ equals $n$ and $\mathcal{G} = \binom{\cup_{G \in \mathcal{G}} G}{k-1}$. □

■

As a corollary, we give a **fourth proof of Theorem 9**. Let $\mathcal{F} \subseteq \binom{[n]}{k}$ be an intersecting family of size at least $\binom{n-1}{k-1}$ and let $\mathcal{F}' \subseteq \mathcal{F}$ be of size $\binom{n-1}{k-1}$. As $\overline{\mathcal{F}'} = \{\overline{F'} : F' \in \mathcal{F}'\}$ has size $\binom{n-1}{k-1} = \binom{n-1}{n-k}$ we can apply Theorem 27 $(n-2k)$ times to obtain that for $\mathcal{G} := \{G \in \binom{[n]}{k} : \exists \overline{F'} \in \overline{\mathcal{F}'}, G \subseteq \overline{F'}\}$ we have $|\mathcal{G}| \geq \binom{n-1}{k}$. As $\mathcal{F}'$ is intersecting we have $\mathcal{F}' \cap \mathcal{G} = \emptyset$ and thus $|\mathcal{F}'| + |\mathcal{G}| \leq \binom{n}{k}$. This is only possible if $|\mathcal{G}| = \binom{n-1}{k}$, $\mathcal{G} = \binom{X}{k}$ for some $(n-1)$-element subset $X$ of $[n]$, and $\mathcal{F} = \mathcal{F}' = \binom{[n]}{k} \setminus \mathcal{G}$. ∎

We continue with a real vertex isoperimetric problem. The *Hamming graph* $H_n$ (also called the *$n$-dimensional cube*) has vertex set $2^{[n]}$ and two sets $F, G \subseteq [n]$ are joined by an edge if and only if $|F \triangle G| = 1$; in other words, if and only if the characteristic vectors $v_F$, $v_G$ differ in exactly one entry. The *Hamming distance* $d(F,G)$ of two sets $F$ and $G$ is the size of $F \triangle G$. If $F, G \subseteq [n]$, then this equals the number of entries in which their characteristic vectors differ, and also their graph distance in $H_n$. The *Hamming distance* $d(\mathcal{F},\mathcal{G})$ of two families $\mathcal{F}$ and $\mathcal{G}$ is the smallest distance between their members, i.e $d(\mathcal{F},\mathcal{G}) = \min\{d(F,G) : F \in \mathcal{F}, G \in \mathcal{G}\}$. The family $\mathcal{A}$ is called a *(Hamming) sphere around* $F$ if $B_r(F) \subseteq \mathcal{A} \subseteq B_{r+1}(F)$ holds for some $r \geq 0$, where $B_k(F) = \{G \in 2^{[n]} : d(F,G) \leq k\}$ is the ball of radius $k$ at the vertex $F$. From the next theorem of Frankl and Füredi the vertex isoperimetric theorem for the graph $H_n$ will easily follow.

**Theorem 30 (Frankl, Füredi, [200])** *For any families $\mathcal{A}, \mathcal{B} \subseteq 2^{[n]}$ there exist spheres $\mathcal{A}_0, \mathcal{B}_0 \subseteq 2^{[n]}$ around $\emptyset$ and $[n]$ with $|\mathcal{A}| = |\mathcal{A}_0|$ and $|\mathcal{B}| = |\mathcal{B}_0|$ such that $d(\mathcal{A},\mathcal{B}) \leq d(\mathcal{A}_0,\mathcal{B}_0)$ holds.*

**Proof.** Let us consider the following two sets of pairs:

$$\{(A,A') : A \in \mathcal{A}, A' \notin \mathcal{A}, |A'| < |A|\}, \qquad \{(B,B') : B \in \mathcal{B}, B' \notin \mathcal{B}, |B'| > |B|\}.$$

If both of them are empty, then $\mathcal{A}$ and $\mathcal{B}$ are spheres centered at $\emptyset$ and $[n]$, respectively. If not, then pick a pair with minimal symmetric difference $|A \triangle A'|$ or $|B \triangle B'|$. Assume $(A_0, A_0')$ is one such pair and define

$$U = A_0 \setminus A_0', \qquad V = A_0' \setminus A_0 \qquad (\text{thus } |U| > |V|).$$

We introduce the following shifting type operations on $\mathcal{A}$ and $\mathcal{B}$:

$$\mu(A) = \begin{cases} (A \setminus U) \cup V & \text{if } U \subseteq A, V \cap A = \emptyset \text{ and } (A \setminus U) \cup V \notin \mathcal{A} \\ A & \text{otherwise} \end{cases}, \tag{1.4}$$

and

$$\nu(B) = \begin{cases} (B \setminus V) \cup U & \text{if } V \subseteq B, U \cap B = \emptyset \text{ and } (B \setminus V) \cup U \notin \mathcal{B} \\ B & \text{otherwise} \end{cases}. \tag{1.5}$$

Note that if $|U| = |V|$, then $\mu$ is the shifting operation $S_{V,U}$ and $\nu$ is $S_{U,V}$.

We decided to use different notation as if $|U| \neq |V|$, then these operations might increase or decrease the size of a set. We let $\mu(\mathcal{A}) = \{\mu(A) : A \in \mathcal{A}\}$ and $\nu(\mathcal{B}) = \{\nu(B) : B \in \mathcal{B}\}$. By definition, we have $|\mu(\mathcal{A})| = |\mathcal{A}|, |\nu(\mathcal{B})| = |\mathcal{B}|$ and $\mu(A_0) = A_0'$. Because of the latter, the expression $\sum_{B\in\mathcal{B}} |B| - \sum_{A\in\mathcal{A}} |A|$ strictly increases whenever such a pair of operations is performed. As it cannot be more than $(n+1)2^{n-1} = \sum_{F\in 2^{[n]}} |F|$ the process will stop and we obtain two spheres.

To complete the proof of the theorem, we need to show that $d(\mu(\mathcal{A}), \nu(\mathcal{B})) \geq d(\mathcal{A}, \mathcal{B})$ holds. If $A \in \mu(\mathcal{A}) \cap \mathcal{A}, B \in \nu(\mathcal{B}) \cap \mathcal{B}$, then by definition $d(A, B) \geq d(\mathcal{A}, \mathcal{B})$ holds. If $A \in \mu(\mathcal{A}) \setminus \mathcal{A}, B \in \nu(\mathcal{B}) \setminus \mathcal{B}$, then $A' = A \cup U \setminus V \in \mathcal{A}, B' = B \cup V \setminus U \in \mathcal{B}$ and $A \triangle B = A' \triangle B'$ and thus $d(A, B) = d(A', B') \geq d(\mathcal{A}, \mathcal{B})$.

Finally we consider the case when one of $A$ and $B$ is new and the other is unchanged, say, $A \in \mu(\mathcal{A}) \cap \mathcal{A}, B \in \nu(\mathcal{B}) \setminus \mathcal{B}$. We write $B' = B \setminus U \cup V$ and as $B \in \nu(\mathcal{B}) \setminus \mathcal{B}$, we have $B' \in \mathcal{B}$. We distinguish three cases:

Case I: $U \subset A, V \cap A = \emptyset$.
This means $A' = A \setminus U \cup V \in \mathcal{A}$ and $d(A, B) = d(A', B') \geq d(\mathcal{A}, \mathcal{B})$.

Case II: $U \not\subset A$ and $V = \emptyset$.
Then for the pair $A_0, A_0'$ we have $A_0' \subset A_0$. If we consider a longest chain of sets from $A_0'$ to $A_0$, there should be a pair of sets next to each other in that chain (thus with Hamming distance 1), such that the smaller one is not in $\mathcal{A}$, and the larger one is in $\mathcal{A}$. So by the minimality of $|A_0 \triangle A_0'|$ we must have $|U| = 1$. Therefore, $A \triangle B = A \triangle (B' \cup U) = (A \triangle B') \cup U$, so we have $d(A, B) \geq d(\mathcal{A}, \mathcal{B}) + 1$.

Case III: ($U \not\subset A$ or $V \cap A \neq \emptyset$), and $1 \leq |V| < |U|$.
Then we can pick $u \in U, v \in V$ such that either $u \notin A$ or $v \in A$ holds (or both). Let us define $B'' := B \cup \{v\} \setminus \{u\}$. As $|B''| = |B| > |B'|$ and $|B'' \triangle B'| = |(U \setminus \{u\} \cup (V \setminus \{v\})| < |U \cup V| = |A_0 \triangle A_0'|$, by the minimality of $|A_0 \triangle A_0'|$ we must have $B'' \in \mathcal{B}$. But then we obtain

$$d(A, B) = d(A, B'' \cup \{u\} \setminus \{v\}) \geq d(A, B'') \geq d(\mathcal{A}, \mathcal{B}),$$

where the first inequality follows from the choice of $u$ and $v$. ■

**Theorem 31 (Harper [303])** *For every $m$ with $0 \leq m \leq 2^n$ there exists a sphere $\mathcal{S} \subseteq 2^{[n]}$ of size $m$ that minimizes the boundary in the Hamming graph $H_n$.*

**Proof.** Let $\mathcal{F} \subseteq 2^{[n]}$ be a family of size $m$. Apply Theorem 30 to $\mathcal{A} := \mathcal{F}$ and $\mathcal{B} := ext(\mathcal{F})$. As $\partial(\mathcal{A}_0) \subseteq 2^{[n]} \setminus (\mathcal{A}_0 \cup \mathcal{B}_0)$, the theorem follows. ■

Theorem 31 gives a complete solution to the vertex isoperimetric problem for $H_n$ if $m = |B_r(F)| = \sum_{i=0}^{r} \binom{n}{i}$. The answer to the question of how to choose the sets at distance $r+1$ if $|B_r(F)| < m < |B_{r+1}(F)|$ is given by the

Kruskal-Katona theorem. Indeed, if $\mathcal{A}$ is a sphere around $[n]$ with $B_r([n]) \subseteq \mathcal{A} \subseteq B_{r+1}([n])$, then the subfamily $\mathcal{A}'$ of sets that are at distance $r+1$ from $[n]$ all have size $n-r-1$. As $\binom{[n]}{n-r} \subseteq \mathcal{A}$ we have $\partial\mathcal{A} = (\binom{[n]}{n-r-1} \setminus \mathcal{A}) \cup \Delta(\mathcal{A}')$. As Theorem 23 tells us how to minimize $|\Delta(\mathcal{A}')|$, together with Theorem 31 they give a complete solution to the vertex isoperimetric problem for $H_n$.

We conclude this section with some remarks on the edge isoperimetric problem for the Johnson graph $H_{n,k}$. The vertex set of $H_{n,k}$ is $\binom{[n]}{k}$ and two sets $F, G$ are connected by an edge if and only if $|F \triangle G| = 2$ or equivalently $|F \cap G| = k-1$. As $H_{n,k}$ is $k(n-k)$-regular, for any family $\mathcal{F} \subseteq \binom{[n]}{k}$ we have $k(n-k)|\mathcal{F}| = 2e(\mathcal{F}) + e(\mathcal{F}, \partial\mathcal{F})$. Therefore the edge isoperimetric problem is equivalent to maximizing the number $e(\mathcal{F})$ of edges joining sets within $\mathcal{F}$ over all $\mathcal{F} \subseteq \binom{[n]}{k}$ with $|\mathcal{F}| = m$ for some fixed $m$. Harper [304] considered a continuous version of this problem; we present his preliminary remarks that involve only the original discrete problem.

**Lemma 32** *For any $\mathcal{F} \subseteq \binom{[n]}{k}$ and $x, y \in [n]$ the inequality $e(\mathcal{F}) \leq e(S_{x,y}(\mathcal{F}))$ holds.*

**Proof.** Let us denote by $d_{\mathcal{F}}(F)$ and $d_{S_{x,y}(\mathcal{F})}(F)$ the degree of $F$ in $H_{n,k}[\mathcal{F}]$ and $H_{n,k}[S_{x,y}(\mathcal{F})]$, respectively. We will show that $\sum_{F\in\mathcal{F}} d_{\mathcal{F}}(F) \leq \sum_{F\in\mathcal{F}} d_{S_{x,y}(\mathcal{F})}(S_{x,y}(F))$ holds.

By definition, if $F$ and $G$ are joined by an edge in $H_{n,k}$, then $F = I \cup \{f\}$, $G = I \cup \{g\}$ for some $(k-1)$-set $I$ and $f \neq g$, $f, g \notin I$. This shows that for an $F \in \mathcal{F}$ if it contains both $x, y$ (or similarly if it contains neither of them), then $S_{x,y}(F) = F$ and $d_{\mathcal{F}}(F) = d_{S_{x,y}(\mathcal{F})}(F)$ holds.

For an $F \in \mathcal{F}$ if $x \in F, y \notin F$ and $F \setminus \{x\} \cup \{y\} \notin \mathcal{F}$, then $S_{x,y}(F) = F$ and for all $G$ with $(F, G)$ being an edge we have $S_{x,y}(G) = G$ thus $d_{\mathcal{F}}(F) \leq d_{S_{x,y}(\mathcal{F})}(F)$.

If $x \notin F, y \in F$ and $F \setminus \{y\} \cup \{x\} \notin \mathcal{F}$, then $S_{x,y}(F) = F \setminus \{y\} \cup \{x\} =: F'$. Let $\mathcal{G} = \{G \in \mathcal{F} : y \notin G, (F, G) \in E(H_{n,k})\}$ and $\mathcal{G}' = \{G \in \mathcal{F} : y \in G, (F, G) \in E(H_{n,k})\}$. Then for any $G \in \mathcal{G}$ we have $S_{x,y}(G) = G$ and $(F', G) \in E(H_{n,k})$, while for any $G \in \mathcal{G}'$ we have $G' = G \setminus \{y\} \cup \{x\} \in S_{x,y}(\mathcal{F})$ and $(F', G') \in E(H_{n,k})$. This shows $d_{\mathcal{F}}(F) \leq d_{S_{x,y}(\mathcal{F})}(F')$.

Finally, if for some $(k-1)$-set $I$ with $x, y \notin I$ both $F = I \cup \{y\}$ and $F' = I \cup \{x\}$ belong to $\mathcal{F}$, then we claim that $d_{\mathcal{F}}(F) + d_{\mathcal{F}}(F') \leq d_{S_{x,y}(\mathcal{F})}(F) + d_{S_{x,y}(\mathcal{F})}(F')$ holds. Indeed, if $I \subset G \in \mathcal{F}$, then $S_{x,y}(G) = G$ and $G$ is connected to both $F$ and $F'$. Otherwise, a set $G \in \mathcal{F}$ can be connected to at most one of $F$ and $F'$ and if it is connected to one of them, then we have $G = J \cup \{z\} \cup \{a\}$ for a $(k-2)$-subset $J$ of $I$, $z \in \{x, y\}$, $a \in [n] \setminus (I \cup \{x, y\})$. For fixed $J$ and $a$, the number of $z$'s for which $J \cup \{a\} \cup \{z\}$ belongs to $\mathcal{F}$ and $S_{x,y}(\mathcal{F})$ is equal. ■

For a set $F \in \binom{[n]}{k}$ let $y_F(j)$ be the number of zero coordinates of its characteristic vector $v_F$ before the $j$th 1-coordinate. It is easy to see that $y_F(j) = i_j - j$, where $i_j$ is the index of the $j$th one entry of $v_F$. Also we have

$0 \leq y_F(1) \leq y_F(2) \leq \cdots \leq y_F(k) \leq n-k$. Let $L_{a,b} = \{x \in [0,b]^a : x(1) \leq x(2) \leq \cdots \leq x(a)\}$ and we define the relation $x \leq_{L_{a,b}} y$ if and only if $x(i) \leq y(i)$ for all $1 \leq i \leq a$. Then the vectors $y_F = (y_F(1), y_F(2), \ldots, y_F(k))$ form the poset $L_{k,n-k}$ under coordinate-wise ordering. For the next observation of Harper, we need the following definition: a set $D$ is a *downset* in a poset $P$ if $d' \leq_P d \in D$ implies $d' \in D$.

**Lemma 33** *A family $\mathcal{F} \subset \binom{[n]}{k}$ is left-shifted if and only if the set $Y_{\mathcal{F}} = \{y_F : F \in \mathcal{F}\}$ is a downset in $L_{k,n-k}$.*

**Proof.** If $\mathcal{F}$ is not left-shifted and $F \in \mathcal{F}$ and $F' = F \setminus \{y\} \cup \{x\} \notin \mathcal{F}$ with $x < y$ holds, then $y_{F'} \prec_{L_{k,n-k}} y_F$ and thus $Y_{\mathcal{F}}$ is not a downset.

Conversely, if $Y_{\mathcal{F}}$ is not a downset, then consider a pair $G \notin \mathcal{F}$, $F \in \mathcal{F}$ with $y_G \prec y_F$ and $\sum_{i=1}^{k}(y_F(i) - y_G(i))$ being minimal among such pairs. Let $i$ be the smallest integer with $y_G(i) < y_F(i)$. Then $x = i + y_G(i) \in G$ and $x \notin F$. Let $y$ be the smallest integer with $y \in F, y \notin G$. Obviously, $x < y$ and by the choice of $F$ and $G$ we must have $G = F \setminus \{y\} \cup \{x\}$. It follows that $\mathcal{F}$ is not $S_{x,y}$-shifted and therefore not left-shifted. ■

If $F, F'$ are endpoints of an edge in $H_{n,k}$, then for some $i \neq j$ we have $v_F(i) = 0, v_{F'}(i) = 1, v_F(j) = 1, v_{F'}(j) = 0$ and $v_F(l) = v_{F'}(l)$ for all $l \in [n]$, $l \neq i, j$. If $i < j$, then this means that $F'$ could be obtained from $F$ by using $S_{i,j}$ and therefore $y_{F'} \leq_{L_{k,n-k}} y_F$ holds. Moreover, the number of edges for which $F$ is the "upper endpoint" is $r(y_F) = \sum_{i=1}^{k} y_F(i)$. If $\mathcal{F}$ is left-shifted and $F \in \mathcal{F}$, then all lower endpoints of such edges belong to $\mathcal{F}$, thus the number of edges spanned by $\mathcal{F}$ in $H_{n,k}$ is $\sum_{F \in \mathcal{F}} r(y_F)$. Therefore, the isoperimetric problem in $H_{n,k}$ is equivalent to maximizing $\sum_{y \in Y} r(y)$ over all downsets $Y \subset L_{k,n-k}$ of size $m$.

**Exercise 34 (Patkós, [458])** *Let $\mathcal{F} \subseteq \binom{[n]}{k}$ be a family of size $m < \binom{\delta n}{\delta n/2}$, with $0 \leq \delta < 1$ and $k > \delta n/2$. Then in $H(n,k)$ we have $e(\mathcal{F}) \leq \delta m n^2$.*

**Exercise 35** *Give a proof of Theorem 27 using induction and ordinary shifting operations $S_{x,y}$.*

- *For a left-shifted family $\mathcal{F} \subseteq 2^{[n]}$ consider $\mathcal{F}_0 = \{F \in \mathcal{F} : 1 \notin F\}$ and $\mathcal{F}_1 = \{F \setminus \{1\} : 1 \in F \in \mathcal{F}\}$.*
- *Prove that $\Delta(\mathcal{F}_0) \subseteq \mathcal{F}_1$ and $\{G \cup \{1\} : G \in \Delta(\mathcal{F}_1)\} \cup \mathcal{F}_1 \subseteq \Delta(\mathcal{F})$ hold.*
- *Distinguish two cases according to whether $|\mathcal{F}_1|$ is greater or smaller than $\binom{x-1}{k-1}$.*

## 1.5 Sunflowers

The aim of this last section of Chapter 1 is to present a very recent application of the polynomial method and also to introduce the important notion of a sunflower. For a short survey paper in the topic that predates the main result presented in this section see [382]. A *sunflower* (also known as a $\Delta$-system) is a family $\mathcal{F}$ of sets such that for any distinct $F, F' \in \mathcal{F}$ we have $F \cap F' = \cap_{F^* \in \mathcal{F}} F^*$. The *kernel* $K$ of the sunflower $\mathcal{F}$ is $K = \cap_{F \in \mathcal{F}} F$ and the *petals* of $\mathcal{F}$ are $\{F \setminus K : F \in \mathcal{F}\}$. A sunflower of cardinality $p$ is called a sunflower with $p$ petals. Families without large sunflowers were first considered by Erdős and Rado [166]. They obtained the following simple lemma.

**Lemma 36 (Sunflower-Lemma, [166])** *Let $\mathcal{F}$ be family of sets each of size $k$. If $\mathcal{F}$ does not contain a sunflower with $p$ petals, then $|\mathcal{F}| \le k!(p-1)^k$ holds.*

**Proof.** We proceed by induction on $k$. For $k = 1$ any family of singletons form a sunflower with empty kernel, so $|\mathcal{F}| \le p - 1$ must hold. Let $k > 2$ and suppose the statement of the lemma holds for $k - 1$. Suppose towards a contradiction that there is a family $\mathcal{F}$ of $k$-sets with $|\mathcal{F}| > k!(p-1)^k$ not containing a sunflower with $p$ petals. Let $\mathcal{M} \subseteq \mathcal{F}$ be a maximal subfamily of pairwise disjoint sets. As pairwise disjoint sets form a sunflower, we have $|\mathcal{M}| \le p - 1$. Let us write $A = \cup_{M \in \mathcal{M}} M$. As $|A| \le k(p-1)$, there exists an $a \in A$ contained in at least $|\mathcal{F}|/|A| > (k-1)!(p-1)^{k-1}$ sets of $\mathcal{F}$. Therefore, by induction, the family $\mathcal{F}_a = \{F \setminus \{a\} : F \in \mathcal{F}, a \in F\}$ contains a sunflower $\{F_1 \setminus \{a\}, F_2 \setminus \{a\}, \dots, F_p \setminus \{a\}\}$ with kernel $K$, and as all $F_i$'s contain $a$, the family $\{F_1, F_2, \dots, F_p\}$ is a sunflower with kernel $K \cup \{a\}$. ■

Erdős and Rado did not find any construction that would show that the upper bound of Lemma 36 is sharp (see their best construction at the end of this section as an exercise). They conjectured that the dependence of their bound from $k$ is only exponential.

**Conjecture 37 (Erdős, Rado, [166])** *For every $p$ there exists a constant $C_p$ such that for every family $\mathcal{F}$ of $k$-sets that does not contain a sunflower with $p$ petals $|\mathcal{F}| \le C_p^k$ must hold.*

There was a weaker conjecture by Erdős and Szemerédi [168] that involved sets of arbitrary size, but in an underlying set of size $n$. They conjectured that there exists a constant $c < 2$ such that if $\mathcal{F} \subseteq 2^{[n]}$ does not contain a sunflower with three petals, then $|\mathcal{F}| \le c^n$ holds. They showed that Conjecture 37 implies their conjecture. It was proved by Alon, Shpilka, and Umans [18] that the so-called Strong Cap Set conjecture also implies the Erdős-Szemerédi conjecture. This conjecture stated that there is $\delta > 0$ such that a subset of $\mathbb{Z}_3^n$ without a solution to $x + y + z = 0$ with $x, y, z$ being distinct, contains at most $(3-\delta)^n$ elements.

The Strong Cap Conjecture was recently proved by Ellenberg and Gijswijt [141] based on ideas of the paper by Croot, Lev, and Pach [110]. Here we present a direct proof of the Erdős-Szemerédi conjecture due to Naslund and Sawin [442]. It also uses the polynomial method with the ideas of Croot, Lev, and Pach.

Before starting the proof, we need to define the *rank of a function* taking values in a field $\mathbb{F}$. A function $F : A \times A \to \mathbb{F}$ is of rank one if it is non-zero and of the form $(x, y) \mapsto f(x)g(y)$ for some $f, g : A \to \mathbb{F}$. Any function $F : A \times A \to \mathbb{F}$ can be expressed as a linear combination of rank one functions. The rank of $F$ is the least number of rank one functions in such a linear combination. For example, let $\delta$ be the *Kronecker-delta* function with $\delta_a(x) = 1$ if and only if $x = a$. Then the function $F = \sum_{a \in A} c_a \delta_a(x)\delta_a(y)$ has rank $|A|$. Indeed, for every function $G : A \times A \to \mathbb{F}$ one can consider the $|A| \times |A|$ matrix $M_G$ with $M_G(a, a') = G(a, a')$. Then the matrix of $M_F$ is the matrix with non-zero entries $c_a$ in its diagonal, therefore the matrix rank of $M_F$ is $|A|$. Clearly, the matrix rank of $M_G$ for a rank one function $G$ is one. Therefore by the subadditivity of the matrix rank we need at least $|A|$ rank one functions to generate $F$ as a linear combination.

The above holds more generally. If $k \geq 2$, rank one functions are of the form $(x_1, \ldots, x_k) \mapsto f(x_i)g(x_1, \ldots, x_{i-1}, x_{i+1}, \ldots, x_k)$ for an integer $i$ with $1 \leq i \leq k$ and some functions $f : A \to \mathbb{F}$, $g : A^{k-1} \to \mathbb{F}$. Again we define the rank of a function $F : A^k \to \mathbb{F}$ to be the least number of rank one functions that are needed to generate $F$ as a linear combination. With an inductive argument one can prove the following lemma.

**Lemma 38** *Let $k \geq 2$, let $A$ be a finite subset of the field $\mathbb{F}$, and for each $a \in A$, let $c_a \in \mathbb{F}$ be a non-zero coefficient. Then the rank of the function $(x_1, \ldots, x_k) \mapsto \sum_{a \in A} c_a \delta_a(x_1) \ldots \delta_a(x_k)$ is $|A|$.*

The proof is from Tao's blogpost [527].

**Proof.** We proceed by induction on $k$. As mentioned above, the case $k = 2$ follows from the subadditivity of matrix rank, so suppose that $k > 2$ and the claim has already been proven for $k - 1$.

For the function in the statement of the lemma, the summands on the right-hand side are rank one functions, therefore its rank is at most the number of non-zero $c_a$'s. So it suffices to establish the lower bound. Now suppose for contradiction that the function has rank at most $|A| - 1$, then we obtain a representation

$$\sum_{a \in A} c_a \delta_a(x_1) \ldots \delta_a(x_k) = \sum_{i=1}^{k} \sum_{\alpha \in I_i} f_{i,\alpha}(x_i) g_{i,\alpha}(x_1, \ldots, x_{i-1}, x_{i+1}, \ldots, x_k) \tag{1.6}$$

for some sets $I_1, \ldots, I_k$ of cardinalities adding up to at most $|A| - 1$, and some functions $f_{i,\alpha} : A \to \mathbb{F}$ and $g_{i,\alpha} : A^{k-1} \to \mathbf{R}$.

Consider the space of functions $h : A \to \mathbb{F}$ that are orthogonal to all the $f_{k,\alpha}$, $\alpha \in I_k$ in the sense that $\sum_{x \in A} f_{k,\alpha}(x)h(x) = 0$ for all $\alpha \in I_k$. This space is a vector space whose dimension $d$ is at least $|A| - |I_k|$. A basis of this space generates a $d \times |A|$ matrix of full rank, which implies that there is at least one non-singular $d \times d$ minor. This implies that there exists a function $h : A \to \mathbf{F}$ in this space which is nowhere vanishing on some subset $A'$ of $A$ of cardinality at least $d$.

If we multiply (1.6) by $h(x_k)$ and sum in $x_k$, we conclude that

$$\sum_{a \in A} c_a h(a)\delta_a(x_1)\dots\delta_a(x_{k-1}) =$$

$$= \sum_{i=1}^{k-1} \sum_{\alpha \in I_i} f_{i,\alpha}(x_i)\tilde{g}_{i,\alpha}(x_1, \dots, x_{i-1}, x_{i+1}, \dots, x_{k-1})$$

where

$$\tilde{g}_{i,\alpha}(x_1, \dots, x_{i-1}, x_{i+1}, \dots, x_{k-1}) := \sum_{x_k \in A} g_{i,\alpha}(x_1, \dots, x_{i-1}, x_{i+1}, \dots, x_k)h(x_k).$$

The right-hand side has rank at most $|A| - 1 - |I_k|$, since the summands are rank one functions. On the other hand, from induction hypothesis the left-hand side has rank at least $d \geq |A| - |I_k|$, giving the required contradiction. ■

**Theorem 39 (Naslund, Sawin [442])** *If $\mathcal{F} \subseteq 2^{[n]}$ does not contain a sunflower with 3 petals, then $|\mathcal{F}| \leq 3n \sum_{k \leq n/3} \binom{n}{k}$ holds.*

**Proof.** It is enough to prove that for any $1 \leq j \leq n$ we have $|\mathcal{F}_j| \leq 3 \sum_{k \leq n/3} \binom{n}{k}$, where $\mathcal{F}_j = \{F \in \mathcal{F} : |F| = j\}$. Let $V_j = \{v_F : F \in \mathcal{F}_j\}$ be the set of characteristic vectors. Note that $\mathcal{F}_j$ does not contain a sunflower with 3 petals if and only if for any three distinct vectors $x, y, z \in V_j$ there exists an index $1 \leq i \leq n$ such that out of $x(i), y(i), z(i)$ exactly two is 1 and one is 0. Hence for $F, G, H \in \mathcal{F}_j$ we cannot have $v_F + v_G + v_H \in \{0, 1, 3\}^n$ unless $F = G = H$. Indeed, if $F, G, H$ are all distinct, then it follows from the sunflower-free property, while if $F = G \neq H$, then there exists $i \in F, G$, $i \notin H$, therefore $v_F(i) + v_G(i) + v_H(i) = 2$ (this is the only place we use that $F, G$, and $H$ have the same size). Let us define the function $T : V_j \times V_j \times V_j \to \mathbb{R}$ by

$$T(x, y, z) = \prod_{i=1}^{n} (2 - (x_i + y_i + z_i)).$$

By the above, $T(x, y, z) \neq 0$ if and only if $x = y = z$, i.e. it is of the form $\sum_{v \in V_j} T(v, v, v)\delta_v(x)\delta_v(y)\delta_v(z)$. Thus, by Lemma 38, its rank is $|\mathcal{F}_j|$. On the other hand, expanding the above product we can write $T(x, y, z)$ as sums of monomials $x_1^{h_1}x_2^{h_2}\dots x_n^{h_n}y_1^{k_1}y_2^{k_2}\dots y_n^{k_n}z_1^{l_1}z_2^{l_2}\dots z_n^{l_n}$ where the sum of the exponents is at most $n$. Therefore, for every such monomial at least one of $\sum_{i=1}^n h_i$, $\sum_{i=1}^n k_i$, $\sum_{i=1}^n l_i$ is at most $n/3$. Hence we can partition these monomials accordingly to write $T(x,y,z)$ as

$$\sum_{h:\sum_{i=1}^n h_i \le n/3} x_1^{h_1}x_2^{h_2}\dots x^{h_n} g_h(y,z) + \sum_{k:\sum_{i=1}^n k_i \le n/3} y_1^{k_1}y_2^{k_2}\dots y^{k_n} g_k(x,z)$$
$$+ \sum_{l:\sum_{i=1}^n l_i \le n/3} z_1^{l_1}z_2^{l_2}\dots z^{l_n} g_l(x,y).$$

Therefore, by definition, we obtained that the rank of $T$ is at most $3\sum_{i=1}^{n/3}\binom{n}{i}$. This shows $|\mathcal{F}_j| \le 3\sum_{i=1}^{n/3}\binom{n}{i}$ as claimed. ■

**Exercise 40 (Erdős, Rado, [166])** *Let $X_1, X_2, \dots, X_k$ be pairwise disjoint sets of size $r-1$. Show that the family $\mathcal{F} = \{F : |F \cap X_i| = 1$ for all $i = 1, 2, \dots, k\}$ does not contain a sunflower with $r$ petals.*

# 2

# *Intersection theorems*

## CONTENTS



## 2.1 Stability of the Erdős-Ko-Rado theorem

Most of the extremal problems we consider in this book ask for the maximum possible size of a family that satisfies a prescribed property. If this maximum size is determined, then one can try to describe all families attaining this size, i.e. the extremal families. If (up to isomorphism) there is a unique extremal family, or there is just a limited number of them, then one can hope for a stability result. Such a theorem would state that every family of almost extremal size, must have very similar structure to (one of) the extremal family(ies). The first such result was obtained by Erdős [153] and Simonovits [513] who proved the following stability version of Turán's famous theorem on graphs with clique number at most $p$: for every $\delta > 0$ there exists an $\varepsilon > 0$ and $n_0$ such that if a graph on $n \geq n_0$ vertices does not contain a clique of size $p+1$ and has at most $\varepsilon n^2$ fewer edges than the extremal graphs, then after the removal of at most $\delta n^2$ edges of $G$ one can obtain a subgraph of an extremal graph.

In this section we prove stability versions of Theorem 9 and state some other related results. First, we prove a theorem of Hilton and Milner that determines the size of the "second largest" intersecting family $\mathcal{F} \subseteq \binom{[n]}{k}$. Note that if $n = 2k$, then all maximal $k$-uniform intersecting families contain exactly one set from every complement pair of $k$-sets and therefore there is no uniqueness, no stability of the extremal families. By Theorem 9, we know that if $2k + 1 \leq n$ holds, then maximal trivially intersecting families are the only maximum size intersecting families. The next theorem gives the maximum size of a non-trivially intersecting family. One could rephrase the result as that any intersecting family, whose size is bigger than the bound below, is a subfamily of an extremal family.

The following two families are the candidates to be the largest non-trivially intersecting families.

$$\mathcal{F}^3 = \left\{ F \in \binom{[n]}{k} : |F \cap [3]| \geq 2 \right\},$$

$$\mathcal{F}_{HM} = \{[2, k+1]\} \cup \left\{ F \in \binom{[n]}{k} : 1 \in F, |F \cap [2, k+1]| \geq 1 \right\}.$$

**Theorem 41 (Hilton, Milner, [309])** *If $\mathcal{F} \subseteq \binom{[n]}{k}$ is a non-trivially intersecting family with $k \geq 2$ and $2k + 1 \leq n$, then we have $|\mathcal{F}| \leq \binom{n-1}{k-1} - \binom{n-k-1}{k-1} + 1$. Moreover, equality holds if and only if $\mathcal{F}$ is isomorphic to $\mathcal{F}_{HM}$ or if $k = 3$ and $\mathcal{F}$ is isomorphic to $\mathcal{F}^3$.*

The proof we present here is due to Frankl and Füredi [204]. It uses the shifting technique and is basically an extension of the first uniqueness proof of Theorem 9. There we stop applying shifting operations when we arrive to a family of size smaller than $\binom{n-1}{k-1}$; here we further examine that family.

**Proof.** We proceed by induction on $k$. The base case $k = 2$ is easy as there is only one non-trivial intersecting family: the triangle $\{\{1, 2\}, \{1, 3\}, \{2, 3\}\}$. To prove the inductive step, let $\mathcal{F} \subseteq \binom{[n]}{k}$ be a non-trivial intersecting family. Our goal is to find a non-trivial intersecting family $\mathcal{H} \subseteq \binom{[n]}{k}$ with $|\mathcal{H}| = |\mathcal{F}|$ and a set $X$ of size at most $2k$ such that for any $H, H' \in \mathcal{H}$ we have $H \cap H' \cap X \neq \emptyset$. Let us start applying shifting operations $S_{x,y}$ with $x < y$.

We consider two cases. If we obtain a left-shifted family $\mathcal{H}$ that is not a star, we can find $X$ by Lemma 11. Otherwise at some point we obtain a family $\mathcal{G}$ and elements $x_1, y_1$ such that $S_{x_1,y_1}(\mathcal{G})$ is a star. Therefore $G \cap \{x_1, y_1\} \neq \emptyset$ for all $G \in \mathcal{G}$. Whenever this occurs, instead of applying $S_{x_1,y_1}$ we let $X_1 = \{x_1, y_1\}$ and from that point we apply only the shifting operations $S_{x',y'}$ with $x' < y'$ and $x', y' \in [n] \setminus X_1$. We continue this procedure; we always have a set $X_i$, and we only apply shifting operations $S_{x',y'}$ with $x' < y'$ and $x', y' \in [n] \setminus X_i$. Whenever we arrive to a family $\mathcal{G}$ such that $S_{x_{i+1},y_{i+1}}(\mathcal{G})$ is a star for some $x_{i+1} < y_{i+1}$ and $x_{i+1}, y_{i+1} \in [n] \setminus X_i$, instead of applying $S_{x_{i+1},y_{i+1}}$ we let $X_{i+1} = X_i \cup \{x_{i+1}, y_{i+1}\}$. We continue this as long as we can change the family this way, i.e. until we obtain a family $\mathcal{H}$ with $S_{x',y'}(\mathcal{H}) = \mathcal{H}$ for all $x' < y'$ such that $x', y' \in [n] \setminus X_j$.

At this point $X_j = \{x_1, y_1, x_2, y_2, \ldots, x_j, y_j\}$. Clearly, we still have $H \cap \{x_i, y_i\} \neq \emptyset$ for every $H \in \mathcal{H}$ and every $i \leq j$. This implies $j \leq k$. Let $X$ be the union of $X_j$ and the first $2k - 2j$ elements of $[n] \setminus X_j$. Similarly to Lemma 12 we can show that for any $H, H' \in \mathcal{H}$ we have $H \cap H' \cap X \neq \emptyset$. If $j = k$, then for any $H \in \mathcal{H}$ we have $H \subseteq X_k = X$. So we can assume $j < k$; let us suppose towards a contradiction that $H, H'$ are members of $\mathcal{H}$ with $H \cap H' \cap X = \emptyset$ and $|H \cap H'|$ minimal. This means that $H$ and $H'$ meet $\{x_i, y_i\}$ in different elements for every $i \leq j$, and as $\mathcal{H}$ is intersecting there exists $y' \in H \cap H' \setminus X$. Also there exists $x' \in (X \setminus X_j) \setminus (H \cup H')$ as $H \setminus X_j$ and $H' \setminus X_j$ are $(k-j)$-sets both containing $y' \notin A$. Then $H \setminus \{y'\} \cup \{x'\} \in \mathcal{H}$ by $S_{x',y'}(\mathcal{H}) = \mathcal{H}$ which contradicts the choice of $H$ and $H'$.

We introduce the families $\mathcal{F}_i = \{F \in \mathcal{F} : |F \cap X| = i\}$ and $\mathcal{A}_i = \{F \cap X : F \in \mathcal{F}_i\}$ for $i = 1, 2, \ldots, k$.

**Lemma 42** *For $i = 1, 2, \ldots, k-1$ we have $|\mathcal{A}_i| \leq \binom{2k-1}{i-1} - \binom{k-1}{i-1}$.*
*Also $|\mathcal{A}_k| \leq \binom{2k-1}{k-1} - \binom{k-1}{k-1} + 1 = \binom{2k-1}{k-1} = \frac{1}{2}\binom{2k}{k}$ holds.*

**Proof of Lemma.** The fact that $\mathcal{A}_1 = \emptyset$ follows from the non-triviality of $\mathcal{H}$. The statement on $\mathcal{A}_k$ follows from the fact that a $k$-subset $A$ and its complement $[2k] \setminus A$ cannot belong to an intersecting family at the same time.

For $i = 2, 3, \ldots, k-1$ if the statement on $|\mathcal{A}_i|$ does not hold, then by the inductive hypothesis of the theorem we obtain that $\mathcal{A}_i$ is trivial, i.e. there exists $x \in X$ with $x \in A$ for all $A \in \mathcal{A}_i$. As $\mathcal{H}$ is non-trivial, there exists $H \in \mathcal{H}$ with $x \notin H$. As $H \cap A \neq \emptyset$, we obtain $|\mathcal{A}_i| \leq \binom{2k-1}{i-1} - \binom{k-1}{i-1}$. □

As any $A \in \mathcal{A}_i$ can be extended to a $k$-subset of $[n]$ in $\binom{n-2k}{k-i}$ ways, we have $|\mathcal{F}_i| \leq |\mathcal{A}_i|\binom{n-2k}{k-i}$. Therefore, by Lemma 42 we obtain

$$\begin{aligned} |\mathcal{F}| = \sum_{i=1}^{k} |\mathcal{F}_i| &\leq \sum_{i=1}^{k} |\mathcal{A}_i| \binom{n-2k}{k-i} \\ &\leq 1 + \sum_{i=1}^{k} \left( \binom{2k-1}{i-1} - \binom{k-1}{i-1} \right) \binom{n-2k}{k-i} \\ &= 1 + \binom{n-1}{k-1} - \binom{n-k-1}{k-1}, \end{aligned}$$

which proves the upper bound of the theorem.

If $|\mathcal{F}| = 1 + \binom{n-1}{k-1} - \binom{n-k-1}{k-1}$, then for the family $\mathcal{H}$ obtained by shifting, we must have equality in Lemma 42 for all $i = 2, 3, \ldots, k$; in particular $|\mathcal{A}_2| = k$ must hold. $\mathcal{A}_2$ is either the triangle and thus $\mathcal{H} \subseteq \mathcal{F}^3$ and $k$ should be 3, or $\mathcal{A}_2$ is a star with $k$ elements and with center $x$. As $\mathcal{H}$ is non-trivial, there exists $H \in \mathcal{H}$ with $x \notin H$. But we know that $H$ is a $k$-set intersecting every $A \in \mathcal{A}_2$, and there is only one such $k$-set, namely $H = \cup_{A \in \mathcal{A}_2} A \setminus \{x\}$. All other sets in $\mathcal{H}$ must meet $H$ and contain $x$; therefore $\mathcal{H}$ is isomorphic to $\mathcal{F}_{HM}$.

All that remains to prove is that $\mathcal{F}^3$ and $\mathcal{F}_{HM}$ "cannot be created by shifting", i.e. if $\mathcal{F}$ is an intersecting family and $S_{x,y}(\mathcal{F})$ is isomorphic to $\mathcal{F}^3$ or $\mathcal{F}_{HM}$, then $\mathcal{F}$ is isomorphic to $\mathcal{F}^3$ or $\mathcal{F}_{HM}$, respectively. Suppose first that $k=3$, $\mathcal{F}$ is not isomorphic to $\mathcal{F}^3$, but $S_{x,y}(\mathcal{F})$ is. We can assume without loss of generality that $S_{3,4}(\mathcal{F})=\mathcal{F}^3$ holds. Observe first that each set in $\mathcal{F}$ must contain 1 or 2. Take $F \in \mathcal{F}\setminus\mathcal{F}^3$ and observe that as $S_{3,4}(F)\in\mathcal{F}^3$ we must have $3\notin F$, $4\in F$, and $|\{1,2\}\cap F|=1$. Suppose without loss of generality $F=\{2,4,x\}$ for some $x>4$, and let $\mathcal{G}=\{\{1,3,y\}: y\neq 1,2,3,4,x\}$. Then, as sets in $\mathcal{G}$ are disjoint from $F$, we have $\mathcal{G}\subseteq\mathcal{F}^3\setminus\mathcal{F}$; thus $\mathcal{G}'=\{\{1,4,y\}: y\neq 1,2,3,4,x\}\subseteq\mathcal{F}$. Similarly every set of the form $\{2,3,z\}$ is in $\mathcal{F}^3\setminus\mathcal{F}$; thus $\{2,4,z\}\in\mathcal{F}$ for every $z>4$. As $\{1,4,z\},\{2,4,z\}\in\mathcal{F}$ for any $z>4$, $z\neq x$; therefore if a set in $\mathcal{F}$ does not contain 4, it has to contain both 1 and 2. As every set in $\mathcal{F}$ intersects $\{1,2\}$, this implies that each set in $\mathcal{F}$ intersects $\{1,2,4\}$ in at least two elements, hence $\mathcal{F}$ is a subfamily of a family isomorphic to $\mathcal{F}^3$.

Suppose next (without loss of generality) that $\mathcal{F}\neq\mathcal{F}_{HM}$ but $S_{x,y}(\mathcal{F})=\mathcal{F}_{HM}$ holds for some $1\le x<y\le n$. Note that $\mathcal{F}_{HM}$ is the only maximal intersecting family such that it contains exactly one set $F$ that does not contain the element with maximum degree. Therefore, $\mathcal{F}$ contains at least two such sets $F,F'$ and by this we know that $x=1$ should hold. Thus we have $F=[2,k+1]\in\mathcal{F}_{HM}\cap\mathcal{F}$. Suppose that $F'\in\mathcal{F}\setminus\mathcal{F}_{HM}$. By the above, $1\notin F', y\in F'$.

Let $z\in[2,k+1]\setminus F'$ and $\mathcal{G}_z=\{G\in\binom{[n]}{k}: 1,z\in G, G\cap F'=\emptyset\}$. By the intersecting property of $\mathcal{F}$ we have $\mathcal{G}_z\subset\mathcal{F}_{HM}\setminus\mathcal{F}$ and thus $\mathcal{G}'_z=\{G\setminus\{1\}\cup\{y\}: G\in\mathcal{G}_z\}\subseteq\mathcal{F}$. As for any $k$-set $S$ with $1\in S$, $y,z\notin S$, $S\cap[2,k+1]\neq\emptyset$ there exists $G\in\mathcal{G}'_z$ with $G\cap S=\emptyset$, we must have $S\setminus\{1\}\cup\{y\}\in\mathcal{F}$. Let us denote the family of these sets by $\mathcal{G}^+$. Finally, for any $k$-set $T$ with $1,z\in T, y\notin T$ and $T\notin\mathcal{G}_z$ there exists $G\in\mathcal{G}^+$ with $T\cap G=\emptyset$. Therefore $T\setminus\{1\}\cup\{y\}\in\mathcal{F}$. Let us denote the family of these sets by $\mathcal{G}^{++}$. We obtained that $\mathcal{G}'_z\cup\mathcal{G}^+\cup\mathcal{G}^{++}=\{G\in\binom{[n]}{k}: y\in G, G\cap[2,k+1]\neq\emptyset\}\subseteq\mathcal{F}$, i.e. $\mathcal{F}$ contains a family isomorphic to $\mathcal{F}_{HM}$. ■

The property of being non-trivially intersecting can be captured by the following often used parameter. A set $T$ is a *cover* or *transversal* of a family $\mathcal{F}$ if $T\cap F\neq\emptyset$ for every $F\in\mathcal{F}$. The minimum size $\tau(\mathcal{F})$ of all covers of $\mathcal{F}$ is the *covering number* of $\mathcal{F}$. The non-triviality condition is equivalent to $\tau(\mathcal{F})\ge 2$. Note that if $\mathcal{F}$ is $k$-uniform intersecting, then $\tau(\mathcal{F})\le k$ holds as any $F\in\mathcal{F}$ is a cover of $\mathcal{F}$ by the intersecting property of $\mathcal{F}$.

If $\mathcal{F}\subseteq\binom{[n]}{k}$ is a maximal intersecting family, then let $\mathcal{B}=\mathcal{B}(\mathcal{F})$ be the family of all minimal covers of $\mathcal{F}$. Clearly $|B|\ge\tau(\mathcal{F})$ holds for all $B\in\mathcal{B}$. Also for every $A\subsetneq B\in\mathcal{B}$ there exists $F\in\mathcal{F}$ with $A\cap F=\emptyset$ as $A$ is not a cover of $\mathcal{F}$. Therefore there exists a $B'\in\mathcal{B}$ with $A\cap B'=\emptyset$, as every $F\in\mathcal{F}$

contains a set $B' \in \mathcal{B}$. This also implies

$$|\mathcal{F}| \leq \sum_{j=\tau(\mathcal{F})}^{k} |\mathcal{B}_j| \binom{n-j}{k-j},$$

where $\mathcal{B}_j = \{B \in \mathcal{B} : |B| = j\}$.

The next lemma is a slight generalization of Lemma 1 from [216]. It gives bounds on the number of covers of $\mathcal{F}$ of size $j$, and therefore can be applied (with many additional ideas and technicalities) to prove upper bounds on the size of intersecting families with large covering number.

**Lemma 43** *For all $j$ with $\tau(\mathcal{F}) \leq j \leq k$ we have $\sum_{i=\tau(\mathcal{F})}^{j} k^{j-i}|\mathcal{B}_i| \leq s_j j^{s_j-1}k^{j-s_j}$, where $s_j = \tau(\cup_{i=\tau(\mathcal{F})}^{j}\mathcal{B}_i)$.*

**Proof of Lemma.** For any set $X$ and family $\mathcal{G}$ let $\mathcal{G}(X) := \{G \in \mathcal{G} : X \subseteq G\}$, $\mathcal{G}_i = \{G \in \mathcal{G} : |G| = i\}$, $w(\mathcal{G}) = \sum_{i=\tau(\mathcal{F})}^{j} k^{j-i}|\mathcal{G}_i|$ and $\mathcal{B}' := \cup_{i=\tau(\mathcal{F})}^{j}\mathcal{B}_i$. We are going to construct sets $X_i$ of size $i$ such that $w(\mathcal{B}'(X_i))$ is large. Let $S_j$ be a cover of $\mathcal{B}'$ of size $s_j$. As $S_j$ is a cover of $\mathcal{B}'$, every member of $\mathcal{B}'$ is contained in one of the families $\mathcal{B}'(\{x\})$ with $x \in S_j$; thus we have

$$\sum_{x \in S_j} w(\mathcal{B}'(\{x\})) \geq w(\mathcal{B}').$$

Let $x_1$ be the element of $S_j$ with the largest $w(\mathcal{B}'(\{x_1\}))$, then we have $w(\mathcal{B}'(X_1)) \geq w(\mathcal{B}')/s_j$, where $X_1 = \{x_1\}$.

We are going to construct $X_{i+1}$ by adding a new element to $X_i$. As long as $i < s_j$ holds, one can extend $X_i$ in the following way: as $X_i$ is not a cover of $\mathcal{B}'$, there exists a $B \in \mathcal{B}'$ with $X_i \cap B = \emptyset$. Similarly to the previous case we choose $x_{i+1} \in B$ such that letting $X_{i+1} = X_i \cup \{x_{i+1}\}$ we have the largest $w(\mathcal{B}'(X_{i+1}))$. Then $w(\mathcal{B}'(X_{i+1})) \geq w(\mathcal{B}'(X_j))/j \geq w(\mathcal{B}')/s_j j^i$, as $|B| \leq j$. If $s_j \geq j$ holds, then rearranging this inequality finishes the proof.

If $s_j < j$, then we start our procedure as above. If $i \geq s_j$ and $X_i \notin \mathcal{B}'$, similarly to the previous case we can find a set $B \in \mathcal{B}$ with $X_i \cap B = \emptyset$, but here we only know that $|B| \leq k$, thus it leads to $w(\mathcal{B}'(X_{i+1}) \geq w(\mathcal{B}'(X_i))/k \geq w(\mathcal{B}')/s_j j^{s_j-1}k^{i+1-s_j}$. Once for some $i > s_j$ we have $X_i \in \mathcal{B}'$, then it implies $w(\mathcal{B}'(X_i)) = k^{j-i}$. As $i$ is the smallest such index, we also have $w(\mathcal{B}')/s_j j^{s_j-1}k^{i-s_j} \leq w(\mathcal{B}'(X_i))$, and we are done by rearranging.

Finally note that if $X_j$ is defined, then $X_j \in \mathcal{B}'$ must hold as the above reasoning shows that $w(\mathcal{B}'(X_j)) > 0$ and the only set that may belong to $\mathcal{B}'(X_j)$ is $X_j$ itself. ■

Note that as long as $j \leq k/2$, the expression $s_j j^{s_j-1}k^{j-s_j}$ takes its maximum when $s_j = 1$, therefore $|w(\mathcal{B}')| \leq k^{j-1}$ holds for $j \leq k/2$ and $|\mathcal{B}_j| \leq k^j$ holds for all $j$. We will see many more problems of this type and use this lemma in Section 2.6.

Let us continue this section by stating two results that determine the maximum size of intersecting families with covering number 3 and 4, respectively. In view of the above reasoning, it is no surprise that the optimal families are defined via their minimal covers. Let

$$\mathcal{G}_3 = \left\{G \in \binom{[n]}{3} : 1 \in G, G \cap [2, k+1] \neq \emptyset, G \cap [k+2, 2k] \neq \emptyset\right\}$$
$$\cup \{\{1, 2, 3\}, [2, k+1], \{2\} \cup [k+2, 2k], \{3\} \cup [k+2, 2k]\}$$

and

$$\mathcal{F}_{\mathcal{G}_3} = \left\{F \in \binom{[n]}{k} : \exists G \in \mathcal{G}_3, G \subseteq F\right\}.$$

**Theorem 44 (Frankl, [189])** *For every $k \geq 3$, there exists an integer $n_0(k)$ such that for every $n \geq n_0(k)$ the following holds: if $\mathcal{F} \subseteq 2^{[n]}$ is an intersecting family with $\tau(\mathcal{F}) \geq 3$, then $|\mathcal{F}| \leq |\mathcal{F}_{\mathcal{G}_3}|$ and equality holds if and only if $\mathcal{F}$ is isomorphic to $\mathcal{F}_{\mathcal{G}_3}$.*

Similarly, Frankl, Ota and Tokushige [216] defined an intersecting family $\mathcal{G}_4 \subseteq 2^{[n]}$ of sets of size at most $k$ with $\tau(\mathcal{G}_4) = 4$. $\mathcal{G}_4$ contains six sets of size $k$:

$$[k-1] \cup \{k+1\}, [k-1] \cup \{k+2\},$$
$$[k+1, 2k-1] \cup \{2k+1\}, [k+1, 2k-1] \cup \{2k+2\},$$
$$[2k+1, 3k-1] \cup \{1\}, [2k+1, 3k-1] \cup \{2\}$$

and $(k-1)^3 + 3(k-1)$ sets of size 4:

$$\{\{n, x, y, z\} : x \in [k-1], y \in [k+1, 2k-1], z \in [2k+1, 3k-1]\},$$
$$\{\{n, 1, 2, y\} : y \in [k+1, 2k-1]\},$$
$$\{\{n, k+1, k+2, z\} : z \in [2k+1, 3k-1]\},$$
$$\{\{n, 2k+1, 2k+2, x\} : x \in [k-1]\}.$$

Finally, let $\mathcal{F}_{\mathcal{G}_4} = \{F \in \binom{[n]}{k} : \exists G \in \mathcal{G}_4 \text{ with } G \subseteq F\}$.

**Theorem 45 (Frankl, Ota, Tokushige, [216])** *For every $k \geq 9$, there exists an integer $n_0(k)$ such that for every $n \geq n_0(k)$ the following holds: if $\mathcal{F} \subseteq 2^{[n]}$ is an intersecting family with $\tau(\mathcal{F}) \geq 4$, then $|\mathcal{F}| \leq |\mathcal{F}_{\mathcal{G}_4}|$ and equality holds if and only if $\mathcal{F}$ is isomorphic to $\mathcal{F}_{\mathcal{G}_4}$.*

Let us finish this section with a stability result that states that every intersecting family $\mathcal{F} \subseteq \binom{[n]}{k}$ that is close to having extremal size can be made trivially intersecting by the removal of a very small number of sets.

**Theorem 46 (Keevash, [347])** *Suppose $1 < k < n/2$, $\delta < 10^{-3}n^{-4}$ and $\mathcal{F} \subseteq \binom{[n]}{k}$ is an intersecting family with $|\mathcal{F}| > (1-\delta)\binom{n-1}{k-1}$. Then there is some vertex $v$ so that all but at most $25n\delta^{1/2}\binom{n-1}{k-1}$ sets in $\mathcal{F}$ contain $v$.*

The first such result was obtained by Friedgut [225], but he needed the extra assumption $\alpha n < k < (1/2 - \alpha)n$ for some fixed positive $\alpha$. The condition $\alpha n < k$ was then eliminated by Dinur and Friedgut [134]. Keevash and Mubayi [352] obtained somewhat stronger results than Theorem 46 with different methods but only for the range $\alpha n < k < n/2$.

## 2.2 $t$-intersecting families

An obvious generalization of the intersecting property is to prescribe a larger intersection. A family $\mathcal{F}$ is *$t$-intersecting* if for any two members $F_1, F_2$ we have $|F_1 \cap F_2| \geq t$. For $t = 1$ this is the same as the intersecting property. For larger $t$ determining the maximum cardinality even in the non-uniform case is non-trivial. It was solved by Katona [332]. We start with the analog of Lemma 10 that we will use in the proofs of both the non-uniform and the uniform case. The proof is left to the reader (Exercise 62).

**Lemma 47** *If $\mathcal{F} \subseteq 2^{[n]}$ is $t$-intersecting, then so is $S_{x,y}(\mathcal{F})$ for any $x, y \in [n]$.*

**Theorem 48 (Katona, [332])** *Let $\mathcal{F} \subset 2^{[n]}$ be a $t$-intersecting family with $1 \leq t \leq n$. Then*

$$|\mathcal{F}| \leq \begin{cases} \sum_{i=\frac{n+t}{2}}^{n} \binom{n}{i} & \text{if } n+t \text{ is even} \\ \binom{n-1}{\frac{n+t-1}{2}} + \sum_{i=\frac{n+t+1}{2}}^{n} \binom{n}{i} & \text{if } n+t \text{ is odd} \end{cases} \tag{2.1}$$

**Proof.** We proceed by induction on $n$ the case $n = 1, 2$ being trivial. Let us assume that the statement of the theorem has already been proved for $n-1$ and arbitrary values of $t$. By Lemma 47 we can restrict our attention to left-shifted families $\mathcal{F}$. Let $\mathcal{F}_0 = \{F \in \mathcal{F} : 1 \notin \mathcal{F}\}$ and $\mathcal{F}_1 = \{F \setminus \{1\} : 1 \in F \in \mathcal{F}\}$. Clearly, we have $|\mathcal{F}| = |\mathcal{F}_0| + |\mathcal{F}_1|$, $\mathcal{F}_0, \mathcal{F}_1 \subseteq 2^{[2,n]}$ and $\mathcal{F}_1$ is $(t-1)$-intersecting by definition.

**Lemma 49** *The family $\mathcal{F}_0$ is $(t+1)$-intersecting.*

**Proof of Lemma.** As $\mathcal{F}_0 \subseteq \mathcal{F}$, we know that $\mathcal{F}_0$ is $t$-intersecting. So suppose that for some sets $F, F' \in \mathcal{F}_0$ we have $|F \cap F'| = t$ and let $x \in F \cap F'$. By the left-shiftedness of $\mathcal{F}$, the set $F'' := F' \setminus \{x\} \cup \{1\}$ belongs to $\mathcal{F}$. This contradicts the $t$-intersecting property of $\mathcal{F}$ as $|F \cap F''| = t - 1$. ■

We can apply the induction hypothesis to $\mathcal{F}_0$ and $\mathcal{F}_1$. We have to distinguish two cases depending on the parity of $n+t$. We consider here $n+t$ being even, the other case is similar. Note that the parity of $n+t$ is the same as that of $(n-1)+(t-1)$ and $(n-1)+(t+1)$. Therefore, we obtain

$$|\mathcal{F}| = |\mathcal{F}_0| + |\mathcal{F}_1| \leq \sum_{i=\frac{(n-1)+(t-1)}{2}}^{n-1} \binom{n-1}{i} + \sum_{i=\frac{(n-1)+(t+1)}{2}}^{n} \binom{n-1}{i} = \sum_{i=\frac{n+t}{2}}^{n} \binom{n}{i}.$$

□

Observe that the bound of Theorem 48 is sharp, as shown by the families $\cup_{i=\frac{n+t}{2}}^{n} \binom{[n]}{i}$ if $n+t$ is even and $\{G \cup \{n\} : G \in \binom{[n-1]}{\frac{n+t-1}{2}}\} \cup \bigcup_{i=\frac{n+t+1}{2}}^{n} \binom{[n]}{i}$ if $n+t$ is odd. As Exercise 15 shows, the extremal family is not unique if $t=1$. The uniqueness of the above extremal families for $t \geq 2$ is given as an exercise at the end of this section.

Let us now turn our attention to uniform $t$-intersecting families. A family $\mathcal{F}$ is said to be *trivially $t$-intersecting* if there exists a $t$-set $T$ such that $T \subseteq F$ for all $F \in \mathcal{F}$, and non-trivially $t$-intersecting if it is $t$-intersecting but $|\cap_{F\in\mathcal{F}} F| < t$. It was proved by Erdős, Ko, and Rado in their seminal paper [163] that for any $1 \leq t < k$, the largest $t$-intersecting families $\mathcal{F} \subseteq \binom{[n]}{k}$ are the maximal trivially $t$-intersecting ones if $n$ is large enough. Therefore, $|\mathcal{F}| \leq \binom{n-t}{k-t}$ holds for large enough $n$. The threshold function $n_0(k,t)$ of this property was determined by Wilson [550] after Frankl [186] settled the cases $t \geq 15$. They proved $n_0 = (t+1)(k-t+1)$. Frankl in his paper [186] introduced the following $t$-intersecting families for $0 \leq i \leq k-t$:

$$\mathcal{A}(n,k,t,i) = \left\{A \in \binom{[n]}{k} : |A \cap [2i+t]| \geq i+t\right\}.$$

Note that the family $\mathcal{A}(n,k,t,0)$ is maximal trivially $t$-intersecting and $|\mathcal{A}(n,k,t,0)| \geq |\mathcal{A}(n,k,t,1)|$ if and only if $n \geq (t+1)(k-t+1)$. Frankl conjectured that the largest possible size of a $t$-intersecting family is always attained at one of the $\mathcal{A}(n,k,t,i)$'s. This was proved almost twenty years later by Ahlswede and Khachatrian.

**Theorem 50 (The Complete Intersection Theorem [5])** *Let $\mathcal{F} \subseteq \binom{[n]}{k}$ be $t$-intersecting with $t \geq 2$.*

***(i)*** *If for some non-negative integer $i$ we have $(k-t+1)(2+\frac{t-1}{i+1}) < n < (k-t+1)(2+\frac{t-1}{i})$, then $|\mathcal{F}| \leq |\mathcal{A}(n,k,t,i)|$ and equality holds if and only if $\mathcal{F}$ is isomorphic to $\mathcal{A}(n,k,t,i)$,*

***(ii)*** *If for some non-negative integer $i$ we have $(k-t+1)(2+\frac{t-1}{i+1}) = n$, then $|\mathcal{F}| \leq |\mathcal{A}(n,k,t,i)| = |\mathcal{A}(n,k,t,i+1)|$ and equality holds if and only if $\mathcal{F}$ is isomorphic to $\mathcal{A}(n,k,t,i)$ or $\mathcal{A}(,n,k,t,i+1)$.*

Before starting the proof of Theorem 50, we need to define generating families and related concepts, and prove some auxiliary statements. We say that $\mathcal{G} \subseteq 2^{[n]}$ is a *generating family of* $\mathcal{F} \subseteq \binom{[n]}{k}$ if $\mathcal{F} = \nabla_k(\mathcal{G})$. For example, as seen before Lemma 43, if $\mathcal{F}$ is a maximal intersecting family, then the family of minimal covers forms a generating family of $\mathcal{F}$. The set of all generating families of $\mathcal{F}$ is denoted by $\Gamma(\mathcal{F})$. Note that $\Gamma(\mathcal{F})$ is not empty, as $\mathcal{F}$ is a generating family of itself. Also, if $\mathcal{G}$ is a generating family, then so is the subfamily $\mathcal{G}_{min}$ of minimal sets. Therefore, we can restrict our attention to generating antichains. For a family $\mathcal{F} \subseteq 2^{[n]}$ we define $\max(\mathcal{F}) = \max_{F \in \mathcal{F}} \max_{i \in F} i$ and for a family $\mathcal{F} \subseteq \binom{[n]}{k}$ we define $ak(\mathcal{F}) = \min_{\mathcal{G} \in \Gamma(\mathcal{F})} \max(\mathcal{G})$.

**Proposition 51** *If $\mathcal{F} \subseteq \binom{[n]}{k}$ is left-shifted, then there exists a Sperner family $\mathcal{G} \in \Gamma(\mathcal{F})$ with $\max(\mathcal{G}) = ak(\mathcal{F})$ such that if $G \in \mathcal{G}$ and $G'$ can be obtained from $G$ by shifting operations, then there exists $G^* \in \mathcal{G}$ with $G^* \subseteq G'$.*

**Proof.** It is enough to prove the existence of a left-shifted, not necessarily Sperner family $\mathcal{G}' \in \Gamma(\mathcal{F})$ with $\max(\mathcal{G}') = ak(\mathcal{F})$, as then the subfamily $\mathcal{G}$ of minimal sets will be as required. We claim that if $\mathcal{G} \in \Gamma(\mathcal{F})$ with $\max(\mathcal{G}) = ak(\mathcal{F})$ (by definition there exists such $\mathcal{G}$) and of maximum size among such families, then $\mathcal{G}$ is left-shifted. Suppose not and for some $G \in \mathcal{G}$ and $i < j$, we have $G' = G \setminus \{j\} \cup \{i\} \notin \mathcal{G}$. Then, as $\mathcal{F}$ is left-shifted and $\{K \in \binom{[n]}{k} : G \subseteq K\} \subseteq \mathcal{F}$, we have $\{K \in \binom{[n]}{k} : G' \subseteq K\} \subseteq \mathcal{F}$. But then $\mathcal{G}' = \mathcal{G} \cup \{G'\}$ is also generating with $\max \mathcal{G}' = \max \mathcal{G} = ak(\mathcal{F})$, contradicting the choice of $\mathcal{G}$. ■

Generating families satisfying the properties of Proposition 51 will be called *AK-families of $\mathcal{F}$*. For $G \subseteq [n]$ let us define $\nabla'_k(G) = \{F \in \binom{[n]}{k} : G \subseteq F, \max G < \min(F \setminus G)\}$, and for a family $\mathcal{G}$ we let $\nabla'_k(\mathcal{G}) = \cup_{G \in \mathcal{G}} \nabla'_k(G)$.

**Proposition 52** *If $\mathcal{F} \subseteq \binom{[n]}{k}$ is left-shifted, and $\mathcal{G}$ is an AK-family of $\mathcal{F}$, then $\cup_{G \in \mathcal{G}} \nabla'_k(G)$ is a partition of $\mathcal{F}$.*

**Proof.** As $\mathcal{G}$ is generating, we have $\nabla'_k(G) \subseteq \mathcal{F}$ for all $G \in \mathcal{G}$. Let $G_1 \neq G_2$ with $\max G_1 \leq \max G_2$ and suppose $F \in \nabla'_k(G_1) \cap \nabla'_k(G_2)$. Then $G_1 \subseteq F \cap [\max G_2] = F \cap G_2$ implies $G_1 \subseteq G_2$ which contradicts the Sperner property of $\mathcal{G}$.

Finally, for any set $F \in \mathcal{F}$, as $\mathcal{G}$ is generating, there exists $G \in \mathcal{G}$ with $G \subseteq F$. Let $G$ be such a set with smallest size and among all such sets with $s = |(F \setminus G) \cap [\max G]|$ minimal. We need to show $s = 0$. Assume not and let $m = \max G$ and $x \in (F \setminus G) \cap [\max G]$; then by Proposition 51 there exists $G^* \in \mathcal{G}$ with $G^* \subseteq G \setminus \{m\} \cup \{x\}$, contradicting the choice of $G$. ■

**Proposition 53** *If $\mathcal{F} \subseteq \binom{[n]}{k}$ is left-shifted and $\mathcal{G}$ is an AK-family of $\mathcal{F}$, then letting $\mathcal{H}_i = \{G \in \mathcal{G} : |G| = i, \max G = ak(\mathcal{F}) = \max \mathcal{G}\}$ and $\mathcal{O}_{\mathcal{H}_i} = \nabla_k(\mathcal{H}_i) \setminus \nabla_k(\mathcal{G} \setminus \mathcal{H}_i)$, we have $\mathcal{O}_{\mathcal{H}_i} = \nabla'_k(\mathcal{H}_i)$ and $|\mathcal{O}_{\mathcal{H}_i}| = |\mathcal{H}_i|\binom{n-ak(\mathcal{F})}{k-i}$.*

**Proof.** The first statement clearly implies the second. If $F \in \nabla'_k(H_i)$ with $H_i \in \mathcal{H}_i$, then $H_i = F \cap [\max H_i]$. If $F \in \nabla_k(G)$ for a $G \in \mathcal{G}$, then $G \subseteq H_i$ as $\max H_i = \max \mathcal{G} \geq \max G$. By the Sperner property of $\mathcal{G}$ we have $G = H_i$. This implies $\nabla'_k(\mathcal{H}_i) \subseteq \mathcal{O}_{\mathcal{H}_i}$.

If $F \in \nabla_k(H_i) \setminus \nabla'_k(H_i)$ with $H_i \in \mathcal{H}_i$, then there exists $f \in F \setminus H_i$ with $f < \max H_i$. Then as $\mathcal{G}$ is an AK-family of $\mathcal{F}$, there exists $G \in \mathcal{G}$ with $G \subseteq H_i \setminus \{\max H_i\} \cup \{f\}$ and thus $F \in \nabla_k(G)$ and $G \notin \mathcal{H}_i$, therefore $F \in \nabla_k(\mathcal{G} \setminus \mathcal{H}_i)$. This implies $\mathcal{O}_{\mathcal{H}_i} \subseteq \nabla'_k(\mathcal{H}_i)$. ■

**Lemma 54** *Let $2k - t < n$, $\mathcal{F} \subseteq \binom{[n]}{k}$ be a left-shifted t-intersecting family and $\mathcal{G}$ is an AK-family of $\mathcal{F}$, then*
***(i)*** *for any $G_1, G_2 \in \mathcal{G}$ we have $|G_1 \cap G_2| \geq t$,*
***(ii)*** *if $i < j$ and for some $G_1, G_2 \in \mathcal{G}$ we have $i \notin G_1 \cup G_2$ and $j \in G_1 \cap G_2$, then $|G_1 \cap G_2| \geq t + 1$.*

**Proof.** For any $G_1, G_2 \in \mathcal{G}$ there exist $F_1 \in \nabla_k(G_1)$ and $F_2 \in \nabla_k(G_2)$ with $F_1 \cap F_2 = G_1 \cap G_2$. This proves **(i)**.

As $\mathcal{G}$ is an AK-family of $\mathcal{F}$, there exists $G \in \mathcal{G}$ with $G \subseteq G_1 \setminus \{j\} \cup \{i\}$ and $|G_1 \cap G_2| \leq t$ would imply $|G \cap G_2| < t$, contradicting **(i)**. This finishes the proof of **(ii)**. ■

The next lemma is the core of the proof of Theorem 50.

**Lemma 55** *If $\mathcal{F} \subseteq \binom{[n]}{k}$ is a left-shifted t-intersecting family of maximum size with $n > 2k - t$, $t \geq 1$ and $(k - t + 1)(2 + \frac{t-1}{r+1}) < n$ for some $r \in [0, k - t]$, then $ak(\mathcal{F}) \leq t + 2r$ holds.*

**Proof.** If $n = 2k - t + 1$, then by the condition $(k - t + 1)(2 + \frac{t-1}{r+1}) < n$, we have $r \geq k - t + 1$, and $t + 2r \geq 2k - t + 1 = n \geq ak(\mathcal{F})$.

Let $n \geq 2k - t + 2$, $\mathcal{G}$ be an AK-family of $\mathcal{F}$, and, towards a contradiction, let us assume $\max \mathcal{G} = t + 2r + x$ for some positive $x$. Let us partition $\mathcal{G}$ into $\mathcal{G}_1 = \{G \in \mathcal{G} : \max \mathcal{G} \notin G\}$ and $\mathcal{G}_2 = \{G \in \mathcal{G} : \max \mathcal{G} \in G\}$. Let $\mathcal{G}'_2 = \{G \setminus \{\max \mathcal{G}\} : G \in \mathcal{G}_2\}$ and note that $\mathcal{G}_1$ and $\mathcal{G}'_2$ are disjoint as $\mathcal{G}$ is an antichain.

**Claim 56** *If $G, G' \in \mathcal{G}_1 \cup \mathcal{G}_2 \cup \mathcal{G}'_2$, then $|G \cap G'| \geq t - 1$, and if equality holds, then $G, G' \in \mathcal{G}'_2$, $|G| + |G'| = 2t + 2r + x - 2$ and $|G|, |G'| \geq k - (n - t - 2r - x)$ or $G \in \mathcal{G}_2$, $G' \in \mathcal{G}'_2$, $|G| + |G'| = 2t + 2r + x - 1$ and $|G'| \geq k - (n - t - 2r - x)$.*

**Proof of Claim.** If $G \in \mathcal{G}_1$ and $G' \in \mathcal{G}_1 \cup \mathcal{G}_2$, then Lemma 54 **(i)** implies $|G \cap G'| \geq t$. If $G' \in \mathcal{G}'_2$, then it was obtained from a set in $\mathcal{G}_2$ by deleting $\max \mathcal{G}$, that is not contained in $G$ anyway, thus the intersection still has size at least $t$. If $G, G' \in \mathcal{G}_2$, then similarly Lemma 54 **(i)** implies $|G \cap G'| \geq t$. If $G \in \mathcal{G}_2$ and $G' \in \mathcal{G}'_2$, or both $G, G' \in \mathcal{G}'_2$ then Lemma 54 **(ii)** implies their intersection contains $t$ elements besides $\max \mathcal{G}$, unless neither contains any element larger than $\max \mathcal{G}$.

If $G, G' \in \mathcal{G}'_2$, then $G \cup G' = [\max \mathcal{G} - 1]$, hence $|G| + |G'| = 2t + 2r + x - 2$ holds. Then $|G|, |G'| \geq k - (n - t - 2r - x)$ follows as otherwise the other set is of size at least $n - k - t + 1 \geq k$, but as members of $\mathcal{G}'_2$ their size is at most $k - 1$. If $G \in \mathcal{G}_2$, $G' \in \mathcal{G}'_2$, then $G \cup G' = [\max \mathcal{G}]$, but $\max \mathcal{G} \notin G'$, hence

$|G| + |G'| = 2t + 2r + x - 1$ holds, and $|G'| \geq k - (n - t - 2r - x)$ follows similarly. ■

Aiming to apply Proposition 53, we recall $\mathcal{H}_i = \{G \in \mathcal{G}_2 : |G| = i\}$ and define $\mathcal{H}'_i = \{G \setminus \{\max \mathcal{G}\} : G \in \mathcal{H}_i\}$. As $\mathcal{F}$ is $t$-intersecting and $ak(\mathcal{F}) = t + 2r + x$, $\mathcal{H}_i \neq \emptyset$ implies $t \leq i \leq t + 2r + x$. If $\mathcal{H}_t \neq \emptyset$, then $\mathcal{F}$ contains a maximal trivially $t$-intersecting family, and therefore $\mathcal{F}$ *is* a maximal $t$-intersecting family; thus $\mathcal{G} = \mathcal{H}_t$ is a singleton. As $\mathcal{F}$ is left-shifted, we have $\mathcal{H}_t = \{[t]\}$ and thus $\max \mathcal{G} = t < t + 2r + x$ a contradiction.

If $\mathcal{H}_{t+2r+x} \neq \emptyset$, then by the Sperner property of $\mathcal{G}$, we have $\mathcal{G} = \{[t+2r+x]\}$. Therefore $\mathcal{F} = \nabla_k([t + 2r + x])$ which is a strict subfamily of a trivially $t$-intersecting family as $x$ is positive. This contradicts that $\mathcal{F}$ is of maximum size.

If $\mathcal{H}_i = \emptyset$ for all $i > k - (n - t - 2r - x)$, then by Claim 56 $\mathcal{G}_1 \cup \mathcal{G}'_2$ is $t$-intersecting and so is $\mathcal{F}' = \nabla_k(\mathcal{G}_1 \cup \mathcal{G}'_2)$. By definition, $\mathcal{F} \subseteq \mathcal{F}'$ and as $\mathcal{F}$ is of maximum size, we have $\mathcal{F} = \mathcal{F}'$. But then $\mathcal{G}_1 \cup \mathcal{G}'_2$ is a generating family of $\mathcal{F}$ with $\max(\mathcal{G}_1 \cup \mathcal{G}'_2) < t + 2r + x = ak(\mathcal{F})$, a contradiction.

By the above, we obtain that there exists an $i$ with $k - (n - t - 2r - x) < i < t + 2r + x$ such that $\mathcal{H}_i \neq \emptyset$ holds. We distinguish two cases.

CASE I: $i \neq \frac{2t+2r+x}{2}$

Then we define the following two families:

$$\mathcal{G}' = \mathcal{G}_1 \cup (\mathcal{G}_2 \setminus (\mathcal{H}_i \cup \mathcal{H}_{2t+2r+x-i})) \cup \mathcal{H}'_i,$$

$$\mathcal{G}'' = \mathcal{G}_1 \cup (\mathcal{G}_2 \setminus (\mathcal{H}_i \cup \mathcal{H}_{2t+2r+x-i})) \cup \mathcal{H}'_{2t+2r+x-i}.$$

By Claim 56, both $\mathcal{G}'$ and $\mathcal{G}''$ are $t$-intersecting and therefore so are $\mathcal{F}' = \nabla_k(\mathcal{G}')$ and $\mathcal{F}'' = \nabla_k(\mathcal{G}'')$. The following claim shows that either $\mathcal{F}'$ or $\mathcal{F}''$ is larger than $\mathcal{F}$, yielding to a contradiction in Case I.

**Claim 57** $\max\{|\mathcal{F}'|, |\mathcal{F}''|\} > |\mathcal{F}|$.

**Proof of Claim.** Observe that $\mathcal{F} \setminus \mathcal{F}'$ consists of those sets that belong only to $\nabla_k(\mathcal{H}_{2t+2r+x-i})$. Therefore by Proposition 53 we obtain

$$|\mathcal{F} \setminus \mathcal{F}'| = |\mathcal{H}_{2t+2r+x-i}| \binom{n - t - 2r - x}{k - 2t - 2r - x + i}.$$

To estimate $|\mathcal{F}' \setminus \mathcal{F}|$ we claim that

$$\left\{ H \cup K : H \in \mathcal{H}'_i, K \in \binom{[t + 2r + x + 1, n]}{k - i + 1} \right\} \subseteq \mathcal{F}' \setminus \mathcal{F}.$$

Indeed, if $H \cup K \in \mathcal{F}$, then as $\max \mathcal{G} = t + 2r + x$ it should be generated by a set $G \in \mathcal{G}$ with $G \subseteq H$. But $\mathcal{G}$ is Sperner and $H \cup \{t + 2r + x\} \in \mathcal{G}_2 \subseteq \mathcal{G}$. Therefore we obtain

$$|\mathcal{F}' \setminus \mathcal{F}| \geq |\mathcal{H}'_i| \binom{n - t - 2r - x}{k - i + 1}.$$

With the same reasoning we obtain

$$|\mathcal{F} \setminus \mathcal{F}''| = |\mathcal{H}_i|\binom{n-t-2r-x}{k-i}$$

and

$$|\mathcal{F}'' \setminus \mathcal{F}| \geq |\mathcal{H}'_{2t+2r+x-i}|\binom{n-t-2r-x}{k-2t-2r-x+i+1}.$$

By the above, the inequalities $|\mathcal{F}'' \setminus \mathcal{F}| \leq |\mathcal{F} \setminus \mathcal{F}''|$ and $|\mathcal{F}' \setminus \mathcal{F}| \leq |\mathcal{F} \setminus \mathcal{F}'|$ would imply

$$|\mathcal{H}'_{2t+2r+x-i}|\binom{n-t-2r-x}{k-2t-2r-x+i+1} \leq |\mathcal{H}_i|\binom{n-t-2r-x}{k-i}$$

and

$$|\mathcal{H}'_i|\binom{n-r-2t-x}{k-i+1} \leq |\mathcal{H}_{2t+2r+x-i}|\binom{n-t-2r-x}{k-2t-2r-x+i}.$$

Note that $|\mathcal{H}'_i| = |\mathcal{H}_i|$ and$|\mathcal{H}'_{2t+2r+x-i}| = |\mathcal{H}_{2t+2r+x-i}|$. As $\mathcal{H}'_i \neq \emptyset$ the second inequality implies that $\mathcal{H}'_{2t+2r+x-i} \neq \emptyset$. Using $\frac{n-t-2r-x-k+i}{k-i+1}\binom{n-t-2r-x}{k-i} = \binom{n-t-2r-x}{k-i+1}$ and the last two inequalities, we obtain

$$\frac{n-t-2r-x-k+i}{k-i+1}\binom{n-r-2r-x}{k-2t-2r-x+i+1} \leq \binom{n-t-2r-x}{k-2t-2r-x+i}.$$

Rearranging yields $(n-t-2r-x-k+i)(n-k+t-i) \leq (k-i+1)(k-2t-2r-x+i+1)$, which is a contradiction as $n \geq 2k-t+2$ implies $n-k+t-x-i > k-i+1$ and $n-t-2r-x-k+i > k-2t-2r-x+i+1$. This shows that either $|\mathcal{F}'' \setminus \mathcal{F}| > |\mathcal{F} \setminus \mathcal{F}''|$ or $|\mathcal{F}' \setminus \mathcal{F}| > |\mathcal{F} \setminus \mathcal{F}'|$ holds and thus $\max\{|\mathcal{F}'|, |\mathcal{F}''|\} > |\mathcal{F}|$ as required. □

CASE II: $i = \frac{2t+2r+x}{2}$.

We will again get a contradiction by constructing a $t$-intersecting family that is larger than $\mathcal{F}$. To this end, for $j \in [t+2r+x-1]$ we define $\mathcal{M}_j = \{H \in \mathcal{H}'_i : j \notin H\}$. As for every set $H$ in $\mathcal{H}'_i$ we have $|H| = t+r+x/2-1$ and $|[t+2r+x-1] \setminus H| = r+x/2$, by averaging, we obtain a $j$ with $|\mathcal{M}_j| \geq \frac{r+x/2}{t+2r+x-1}|\mathcal{H}_i|$. It follows from Lemma 54 **(ii)** that $\mathcal{M}_j$ is $t$-intersecting and Claim 56 shows that so are $(\mathcal{G} \setminus \mathcal{H}_i) \cup M_j$ and $\mathcal{F}' = \nabla_k((\mathcal{G} \setminus \mathcal{H}_i) \cup \mathcal{M}_j)$. The next claim will finish the proof of Case II.

**Claim 58** $|\mathcal{F}'| > |\mathcal{F}|$.

**Proof of Claim.** Let $\mathcal{F}_1 = \nabla_k(\mathcal{G} \setminus \mathcal{H}_i)$, $\mathcal{F}_2 = \nabla_k(\mathcal{H}_i) \setminus \nabla_k(\mathcal{G} \setminus \mathcal{H}_i)$, $\mathcal{F}_3 = \nabla_k(\mathcal{M}_j) \setminus \nabla_k(\mathcal{G} \setminus \mathcal{H}_i)$. Then we have $\mathcal{F} = \mathcal{F}_1 \cup \mathcal{F}_2$ and $\mathcal{F}' = \mathcal{F}_1 \cup \mathcal{F}_3$. We need to prove $|\mathcal{F}_3| > |\mathcal{F}_2|$. By Proposition 53, we have

$$|\mathcal{F}_2| = |\mathcal{H}_i|\binom{n-t-2r-x}{k-t-r-x/2}.$$

Just as in Case I, we see that $\{H \cup K : H \in \mathcal{M}_j, K \in \binom{[n-t-2r-x,n]}{k-i+1}\} \subseteq \mathcal{F}_3$ and thus

$$|\mathcal{F}_3| \geq |\mathcal{M}_j| \binom{n-t-2r-x+1}{k-i+1}.$$

Using the lower bound on $|\mathcal{M}_j|$ and plugging back $i = r+t+x/2$ we obtain that if $|\mathcal{F}_3| \leq |\mathcal{F}_2|$ holds, then

$$\frac{r+x/2}{t+2r+x-1}\binom{n-t-2r-x+1}{k-t-r-x/2+1} \leq \binom{n-t-2r-x}{k-t-r-x/2}$$

and rearranging gives

$$n \leq \frac{(k-t+1)(t+2r+x-1)}{r+x/2}.$$

As $i$ and therefore $x/2$ is an integer, we have $x \geq 2$. Hence the right hand side is at least $\frac{(k-t+1)(t+2r+1)}{r+1}$ which contradicts the assumption of the lemma. This finishes the proof of $|\mathcal{F}_3| > |\mathcal{F}_2|$ and the claim. □

As both cases led to a contradiction, the proof of Lemma 55 is finished. ■

We need one more simple lemma before the proof of Theorem 50.

**Lemma 59** *Let $\mathcal{F} \subseteq \binom{[n]}{k}$ be a $t$-intersecting family with $n > 2k-t$, $t \geq 1$ and let $\overline{\mathcal{F}} = \{\overline{F} : F \in \mathcal{F}\}$. If $\mathcal{G}$ is a generating family of $\mathcal{F}$, and $\mathcal{H}$ is a generating family of $\overline{\mathcal{F}}$, then for any $G \in \mathcal{G}, H \in \mathcal{H}$ we have $|G \cup H| \geq n-k+t$.*

**Proof.** If not, then for some $G \in \mathcal{G}, H \in \mathcal{H}$ we have $G \cup H \subseteq Z$ for some $(n-k+t-1)$-subset $Z$ of $[n]$. Then as $\max\{k, n-k\} \leq n-k+t-1$ there exist $F, F' \in \binom{[n]}{k}$ such that $F \in \nabla_k(G) \cap \binom{Z}{k}$ and $\overline{F'} \in \nabla_{n-k}(H) \cap \binom{Z}{n-k}$. As $\mathcal{G}$ is a generating family of $\mathcal{F}$ and $\mathcal{H}$ is a generating family of $\overline{\mathcal{F}}$, we have $F, F' \in \mathcal{F}$. But $|F \cap F'| = |F \cup \overline{F'}| - |\overline{F'}| \leq n-k+t-1-(n-k) = t-1$ contradicts the $t$-intersecting property of $\mathcal{F}$. ■

We say that a family $\mathcal{F}$ is *right-shifted* if $S_{x,y}(\mathcal{F}) = \mathcal{F}$ for all $1 \leq y < x \leq n$, or equivalently if the complement family $\overline{\mathcal{F}}$ is left-shifted. If $\mathcal{F} \subseteq \binom{[n]}{k}$ is a left-shifted $t$-intersecting family, then $\overline{\mathcal{F}}$ is obviously right-shifted, $(n-k)$-uniform and $(n-2k+t)$-intersecting. By symmetry, for a statement about left-shifted families, an appropriately modified version of it holds for right-shifted families. In particular, if the conditions of Lemma 55 hold for a right-shifted family $\mathcal{F}'$, then the conclusion is the existence of a generating family $\mathcal{H}$ of $\mathcal{F}'$ with $\min(\mathcal{H}) \geq n+1-t-2r$.

**Proof of Theorem 50.** Let $\mathcal{F} \subseteq \binom{[n]}{k}$ be a left-shifted $t$-intersecting family of maximum size. Let us set $k' = n-k$, $t' = n-2k+t$ and $r' = k-t-r$. Simple calculation shows that the condition $n < (k-t+1)(2+\frac{t-1}{r})$ implies

$(k'-t'+1)(2+\frac{t'-1}{r'+1}) < n$. Therefore, using Lemma 55, we can assume the existence of a generating family $\mathcal{H}$ of $\overline{\mathcal{F}}$ with $\mathcal{H} \subseteq 2^{[t+2r+1,n]}$ as $n+1-t'-2r' = n+1-(n-2k+t)-2(k-t-r) = t+2r+1$.

We still have to distinguish two cases according to whether both inequalities on $n$ in the condition of the theorem are strict. If yes, then Lemma 55 shows the existence of a generating family $\mathcal{G}$ of $\mathcal{F}$ with $\max(\mathcal{G}) \leq t+2r$, while if not, then we only know $\max(\mathcal{G}) \leq t+2r+2$.

CASE I: Both inequalities on $n$ in the condition of the theorem are strict.

Then Lemma 55 shows that $\mathcal{G} \subseteq 2^{[t+2r]}$ and $\mathcal{H} \subseteq 2^{[t+2r+1,n]}$. If for all $G \in \mathcal{G}$ we have $|G| \geq t+r$, then $\mathcal{F}$ is $\mathcal{A}(n,k,t,r)$. Similarly, if for all $H \in \mathcal{H}$ we have $|H| \geq t'+r' = n-k-r$, then again $\mathcal{F}$ is $\mathcal{A}(n,k,t,r)$. If there exist $G \in \mathcal{G}, H \in \mathcal{H}$ with $|G| \leq t+r-1$, $|H| \leq n-k-r-1$, then $|G \cup H| \leq n-k+t-2$ contradicts Lemma 59.

CASE II: The first inequality is an equality.

Then Lemma 55 shows that $\mathcal{G} \subseteq 2^{[t+2r+2]}$ and $\mathcal{H} \subseteq 2^{[t+2r+1,n]}$. If for all $G \in \mathcal{G}$ we have $|G| \geq t+r+1$, then $\mathcal{F}$ is $\mathcal{A}(n,k,t,r+1)$. As in Case I, if for all $H \in \mathcal{H}$ we have $|H| \geq t'+r' = n-k-r$, then $\mathcal{F}$ is $\mathcal{A}(n,k,t,r)$. If there exist $G \in \mathcal{G}, H \in \mathcal{H}$ with $|G| \leq t+r$, $|H| \leq n-k-r-1$, then $|G \cup H| \leq n-k+t-1$ contradicts Lemma 59. ■

Earlier in this chapter, we saw several stability versions of the Erdős-Ko-Rado theorem. Theorem 41 gave the largest size of a non-trivially intersecting family $\mathcal{F} \subseteq \binom{[n]}{k}$. Since when $n \leq (t+1)(k-t+1)$, the largest $t$-intersecting family is non-trivial, the analogous problem for $t$-intersecting families becomes interesting when $n > (t+1)(k-t+1)$. The two non-trivially intersecting families that are natural candidates to have largest possible size in this range are $\mathcal{A}(n,k,t,1)$ and

$$\mathcal{A}^*(n,k,t) = \left\{A \in \binom{[n]}{k} : [t] \subseteq A, [t+1,k+1] \cap A \neq \emptyset\right\} \cup \{[k+1] \setminus \{i\} : i \in [t]\}.$$

Note that for $t=1$ the family $\mathcal{A}^*(n,k,1)$ is $\mathcal{F}_{HM}$, the extremal example from the result of Hilton and Milner (Theorem 41).

Ahlswede and Khachatrian managed to determine the largest non-trivially intersecting families.

**Theorem 60 (The Complete Nontrivial Intersection Theorem [4])** *If $\mathcal{F} \subseteq \binom{[n]}{k}$ is non-trivially $t$-intersecting with $t \geq 2$, $(t+1)(k-t+1) < n$ and*

***(i)*** *$t+1 \leq k \leq 2t+1$, then $|\mathcal{F}| \leq |\mathcal{A}(n,k,t,1)|$ and equality holds if and only if $\mathcal{F}$ is isomorphic to $\mathcal{A}(n,k,t,1)$,*

***(ii)*** *$2t+1 < k$, then $|\mathcal{F}| \leq |\mathcal{A}^*(n,k,t)|$ and equality holds if and only if $\mathcal{F}$ is isomorphic to $\mathcal{A}^*(n,k,t)$.*

Prior to the work of Ahlswede and Khachatrian, Frankl [188] obtained partial results if $n$ is large enough. Later Balogh and Mubayi [45] gave a simple and short proof of part **(i)**.

Let us finish this section by stating a result of Milner. In Chapter 1, we gave several proofs of Theorem 16 that determined the maximum possible size of an intersecting Sperner family $\mathcal{F} \subseteq 2^{[n]}$. Milner's result was more general: he obtained the maximum possible size of $t$-intersecting antichains.

**Theorem 61 (Milner, [419])** *If $\mathcal{F} \subseteq 2^{[n]}$ is $t$-intersecting and Sperner, then $|\mathcal{F}| \le \binom{n}{\lfloor \frac{n+t+1}{2} \rfloor}$ holds.*

**Exercise 62** *Prove Lemma 47.*

**Exercise 63** *Prove that if $t \ge 2$, then the only families achieving the bound of Theorem 48 are $\cup_{i=\frac{n+t}{2}}^{n} \binom{[n]}{i}$ if $n+t$ is even and $\{G \cup \{n\} : G \in \binom{[n-1]}{\frac{n+t-1}{2}}\} \cup \bigcup_{i=\frac{n+t+1}{2}}^{n} \binom{[n]}{i}$ if $n+t$ is odd.*

**Exercise 64** *Prove that if $\mathcal{F} \subseteq \binom{[n]}{k}$ is a non-trivially $t$-intersecting family, then for any $t$-set $T$ we have $deg_{\mathcal{F}}(T) \le k\binom{n-k-t}{k-t-1}$. As a corollary, deduce that there exists $n_0$ such that for any $n \ge n_0$ and $\mathcal{F} \subseteq \binom{[n]}{k}$ non-trivially $t$-intersecting family, the inequality $|\mathcal{F}| < \binom{n-t}{k-t}$ holds.*

## 2.3 Above the Erdős-Ko-Rado threshold

The Erdős-Ko-Rado theorem states that if $n \ge 2k$ and a family $\mathcal{F} \subseteq \binom{[n]}{k}$ has size larger than $\binom{n-1}{k-1}$, then $\mathcal{F}$ cannot be intersecting. In this section we will consider families of size from this range and using different concepts we will examine how "close" they can be to being intersecting.

### 2.3.1 Erdős's matching conjecture

A collection $\mathcal{M}$ of pairwise disjoint sets is called a *matching*. The *matching number* $\nu(\mathcal{F})$ of $\mathcal{F}$ is the size of the largest matching that $\mathcal{F}$ contains. Note that a family is intersecting if $\nu(\mathcal{F}) = 1$. Therefore one can weaken the condition of the Erdős-Ko-Rado theorem if one is interested in the maximum size of a family $\mathcal{F} \subseteq \binom{[n]}{k}$ with $\nu(\mathcal{F}) \le s$. There are two trivial examples of such families: $\mathcal{A}(k,s) = \binom{[(k(s+1)-1]}{k}$ and $\mathcal{A}(n,k,s) = \{A \in \binom{[n]}{k} : [s] \cap A \neq \emptyset\}$ both have matching number $s$ by the pigeon-hole principle. Erdős conjectured that the largest $k$-uniform family on an underlying set of size $n$ is always one of these two families.

**Conjecture 65 (Erdős [152])** *Let $\mathcal{F} \subseteq \binom{[n]}{k}$ with $\nu(\mathcal{F}) \leq s$ and $n \geq (s+1)k-1$. Then $|\mathcal{F}| \leq \max\{\binom{(s+1)k-1}{k}, \binom{n}{k} - \binom{n-s}{k}\}$*

Erdős verified Conjecture 65 for large $n$. Later the threshold was improved in several papers by Bollobás, Daykin, Erdős [68], Huang, Loh, Sudakov [312], Frankl, Łuczak, Mieczkowska [215]. The current best threshold is $n \geq 5sk/3 - 2s/3$ for $s$ large enough, due to Frankl and Kupavskii [213]. Here we present a proof by Frankl that gives linear bound for the threshold both in $s$ and $k$.

**Theorem 66 (Frankl [197])** *Let $\mathcal{F} \subseteq \binom{[n]}{k}$ with $\nu(\mathcal{F}) \leq s$ and $n \geq (2s+1)k-s$. Then $|\mathcal{F}| \leq \binom{n}{k} - \binom{n-s}{k}$.*

Before starting the proof of Theorem 66, we need some preliminary results. The first lemma should be routine to the reader by now. (It is given as an exercise at the end of the section.)

**Lemma 67** *If $\mathcal{F} \subseteq 2^{[n]}$ is a family with $\nu(\mathcal{F}) = s$, then for any $i,j \in [n]$ we have $\nu(S_{i,j}(\mathcal{F})) \leq s$.*

The first auxiliary result Frankl needed is the following theorem. Its special case $s = 1$ was proved by Katona [332].

**Theorem 68** *If $\mathcal{F} \subseteq \binom{[n]}{k}$ with $\nu(\mathcal{F}) = s$, then $s|\Delta(\mathcal{F})| \geq |\mathcal{F}|$ holds.*

**Proof.** Let us first assume that $n \leq k(s+1) - 1$ holds and consider the bipartite graph $G = (\mathcal{F}, \Delta(\mathcal{F}), E)$ with $(F,G) \in E$ if and only if $G \subseteq F$. The degree of every set $F \in \mathcal{F}$ is $k$, while the degree of every set $G \in \Delta(\mathcal{F})$ is at most $n-k+1 \leq sk$ by the assumption on $n$. Therefore $sk|\Delta(\mathcal{F})| \geq |E| = k|\mathcal{F}|$ as required.

If $n \geq k(s+1)$ we proceed by induction on $n+k$. (The case $k=1$ is trivial, and for any $k$ and $s$, small values of $n$ are dealt with in the previous paragraph.) By Lemma 67 we can assume that $\mathcal{F}$ is left-shifted. Let $\mathcal{F}_1 = \{F \in \mathcal{F} : n \notin F\}$ and $\mathcal{F}_2 = \{F \setminus \{n\} : n \in F \in \mathcal{F}\}$. As $\mathcal{F}_1 \subseteq \mathcal{F}$ we have $\nu(\mathcal{F}_1) \leq s$. We claim that $\nu(\mathcal{F}_2) \leq s$ holds as well. If not, then there exist pairwise disjoint $F_1, F_2, \ldots, F_{s+1} \in \mathcal{F}_2$ and $F_i \cup \{n\} \in \mathcal{F}$ for all $1 \leq i \leq s+1$. As $|\cup_{i=1}^{s+1} F_i| = (k-1)(s+1)$ and $n \geq k(s+1)$, there exist distinct elements $x_1, x_2, \ldots, x_s, x_{s+1} \in [n]$ with $x_j \notin \cup_{i=1}^{s+1} F_i$ for all $1 \leq j \leq s+1$. But as $\mathcal{F}$ is left-shifted and $F_i \cup \{n\} \in \mathcal{F}$, the sets $F_i' = F_i \cup \{x_i\} \in \mathcal{F}$ contradicting $\nu(\mathcal{F}) \leq s$.

Note that $\Delta(\mathcal{F}_1) \subseteq \Delta(\mathcal{F})$, $\{G \cup \{n\} : G \in \Delta(\mathcal{F}_2)\} \subseteq \Delta(\mathcal{F})$ and $\Delta(\mathcal{F}_1) \cap \{G \cup \{n\} : G \in \Delta(\mathcal{F}_2)\} = \emptyset$. Therefore $|\Delta(\mathcal{F})| \geq |\Delta(\mathcal{F}_1)| + |\Delta(\mathcal{F}_2)|$, and applying the inductive hypothesis to $\mathcal{F}_1$ and $\mathcal{F}_2$, we obtain

$$|\mathcal{F}| = |\mathcal{F}_1| + |\mathcal{F}_2| \leq s|\Delta(\mathcal{F}_1)| + s|\Delta(\mathcal{F}_2)| \leq s|\Delta(\mathcal{F})|.$$

■

We will need one more auxiliary result, Theorem 71. Its proof uses a theorem of König on bipartite graphs.

**Theorem 69 (König [374])** *If $G = (A, B, E)$ is a bipartite graph, then $\nu(E) = \tau(E)$.*

Before continuing with the proof of Theorem 66, let us mention another version of the above theorem, that we will use later.

**Theorem 70 (Hall's marriage theorem, [299])** *If $G = (A, B, E)$ is a bipartite graph, then there is a matching covering $A$ if and only if for every $A' \subseteq A$ we have $N(A')| \geq |A'|$.*

A sequence $\mathbf{F} = \mathcal{F}_1, \mathcal{F}_2, \ldots, \mathcal{F}_{s+1}$ of families is *nested* if $\mathcal{F}_{s+1} \subseteq \mathcal{F}_s \subseteq \cdots \subseteq \mathcal{F}_1$ holds and *cross-dependent* if there does not exist a matching $\{F_1, F_2, \ldots, F_{s+1}\}$ with $F_i \in \mathcal{F}_i$.

**Theorem 71** *If the sequence $\mathbf{F} = \mathcal{F}_1, \mathcal{F}_2, \ldots, \mathcal{F}_{s+1} \subseteq \binom{X}{l}$ is nested and cross-dependent with $|X| \geq tl$ and $t \geq 2s + 1$, then*

$$\sum_{i=1}^{s} |\mathcal{F}_i| + (s+1)|\mathcal{F}_{s+1}| \leq s\binom{|X|}{l}.$$

**Proof.** Let us pick a matching $\mathcal{M}$ of size $t$ uniformly at random among all such matchings in $\binom{X}{l}$. Clearly, we have $\mathbb{E}(|\mathcal{F}_i \cap \mathcal{M}|) = \frac{t|\mathcal{F}_i|}{\binom{|X|}{l}}$.

**Lemma 72** *For every $\mathcal{M}$ we have $(s+1)|\mathcal{F}_{s+1} \cap \mathcal{M}| + \sum_{i=1}^{s} |\mathcal{F} \cap \mathcal{M}| \leq st$.*

**Proof of Lemma.** Let us consider the bipartite graph $G = (\mathcal{M}, \mathbf{F}, E)$ with $(M, \mathcal{F}_i) \in E$ if and only if $M \in \mathcal{F}_i$. As $\mathcal{F}_1, \mathcal{F}_2, \ldots, \mathcal{F}_{s+1}$ are cross-dependent, we have $\nu(G) \leq s$ and thus by Theorem 69 there exists a set $S$ of $s$ vertices such that $S \cap e \neq \emptyset$ for all $e \in E$. Let $|S \cap \mathcal{M}| = x$ and hence $|S \cap \mathbf{F}| = s - x$. Then an $\mathcal{F}_i \in S$ is incident to at most $|\mathcal{M}| = t$ edges of $G$; thus elements of $S \cap \mathbf{F}$ are incident to at most $(s - x)t$ edges in $G$. Every other edge of $G$ is incident to an $M \in \mathcal{M} \cap S$. Such an $M$ is adjacent to at most $(s+1)-(s-x)$ distinct $\mathcal{F}_j \notin S$. This implies $|E| \leq (s-x)t + x(x+1) = x^2 - x(t-1) + st$.

As $\mathbf{F}$ is nested, if $M \in \mathcal{F}_{s+1}$, then $M \in \mathcal{F}_i$ for all $i$, i.e. $deg_G(M) = s + 1$. Therefore as $|S| = s$, we have $\mathcal{F}_{s+1} \cap \mathcal{M} \subseteq S$. Let $b = |\mathcal{M} \cap \mathcal{F}_{s+1}|$; then, clearly, we have $b \leq x$. In the range $b \leq x \leq s$ the expression $x^2 - x(t-1) + st$ takes its maximum either at $x = b$ or at $x = s$. Therefore we have

$$\sum_{i=1}^{s+1} |\mathcal{M} \cap \mathcal{F}_i| = |E| \leq \max\{b^2 - (t-1)b + st, s^2 - (t-1)s + st\}. \tag{2.2}$$

We claim that the right-hand-side of (2.2) is at most $st - sb$. Indeed $b^2 - (t-1)b + st \leq st - sb$ is equivalent to $b(b - t + 1 + s) \leq 0$ which follows from $0 \leq b \leq s$

and the assumption $t \geq 2s+1$. The inequality $s^2 - (t-1)s + st \leq st - sb$ holds similarly.

Adding $sb = s|\mathcal{M} \cap \mathcal{F}_{s+1}|$ to both sides of (2.2) we obtain the statement of the Lemma. □

Applying Lemma 72 to the expected values of the random variables of the left-hand-side we obtain $(s+1)\frac{t|\mathcal{F}_{s+1}|}{\binom{|X|}{l}} + \sum_{i=1}^{s} \frac{t|\mathcal{F}_i|}{\binom{|X|}{l}} \leq st$. Multiplying by $\binom{|X|}{l}/t$ finishes the proof. ■

**Proof of Theorem 66.** Let $\mathcal{F} \subseteq \binom{[n]}{k}$ with $\nu(\mathcal{F}) \leq s$ and $n \geq (2s+1)k - s$. By Lemma 67 it is enough to prove $|\mathcal{F}| \leq \binom{n}{k} - \binom{n-s}{k} = |\mathcal{A}(n,k,s)|$ for left-shifted $\mathcal{F}$. Let us write $\mathcal{A} = \mathcal{A}(n,k,s)$ and for any $S \subseteq [s+1]$ let

$$\mathcal{F}(S) = \{F \in \mathcal{F} : F \cap [s+1] = S\} \quad \text{and} \quad \mathcal{A}(S) = \{A \in \mathcal{A} : A \cap [s+1] = S\}.$$

By definition, we have $|\mathcal{A}(S)| = \binom{n-s-1}{k-|S|} \geq |\mathcal{F}(S)|$ provided $|S| \geq 2$. Furthermore, $\mathcal{A}(\emptyset) = \mathcal{A}(\{s+1\}) = \emptyset$ and for any $i \in [s]$ we have $|\mathcal{A}(\{i\})| = \binom{n-s-1}{k-1}$. Therefore it is enough to show

$$|\mathcal{F}(\emptyset)| + \sum_{i=1}^{s+1} |\mathcal{F}(\{i\})| \leq s\binom{n-s-1}{k-1}. \tag{2.3}$$

Observe that as $\mathcal{F}$ is left-shifted, $H \in \Delta(\mathcal{F}(\emptyset))$ implies $H \cup \{s+1\} \in \mathcal{F}(\{s+1\})$. We obtained $|\mathcal{F}(\{s+1\})| \geq |\Delta(\mathcal{F}(\emptyset))| \geq \frac{1}{s}|\mathcal{F}(\emptyset)|$, where the last inequality follows from Theorem 68 and $\nu(\mathcal{F}(\emptyset)) \leq \nu(\mathcal{F}) \leq s$.

For $1 \leq i \leq s+1$, let us define $\mathcal{F}_i = \{F \setminus \{i\} : F \in \mathcal{F}(\{i\})\}$. Clearly, we have $|\mathcal{F}_i| = |\mathcal{F}(\{i\})|$ and $\mathcal{F}_i \subseteq \binom{[s+2,n]}{k-1}$ for all $1 \leq i \leq s+1$. We claim that the collection $\mathcal{F}_1, \mathcal{F}_2, \ldots, \mathcal{F}_{s+1}$ is nested and cross-dependent. Indeed, if $F_i \in \mathcal{F}_i$ are pairwise disjoint, then so are $F_i \cup \{i\} \in \mathcal{F}(\{i\}) \subseteq \mathcal{F}$ contradicting $\nu(\mathcal{F}) \leq s$. If $i < j$ then $F \in \mathcal{F}_j$ implies $F \cup \{j\} \in \mathcal{F}(\{j\})$, $i \notin F \cup \{j\}$. Therefore as $\mathcal{F}$ is left-shifted, $F \cup \{i\} \in \mathcal{F}(\{i\})$ and hence $F \in \mathcal{F}_i$. Applying Theorem 71 we obtain

$$(s+1)|\mathcal{F}(\{s+1\})| + \sum_{i=1}^{s} |\mathcal{F}(\{i\})| \leq s\binom{n-s-1}{k-1}.$$

This implies (2.3) using $|\mathcal{F}(\emptyset)| \leq s|\mathcal{F}(\{s+1\})|$. ■

Let us finish this subsection by mentioning the special cases of Conjecture 65 that have been completely settled. The case of graphs, i.e. $k=2$ was proved by Erdős and Gallai [159]. The case $k=3$ was settled by Frankl [198] after partial results by Frankl, Rödl, Rucinski [220] and Łuczak, Mieczkowska [398].

**Exercise 73** *Prove Lemma 67.*

### 2.3.2 Supersaturation - minimum number of disjoint pairs

If a property $P$ of some families is defined via a forbidden subfamily, and $M$ is the maximum size that a family satisfying $P$ can have, then we know that a family of size $M+e$ will contain at least $e$ copies of the forbidden subfamily. In most cases the lower bound $e$ is far from being sharp. Problems asking for the minimum number of copies are called *supersaturation* or *Erdős-Rademacher type problems* as the first such results are due to them. Mantel [405] proved that a triangle-free graph on $n$ vertices can have at most $\lfloor \frac{n^2}{4} \rfloor$ edges and the only triangle-free graph with that many edges is the complete bipartite graph $G$ with parts of size $\lfloor n/2 \rfloor$ and $\lceil n/2 \rceil$. If we add one more edge to $G$, then it will contain (at least) $\lfloor n/2 \rfloor$ triangles. Rademacher proved (unpublished) that this is best possible: any graph on $n$ vertices with $\lfloor \frac{n^2}{4} \rfloor + 1$ edges contains at least $\lfloor n/2 \rfloor$ triangles. This was improved by Erdős [147] who showed that any graph on $n$ vertices with $\lfloor \frac{n^2}{4} \rfloor + e$ edges contains at least $e\lfloor n/2 \rfloor$ triangles if $e \le cn$ for some sufficiently small constant $c > 0$.

A family $\mathcal{F}$ is intersecting if there does not exist a disjoint pair $F, F' \in \mathcal{F}$. Therefore the corresponding Erdős-Rademacher problem is to minimize the number $D(\mathcal{F})$ of disjoint pairs that $\mathcal{F}$ contains over all families $\mathcal{F} \subseteq 2^{[n]}$ or $\mathcal{F} \subseteq \binom{[n]}{k}$ of size $s$. For a set $F$ and families $\mathcal{F}, \mathcal{G}$ we introduce $\mathcal{D}_{\mathcal{G}}(F) = \{G \in \mathcal{G} : F \cap G = \emptyset\}$, $\mathcal{D}(\mathcal{F}, \mathcal{G}) = \{(F, G) : F \in \mathcal{F}, G \in \mathcal{G}, F \cap G = \emptyset\}$, so $D(\mathcal{F}) = \frac{1}{2}|\mathcal{D}(\mathcal{F}, \mathcal{F})|$ unless $\emptyset \in \mathcal{F}$. In case of the $t$-intersecting property, the corresponding Erdős-Rademacher problem is to minimize the number $D_t(\mathcal{F})$ of pairs $(F, F')$ in $\mathcal{F}$ with $|F \cap F'| < t$, while for the property of $\nu(\mathcal{F}) < q$ it is to minimize the number $D_{q,\times}(\mathcal{F})$ of pairwise disjoint $q$-tuples $(F_1, F_2, \ldots, F_q)$. By Proposition 8 and Theorem 9, the problem to minimize $D(\mathcal{F})$ becomes interesting when $m > 2^{n-1}$ in the non-uniform case, and when $m > \binom{n-1}{k-1}$ in the uniform case. We first consider non-uniform families.

**Theorem 74 (Ahlswede [1], Frankl [185])** *Let $s$, $m$ and $n$ be positive integers with $|\binom{[n]}{\ge m}| \le s < |\binom{[n]}{\ge m-1}|$. Then there exists a family $\mathcal{F} \subseteq 2^{[n]}$ that minimizes $D(\mathcal{F})$ over all families of size $s$ such that $\binom{[n]}{\ge m} \subseteq \mathcal{F} \subseteq \binom{[n]}{\ge m-1}$ holds.*

**Proof.** Let $\mathcal{F} \subseteq 2^{[n]}$ be a family of size $s$ that minimizes $D(\mathcal{F})$ and among all such families have maximum *volume* $\sum_{F \in \mathcal{F}} |F|$. It is enough to prove that $\mathcal{F}$ has the property that $F \in \mathcal{F}, G \in 2^{[n]}$, and $|F| < |G|$ imply $G \in \mathcal{F}$. Suppose not and $F \in \mathcal{F}, G \in 2^{[n]} \setminus \mathcal{F}$ with $|F| < |G|$. If $F \subset G$, then $\mathcal{F}' := \mathcal{F} \setminus \{F\} \cup \{G\}$ has size $s$ and larger volume than $\mathcal{F}$, and clearly $D(\mathcal{F}) \ge D(\mathcal{F}')$, contradicting the choice of $\mathcal{F}$.

So we may suppose $F \not\subset G$ for any pair $F \in \mathcal{F}, G \in 2^{[n]}$ with $|F| < |G|$. Let us pick $F, G$ with $G \setminus F$ being minimal among these pairs. We will use one of the shifting operations from the proof of Theorem 30. Let $F' = F \setminus G$,

$G' = G \setminus F$ and for any $H \in \mathcal{F}$ let

$$\mu(H) = \begin{cases} (H \setminus F') \cup G' & \text{if } F' \subseteq H, G' \cap H = \emptyset \text{ and } (H \setminus F') \cup G' \notin \mathcal{F} \\ H & \text{otherwise.} \end{cases}$$

As $|G'| > |F'|$, the family $\mu(\mathcal{F}) = \{\mu(H) : H \in \mathcal{F}\}$ has strictly larger volume than $\mathcal{F}$ and is of size $s$. Therefore to get our contradiction it is enough to prove $D(\mu(\mathcal{F})) \leq D(\mathcal{F})$. We have

$$D(\mathcal{F}) = D(\mathcal{F} \cap \mu(\mathcal{F})) + D(\mathcal{F} \setminus \mu(\mathcal{F})) + \sum_{H \in \mathcal{F} \setminus \mu(\mathcal{F})} |\mathcal{D}_{\mathcal{F} \cap \mu(\mathcal{F})}(H)|$$

and

$$D(\mu(\mathcal{F})) = D(\mathcal{F} \cap \mu(\mathcal{F})) + D(\mu(\mathcal{F}) \setminus \mathcal{F}) + \sum_{H \in \mu(\mathcal{F}) \setminus \mathcal{F}} |\mathcal{D}_{\mathcal{F} \cap \mu(\mathcal{F})}(H)|.$$

As $\mu(\mathcal{F}) \setminus \mathcal{F}$ is intersecting (all its sets contain $G'$), we have $D(\mu(\mathcal{F}) \setminus \mathcal{F}) = 0$. Since $\mu(\mathcal{F}) \setminus \mathcal{F} = \{\mu(H) : H \in \mathcal{F} \setminus \mu(\mathcal{F})\}$, it is enough to prove that for every set $H \in \mathcal{F} \setminus \mu(\mathcal{F})$ the inequality $|\mathcal{D}_{\mathcal{F} \cap \mu(\mathcal{F})}(H)| \geq |\mathcal{D}_{\mathcal{F} \cap \mu(\mathcal{F})}(\mu(H))|$ holds. To this end we define an injective mapping $f$ from $\mathcal{D}_{\mathcal{F} \cap \mu(\mathcal{F})}(\mu(H))$ to $\mathcal{D}_{\mathcal{F} \cap \mu(\mathcal{F})}(H)$. Let us fix $g \in G'$ and an injective mapping $i : F' \to G' \setminus \{g\}$ (as $|G'| > |F'|$, there exists such a mapping). For any $E \in \mathcal{D}_{\mathcal{F} \cap \mu(\mathcal{F})}(\mu(H))$ let $f(E) = (E \setminus (E \cap F')) \cup i(E \cap F') \cup \{g\}$. It is easy to see that the mapping $f$ is injective, as $E$ is disjoint from $G'$. We have to prove that $f(E) \in \mathcal{D}_{\mathcal{F} \cap \mu(H)}(H)$, i.e. $f(E) \cap H = \emptyset$ and $f(E) \in \mathcal{F} \cap \mu(\mathcal{F})$.

The fact $f(E) \cap H = \emptyset$ follows from $E \cap \mu(H) = \emptyset$ and the definition of $f$.

As $g \in f(E)$, if $f(E) \in \mathcal{F}$, then $f(E) = \mu(f(E)) \in \mu(\mathcal{F})$ by the definition of $\mu$; therefore it is left to prove that $f(E) \in \mathcal{F}$. If $|G'| = |F'| + 1$ and $F' \subseteq E$; then $E \in \mathcal{F} \cap \mu(\mathcal{F})$ and $f(E) = (E \setminus F') \cup G'$ imply $f(E) \in \mathcal{F}$. If $f(E) \notin \mathcal{F}$and either $|G'| \geq |F'| + 2$ or $F' \not\subseteq E$, then the pair $E, f(E)$ would have been chosen instead of the pair $F, G$. Indeed, $|E| < |f(E)| = |E| + 1$ and $E \in \mathcal{F}$ makes $E, f(E)$ a possible choice, and we have $f(E) \setminus E \subseteq i(E \cup F') \cup \{g\} \subsetneq G'$. This implies $|f(E) \setminus E| < |G \setminus F|$, contradicting the choice of $F$ and $G$, finishing the proof. ∎

As shown by the following example of Frankl [185], the statement of Theorem 74 does not stay true if instead of $D(\mathcal{F})$ one wants to minimize $D_t(\mathcal{F})$ (See Exercise 84).

**Construction 75** *Let* $n = 2m - t$, $m \geq 2t$ *and* $X \in \binom{[n]}{m}$ *be a fixed set. Let* $\mathcal{F} = (\binom{[n]}{\geq m} \setminus \{X\}) \cup \{Y \in \binom{[n]}{m-1} : \overline{X} \subseteq Y\}$.

However, when Bollobás and Leader gave a new proof to the special case $s = |\binom{[n]}{\geq m}|$ of Theorem 74, they used fractional set systems and their method can be applied to minimize $D_t(\mathcal{F})$ or $D_{q,\times}(\mathcal{F})$ over all families of fixed size. Somewhat surprisingly the statement of Theorem 74 does remain valid in this special case.

**Theorem 76 (Bollobás, Leader [71])** *Let $\mathcal{F} \subseteq 2^{[n]}$ be a family with $|\mathcal{F}| = |\binom{[n]}{\geq m}|$; then for any $t$ and $k$ we have $D_t(\mathcal{F}) \geq D_t(\binom{[n]}{\geq m})$ and $D_{q,\times}(\mathcal{F}) \geq D_{q,\times}(\binom{[n]}{\geq m})$.*

Note that Theorem 74 gives the minimum value of $D(\mathcal{F})$ if and only if $s = |\binom{[n]}{\geq m}|$ for some $0 \leq m \leq n/2$. If $s$ is not of that form, then the problem reduces to the uniform case. Indeed, as $|\mathcal{D}_{\binom{[n]}{\geq m}}(X)|$ is the same for all $(m-1)$-subsets $X \subset [n]$, minimizing $D(\mathcal{F})$ over all families $\mathcal{F} \subseteq 2^{[n]}$ of size $s$ is equivalent to minimizing $D(\mathcal{G})$ over all families $\mathcal{G} \subseteq \binom{[n]}{m-1}$ of size $s - |\binom{[n]}{\geq m}|$.

The uniform case was first addressed by Ahlswede and Katona [3]. They considered the special case of graphs, i.e. $k = 2$, when minimizing the number of disjoint edges is equivalent to maximizing $\sum_{v \in V(G)} \binom{deg_G(v)}{2}$ under the condition $\sum_{v \in V(G)} deg_v(G) = 2e$ is fixed. They showed that this maximum is always attained at either the *quasi-star* $S_{n,e}$ or its complement. $S_{n,e} = (V, E)$ has $n$ vertices and $e$ edges such that there exists a set of vertices $U \subseteq V$ and $v \in V \setminus U$ with $\{f \in \binom{V}{2} : f \cap U \neq \emptyset\} \subseteq E \subseteq \{f \in \binom{V}{2} : f \cap (U \cup \{v\}) \neq \emptyset\}$.

To state a result in the general case we need the following definition. Recall that the lexicographic ordering on $\binom{[n]}{k}$ is defined as $A < B$ if and only if $\min A \triangle B \in A$. Let us denote initial segment of size $s$ of this ordering by $\mathcal{L}_{n,k}(s)$. Let $r$ be the integer with $\binom{n}{k} - \binom{n-r+1}{k} < s \leq \binom{n}{k} - \binom{n-r}{k}$, then counting disjoint pairs $F, F' \in \mathcal{L}_{n,k}(s)$ according to $i = \max\{\min F, \min F'\}$, we obtain that $D(\mathcal{L}_{n,k}(s)) = \sum_{i=2}^{r-1} \binom{n-i}{k-1} \sum_{j=1}^{i-1} \binom{n-k-j}{k-1} + (s - \sum_{i=1}^{r-1} \binom{n-i}{k-1}) \sum_{j=1}^{r-1} \binom{n-j-k}{k-1}$.

**Theorem 77 (Das, Gan, Sudakov [116])** *Let $n, k, l, s$ be integers satisfying $n \geq 108k^2 l(k+l)$ and $0 \leq s \leq \binom{n}{k} - \binom{n-l}{k}$. Then if $\mathcal{F} \subseteq \binom{[n]}{k}$ with $|\mathcal{F}| = s$, then $D(\mathcal{F}) \geq D(\mathcal{L}_{n,k}(s))$ holds.*

**Proof.** Let $r$ be the integer with $\binom{n}{k} - \binom{n-r+1}{k} < s \leq \binom{n}{k} - \binom{n-r}{k}$. By assumption, we have $1 \leq r \leq l$. Let us start by an upper bound on $D(\mathcal{L}_{n,k}(s))$ and a lower bound on $s$ that will be useful and easier to calculate with. Obviously $\min F \in [r]$ for all $F \in \mathcal{L}_{n,k}(s)$ and if $F \cap F' = \emptyset$, then $\min F \neq \min F'$. Thus $\min F$ partitions $\mathcal{L}_{n,k}(s)$ into $r$ intersecting subfamilies $\mathcal{A}_1, \ldots, \mathcal{A}_r$. This implies

$$D(\mathcal{L}_{n,k}(s)) \leq \sum_{1 \leq i < j \leq r} |\mathcal{A}_i||\mathcal{A}_j| \leq \frac{1}{2}\binom{r}{2}\left(\frac{s}{r}\right)^2 = \frac{1}{2}\left(1 - \frac{1}{r}\right)s^2 \tag{2.4}$$

As $\mathcal{L}_{n,k}(s)$ contains all sets that intersect $[r-1]$ we have

$$s \geq (r-1)\binom{n-1}{k-1} - \binom{r-1}{2}\binom{n-2}{k-2} \geq \frac{rn}{3k}\binom{n-2}{k-2}, \tag{2.5}$$

where we used the assumption that $n$ is large compared to $r$.

To prove the theorem we proceed by induction on $n$ and $s$. If $s \leq \binom{n-1}{k-1}$, then $\mathcal{L}_{n,k}(s)$ is a trivially intersecting family, thus $D(\mathcal{L}_{n,k}(s)) = 0$. So we may suppose $r \geq 2$. Let $\mathcal{F} \subseteq \binom{[n]}{k}$ be a family of size $s$ minimizing $D(\mathcal{F})$. If there exists $x \in [n]$ with $deg_{\mathcal{F}}(x) = \binom{n-1}{k-1}$, say $x = 1$, then letting $\mathcal{F}_x = \{F \in \mathcal{F} : x \in F\}$ we have $D(\mathcal{F}) = D(\mathcal{F} \setminus \mathcal{F}_1) + |\mathcal{D}(\mathcal{F} \setminus \mathcal{F}_1, \mathcal{F}_1)|$ and $\mathcal{F}_1 \subseteq \mathcal{F} \cap \mathcal{L}_{n,k}(s)$. Note that $|\mathcal{D}_{\mathcal{F}_1}(G)|$ is the same for any $G \in \binom{[n]}{k} \setminus \mathcal{F}_1$, so minimizing $D(\mathcal{F})$ is equivalent to minimizing $D(\mathcal{F} \setminus \mathcal{F}_1)$ among families containing $\mathcal{F}_1$. By induction $\mathcal{F} \setminus \mathcal{F}_1$ should be the initial segment $\mathcal{L}$ of size $s - \binom{n-1}{k-1}$ of the lexicographic order on $[2, n]$. But $\mathcal{F}_1 \cup \mathcal{L} = \mathcal{L}_{n,k}(s)$.

So we may assume that $deg_{\mathcal{F}}(x) < \binom{n-1}{k-1}$ for all $x \in [n]$ and therefore we can "replace a set $F \in \mathcal{F}$ with a set containing $x$", i.e. for a set $x \in G \notin \mathcal{F}$ consider $(\mathcal{F} \setminus \{F\}) \cup \{G\}$. The rest of the proof consists of three steps: the first shows the existence of a vertex of high degree, the second shows that the covering number of $\mathcal{F}$ is at most $r$, and the last one yields the statement of the theorem.

STEP 1: There exists $x \in [n]$ with $deg_{\mathcal{F}}(x) \geq \frac{s}{3r}$.

We start by a claim showing that there cannot exist too many vertices of relatively high degree.

**Claim 78** $|\{x \in [n] : deg_{\mathcal{F}}(x) \geq \frac{s}{3kr}\}| < 6kr$.

**Proof of Claim.** If not, then for a $6kr$-subset $X$ of $\{x \in [n] : deg_{\mathcal{F}}(x) \geq \frac{s}{3kr}\}$ we have

$$s \geq \sum_{x \in X} |\mathcal{F}_x| - \sum_{x,y \in X} |\mathcal{F}_x \cap \mathcal{F}_y| \geq |X| \frac{s}{3kr} - \binom{|X|}{2}\binom{n-2}{k-2} \geq \left(2 - \frac{54k^3 r}{n}\right) s,$$

where we used (2.5) for the last inequality. This is a contradiction as $n > 108k^2 l(k + l) > 54k^3 r$. □

Note that $D(\mathcal{F}) = \frac{1}{2}\sum_{F \in \mathcal{F}} |\mathcal{D}_{\mathcal{F}}(F)|$ and $|\mathcal{D}_{\mathcal{F}}(F)| = s - |\cup_{x \in F} \mathcal{F}_x|$, therefore $2D(\mathcal{F}) \geq s^2 - \sum_{x \in [n]} |\mathcal{F}_x|^2$. As $\mathcal{F}$ minimizes $D(\mathcal{F})$, using (2.4), we obtain $(1 - \frac{1}{r})s^2 \geq s^2 - \sum_{x \in [n]} |\mathcal{F}_x|^2$.

Let $X = \{x \in [n] : deg_{\mathcal{F}}(x) \geq \frac{s}{3kr}\}$ and by Claim 78 we have $|X| < 6kr$. Then assuming that 1 has the highest degree we obtain

$$\frac{1}{r}s^2 \leq \sum_{x \in X} |\mathcal{F}_x|^2 + \sum_{x \notin X} |\mathcal{F}_x|^2 \leq |\mathcal{F}_1| \sum_{x \in X} |\mathcal{F}_x| + \frac{s}{3kr} \sum_{x \notin X} |\mathcal{F}_x|. \tag{2.6}$$

Using (2.5) we bound the first term of the sum by

$$\sum_{x \in X} |\mathcal{F}_x| \leq |\mathcal{F}| + \sum_{x,y \in X} |\mathcal{F}_x \cap \mathcal{F}_y| \leq s + \binom{|X|}{2}\binom{n-2}{k-2} \leq \left(1 + \frac{54k^3 r}{n}\right) s \leq 2s$$

as $n$ is large enough. For the second term of the sum we have $\sum_{x \notin X} |\mathcal{F}_x| \leq \sum_{i=1}^{n} |\mathcal{F}_i| \leq ks$. Plugging back these bounds to (2.6) we obtain $\frac{1}{r}s^2 \leq 2s|\mathcal{F}_1| + \frac{s^2}{3r}$. Rearranging finishes the proof of Step 1.

STEP 2: $\tau(\mathcal{F}) \leq r$.

We accomplish this step in a series of claims getting better and better bounds on the covering number of $\mathcal{F}$. The first claim together with Claim 78 will imply $\tau(\mathcal{F}) \leq 6kr$.

**Claim 79** *The set* $X = \{x \in [n] : deg_{\mathcal{F}}(x) \geq \frac{s}{3kr}\}$ *is a cover of* $\mathcal{F}$.

**Proof of Claim.** If not, then there exists $F \in \mathcal{F}$ such that $|\mathcal{F}_x| < \frac{s}{3kr}$ for all $x \in F$. This shows $|\mathcal{D}_{\mathcal{F}}(F)| \geq s - k\frac{s}{3kr} = s(1 - \frac{1}{3r})$. On the other hand, by Step 1, for any $1 \in G$ we have $|\mathcal{D}_{\mathcal{F}}(G)| \leq s - \frac{s}{3r}$. That means replacing $F$ with $G$ would yield a family $\mathcal{F}'$ with $D(\mathcal{F}') < D(\mathcal{F})$, a contradiction. □

Let us consider a minimal subcover of $X$. By Claim 78 and rearranging elements, we may suppose that this subcover is $[m]$ for some integer $m$ with $r \leq m < 6kr$. Let us partition $\mathcal{F}$ into the families $\mathcal{F}_i^* = \{F \in \mathcal{F} : \min F = i\}$. Our next claim shows that these families cannot differ much in size.

**Claim 80** *For every* $i, j \in [m]$ *we have* $|\mathcal{F}_i^*| \geq |\mathcal{F}_j^*| - \frac{3mk^2}{rn}s$.

**Proof of Claim.** First note that for every $i \in [m]$ there exists a set $F_i \in \mathcal{F}_i^*$ with $F \cap [m] = \{i\}$ by the minimality of the subcover.

For any $j \in [m] \setminus \{i\}$ there exist at most $k\binom{n-2}{k-2}$ sets $F \in \mathcal{F}_j$ such that $F_i \cap F \neq \emptyset$ ($F$ contains $j$ and an element of $F_i$). Using (2.5), this yields

$$|\mathcal{D}_{\mathcal{F}}(F_i)| = \sum_{1 \leq j \leq m, i \neq j} |\mathcal{D}_{\mathcal{F}_j^*}(F_i)| \geq \sum_{1 \leq j \leq m, i \neq j} \left(|\mathcal{F}_j^*| - k\binom{n-2}{k-2}\right)$$
$$= s - |\mathcal{F}_i^*| - (m-1)k\binom{n-2}{k-2} \geq s - |\mathcal{F}_i^*| - \frac{3mk^2}{rn}s.$$

By minimality of $D(\mathcal{F})$ the replacement of $F_i$ by a set $G$ containing $j$ cannot decrease the number of disjoint pairs. Therefore we must have

$$s - |\mathcal{F}_i^*| - \frac{3mk^2}{rn}s \leq |\mathcal{D}_{\mathcal{F}}(F_i)| \leq |\mathcal{D}_{\mathcal{F}}(G)| \leq s - |\mathcal{F}_j^*|$$

as $j \in G \cap F$ for all $F \in \mathcal{F}_j^*$. Rearranging completes the proof of the Claim. □

We have one more intermediate claim before establishing the statement of Step 2.

**Claim 81** $m \leq 6r$.

**Proof of Claim.** By Step 1, we have $|\mathcal{F}_1| = |\mathcal{F}_1^*| \geq \frac{s}{3r}$. Therefore by Claim 80, $n > 108k^3r$ and $m < 6kr$, we have $|\mathcal{F}_i^*| \geq |\mathcal{F}_1^*| - \frac{3mk^2}{rn}s \geq \frac{s}{3r} - \frac{18k^3}{n}s \geq \frac{s}{6r}$ for all $i \in [m]$. As the $\mathcal{F}_i^*$'s partition $\mathcal{F}$, we obtain $m\frac{s}{6r} \leq s$. □

Now we are ready for our final argument in Step 2. Observe that

- $D(\mathcal{F}) = \sum_{i<j} |\mathcal{D}(\mathcal{F}^*_i, \mathcal{F}^*_j)|$ as the $\mathcal{F}^*_i$'s are intersecting and partition $\mathcal{F}$,
- $|\mathcal{D}(\mathcal{F}^*_i, \mathcal{F}^*_j)| \geq (|\mathcal{F}^*_i| - k\binom{n-2}{k-2})|\mathcal{F}^*_j|$ as an $F \in \mathcal{F}^*_j$ can intersect at most $k\binom{n-2}{k-2}$ sets in $\mathcal{F}_i$,
- $|\mathcal{F}^*_1| \leq \frac{s}{m} + \frac{3mk^2}{rn}s$ by Claim 80 as the $\mathcal{F}^*_i$'s partition $\mathcal{F}$ so there exists $i$ with $|\mathcal{F}^*_i| \leq \frac{s}{m}$,
- $|\mathcal{F}^*_i| \geq |\mathcal{F}^*_1| - \frac{3mk^2}{rm}s$ for every $i \in [m]$ by Claim 80.

Using these, we obtain

$$\begin{aligned}
D(\mathcal{F}) &\geq \sum_{1\leq i<j\leq m} \left(|\mathcal{F}^*_i| - k\binom{n-2}{k-2}\right)|\mathcal{F}^*_j| \\
&= \sum_{1\leq i<j\leq m} |\mathcal{F}^*_i||\mathcal{F}^*_j| - k\binom{n-2}{k-2}\sum_{2\leq j\leq m}(j-1)|\mathcal{F}^*_j| \\
&\geq \frac{1}{2}\left(\left(\sum_{i=1}^{m}|\mathcal{F}^*_i|\right)^2 - \sum_{i=1}^{m}|\mathcal{F}^*_i|^2\right) - mk\binom{n-2}{k-2}\sum_{2\leq j\leq m}|\mathcal{F}^*_j| \\
&\geq \frac{1}{2}\left(s^2 - |\mathcal{F}^*_1|\sum_{i=1}^{m}|\mathcal{F}^*_i|\right) - mk\binom{n-2}{k-2}s \\
&\geq \frac{1}{2}\left(s^2 - \left(\frac{s}{m} + \frac{3mk^2}{rn}s\right)s\right) - \frac{3mk^2}{rn}s^2 \\
&= \frac{1}{2}\left(1 - \frac{1}{m} - \frac{9mk^2}{rn}\right)s^2.
\end{aligned}$$

As $\mathcal{F}$ minimizes $\mathcal{D}$, we have $\mathcal{D}(\mathcal{F}) \leq \mathcal{D}(\mathcal{L}_{n,k}(s)) \leq \frac{1}{2}(1-\frac{1}{r})s^2$ by (2.4). Therefore we obtain $\frac{1}{r} \leq \frac{1}{m} + \frac{9mk^2}{rn} \leq \frac{1}{m} + \frac{54k^2}{n}$. Using $n > 54k^2r(k+r)$ we have $\frac{54k^2}{n} < \frac{1}{r} - \frac{1}{r+1}$, which implies $m \leq r$. This finishes the proof of Step 2.

STEP 3: $D(\mathcal{F})$ is minimized by $\mathcal{L}_{n,k}(s)$.

Let $\mathcal{A} := \{A \in \binom{[n]}{k} : A \cap [r] \neq \emptyset\}$. By Step 2, we know that $\mathcal{F} \subseteq \mathcal{A}$. Writing $\mathcal{G} = \mathcal{G}_{\mathcal{F}} = \mathcal{A} \setminus \mathcal{F}$ we have

$$D(\mathcal{F}) = D(\mathcal{A}) - D(\mathcal{G}) - |\mathcal{D}(\mathcal{G}, \mathcal{A})|.$$

Therefore minimizing $D(\mathcal{F})$ is equivalent to minimizing $D(\mathcal{G}) - |\mathcal{D}(\mathcal{G}, \mathcal{A})|$. As $s = |\mathcal{F}|$ is fixed, so is $|\mathcal{G}|$. For any $G \in \mathcal{G}$ the number $|\mathcal{D}_{\mathcal{A}}(G)|$ is maximized when $|G \cap [r]|$ is as small as possible. As $G \in \mathcal{A}$ by definition, $|G \cap [r]| \geq 1$ holds for all $G \in \mathcal{G}$. Note that for any $G \in \mathcal{G}_{\mathcal{L}_{n,k}(s)}$ we have $|G \cap [r]| = 1$, therefore $|\mathcal{D}(\mathcal{G}, \mathcal{A})|$ is maximized by $\mathcal{G}_{\mathcal{L}_{n,k}(s)}$. Obviously, $D(\mathcal{G}) \geq 0$ for any family $\mathcal{G}$ and we have $D(\mathcal{G}_{\mathcal{L}_{n,k}(s)}) = 0$ as all sets $G \in \mathcal{G}_{\mathcal{L}_{n,k}(s)}$ contain $r$. This means that $\mathcal{G}_{\mathcal{L}_{n,k}(s)}$ minimizes $D(\mathcal{G}) - |\mathcal{D}(\mathcal{G}, \mathcal{A})|$ which completes the proof of Step 3 and the theorem. ■

We finish this section by stating the results of Das, Gan, and Sudakov on $D_t(\mathcal{F})$ and $D_{q,\times}(\mathcal{F})$.

**Theorem 82 (Das, Gan, Sudakov [116])** *For every $t,k,l$ there exists an integer $n_0$ such that if $n \geq n_0$ and $0 \leq s \leq \binom{n}{k} - \binom{n-l}{k}$, then $D_t(\mathcal{F}) \geq D_t(\mathcal{L}_{n,k}(s))$ holds for any $\mathcal{F} \subseteq \binom{[n]}{k}$ with $|\mathcal{F}| = s$.*

**Theorem 83 (Das, Gan, Sudakov [116])** *For every $t,k,l$ there exists an integer $n_0$ such that if $n \geq n_0$ and $0 \leq s \leq \binom{n}{k} - \binom{n-l}{k}$, then $D_{q,\times}(\mathcal{F}) \geq D_{q,\times}(\mathcal{L}_{n,k}(s))$ holds for any $\mathcal{F} \subseteq \binom{[n]}{k}$ with $|\mathcal{F}| = s$.*

**Exercise 84** *Prove that for the family $\mathcal{F}$ defined in Construction 75 and any $\mathcal{G}$ with $\binom{[n]}{\geq m} \subset \mathcal{G} \subset 2^{[n]}$ we have $D_t(\mathcal{F}) < D_t(\mathcal{G})$.*

**Exercise 85 (Das, Gan, Sudakov [116])** *Using Theorem 77, prove that if $n,k,l,s$ are integers satisfying $n \geq 108k^2l(k+l)$ and $\binom{n-l}{k} \leq s \leq \binom{n}{k}$, then $\binom{[n]}{k} \setminus \mathcal{L}_{n,k}(\binom{n}{k} - s)$ minimizes $D(\mathcal{F})$ over all $\mathcal{F} \subseteq \binom{[n]}{k}$ with $|\mathcal{F}| = s$.*

### 2.3.3 Most probably intersecting families

The following property measuring how close a family is to being intersecting was introduced by Katona, Katona, and Katona [343]. For a family $\mathcal{F}$ of sets let $\mathcal{F}(p)$ denote the random subfamily that contains every set $F \in \mathcal{F}$ with probability $p$ independently of all other sets. For given positive integer $m$ and real $0 < p < 1$, they posed the problem of finding the families for which

$$I_{n,p}(m) := \max_{\mathcal{F} \subseteq 2^{[n]}, |\mathcal{F}| = 2^{n-1}+m} \mathbb{P}(\mathcal{F}(p) \text{ is intersecting})$$

and

$$I_{n,p,k}(m) := \max_{\mathcal{F} \subseteq \binom{[n]}{k}, |\mathcal{F}| = \binom{n-1}{k-1}+m} \mathbb{P}(\mathcal{F}(p) \text{ is intersecting})$$

are attained. These families are said to be *most probably intersecting.*

In the non-uniform case, Katona, Katona, and Katona determined $I_{n,p}(m)$ for small values of $m$ and conjectured that there exists a sequence $F_1, F_2, \ldots, F_{2^n}$ of sets such that $\mathcal{F}_i = \{F_1, F_2, \ldots, F_i\}$ is most probably intersecting for all values of $i$. Russell proved the following theorem.

**Theorem 86 (Russell [488])** *For any $0 < p < 1$ and $0 < m \leq 2^{n-1}$ if $r$ is the integer with the property $\sum_{i=r}^{n} \binom{n}{i} \leq 2^{n-1} + m < \sum_{i=r-1}^{n} \binom{n}{i}$, then there exists a family $\mathcal{F}$ with $I_{n,p}(m) = \mathbb{P}(\mathcal{F}(p)$ is intersecting$)$ and $\binom{[n]}{\geq r} \subseteq \mathcal{F} \subseteq \binom{[n]}{\geq r-1}$.*

However, a year later Russell and Walters disproved the conjecture by the following result.

**Theorem 87 (Russell and Walters [489])** *Let $n \geq 21$ and $2 \leq s \leq 2^{n-1}$. Then*

*(i)* $\mathcal{F} = \binom{[n]}{\geq 3} \cup \{F \in \binom{[n]}{2} : 1 \in F\}$ *is the unique family of size* $\sum_{i=3}^{n} \binom{n}{i} + n - 1$ *that maximizes the number of intersecting subfamilies of size $s$,*

*(ii)* $\mathcal{G} = \binom{[n]}{\geq 3} \cup \{G \in \binom{[n]}{2} : 1 \notin F\}$ *is the unique family of size* $\sum_{i=3}^{n} \binom{n}{i} + \binom{n-1}{2}$ *that maximizes the number of intersecting subfamilies of size $s$.*

Theorem 87 implies that the conjecture is false, as one can generate $\mathcal{F}(p)$ in two steps: first pick a random number $s$ with binomial distribution $Bi(|\mathcal{F}|, p)$ and then pick a subfamily of $\mathcal{F}$ of size $s$ uniformly at random.

In the uniform case Katona, Katona, and Katona [343] determined $I_{n,p,k}(1)$ and Russell [488] showed that for any $m$ and $p$ there exists a most probably intersecting *left-shifted* family $\mathcal{F}$ of size $\binom{n-1}{k-1} + m$. Using the methods developped in [116], Das and Sudakov obtained the following result. Recall $\mathcal{L}_{n,k}(m)$ is the family of the first $m$ sets in $\binom{[n]}{k}$ in the lexicographic order.

**Theorem 88 (Das, Sudakov [117])** *For any $l$ and $k$, there exists an $n_0 = n_0(l,k)$ such that if $n \geq n_0$ and $\binom{n}{k} - \binom{n-l}{k} \leq m \leq \binom{n}{k} - \binom{n-l}{k} + n - l - k + 1$ hold, then $\mathcal{L}_{n,k}(m)$ is the most probably intersecting family among all families $\mathcal{F} \subseteq \binom{[n]}{k}$ and $|\mathcal{F}| = m$ (independently of $p$).*

### 2.3.4 Almost intersecting families

The Erdős-Rademacher problem for intersecting families is to minimize the number $\mathcal{D}(\mathcal{F})$ of disjoint pairs in an intersecting family $\mathcal{F}$ of given size. Obtaining the function $d(m) = \min_{\mathcal{F}:|\mathcal{F}|=m} D(\mathcal{F})$ for all values of $m$ is equivalent to determining the function $f(d) = \max_{\mathcal{F}:D(\mathcal{F}) \leq d} |\mathcal{F}|$. Here no condition is imposed on the structure of disjoint pairs. One can require that these pairs should be roughly evenly distributed among the sets of $\mathcal{F}$. Gerbner, Lemons, Palmer, Patkós, and Szécsi [259] introduced the following notion: a family $\mathcal{F}$ is said to be $(\leq l)$-*almost intersecting* if for every $F \in \mathcal{F}$ the set $\mathcal{D}_{\mathcal{F}}(F) = \{F' \in \mathcal{F} : F \cap F' = \emptyset\}$ has size at most $l$, and $\mathcal{F}$ is said to be $l$-*almost intersecting* if $|\mathcal{D}_{\mathcal{F}}(F)| = l$ for all $F \in \mathcal{F}$. The two notions are rather different. An intersecting family is $(\leq l)$-almost intersecting for any $l$, but not $l$-almost intersecting. Gerbner et al. proved that for any $k$ and $l$ there exists $n_0 = n_0(k,l)$ such that if $\mathcal{F} \subseteq \binom{[n]}{k}$ is $(\leq l)$-almost intersecting with $n \geq n_0$, then $|\mathcal{F}|$ cannot exceed the Erdős-Ko-Rado bound $\binom{n-1}{k-1}$ (and determined $n_0(k,1) = 2k+2$). They conjectured that something similar happens in the non-uniform case: allowing 1 disjoint set does increase the maximum size, but allowing more does not help if $n$ is large enough.

**Theorem 89** *For $l = 1$ and $l = 2$ there exists $n_0 = n_0(l)$ such that if $n \geq n_0$*

*and $\mathcal{F} \subset 2^{[n]}$ is an $(\leq l)$-almost intersecting family, then*

$$|\mathcal{F}| \leq \begin{cases} \sum_{i=n/2}^{n} \binom{n}{i} & \text{if } n \text{ is even} \\ \binom{n-1}{\lfloor n/2 \rfloor} + \sum_{i=\lceil n/2 \rceil}^{n} \binom{n}{i} & \text{if } n \text{ is odd,} \end{cases}$$

*and equality holds if and only if $\mathcal{F}$ is the family of sets of size at least $n/2$ and (if $n$ is odd) the sets of size $\lfloor n/2 \rfloor$ not containing a fixed element of $[n]$.*

They conjectured that the same upper bound holds for larger $l$, but we can disprove it. Let us assume $n$ is even, and consider the following family $\mathcal{F}$. It contains the $(n/2-1)$-element sets that avoid $n$ and $n-1$, the $n/2$-element sets that contain at most one of $n$ and $n-1$, and all the sets of size greater than $n/2$. It is easy to see that $\mathcal{F}$ is a $(\leq 3)$-almost intersecting family of size greater than $\sum_{i=n/2}^{n} \binom{n}{i}$.

As we mentioned earlier, the notion of $l$-almost intersecting is very different from $(\leq l)$-almost intersecting. If $\mathcal{F}$ is 1-almost intersecting, then $\mathcal{F}$ can be partitioned into disjoint pairs $F_i, F_i'$, such that these pairs form an ISPS. Using this observation they were able to prove that for any $l$ the maximum size of an $l$-almost intersecting family $\mathcal{F} \subseteq 2^{[n]}$ is $\Theta_l(\binom{n}{\lfloor n/2 \rfloor})$, while a $k$-uniform $l$-almost intersecting family cannot have size greater than some constant $f(k)$ independently of the size of the underlying set. Their conjecture about the correct value of $f(k)$ was verified by Scott and Wilmer in a stronger form. To state their result we introduce the following definition: $\mathcal{F}$ is *$[a,b]$-almost intersecting* if $|\mathcal{D}_{\mathcal{F}}(F)| \in [a,b]$ for all $F \in \mathcal{F}$.

**Theorem 90 (Scott, Wilmer [500])** *For any $k > 2$ there exist constants $R = R(k)$ and $s_0 = s_0(k)$ such that if $\mathcal{F}$ is a $k$-uniform $[R,s]$-almost intersecting family with $s \geq s_0$, then $|\mathcal{F}| \leq (s+1)\binom{2k-2}{k-1}$.*

**Exercise 91 (Gerbner et al. [259])** *If $\mathcal{F} \subseteq 2^{[n]}$ is a 1-almost intersecting family, then*

$$|\mathcal{F}| \leq \begin{cases} \binom{n}{n/2} & \text{if } n \text{ is even} \\ 2\binom{n-1}{\lfloor n/2 \rfloor - 1} & \text{if } n \text{ is odd,} \end{cases}$$

*and equality holds if and only if $\mathcal{F} = \binom{[n]}{n/2}$ provided $n$ is even and $\mathcal{F} = \{F \in \binom{[n]}{\lfloor n/2 \rfloor} : x \in F\} \cup \{F \in \binom{[n]}{\lceil n/2 \rceil} : x \notin F\}$ for some fixed $x \in [n]$ provided $n$ is odd.*

## 2.4 $L$-intersecting families

In previous sections we considered families with the property that every pairwise intersection should have large size. In this section we will prescribe all possible pairwise intersection sizes. Formally, for a subset $L$ of non-negative integers we say that a family $\mathcal{F}$ is *$L$-intersecting* if for any distinct $F, F' \in \mathcal{F}$ we have $|F \cap F'| \in L$. Note that $\mathcal{F}$ is $t$-intersecting if and only if it is $L$-intersecting with $L = L_t = \{t, t+1, t+2, \dots\}$.

The first result of this type considers the $L$-intersection problem when $L$ consists of a single element. Fisher proved [182] that a uniform family with this property cannot contain more sets than the the size of its underlying set.

**Exercise 92 (Fisher's inequality)** *For any non-negative integer $\lambda$, if $\mathcal{F} \subseteq \binom{[n]}{k}$ is $\{\lambda\}$-intersecting, then $|\mathcal{F}| \le n$ holds. (Hint (Bose [84]: consider the matrix $B = MM^T$ with $M$ being the incidence matrix of $\mathcal{F}$.)*

Another famous early result of this type is due to deBruijn and Erdős who solved the non-regular case of $L = \{1\}$.

**Exercise 93 (deBruijn, Erdős, [122])** *Prove that if $\mathcal{F} \subseteq 2^{[n]}$ is an antichain, and $L$-intersecting with $|L| = 1$, then the characteristic vectors $\{v_F : F \in \mathcal{F}\}$ are independent over the reals, in particular $|\mathcal{F}| \le n$.*

Finally, the problem of non-uniform $\{\lambda\}$-intersecting families was settled by Majumdar.

**Theorem 94 (Majumdar, [402])** *For any non-negative integer $\lambda$, if $\mathcal{F} \subseteq 2^{[n]}$ is $\{\lambda\}$-intersecting, then $|\mathcal{F}| \le n$ holds.*

The systematic study of $L$-intersecting families was started in the 1970's by Ray-Chaudhuri, Wilson and others. Soon, the modular version of these problems was also investigated. We say that a family $\mathcal{F}$ is $L$-intersecting mod $p$ if $|F \cap F'| \equiv l \pmod{p}$ for some $l \in L$ and for any distinct pair $F, F' \in \mathcal{F}$. Below we state and prove some basic results concerning $L$-intersecting families. A much more detailed introduction to the topic (including several proofs for these results as well) can be found in the excellent book by Babai and Frankl [28].

**Theorem 95 (Frankl, Wilson [224])** *If $\mathcal{F} \subseteq 2^{[n]}$ is $L$-intersecting, then $|\mathcal{F}| \le \sum_{i=0}^{|L|} \binom{n}{i}$ holds.*

**Proof.** The following short proof using the polynomial method is due to Babai [27]. Let us enumerate the sets $F_1, F_2, \dots, F_{|\mathcal{F}|}$ of $\mathcal{F}$ in increasing order according to their sizes. For every $F \in \mathcal{F}$ let $v_F$ be its characteristic vector and let us introduce the $n$-variable polynomials

$$f_{F_i} = \prod_{l \in L, l < |F_i|} (x \cdot v_{F_i} - l).$$

As $0^2 = 0$ and $1^2 = 1$, we can obtain multilinear polynomials $f'_{F_i}(x)$ such that $f_{F_i}(x) = f'_{F_i}(x)$ holds for all $x \in \{0,1\}^n$. The assumption on $\mathcal{F}$ implies

$$f'_{F_i}(v_{F_j}) \begin{cases} \neq 0 & \text{if } j = i \\ = 0 & \text{if } j < i, \end{cases}$$

Lemma 14 implies that the set $\{f'_F(X) : F \in \mathcal{F}\}$ of polynomials is independent over $\mathbb{F}_p$. As discussed after Lemma 14, the dimension of $n$-variable multilinear polynomials of degree at most $s$ is $\sum_{i=0}^{s} \binom{n}{i}$. This finishes the proof. ■

**Theorem 96 (Deza, Frankl, Singhi, [133])** *Let $\mathcal{F} \subseteq 2^{[n]}$ be $L$-intersecting mod $p$ and assume that for every $F \in \mathcal{F}$ and $l \in L$ we have $|F| \not\equiv l \pmod p$. Then $|\mathcal{F}| \le \sum_{i=0}^{|L|} \binom{n}{i}$ holds.*

**Proof.** The proof we present here is due to Alon, Babai, and Suzuki [12] and is very similar to the proof we gave for Theorem 95. For every $F \in \mathcal{F}$ let $v_F$ be its characteristic vector and let us introduce the $n$-variable polynomials $f_F = \prod_{l \in L}(x \cdot v_F - l) \in \mathbb{F}_p^n(x)$. We can obtain multilinear polynomials $f'_F(x)$ such that $f_F(x) = f'_F(x)$ holds for all $x \in \{0,1\}^n$. The assumption on $\mathcal{F}$ implies

$$f'_F(v_G) \begin{cases} \neq 0 & \text{if } F = G \\ = 0 & \text{if } F \neq G, \end{cases}$$

Again, using Lemma 14 and the discussion afterward, we obtain that the set $\{f'_F(X) : F \in \mathcal{F}\}$ of polynomials is independent over $\mathbb{F}_p$, and has cardinality $\sum_{i=0}^{s} \binom{n}{i}$. This finishes the proof. ■

Let us state a strengthening of the bound of Theorem 95 if $L$ contains only positive integers.

**Theorem 97 (Snevily [516])** *If $\mathcal{F} \subseteq 2^{[n]}$ is $L$-intersecting with $0 \notin L$, then $|\mathcal{F}| \le \sum_{i=0}^{|L|} \binom{n-1}{i}$ holds.*

Now we turn our attention to uniform $L$-intersecting families.

**Theorem 98 (Ray-Chaudhuri, Wilson, [473])** *If $\mathcal{F} \subseteq \binom{[n]}{k}$ is $L$-intersecting, then $|\mathcal{F}| \le \binom{n}{|L|}$ holds.*

**Proof.** The proof we present is due to Alon, Babai, and Suzuki [12]. For a set $I \subset [n]$ let $x_I = \prod_{i \in I} x_i$. Then $x_I(F) = x_I(v_F)$ is the product of the appropriate entries of the characteristic vector of $F$, hence $x_I(F) \neq 0$ if and only if $I \subseteq F$. First we prove the following lemma.

**Lemma 99** *Let $f : \{0,1\}^n \to \mathbb{R}$ be a function such that $f(x) \neq 0$ whenever $x \cdot \mathbf{1} \le r$ holds. Then the set $\{x_I f : I \in \binom{[n]}{\le r}\}$ is independent.*

**Proof of Lemma.** Let us enumerate the sets of $\binom{[n]}{\le r}$ in increasing order according to their sizes, and let $v_1, v_2, \dots, v_{\binom{n}{\le r}}$ be their characteristic vectors in this order. Then we have

$$x_{I_i}(I_j)f(I_j)\begin{cases} \neq 0 & \text{if } j = i \\ = 0 & \text{if } j < i. \end{cases}$$

Lemma 14 finishes the proof of the lemma. □

Let $\mathcal{F} \subseteq \binom{[n]}{k}$ be $L$-intersecting. Just as in the proof of Theorem 96, we introduce the polynomials

$$f_F(x) = \prod_{l \in L}(x \cdot v_F - l),$$

and observe that there exist multilinear polynomials $f'_F$ such that $f'_F(x) = f_F(x)$ for all $x \in \{0,1\}^n$ and these polynomials are linearly independent. Let us also consider the polynomials $x_I(\sum_{i=1}^n x_i - k)$ for all $I \in \binom{n}{\le k-1}$. Lemma 99 implies that these polynomials also form an independent set. We claim that the union of these two sets of polynomials is also linearly independent. Indeed, if $\sum_{F \in \mathcal{F}} \lambda_F f'_F + \sum_{I \in \binom{n}{\le k-1}} \kappa_I x_I(\sum_{i=1}^n x_i - k) = 0$, then substituting one particular $v_F$ shows $\lambda_F = 0$; thus we obtain that all $\lambda_F$'s are equal to 0. Then the independence of $\{x_I(\sum_{i=1}^n x_i - k) : I \in \binom{[n]}{\le k-1}\}$ implies that all $\kappa_I$'s are 0.

We can change $x_I(\sum_{i=1}^n x_i - k)$ to a multilinear $g_I$ with $x_I(\sum_{i=1}^n x_i - k)(y) = g_I(y)$ for any $y \in \{0,1\}^n$ and the set $\{f'_F : F \in \mathcal{F}\} \cup \{g_I : I \in \binom{n}{\le k-1}\}$ is independent with the same proof as above. As the dimension of $n$-variable multilinear polynomials of degree at most $k$ is $\sum_{i=0}^k \binom{n}{i}$, we obtain

$$|\mathcal{F}| + \sum_{i=0}^{k-1}\binom{n}{i} \le \sum_{i=0}^{k}\binom{n}{i},$$

which finishes the proof of the theorem. ■

**Theorem 100 (Frankl, Wilson [224])** *If $\mathcal{F} \subseteq \binom{[n]}{k}$ is $L$-intersecting mod $p$, then $|\mathcal{F}| \le \binom{n}{|L|}$ holds.*

To bridge the gap between uniform and non-uniform families, one can introduce a set $K$ of allowed set sizes, i.e. one may consider families $\mathcal{F}$ such that $|F| \in K$ for all sets in $\mathcal{F}$ and $|F \cap F'| \in L$ for all distinct pairs $F, F' \in \mathcal{F}$. The following conjecture of Alon, Babai and Suzuki [12] about such families has been proved recently.

**Theorem 101 (Hwang, Kim [316])** *Let $K = \{k_1, k_2, \dots, k_r\}$ and $L = \{l_1, l_2, \dots, l_s\}$ be two disjoint subsets of $\{0, 1, \dots, p-1\}$, where $p$ is a prime, and let $\mathcal{F} \subseteq \cup_{i=1}^r \binom{[n]}{k_i}$ be an $L$-intersecting family mod $p$. If $n \ge s + \max_{1 \le i \le r} k_i$, then $|\mathcal{F}| \le \binom{n}{s} + \binom{n}{s-1} + \dots + \binom{n}{s-r+1}$ holds.*

In some cases even better bounds are possible. Observe that $\sum_{i=0}^{2r-1} \binom{n-1}{s-i} = \sum_{j=0}^{r-1} \binom{n}{s-2j}$ holds by Pascal's identity and $\binom{n}{s-2i} < \binom{n}{s-i}$ if $n \geq 2s-2$ holds. Therefore, the next theorem improves the bound of Theorem 101 whenever $n \geq 2s-2$ holds.

**Theorem 102 (Wang, Wei, Ge [543])** *Let $K = \{k_1, k_2, \dots, k_r\}$ and $L = \{l_1, l_2, \dots, l_s\}$ be two disjoint subsets of $\{0, 1, \dots, p-1\}$, where $p$ is a prime, and let $\mathcal{F} \subseteq \cup_{i=1}^{r} \binom{[n]}{k_i}$ be an $L$-intersecting family mod $p$. If $n \geq s + \max_{1 \leq i \leq r} k_i$, then $|\mathcal{F}| \leq \binom{n-1}{s} + \binom{n-1}{s-1} + \dots + \binom{n-1}{s-2r+1}$ holds.*

An even stronger conjecture on $L$-intersecting families mod $p$ is still open.

**Conjecture 103 (Snevily [516])** . *Let $p$ be a prime, and let $K = \{k_1, k_2, \dots, k_r\}$ and $L = \{l_1, l_2, \dots, l_s\}$ be two disjoint subsets of $[0, p-1]$. Let $\mathcal{F} \subset \cup_{i=1}^{r} \binom{[n]}{k_i}$ be an $L$-intersecting family mod $p$. Then $|\mathcal{F}| \leq \binom{n}{s}$ holds.*

We finish this section with the following two exercises.

**Exercise 104** *Prove that if every set in a family $\mathcal{F} \subseteq 2^{[n]}$ has odd size and $\mathcal{F}$ is $L$-intersecting with $L = 2\mathbb{Z}^+$, then the characteristic vectors $\{v_F : F \in \mathcal{F}\}$ are independent over $\mathbb{F}_2$.*

**Exercise 105 (Berlekamp [52])** *Prove that if a family $\mathcal{F} \subseteq 2^{[n]}$ is $L$-intersecting with $L = 2\mathbb{Z}^+$, then $|\mathcal{F}| \leq 2^{\lfloor n/2 \rfloor} + \delta_n$, where $\delta_n = 1$ if $n$ is odd and 0 otherwise. Show that this bound is sharp. Furthermore, prove that if every set in a family $\mathcal{F} \subseteq 2^{[n]}$ has even size, then $|\mathcal{F}| \leq 2^{\lfloor n/2 \rfloor}$.*

## 2.5 $r$-wise intersecting families

In this section we consider yet another natural generalization of intersection problems. Instead of posing restrictions on pairwise intersection sizes, one might consider the $r$-wise case. We say that a family $\mathcal{F}$ of sets is $r$-wise $t$-intersecting if for any $F_1, F_2, \dots, F_r \in \mathcal{F}$ we have $|F_1 \cap F_2 \cap \dots \cap F_r| \geq t$. If $t = 1$, then $\mathcal{F}$ is said to be $r$-wise intersecting. If $L$ is a set of non-negative integers, then $\mathcal{F}$ is $r$-wise $L$-intersecting if for any distinct sets $F_1, F_2, \dots, F_r \in \mathcal{F}$ we have $|F_1 \cap F_2 \cap \dots \cap F_r| \in L$. If $r$ divides $n$, an $r$-*equipartition* of $n$ is a partition into $r$ parts of size $n/r$.

**Theorem 106 (Frankl [183])** *If $\mathcal{F} \subseteq \binom{[n]}{k}$ is $r$-wise intersecting and $rk \leq (r-1)n$ holds, then $|\mathcal{F}| \leq \binom{n-1}{k-1}$.*

**Proof.** We proceed by induction on $(r-1)n-rk$. Observe that $\mathcal{F}$ is $r$-wise intersecting if and only if $\overline{\mathcal{F}} = \{[n] \setminus F : F \in \mathcal{F}\}$ is *r-wise union*, i.e. for any $G_1, G_2, \dots, G_r \in \overline{\mathcal{F}}$ we have $\cup_{i=1}^r G_i \neq [n]$. If $(r-1)n - rk = 0$, then sets of $\overline{\mathcal{F}}$ are of size $n/r$ and thus $\overline{\mathcal{F}}$ can contain at most $r-1$ parts of any $r$-equipartition of $[n]$. The number $M_{n,r}$ of $r$-equipartitions is $\frac{\binom{n}{n/r}\binom{n-n/r}{n/r}\dots\binom{n/}{n/r}}{r!}$ and every $(n/r)$-subset of $[n]$ belongs to $M_{n-n/r,r-1}$ many $r$-equipartitions. Therefore counting pairs $(G, \mathcal{P})$ with $G \in \overline{\mathcal{F}} \cap \mathcal{P}$ and $\mathcal{P}$ being an $r$-equipartition of $[n]$, we obtain

$$|\mathcal{F}| = |\overline{\mathcal{F}}| \leq \frac{(r-1)M_{n,r}}{M_{n-n/r,r-1}} = \frac{r-1}{r}\binom{n}{n/r} = \frac{r-1}{r}\binom{n}{n-n/r} = \binom{n-1}{n-k-1}.$$

Now let us assume that we have proved the theorem for all values of $n$, $r$, $k$ with $(r-1)n - rk = p$ and let $\mathcal{F} \subseteq \binom{n}{k}$ be a maximum size $r$-wise intersecting family with $(r-1)n - rk = p+1$. In particular, we have $|\mathcal{F}| \geq \binom{n-1}{k-1}$ as the family of all $k$-sets containing a fixed point is $r$-wise intersecting. Consider the family $\mathcal{G} = \{G \in \binom{[n+1]}{k+1} : \exists F \in \mathcal{F} \text{ with } F \subset G\}$. As $\mathcal{F}$ is $r$-wise intersecting, so is $\mathcal{G}$. Moreover, $(r-1)(n+1) - r(k+1) = p$ holds; thus the inductive hypothesis implies $|\mathcal{G}| \leq \binom{n+1-1}{k+1-1} = \binom{n}{k}$.

We partition $\mathcal{G}$ into two subfamilies: $\mathcal{G}_1 = \{G \in \mathcal{G} : n+1 \in G\}$ and $\mathcal{G}_2 = \mathcal{G} \setminus \mathcal{G}_1$. Clearly, $\mathcal{G}_1 = \{F \cup \{n+1\} : F \in \mathcal{F}\}$ and thus $|\mathcal{G}_1| = |\mathcal{F}|$. On the other hand $\overline{\Delta(\overline{\mathcal{F}})} = \mathcal{G}_2$. As $\binom{n-1}{n-k} = \binom{n-1}{k-1} \leq |\mathcal{F}| = |\overline{\mathcal{F}}|$, Theorem 23 implies $|\mathcal{G}_2| = |\Delta(\overline{\mathcal{F}})| \geq \binom{n-1}{n-k-1} = \binom{n-1}{k}$. We obtained

$$|\mathcal{F}| + \binom{n-1}{k} \leq |\mathcal{G}_1| + |\mathcal{G}_2| = |\mathcal{G}| \leq \binom{n}{k}.$$

Rearranging yields $|\mathcal{F}| \leq \binom{n-1}{k-1}$ as required. ■

**Theorem 107 (Brace, Daykin [87])** *Suppose that $\mathcal{F} \subseteq 2^{[n]}$ is r-wise intersecting with $\cap_{F \in \mathcal{F}} F = \emptyset$. Then $|\mathcal{F}| \leq \frac{r+2}{2^{r+1}} 2^n$ holds and this is sharp as shown by the family $\mathcal{F}_{n,r,1,1} = \{F \subseteq [n] : |F \cap [r+1]| \geq r\}$.*

**Proof.** We proceed by induction on $r+n$. The case $r=2$ is covered by Proposition 8, while the case $r = n-1$ follows from the observation that the intersecting property of $\mathcal{F}$ and $\cap_{F \in \mathcal{F}} F = \emptyset$ imply all sets in $\mathcal{F}$ have size at least $n-1$.

Let $\mathcal{F}$ be a maximum size $r$-wise intersecting family. Therefore $G \supseteq F \in \mathcal{F}$ implies $G \in \mathcal{F}$. Furthermore, we can also assume that there exists a set of size at most $n-2$; otherwise $|\mathcal{F}| \leq n+1 \leq \frac{r+2}{2^{r+1}} 2^n$ holds. By symmetry, we can assume $[n] \setminus \{1,2\} \in \mathcal{F}$. We will make use of a partition of $\mathcal{F}$. As a first step we define

$$\mathcal{B} := \{F \in \mathcal{F} : F \setminus \{1,2\} \notin \mathcal{F}, F \setminus \{1\} \cup \{2\} \in \mathcal{F}, F \setminus \{2\} \cup \{1\} \in \mathcal{F}\}.$$

Now let us consider the partition $\mathcal{F} = \mathcal{A} \cup \mathcal{B}_{1,2} \cup \mathcal{B}_1 \cup \mathcal{B}_2 \cup \mathcal{C}_1 \cup \mathcal{C}_2 \cup \mathcal{D}$ with

$$\mathcal{A}=\{F \in \mathcal{F}: F\setminus\{1,2\} \in \mathcal{F}\},\ \mathcal{D}=\{F \in \mathcal{F}: F\setminus\{1\}\cup\{2\} \notin \mathcal{F}, F\setminus\{2\}\cup\{1\} \notin \mathcal{F}\},$$

$$\mathcal{B}_{1,2} = \{F \in \mathcal{B} : \{1,2\} \subseteq F\},$$

$$\mathcal{B}_1 = \{F \in \mathcal{B} : 1 \notin F, 2 \in F\},\ \mathcal{B}_2 = \{F \in \mathcal{B} : 1 \in F, 2 \notin F\},$$

$$\mathcal{C}_1 = \{F \in \mathcal{F} : F \setminus \{1\} \cup \{2\} \in \mathcal{F}, F \setminus \{2\} \cup \{1\} \notin \mathcal{F}\},$$

$$\mathcal{C}_2 = \{F \in \mathcal{F} : F \setminus \{1\} \cup \{2\} \notin \mathcal{F}, F \setminus \{2\} \cup \{1\} \in \mathcal{F}\}.$$

**Claim 108** *(i)* $F \in \mathcal{C}_1$ *implies* $F \setminus \{1\} \in \mathcal{F}$ *and* $F \cup \{1\} \in \mathcal{F}$.
*(ii)* $|\mathcal{B}_{1,2}| = |\mathcal{B}_1| = |\mathcal{B}_2|$.

**Proof of Claim.** To prove **(i)** observe that $F \cup \{1\} \in \mathcal{F}$ is true as $\mathcal{F}$ is upward closed. To prove $F \setminus \{1\} \in \mathcal{F}$ we consider two cases. If $1 \notin F$, then $F \setminus \{1\} = F \in \mathcal{F}$. If $1 \in F$, then $F \setminus \{2\} \cup \{1\} \notin \mathcal{F}$ implies $2 \in F$, so $F \setminus \{1\} \cup \{2\} = F \setminus \{1\} \in \mathcal{F}$.

The definition of $\mathcal{B}$ immediately implies **(ii)**. □

CASE I: $|\mathcal{B}_{1,2}| \leq |\mathcal{D}|$.

Let us define

$$\mathcal{C}_2' = \{F \in \mathcal{C}_2 : 2 \in F\} \cup \{F \setminus \{1\} \cup \{2\} : 2 \notin F \in \mathcal{C}_2\}, \mathcal{D}' = \{F \setminus \{1\} : F \in \mathcal{D}\}$$

and

$$\mathcal{F}' = \mathcal{A} \cup \mathcal{B}_{1,2} \cup \mathcal{B}_1 \cup \mathcal{D}' \cup \mathcal{C}_1 \cup \mathcal{C}_2' \cup \mathcal{D}.$$

The definitions of $\mathcal{C}_2$ and $\mathcal{D}$ imply that $\mathcal{C}_2'$ and $\mathcal{D}'$ are disjoint from each other and from $\mathcal{F} \setminus (\mathcal{C}_2 \cup \mathcal{B})$. Indeed, for example $F \in \mathcal{D}$ implies $F \setminus \{1\} \subseteq F \setminus \{1\} \cup \{2\} \notin \mathcal{F}$, so $F \setminus \{1\} \notin \mathcal{F}$ by the upward closed property of $\mathcal{F}$. All other cases can be checked similarly.

We obtained

- $|\mathcal{F}'| = |\mathcal{F}| - |\mathcal{B}_{1,2}| + |\mathcal{D}| \geq |\mathcal{F}|$,
- the sets of $\mathcal{F}'$ can be partitioned into pairs $F, F'$ such that $F' = F \cup \{1\}$. This can be seen similarly to **(i)** of Claim 108.

Let us enumerate the sets of $\mathcal{F}'$ as $F_1, F_1', F_2, F_2', \ldots, F_m, F_m'$ such that $F_i' = F_i \cup \{1\}$ and let us write $\mathcal{F}'' = \{F_1, F_2, \ldots, F_m\} \subseteq 2^{[2,n]}$. Observe that $\cap_{F \in \mathcal{F}''} F = \emptyset$. Indeed, we have $\cap_{F \in \mathcal{F}} F = \emptyset$ and the change from $\mathcal{F}$ to $\mathcal{F}'$ affected the sets only in containing 1 and 2. Observe that $[n] \setminus \{1,2\} \in \mathcal{A}$ and therefore $[n] \setminus \{1,2\} \in \mathcal{F}''$. These imply $\cap F \in \mathcal{F}'' F = \emptyset$. We claim that $\mathcal{F}''$ is $r$-wise intersecting. Assume to the contrary that for $F_{i_1}, F_{i_2}, \ldots, F_{i_r} \in \mathcal{F}''$ we have $\cap_{j=1}^r F_{i_j} = \emptyset$. Then there exists a $j$ such that $2 \notin F_{i_j}$ and we can assume $j = 1$. Observe that $F_{i_1} \in \mathcal{A}$ must hold as all sets in $\mathcal{F}' \setminus \mathcal{A}$ contain 2. Also, for every $2 \leq j \leq r$ there exists a set $G_j \in \mathcal{F}$ that differ from $F_{i_j}$ only in 1 and

2. But then, as $\mathcal{A} \subseteq \mathcal{F}$, we have $F_{i_1}, G_2, \dots, G_r \in \mathcal{F}$ and $F_{i_1} \cap (\cap_{j=2}^r G_j) = \emptyset$, contradicting the $r$-wise intersecting property of $\mathcal{F}$.

Therefore we can apply the inductive hypothesis to $\mathcal{F}''$ to obtain

$$|\mathcal{F}| \leq |\mathcal{F}'| = 2|\mathcal{F}''| \leq 2 \cdot \frac{r+2}{2^{r+1}} 2^{n-1} = \frac{r+2}{2^{r+1}} 2^n.$$

CASE II: $|\mathcal{B}_{1,2}| < |\mathcal{D}|$.

Let us define $\mathcal{B}_0 = \{F \setminus \{1,2\} : F \in \mathcal{B}_{1,2}\}$ and

$$\mathcal{M} = \mathcal{A} \cup B_{1,2} \cup \mathcal{B}_1 \cup \mathcal{B}_2 \cup \mathcal{C}_1 \cup \mathcal{C}_2 \cup \mathcal{B}_0.$$

Let $\mathcal{L} = \{[n] \setminus \{i\} : i \in \cap_{M \in \mathcal{M}} M\}$ and set $\mathcal{F}'' = \mathcal{M} \cup \mathcal{L}$.

If $\cap_{M \in \mathcal{M}} M = \emptyset$, then $\mathcal{L}$ is empty; otherwise $\mathcal{L} \subseteq \mathcal{D}$. Indeed, $\cap_{F \in \mathcal{F}} F = \emptyset$ and the only sets that can cause the difference in the intersection are those in in $\mathcal{F} \setminus \mathcal{M} = \mathcal{D}$. Observe that $|\mathcal{F}''| = |\mathcal{F}| + |\mathcal{B}_0| - |\mathcal{D}| = |\mathcal{F}| + |\mathcal{B}_{1,2}| - |\mathcal{D}| > |\mathcal{F}|$, so $\mathcal{F}''$ cannot be $r$-wise intersecting, i.e. there must exist $G_1, G_2, \dots, G_r \in \mathcal{F}''$ with $\cap_{j=1}^r G_j = \emptyset$. All sets in $\mathcal{L}$ are among the $G_j$'s as if $L_i = [n] \setminus \{i\} \in \mathcal{L}$, then $L_i$ is the only set in $\mathcal{F}''$ that does not contain $i$.

Next we claim that none of the $G_j$'s belong to $\mathcal{A}$. Otherwise, for every $1 \leq j \leq t$ we can define $F_j \in \mathcal{F}$ as

- $G_j \setminus \{1,2\}$ if $G_j \in \mathcal{A}$,
- $G_j \cup \{1,2\} \in \mathcal{B}_{1,2}$ if $G_j \in \mathcal{B}_0$,
- $G_j = F_j$ in all other cases.

Then $\cap_{j=1}^r F_j = \emptyset$, contradicting the $r$-wise intersecting property of $\mathcal{F}$.

Now we prove that at most one $G_j$ might belong to $\mathcal{M}' = \mathcal{B}_{1,2} \cup \mathcal{B}_1 \cup \mathcal{B}_2 \cup \mathcal{C}_1 \cup \mathcal{C}_2 \cup \mathcal{B}_0$. Observe that for every $F \in \mathcal{D}$ we have $1, 2 \in F$; thus there is a set avoiding 1 and a set avoiding 2 in $\mathcal{F} \setminus \mathcal{D}$. Those sets are in $\mathcal{M}$; thus $1, 2 \notin \cap_{M \in \mathcal{M}} M$; hence $\{1,2\} \subset L$ for all $L \in \mathcal{L}$. This implies that the sets avoiding 1 and 2 among the $G_j$'s are in $\mathcal{M}'$. Suppose $G_1, G_2 \in \mathcal{M}'$ and 1 and 2 are avoided by at least one of them. If $G_1, G_2 \in \mathcal{B}_0 \cup \mathcal{B}_{1,2}$, then $F_1 = G_1 \setminus \{2\} \cup \{1\}, F_2 = G_2 \setminus \{1\} \cup \{2\}$ are in $\mathcal{B}_1 \cup \mathcal{B}_2 \subseteq \mathcal{F}$ and 1 and 2 are avoided by exactly one of them. If $G_1, G_2 \notin \mathcal{B}_0 \cup \mathcal{B}_{1,2}$, then they belong to $\mathcal{F}$. Finally, if $G_1 \in \mathcal{B}_0 \cup \mathcal{B}_{1,2}$, $G_2 \notin \mathcal{B}_0 \cup \mathcal{B}_{1,2}$, then we define $F_2 = G_2 \in \mathcal{F}$ and $F_1 = G_1 \setminus \{1,2\} \cup \{l\} \in \mathcal{F}$ with $l$ being the element of $\{1,2\}$ avoided by $F_2$. In all cases, $G_3, G_4, \dots, G_r$ can be transformed as in the previous paragraph to obtain $r$ sets in $\mathcal{F}$ with empty intersection - a contradiction with the $r$-wise intersecting property of $\mathcal{F}$.

So we obtained that exactly one of the $G_j$'s belong to $\mathcal{M}'$, say $G_1$, and the others to $\mathcal{L}$. Observe that $G_1 \in \mathcal{B}_0$, as the other $G_j$'s are in $\mathcal{F}$. We can assume that $\mathcal{L} = \{[n] \setminus \{n-r+2\}, [n] \setminus \{n-t+3\}, \dots, [n] \setminus \{n\}\}$. Let us make the following two observations:

- there cannot exist $F \in \mathcal{F}$ and $n-r+1 \leq l < l' \leq n$ with $F \subseteq [n] \setminus \{l, l'\}$. Indeed, otherwise consider the $r-2$ sets in $\mathcal{F}$ that all together avoid $n-r+1, n-r+2, \ldots, n$, $F_1 = G_1 \cup \{1,2\}$ and $[n] \setminus \{1,2\} \in \mathcal{F}$. These $r$ sets contradict the $r$-wise intersecting property of $\mathcal{F}$.

- Similarly, there cannot exist $F \in \mathcal{F}$, $m \in \{1,2\}$ and $n-r+1 \leq l' \leq n$ with $F \subseteq [n] \setminus \{l, m\}$.

It follows that for every $n \notin F \in \mathcal{F}$ we have $\{1, 2, n-r+1, n-r+2, \ldots, n-1\} \subseteq F$; therefore $|\{F \in \mathcal{F} : n \notin F\}| \leq 2^{n-r-1}$ holds. On the other hand $\mathcal{F}_n = \{F \setminus \{n\} : n \in F \in \mathcal{F}\}$ is $(r-1)$-wise intersecting, and, by the assumption that $\mathcal{F}$ is upward closed, we obtain $\cap_{F \in \mathcal{F}_n} F = \cap_{F \in \mathcal{F}} F = \emptyset$. Applying the inductive hypothesis, we obtain $|\mathcal{F}| \leq 2^{n-r-1} + (r-1+2)2^{n-1-(r-1)-1} = (r+2)2^{n-r-1}$. ■

**Conjecture 109 (Erdős, Frankl)** *For any $n, t, r, h$ let*

$$\mathcal{F}_{n,r,h,t} = \{F \in 2^{[n]} : |F \cap [t+rh]| \geq t + (r-1)h\}.$$

*Observe that $\mathcal{F}_{n,r,h,t}$ is $r$-wise $t$-intersecting. Suppose that $\mathcal{F} \subseteq 2^{[n]}$ is $r$-wise $t$-intersecting; then $|\mathcal{F}| \leq \max_h\{|\mathcal{F}_{n,r,h,t}|\}$ holds.*

Observe that $\mathcal{F}_{n,r,0,t}$ is at least the size of $\mathcal{F}_{n,r,1,t}$ if and only if $2^r - r - 1 \geq t$. Frankl proved that if this inequality is satisfied, then $\mathcal{F}_{n,r,0,t} = \{F \in 2^{[n]} : [t] \subseteq F\}$ is the largest $r$-wise $t$-intersecting family. This statement was proved mainly in [196], but the case $r = 3$, $t = 4$ was only proved very recently in [199].

**Theorem 110 (Frankl [196, 199])** *If $\mathcal{F} \subseteq 2^{[n]}$ is $r$-wise $t$-intersecting and $n < r + t$ or $t \leq 2^r - r - 1$, then $|\mathcal{F}| \leq 2^{n-t}$ holds.*

In [196] it is also shown that the second largest maximal $r$-wise $t$-intersecting family is $\mathcal{F}_{n,r,1,t}$, provided $t \leq 2^r - r - 1$ and $r \geq 5$.

**Exercise 111** *Reprove Theorem 106 using the cycle method: if $\mathcal{F} \subseteq \binom{[n]}{k}$ is $r$-wise intersecting and $rk \leq (r-1)n$ holds, then $|\mathcal{F}| \leq \binom{n-1}{k-1}$. Moreover, unless $r = 2$ and $n = 2k$, then equality holds if and only if $\mathcal{F} = \{F \in \binom{[n]}{k} : x \in F\}$ for some $x \in [n]$.*

Strengthening a result of Grolmusz and Sudakov [283], Füredi and Sudakov proved the following asymptotically tight bound

**Theorem 112 (Füredi, Sudakov [254])** *Let $L$ be a set of $s$ non-negative integers. Then for any $r \geq 2$ there exists an $n_0 = n_0(r, s)$ such that whenever $n \geq n_0$ holds and $\mathcal{F} \subseteq 2^{[n]}$ is $r$-wise $L$-intersecting, then we have*

$$|\mathcal{F}| \leq \frac{k+s-1}{s+1}\binom{n}{s} + \sum_{i=0}^{s-1}\binom{n}{i}.$$

Note that the bound of Theorem 112 cannot be improved in general. Indeed, let $L = \{0, 1, \ldots, s-1\}$ and $\mathcal{G} \subseteq \binom{[n]}{s+1}$ such that for any $S \in \binom{[n]}{s}$ there are at most $r-2$ sets $G \in \mathcal{G}$ with $S \subset G$. Then $\mathcal{F} := \binom{[n]}{\le s} \cup \mathcal{G}$ is $r$-wise $L$-intersecting. As every $(s+1)$-set contains $s+1$ many $s$-subsets we obtain that $|\mathcal{G}| \le \frac{r-2}{s+1}\binom{n}{s}$. The existence of families achieving this bound asymptotically was first showed by Rödl [481]. Keevash [349] obtained even stronger results.

A recent common generalization of Theorem 112 and Theorem 97 is the following.

**Theorem 113 (Liu, Liu, [390])** *Let $r$ be an integer at least 2 and $L$ be a set of positive integers of size $s$. Then there exists an $n_0 = n_0(r, s)$ such that if $\mathcal{F} \subseteq 2^{[n]}$ is $r$-wise $L$-intersecting, then we have*

$$|\mathcal{F}| \le \frac{r+s-1}{s+1}\binom{n-1}{s} + \sum_{i=0}^{s-1}\binom{n-1}{i}.$$

Szabó and Vu [521] obtained the following exact result on the $r$-wise version of Exercise 105.

**Theorem 114** *Let $r = 2^l + 1 \ge 3$ and $n \ge 3\frac{r-1}{r-2}\log_2(r-1)$. If $\mathcal{F} \subseteq 2^{[n]}$ contains sets only of odd size such that for any distinct $F_1, F_2, \ldots, F_r \in \mathcal{F}$, the intersection $\cap_{i=1}^{r} F_i$ is of even size, then $|\mathcal{F}| \le (r-1)(n - 2\log_2(r-1))$. Moreover, for any $r$ and $n \ge 3(k-1) + 2\lceil \log_2(k-1)\rceil - 2$ a maximum sized family $\mathcal{F}$ with the above property, we have*

$$(k-1)(n - 2\lceil \log_2(k-1)\rceil) \le |\mathcal{F}| \le (k-1)(n - \lceil \log_2(k-1)\rceil - \lfloor \log_2(k-1)\rfloor).$$

## 2.6 $k$-uniform intersecting families with covering number $k$

Uniform intersecting families with the largest possible covering number have a special property among all uniform intersecting families: as Lemma 43 shows the size of such a family can be bounded independently of the size of its ground set. Families of this kind were first studied by Erdős and Lovász [164]. They arrived at this notion from considering colorings of set systems. The *chromatic number* $\chi(\mathcal{F})$ of a family $\mathcal{F}$ is the minimum positive integer $k$ such that there exists a coloring $c : \cup_{F \in \mathcal{F}} F \to [k]$ such that no set $F \in \mathcal{F}$ is monochromatic, i.e. $|\{c(i) : i \in F\}| \ge 2$ for all $F \in \mathcal{F}$ (colorings with this property are called *proper*). Erdős and Lovász were interested in the number of sets a $k$-uniform family must contain if it has chromatic number at least 3. If $\mathcal{F}$ is intersecting then $\chi(\mathcal{F}) \le 3$ as one can two-color a set $F_0 \in \mathcal{F}$ and color every other vertex with the third color. They observed that if a $k$-uniform intersecting family $\mathcal{F}$ has chromatic number 3, then $\tau(\mathcal{F}) = k$ must hold. Indeed, if $\tau(\mathcal{F}) < k$, then one can color a minimum cover with one color, and every other vertex with another color to obtain a proper two-coloring. They posed the problem of finding the most/least number of sets/vertices in a $k$-uniform family $\mathcal{F}$ with $\tau(\mathcal{F}) = k$. (It is trivial that the minimum number of vertices is $2k-1$. Indeed, $\binom{[2k-1]}{k}$ is intersecting and $\tau(\binom{[2k-1]}{k}) = k$. On the other hand if the ground set $V$ of $\mathcal{F}$ has size at most $2k-2$, then any $(k-1)$-subset $T$ of $V$ intersects every $F \in \mathcal{F}$.)

Concerning the minimum number $q(k)$ of sets in a $k$-uniform intersecting family $\mathcal{F}$ with $\tau(\mathcal{F}) = k$, they proved $\frac{8}{3}k - 3 \leq q(k)$ which is still the best known lower bound. For the upper bound Erdős and Lovász, and later Kahn [324], considered families of randomly chosen lines of projective planes and proved that $q(k)$ is at most $4k^{3/2}\log k$ and $O(k \log k)$, respectively. Finally, Kahn obtained the following result verifying a conjecture of Erdős and Lovász.

**Theorem 115 (Kahn [325])** *There exists a $k$-uniform intersecting family $\mathcal{F}$ with $\tau(\mathcal{F}) = k$ and $|\mathcal{F}| = O(k)$, i.e. $q(k) = \Theta(k)$.*

Let us now turn our attention to

$$M(k) = \max\{|\mathcal{F}| : \mathcal{F} \text{ is } k\text{-uniform intersecting with } \tau(\mathcal{F}) = k\}$$

and

$$N(k) = \max\{|\bigcup_{F \in \mathcal{F}} F| : \mathcal{F} \text{ is } k\text{-uniform intersecting with } \tau(\mathcal{F}) = k\}.$$

Erdős and Lovász proved $M(k) \leq k^k$ (this is implied by Lemma 43 as mentioned before) and gave an inductive constriction of size $\lfloor k!(e-1) \rfloor$ based on the following lemma.

**Lemma 116** $M(k) \geq 1 + kM(k-1)$ *holds for all* $k \geq 1$.

**Proof.** Let $\mathcal{F}$ be a $(k-1)$-uniform intersecting family with $\tau(\mathcal{F}) = k-1$ and $|\mathcal{F}| = M(k-1)$. If $G$ is a $k$-set disjoint from $\cup_{F \in \mathcal{F}} F$, then $\mathcal{F}' = \{G\} \cup \{F \cup \{x\} : F \in \mathcal{F}, x \in G\}$ is a $k$-uniform intersecting family with $\tau(\mathcal{F}') = k$. Indeed, if $T$ covers $\mathcal{F}'$, then either $G \subseteq T$ and thus $|T| \geq k$ or $G \cap T \neq \emptyset$ and $G \setminus T \neq \emptyset$. From the latter it follows that $T \cap (\cup_{F \in \mathcal{F}} F)$ covers $\mathcal{F}$ and therefore $|T| \geq |T \cap G| + |T \cap (\cup_{F \in \mathcal{F}} F)| \geq 1 + k - 1 = k$. ■

As $M(1) = 1$, Lemma 116 indeed yields $M(k) \geq \lfloor k!(e-1) \rfloor$. Lovász conjectured [391] that this bound is tight, but it was disproved by Frankl, Ota, and Tokushige [217]. Here we present a recent construction of Majumder [403] that is also larger than the one of Erdős and Lovász. It has the same asymptotic size as the construction by Frankl, Ota and Tokushige, but is much simpler than that.

**Construction 117** *For an even number $k$ and $0 \leq i \leq k-2$ let $X_i$ be pairwise disjoint sets of size $1+k/2$. Let $y \notin \cup_{i=0}^{k-2} X_i$ and define $\mathcal{F}_{1,i} = \{X_i \cup \{x_{i+1}, x_{i+2}, \dots, x_{i+k/2-1}\} : x_{i+j} \in X_{i+j}\}$ where addition in the suffix from here on is modulo $k-1$ . Let $\mathcal{F}_2 = \{y\} \times X_0 \times X_1 \times \dots \times X_{k-2}$ and $\mathcal{F} = \cup_{i=0}^{k-2} \mathcal{F}_{1,i} \cup \mathcal{F}_2$.*

We show that $\mathcal{F}$ is a $k$-uniform intersecting family with $\tau(\mathcal{F}) = k$. Clearly any $F \in \mathcal{F}_2$ meets all $F' \in \mathcal{F}$. Also, if $F, F' \in \mathcal{F}_{1,i}$, then $X_i \subset F \cap F' \neq \emptyset$. Let $F \in \mathcal{F}_{1,i}$ and $F' \in \mathcal{F}_{1,j}$ with $0 \leq i < j \leq k-2$. Note that either $j \in \{i+1, i+2, \dots, i+k/2-1\}$ or $i \in \{j+1, j+2, \dots, j+k/2-1\}$ (modulo $k-1$), as the distance between $i$ and $j$ on the circle is at most $(k-1)/2$. In the former case we have $X_j \subset F'$ and $F$ contains an element from $X_j$. Similarly in the latter case we have $X_i \subset F$ and $F'$ contains an element from $X_i$. This proves that $\mathcal{F}$ is intersecting.

To prove $\tau(\mathcal{F}) = k$ let $C$ be a cover of $\mathcal{F}$ of size at most $k$. We distinguish two cases.

CASE I: $y \notin C$.

Then $C$ must contain one of the $X_i$'s to intersect all sets in $\mathcal{F}_2 \subseteq \mathcal{F}$. As the $X_i$'s have size $1+k/2$ there exists a unique $j$ such that $X_j \subset C$ holds. For any $i \neq j$ let $a_i \in X_i \setminus C$. For any $h = j+l$ with $1 \leq l \leq k/2-1$ we must have $C \cap X_h \neq \emptyset$ as otherwise $C$ does not meet $X_h \cup \{a_{h+i} : 1 \leq i \leq k/2-1\} \in \mathcal{F}_1$. This shows that $|C| \geq k$; moreover $C \in \mathcal{F}_{1,j}$.

CASE II: $y \in C$.

If $C$ contains some $X_i$, then $C$ is disjoint from at least $k/2$ many $X_j$'s; in particular $C$ is disjoint from $X_{i+h}$ with some $h \leq k/2-1$. But then as in the previous case we could find a set $F \in \mathcal{F}_{1,i+h}$ disjoint from $C$. Therefore for *every* $i$ there exists $a_i \in X_i \setminus C_i$. This allows us to use the reasoning of the previous case to conclude that $C$ must meet all $X_i$'s and thus $C \in \mathcal{F}_2$, in particular $|C| \geq k$.

The size of $\mathcal{F}$ is $(k-1)(k/2+1)^{k/2-1} + (k/2+1)^{k-1} = (e^2+o(1))(k/2)^{k-1}$, while the construction of Erdős and Lovász has size $\lfloor k!(e-1) \rfloor = O((k/e)^k)$. To obtain the same asymptotics for odd values of $k$, one may modify the construction by letting some of the $X_i$'s have size $\frac{k+1}{2}$ and the others have size $\frac{k+1}{2}+1$, and adjusting the definition of sets in $\mathcal{F}_{1,i}$ so that they have size $k$. The proof that the resulting family is intersecting with covering number $k$ is similar, but a bit more complicated. An easier way to get the same order of magnitude is to use the even case result and apply Lemma 116.

Let us continue with $N(k)$. The following example by Erdős and Lovász gives a lower bound.

**Construction 118** *Let $|Y| = 2k-2$. For each 2-equipartition $Y$ as $E \cup E' = Y$, $|E| = |E'| = k-1$, we take a new point $x_{E,E'}$, and set $E \cup \{x_{E,E'}\}$, $E' \cup \{x_{E,E'}\}$. This way we obtain $\binom{2k-2}{k-1}$ $k$-element sets, forming an intersecting* *family with covering number $k$, such that the union of these sets consists of $2k-2+\frac{1}{2}\binom{2k-2}{k-1}$ points.*

Tuza [536] observed that there is an important connection between certain types of $k$-uniform families and ISPS's. Before presenting his observation let us recall the definition of an ISPS and introduce some related notions. A family of pairs $\{A_i, B_i)\}_{i=1}^m$ forms an intersecting set pair system (ISPS) if

1. $A_i \cap B_i = \emptyset$ for all $1 \le i \le m$,
2. $A_i \cap B_j \neq \emptyset$ for all $1 \le i \neq j \le m$.

If $|A_i| \le a$ and $|B_i| \le b$ hold for all $1 \le i \le m$, then we call $\{A_i, B_i)\}_{i=1}^m$ an $(a,b)$-ISPS. Theorem 6 states that if $\{A_i, B_i)\}_{i=1}^m$ is an $(a,b)$-ISPS, then $m \le \binom{a+b}{a}$. Let us introduce the functions

$$n(a,b) = \max\{| \cup_{i=1}^m A_i \cup B_i| : \{(A_i, B_i)\}_{i=1}^m \text{ is } (a,b)\text{-ISPS}\}$$

and

$$n'(a,b) = \max\{| \cup_{i=1}^m A_i| : \{(A_i, B_i)\}_{i=1}^m \text{ is } (a,b)\text{-ISPS}\}.$$

By definition, we have $n'(a,b) \le n(a,b) = n(b,a)$.

**Theorem 119 (Tuza [536])** *If $a \ge b$, then $\frac{1}{4}\binom{a+b-1}{a-1} < n'(a,b) \le n(a,b) \le \sum_{i=1}^{2b-2}\binom{i}{\lfloor i/2 \rfloor} + \sum_{i=2b-1}^{a+b-1}\binom{i}{b} < \binom{a+b-1}{a-1}$.*

**Proof.** The lower bound is given by the following construction. Let $a' = \lfloor \frac{ab}{b+1} \rfloor < a$. Consider an $a'+b$-element set $M$ and let

$$\{B_1, \ldots, B_{\binom{a'+b}{b}}\} = \binom{M}{b}.$$

Let us consider $\binom{a'+b}{b}$ pairwise disjoint sets $C_i$ of size $a-a'$ $(1 \le i \le \binom{a'+b}{b})$ such that they are also disjoint from $M$. Let $A_i = (M \setminus B_i) \cup C_i$ for $1 \le i \le \binom{a'+b}{b}$. It is obvious that, for every $i$, $A_i$ is disjoint from $B_i$, and intersects $B_j$ $(j \neq i)$ inside $M$. A straightforward calculation gives the lower bound.

For the upper bound let $A_1, A_2, \ldots, A_m$ and $B_1, B_2, \ldots, B_m$ form an $(a,b)$-ISPS with $n(a,b) = |\bigcup_{i=1}^m (A_i \cup B_i)|$. For $j = 0, 1, \ldots, a+b-1$ we define an index set $M_j$ and an ISPS $\mathcal{S}_j = \{(A_i^j, B_i^j) : i \in M_j\}$ in the following way: let $M_0 = [m]$ and $\mathcal{S}_0 = \{(A_i^0, B_i^0) : i \in M_0\}$ with $A_i^0 = A_i, B_i^0 = B_i$. If for $j \le a+b-2$ the index set $M_j$ and the ISPS $\mathcal{S}_j$ are defined, then let $M_{j+1}$ be a minimal subset of $M_j$ such that $\bigcup_{i \in M_{j+1}}(A_i \cup B_i) = \bigcup_{i \in M_j}(A_i \cup B_i)$ holds. The minimality of $M_{j+1}$ implies, that for every $i \in M_{j+1}$, there exists an element $x_i \in A_i^j \cup B_i^j$ such that $x_i \notin A_l^j \cup B_l^j$ for any $l \in M_{j+1} \setminus \{i\}$. Therefore setting $A_i^{j+1} = A_i^j \setminus \{x_i\}$ and $B_i^{j+1} = B_i^j \setminus \{x_i\}$ we obtain that $\mathcal{S}_{j+1}$ is indeed an ISPS. Furthermore, we have $n(a,b) = |\bigcup_{i=1}^m (A_i \cup B_i)| = \sum_{j=0}^{a+b-1} |M_j|$.

For any $j = 0, 1 \ldots, a+b-1$ and $i \in M_j$ we have $|A_i^j \cup B_i^j| \le a+b-j$

and therefore by Theorem 6 we obtain $|M_j| \leq \binom{a+b-j}{\lfloor\frac{a+b-j}{2}\rfloor}$ for arbitrary $j$ and $|M_j| \leq \binom{a+b-j}{b}$ if $j \leq a-b$. This gives the upper bound on $n(a,b)$. The last inequality can be easily seen by induction. ■

If we replace Condition 2 in the definition of an ISPS by

(2') $A_i \cap B_j \neq \emptyset$ for all $1 \leq i < j \leq m$,

then we obtain the notion of *skew ISPS*. If $|A_i| \leq a$ and $|B_i| \leq b$ hold for all $1 \leq i \leq m$, then we call $\{A_i, B_i)\}_{i=1}^m$ a skew $(a,b)$-ISPS. Frankl and Kalai showed that even these weaker conditions imply the same upper bound.

**Theorem 120 (Frankl [191], Kalai [327])** *If $\{(A_i, B_i)\}_{i=1}^m$ form a skew $(a,b)$-ISPS, then $m \leq \binom{a+b}{a}$ holds.*

Similarly to $n(a,b)$ and $n'(a,b)$ one can define

$$n_1(a,b) = \max\{|\cup_{i=1}^m A_i \cup B_i| : \{(A_i, B_i)\}_{i=1}^m \text{ is skew } (a,b)\text{-ISPS}\}$$

and

$$n'_1(a,b) = \max\{|\cup_{i=1}^m A_i| : \{(A_i, B_i)\}_{i=1}^m \text{ is skew } (a,b)\text{-ISPS}\}.$$

**Construction 121 (Nagy, Patkós [441])** *Let $Y$ be the set $Y = [a+b]$. Consider the colex ordering $A_1, \ldots, A_{\binom{a+b}{a}}$ of all the $a$-element subsets of $Y$. Let $A_i = \{x_{i,1}, x_{i,2}, \ldots, x_{i,a}\}$ $(i = 1 \ldots \binom{a+b}{a})$, and let $B_i$ be defined as follows. $B_i \cap Y = [x_{i,a}] \setminus A_i$, and let all the sets $B_i \setminus Y$ be pairwise disjoint for all $i$ such that $|B_i| = b$.*

A short calculation shows that the above construction contains $a+b+\binom{a+b}{a+1}$ points, giving the lower bound of the next theorem. The proof of the upper bound is a slight modification of Tuza's argument for Theorem 119.

**Theorem 122 (Nagy, Patkós [441])** *Let $a \leq b$ be positive integers. Then $a+b+\binom{a+b}{a+1} \leq n_1(a,b) \leq \binom{a+b+2}{a+1} - \binom{a+b}{a} - 2$ holds.*

Füredi found the following variant of Theorem 120.

**Theorem 123 (Füredi, [226])** *Let $t$ be a positive integer and $A_1, A_2, \ldots, A_m$ be sets of size $a$ and $B_1, B_2, \ldots, B_m$ be sets of size $b$ such that for every $1 \leq i \leq m$, we have $|A_i \cap B_i| < t$, while for every $1 \leq i < j \leq m$ we have $|A_i \cap B_j| \geq t$. Then $m \leq \binom{a+b-2t+2}{a-t+1}$ holds.*

In [538], Tuza proposed the investigation of the so-called *weak cross intersecting set pair systems*. Instead of Condition 2 in the definition of an ISPS, they satisfy

2" $A_i \cap B_j \neq \emptyset$ or $A_j \cap B_i \neq \emptyset$ for all $1 \leq i, j \leq m$, $i \neq j$.

If $|A_i| \leq a, |B_i| \leq b$ for all $1 \leq i \leq j$, then $\{(A_i, B_i)\}_{i=1}^m$ is called an $(a,b)$-weak ISPS. Let $m_{\max}(a,b)$ denote the largest $m \in \mathbb{Z}$ for which a $(a,b)$-weak ISPS $\{(A_i, B_i)\}_{i=1}^m$ exists. We also introduce $n_2(a,b) = \max\{|\cup_{i=1}^m A_i \cup B_i| : \{(A_i, B_i)\}_{i=1}^m$ is $(a,b)$-weak ISPS$\}$.

Not too much is known about the $m_{\max}(a,b)$, even its order of magnitude is yet to be determined.

**Theorem 124 (Tuza [538])** $m_{max}(a,b) \leq \frac{(a+b)^{a+b}}{a^b a^b}$.

The best lower bound is due to Király, Nagy, Pálvölgyi and Visontai [360] implying $\liminf_{a+b\to\infty} m_{max}(a,b) \geq (2-o(1))\binom{a+b}{b}$. They gave two constructions to achieve their bound. One (using lattice paths) works for arbitrary $a$ and $b$ tending to infinity, and the following simple one for the case $a = b$: consider the bipartite graph $G = (\binom{[2a+1]}{a}, \binom{[2a+1]}{a}, E)$ with $(X,Y) \in E$ if and only if $X \cap Y = \emptyset$. As $G$ is regular, by Hall's theorem (Theorem 70) it contains a perfect matching $\mathcal{M}$. The pairs of the endpoints of edges in $\mathcal{M}$ form a $(a,a)$-weak ISPS.

A lower bound for $n_2(a,b)$ follows from a construction of Nagy and Patkós [441].

**Construction 125** *Let $Y$ be a set of $a+b-1$ elements with $a \leq b$. Assign a subset $B'_i \subset (Y \setminus A'_i)$ of size $b-1$ to each $(a-1)$-element subset $A'_i \subset Y$, in such a way that the sets $B'_i$ are distinct. This can be done due to the Hall's theorem (Theorem 70) and the fact that $a \leq b$. For each $i$ consider three distinct elements $x_i, y_i, z_i \notin Y$. Take the set pairs $(A'_i \cup \{x_i\}, B'_i \cup \{y_i\})$, $(A'_i \cup \{y_i\}, B'_i \cup \{z_i\})$, $(A'_i \cup \{z_i\}, B'_i \cup \{x_i\})$ for all $i$. This way we obtain $3\binom{a+b-1}{a-1}$ set pairs such that the union of these sets consists of $a+b-1+3\binom{a+b-1}{a-1}$ points.*

Let us return to the connection between ISPS's and $k$-uniform intersecting families $\mathcal{F}$ with $\tau(\mathcal{F}) = k$. Note that if $G \notin \mathcal{F}$ is a $k$-set such that $G \cap F \neq \emptyset$ for all $F \in \mathcal{F}$, then as $\tau(\mathcal{F}) = k$, we must have $G \subseteq \cup_{F \in \mathcal{F}} F$. Therefore, when bounding $N(k)$, we can restrict ourselves to $k$-uniform intersecting families that are minimal with respect to the property $\tau(\mathcal{F}) = k$. In general, a family $\mathcal{F}$ is *$\tau$-critical* if $\tau(\mathcal{F} \setminus \{F\}) < \tau(\mathcal{F})$ holds for all $F \in \mathcal{F}$. This means that for every set $F_i \in \mathcal{F}$ there exists a $(\tau(\mathcal{F}) - 1)$-set $T_i$ such that $T_i \cap F_i = \emptyset$ but $T_i \cap F_j \neq \emptyset$ for all $F_j \in \mathcal{F} \setminus \{F_i\}$, i.e. $\{(T,F)\}_{i=1}^{|\mathcal{F}|}$ is an ISPS. We obtained the following proposition.

**Proposition 126 (Tuza [536])** *If $\mathcal{F}$ is a $\tau$-critical $k$-uniform family with $\tau(\mathcal{F}) = t$, then $|\cup_{F \in \mathcal{F}} F| \leq n'(k, t-1)$ holds. In particular, $N(k) \leq n'(k, k-1)$.*

Theorem 119 and Proposition 126 imply that $N(k)$ is asymptotically at most twice as large as the lower bound given by Construction 118. This upper bound was improved by Majumder.

**Theorem 127 (Majumder [404])** $N(k) \leq \frac{1}{2}\binom{2k-2}{k-2} + n(k, k-2)$.

**Proof.** For a family $\mathcal{F}$ let $\mathcal{F}^T$ denote the family of all minimum size covers of $\mathcal{F}$ (i.e. those of size $\tau(\mathcal{F})$). Let us introduce the function

$$N^T(k,t) = \max\{|\cup_{T \in \mathcal{F}^T} T| : \mathcal{F} \text{ is } k\text{-uniform}, \ \tau(\mathcal{F}) = t\}.$$

**Lemma 128** $N^T(k,t) \leq n(k, t-1)$.

**Proof of Lemma.** Let $\mathcal{F}$ be a $k$-uniform family with $\tau(\mathcal{F}) = t$. Let $\mathcal{E}$ be a $\tau$-critical subfamily of $\mathcal{F}$ with $\tau(\mathcal{E}) = t$. Then for any $E_i \in \mathcal{E}$, there exists a $(t-1)$-set $T_i$ that covers $\mathcal{E} \setminus \{E_i\}$. Clearly, $\{(E_i, T_i)\}_{i=1}^{|\mathcal{E}|}$ is a $(k, t-1)$-ISPS. We need to show $\cup_{T \in \mathcal{F}^T} T \subseteq \cup_{E \in \mathcal{E}} E$. If $x \in T$ for some $T \in \mathcal{F}^T$, then $x \in \cup_{E \in \mathcal{E}} E$ as otherwise $T \setminus \{x\}$ would be a cover of $\mathcal{E}$ contradicting $\tau(\mathcal{E}) = t$. □

**Lemma 129** *Let $\mathcal{F}$ be a maximal $k$-uniform intersecting family with $\tau(\mathcal{F}) = k$, and let $x, y \in \cup_{F \in \mathcal{F}} F$ such that no $F \in \mathcal{F}$ contains both $x$ and $y$. Then writing $\mathcal{G} = \{F \in \mathcal{F} : x, y \notin F\}$, the family*

$$\mathcal{F}_{x \mapsto y} = \mathcal{G} \cup \{T \cup \{y\} : T \in \mathcal{G}^T\}$$

*is maximal $k$-uniform intersecting with $\tau(\mathcal{F}_{x \mapsto y}) = k$ and $\cup_{F' \in \mathcal{F}_{x \mapsto y}} F' = \cup_{F \in \mathcal{F}} F \setminus \{x\}$.*

**Proof of Lemma.** We first show that $\tau(\mathcal{G}) = k - 1$ and thus $\mathcal{F}_{x \mapsto y}$ is $k$-uniform. As for any $F \in \mathcal{F}$ with $x \in F$, the set $F \setminus \{x\}$ covers $\mathcal{G}$, we have $\tau(\mathcal{G}) \leq k - 1$. If $T$ covers $\mathcal{G}$ with $|T| \leq k-2$, then $T \cup \{x, y\}$ covers $\mathcal{F}$, so, by the maximality of $\mathcal{F}$, we have $T \cup \{x, y\} \in \mathcal{F}$, contradicting the assumption that no set of $\mathcal{F}$ contains both $x$ and $y$.

Similarly, if $x \notin F \in \mathcal{F}$, then $F \in \mathcal{F}_{x \mapsto y}$, as if $y \notin F$, then $F \in \mathcal{G}$, while if $y \in F$, then $F \setminus \{y\}$ is a cover. On the other hand if $x \in F \in \mathcal{F}$, then $(F \setminus \{x\}) \cup \{y\} \in \mathcal{F}_{x \mapsto y}$.

$\mathcal{F}_{x \mapsto y}$ is intersecting since (i) $G, G' \in \mathcal{G}$ intersect as $\mathcal{G} \subseteq \mathcal{F}$, (ii) $G, T \cup \{y\}$ intersect as $T$ is a cover of $\mathcal{G}$, and (iii) $T_1 \cup \{y\}, T_2 \cup \{y\}$ intersect because of $y$.

We claim that for any $T \in \mathcal{G}^T$, there exists $T' \in \mathcal{G}^T$ with $T \cap T' = \emptyset$. If not, then there exists a $T \in \mathcal{G}^T$ that covers $\mathcal{G}^T$. So $T$ would cover $\mathcal{G} \cup \mathcal{G}^T$. Therefore $T$ would cover $\mathcal{F}$, as if $x \in F \in \mathcal{F}$, then $F \setminus \{x\} \in \mathcal{G}^T$, while if $y \in F \in \mathcal{F}$ then $F \setminus \{y\} \in \mathcal{G}^T$. This would contradict $\tau(\mathcal{F}) = k$, as we have already shown $\tau(\mathcal{G}) = k - 1$.

As $\mathcal{F}_{x \mapsto y}$ is intersecting, we have $\tau(\mathcal{F}_{x \mapsto y}) \leq k$. As $\mathcal{G} \subseteq \mathcal{F}_{x \mapsto y}$ and $\tau(\mathcal{G}) = k - 1$ we have $\tau(\mathcal{F}_{x \mapsto y}) \geq k - 1$. Suppose $C$ is a cover of $\mathcal{F}_{x \mapsto y}$ of size $k - 1$. Then $C$ also covers $\mathcal{G}$; hence $C \in \mathcal{G}^T$, which implies $y \notin C$. But then there exists $T \in \mathcal{G}^T$ with $C \cap T = \emptyset$, so $T \cup \{y\} \in \mathcal{F}_{x \mapsto y}$ is disjoint from $C$. This contradiction shows $\tau(\mathcal{F}_{x \mapsto y}) = k$.

To prove that $\mathcal{F}_{x \mapsto y}$ is maximal intersecting, we need to show that every cover $T$ of $\mathcal{F}_{x \mapsto y}$ with $|T| = k$ belongs to $\mathcal{F}_{x \mapsto y}$. If such a $T$ contains $y$, then we have $T \setminus \{y\} \in \mathcal{G}^T$. But then by the definition of $\mathcal{F}_{x\mapsto y}$, we have $T = (T \setminus \{y\}) \cup \{y\} \in \mathcal{F}_{x\mapsto y}$. If $y \notin T$, then $T$ covers $\mathcal{F}$ as if $x \notin F \in \mathcal{F}$, then $F \in \mathcal{F}_{x\mapsto y}$, while if $x \in F \in \mathcal{F}$, then $(F \setminus \{x\}) \cup \{y\} \in \mathcal{F}_{x\mapsto y}$. By the maximality of $\mathcal{F}$, we have $T \in \mathcal{F}$ and thus $T$ is in $\mathcal{G} \subseteq \mathcal{F}_{x\mapsto y}$.

Finally, we need to show $\cup_{F' \in \mathcal{F}_{x\mapsto y}} F' = \cup_{F\in\mathcal{F}} F \setminus \{x\}$. $\cup_{F' \in \mathcal{F}_{x\mapsto y}} F' \subseteq \cup_{F\in\mathcal{F}} F \setminus \{x\}$ is true by definition. Let $z \in F \in \mathcal{F}$ with $z \neq x$. If $x \notin F$, then $z \in F \in \mathcal{F}_{x\mapsto y}$, while if $x \in F$, then $z \in (F \setminus \{x\}) \cup \{y\} \in \mathcal{F}_{x\mapsto y}$. □

Having done all preparations, we are ready to prove Theorem 127. Let $\mathcal{F}$ be a $k$-uniform maximal intersecting family with $\tau(\mathcal{F}) = k$, and let us fix an arbitrary $y \in \cup_{F\in\mathcal{F}} F$. We define two sequences $x_1, x_2, \dots, x_{N-1}$ and $\mathcal{F}_1, \mathcal{F}_2, \dots, \mathcal{F}_N$ in the following way: let $\mathcal{F}_1 = \mathcal{F}$ and whenever $\mathcal{F}_i$ is defined, then let $x_i \in \cup_{F\in\mathcal{F}_i} F$ such that no $F \in \mathcal{F}_i$ contains both $x_i$ and $y$, and let $\mathcal{F}_{i+1} = \mathcal{F}_{x_i \mapsto y}$. If there is no such $x_i$, then the process ends and we set $i = N$. As $\cup_{F\in\mathcal{F}} F \leq \binom{2k-2}{k-1}$ by Proposition 126 and Theorem 119, the process terminates after a finite number of iterations.

$\mathcal{F}_N$ has the property that for every $x \in \cup_{F\in\mathcal{F}_N} F$ there exists $F_x \in \mathcal{F}_N$ with $x, y \in F_x$. Let $\mathcal{G} = \{F \in \mathcal{F}_N : y \notin F\}$. We claim $\tau(\mathcal{G}) = k-1$. As $\mathcal{F}_N$ is intersecting by Lemma 129, for any $F \in \mathcal{F}_N$ with $y \in F$, the set $F \setminus \{y\}$ covers $\mathcal{G}$, and therefore $\tau(\mathcal{G}) \leq k-1$. On the other hand, if $T$ covers $\mathcal{G}$ with $|T| \leq k-1$, then $T \cup \{y\}$ covers $\mathcal{F}$ and as $\tau(\mathcal{F}_N) = k$, we must have $|T| = k-1$. It also shows that $\mathcal{G}^T = \{F \setminus \{y\} : y \in F \in \mathcal{F}_N\}$. Therefore, again by Lemma 129, we have $\cup_{G\in\mathcal{G}} G = \cup_{T\in\mathcal{G}^T} T = \cup_{F\in\mathcal{F}} F \setminus \{y, x_1, x_2, \dots, x_{N-1}\}$. Using Lemma 128, we obtain

$$| \cup_{F\in\mathcal{F}} F| = N + | \cup_{T\in\mathcal{G}^T} T| \leq N + N^T(k, k-1) \leq N + n(k, k-2).$$

So all that is left to prove is $N \leq \frac{1}{2}\binom{2k-2}{k-1}$. Note that if $\mathcal{G}$ is a $k$-uniform maximal intersecting family with $\tau(\mathcal{G}) = k$, then for any $x \in \cup_{G\in\mathcal{G}} G$ there exist $G, G' \in \mathcal{F}$ with $G \cap G' = \{x\}$. Indeed, otherwise for any $G$ with $x \in G$ the set $G \setminus \{x\}$ would cover $\mathcal{G}$, contradicting $\tau(\mathcal{G}) = k$. Let us choose $F_0, F_0' \in \mathcal{F} = \mathcal{F}_1$ with $F \cap F' = \{y\}$, and for any $1 \leq i \leq N-1$, let us pick $F_i, F_i' \in \mathcal{F}_i$ with $F_i \cap F_i' = \{x_i\}$. Let $T_0 = F_0 \setminus \{y\}$, $T_0' = F_0' \setminus \{y\}$ and for $1 \leq i \leq N-1$ let $T_i = F_i \setminus \{x_i\}$, $T_i' = F_i' \setminus \{x_i\}$. By the above, $T_i \cap T_i' = \emptyset$ for all $0 \leq i \leq N-1$.

**Claim 130** *For any* $0 \leq i < j \leq N-1$, *we have* $T_i \cup \{y\}, T_i' \cup \{y\} \in \mathcal{F}_j$.

**Proof of Claim.** We use induction on $j$. If $j = 1$ (and thus $i = 0$), the claim holds by definition. The inductive step follows from the property that $\mathcal{G}_{a\mapsto b}$ contains all sets $G \in \mathcal{G}$ with $b \in G$ and all sets $G' \setminus \{a\} \cup \{b\}$ with $a \in G', b \notin G', G' \in \mathcal{G}$. □

By Claim 130, we have $T_i \cup \{y\}, T_i' \cup \{y\} \in \mathcal{F}_j$ for $i < j$, and by definition we have $T_j \cup \{x_j\}, T_j' \cup \{x_j\} \in \mathcal{F}_j$. As $\mathcal{F}_j$ is intersecting, we have $T_i \cap T_j \neq \emptyset$, $T_i \cap T_j' \neq \emptyset$, $T_i' \cap T_j \neq \emptyset$, $T_i' \cap T_j' \neq \emptyset$. This means that $\{(A_h, B_h)\}_{h=1}^{2N}$ with $A_h = T_h, B_h = T_h'$ and $A_{N+h} = T_i', B_{N+h} = T_h$ for $1 \leq h \leq N$ form an ISPS. Therefore by Theorem 6, we have $2N \leq \binom{2k-2}{k-1}$ as required. ■

**Corollary 131** $N(k) \leq (1+o(1))\frac{3}{2}\binom{2k-2}{k-1}$.

**Exercise 132 (Tuza [536])** *A family $\mathcal{F}$ is $\nu$-critical if for any $x \in F \in \mathcal{F}$ and the family $\mathcal{F}_{x,F} = (\mathcal{F} \setminus \{F\}) \cup \{F \setminus \{x\}\}$ we have $\nu(\mathcal{F}) < \nu(\mathcal{F}_{x,F})$. Prove that if $\mathcal{F}$ is $k$-uniform, intersecting and $\nu$-critical, then $|\cup_{F \in \mathcal{F}} F| \leq n(k, k-1)$ holds.*

**Exercise 133 (Tuza [538])** *We say that $X$ is a kernel of the pair of families $\mathcal{A}, \mathcal{B}$ if for every $A \in \mathcal{A}, B \in \mathcal{B}$ we have $A \cap B \cap X \neq \emptyset$. Let $m(a,b) = \max_{\mathcal{A},\mathcal{B}} \min\{|X| : X$ is a kernel$\}$, where $\mathcal{A}, \mathcal{B}$ are such that for all $A \in \mathcal{A}, B \in \mathcal{B}$ we have $|A| \leq a$, $|B| \leq b$ and $A \cap B \neq \emptyset$. Show that $m(a,b) = n'(a, b-1)$ holds.*

## 2.7 The number of intersecting families

To determine the number of certain families of sets does not seem to be an extremal problem at first sight. However, being an intersecting family is a monotone property. We say a property $P$ of families is monotone, if whenever a family $\mathcal{F}$ possesses property $P$, then so do all subfamilies of $\mathcal{F}$. For such properties, determining the number of families satisfying $P$ can often be considered as a very strong version of stability. If $\mathcal{F}$ is a family possessing $P$, then by monotonicity we already have $2^{|\mathcal{F}|}$ such families. Clearly, we get the best lower bound if we take $\mathcal{F}$ to be an extremal family $\mathcal{F}_{ext}$. A result that states that the number of families with property $P$ is not much larger than $2^{|\mathcal{F}_{ext}|}$ means almost all families with property $P$ are subfamilies of extremal ones. It tells us that non-extremal but maximal families with property $P$ are either very few, or so much smaller than $\mathcal{F}_{ext}$, that even adding up the number of their subfamilies will be negligible, compared to $2^{|\mathcal{F}_{ext}|}$.

The problem of determining $\Lambda(n)$, the number of non-uniform intersecting families in $2^{[n]}$ and $\lambda(n)$, the number of non-uniform *maximal* intersecting families in $2^{[n]}$ was posed by Erdős [148]. As by Exercise 15 there is no uniqueness of the extremal family, we cannot expect a result as above. In fact, there is no conjecture for the asypmtotics of these functions. The next proposition determines the asymptotics of $\log_2 \Lambda(n)$ and $\log_2 \lambda(n)$.

**Proposition 134 (Erdős, Hindman [160])** ***(i)*** *For the number $\lambda(n)$ of maximal intersecting families in $2^{[n]}$ we have $\log_2 \lambda(n) = (1+o(1))\binom{n-1}{\lfloor \frac{n-1}{2} \rfloor}$.*

***(ii)*** *For the number $\Lambda(n)$ of intersecting families in $2^{[n]}$ we have $\log_2 \Lambda(n) = (1+o(1))2^{n-1}$.*

**Proof of Proposition.** To obtain the lower bound in **(i)** on $\lambda(n)$, consider the pairs $\{\{A, \overline{A}\} : A \in \binom{n}{\lfloor n/2 \rfloor}, 1 \in A\}$ and pick one set from each pair. For any choice, the sets form an intersecting family and if we extend them to maximal ones, we obtain different families.

To get the upper bound, consider the following map from the set of maximal intersecting families of $2^{[n]}$ to the set of intersecting antichains in $2^{[n-1]}$: $\mathcal{F} \mapsto \mathcal{G}_{\mathcal{F}} = \{F \text{ is minimal in } \mathcal{F}, n \notin F\}$. (The family $\{\emptyset\}$ can also be an image.) We claim that this mapping is injective. Indeed, a maximal intersecting family $\mathcal{F} \subseteq 2^{[n]}$ is clearly determined by the family $\mathcal{F}_{min}$ of its minimal sets. But $\mathcal{F}_{min} = \mathcal{G}_{\mathcal{F}} \cup \mathcal{G}_{n,min}$, where $\mathcal{G}_{n,min}$ is the family of minimal sets in $\{H \in 2^{[n]} : n \in H, \forall G \in \mathcal{G}_{\mathcal{F}}\ G \cap H \neq \emptyset\}$. Therefore, $\lambda(n) \leq \phi(n-1)$, where $\phi(n)$ is the number of antichains in $2^{[n]}$. Kleitman [363] proved that $\log_2 \phi(n) = (1+o(1))\binom{n}{\lfloor n/2 \rfloor}$ (see Section 3.3 for more details). This finishes the proof of **(i)**.

To prove the lower bound of **(ii)** consider any maximal intersecting family $\mathcal{F}$. As by Exercise 15 it has size $2^{n-1}$, there are $2^{2^{n-1}}$ subfamilies of $\mathcal{F}$. On the other hand, every intersecting family is a subfamily of a maximal one; therefore $\Lambda(n) \leq \lambda(n)2^{2^{n-1}}$, and, using **(i)**, we obtain $\log_2 \Lambda(n) \leq (1+o(1))\binom{n-1}{\lfloor \frac{n-1}{2} \rfloor} + 2^{n-1}$. ∎

We now turn our attention to uniform intersecting families. We start our investigations with maximal intersecting families. Let $M(n,k)$ denote the number of maximal intersecting families in $\binom{[n]}{k}$. We will use intersecting set pair systems to give bounds on $M(n,k)$.

$$g(k) = \max\{|\cup_{i=1}^{s} A_i| : \{(A_i, B_i)\}_{i=1}^{s} \text{ is } (k,k)\text{-ISPS and } \{A_i\}_{i=1}^{s} \text{ is intersecting}\}.$$

Recall that the similar notion without the additional property that $\{A_i\}_{i=1}^{s}$ is intersecting was denoted by $n'(k,k)$. Thus by definition we have $g(k) \leq n'(k,k)$. We continue with a couple simple statements about $M(n,k)$ due to Balogh, Das, Delcourt, Liu and Sharifzadeh [40] and Nagy and Patkós [441].

**Lemma 135** *[40] (i)*

$$M(n,k) \leq \sum_{i=1}^{\frac{1}{2}\binom{2k}{k}} \binom{\binom{n}{k}}{i} \leq \binom{n}{k}^{\frac{1}{2}\binom{2k}{k}}.$$

*[441] (ii)*

$$M(n,k) \leq 2^{2^{g(k)}} \binom{n}{g(k)}.$$

**Proof of Lemma.** For a family $\mathcal{F} \subseteq \binom{[n]}{k}$ let us define $\mathcal{I}(\mathcal{F}) = \{G \in \binom{[n]}{k} : \forall F \in \mathcal{F}\ F \cap G \neq \emptyset\}$, the family of $k$-covers. A family is intersecting if and only if $\mathcal{F} \subseteq \mathcal{I}(\mathcal{F})$ and is maximal intersecting if and only if $\mathcal{I}(\mathcal{F}) = \mathcal{F}$. Let us consider a function $f$, that maps any maximal intersecting $k$-uniform family $\mathcal{F}$ to one of its subfamilies $\mathcal{F}_0$, that is minimal with respect to the property that $\mathcal{I}(\mathcal{F}_0) = \mathcal{F}$. Note that the function $f$ is not unique. Clearly, $f$ is injective, $\mathcal{F}_0$

is intersecting and by the minimality of $\mathcal{F}_0$, for every set $F_i \in \mathcal{F}_0$ there exists a set $G_i \in \mathcal{I}(\mathcal{F}_0 \setminus \{F\}) \setminus \mathcal{F}$. This means that the pairs of $k$-sets $\{(F_i, G_i)\}_{i=1}^{|\mathcal{F}_0|}$ form an ISPS. Thus, by definition, $|\cup_{F\in\mathcal{F}_0} F| \leq g(k)$ holds. Therefore, the families, that can be the image of a maximal intersecting $k$-uniform family with respect to $f$, are subfamilies of $2^X$ for some $X \in \binom{[n]}{g(k)}$. The number of such families is not more than $2^{2^{g(k)}}\binom{n}{g(k)}$. This proves **(ii)**. (Actually, the upper bound on $M(n,k)$ could be $\binom{n}{g(k)}\Lambda(g(k))$ which by Proposition 134 is roughly the square root of the bound claimed.)

To see **(i)**, consider the pairs $\{(A_i, B_i)\}_{i=1}^{|2\mathcal{F}_0|}$ with $A_i = F_i$, $B_i = G_i$ for $i = 1, 2, \ldots, |\mathcal{F}_0|$ and $A_i = G_{i-|\mathcal{F}_0|}$, $B_i = F_{i-|\mathcal{F}_0|}$ for $i = |\mathcal{F}_0|+1, |\mathcal{F}_0|+2, \ldots, 2|\mathcal{F}_0|$. They form a skew ISPS; therefore by Theorem 120 we obtain $|\mathcal{F}_0| \leq \frac{1}{2}\binom{2k}{k}$. Therefore $M(n,k)$ is at most the number of all subfamilies of $\binom{[n]}{k}$ of size at most $\frac{1}{2}\binom{2k}{k}$. ■

Note that by Theorem 119 we have $g(k) \leq n'(k,k) \leq n(k,k) = O(\binom{2k}{k})$; therefore, if $n$ is large enough, then **(ii)** gives a better bound than **(i)**, but for small values of $n$ it is better to use **(i)**.

**Theorem 136 (Nagy, Patkós, [441])** *For any fixed integer $k$, as $n$ tends to infinity, the number $M(n,k)$ satisfies*

$$M(n,k) = n^{\Theta(\binom{2k}{k})}.$$

*Moreover,*

$$\frac{1}{8} \leq \limsup_{n} \frac{\log M(n,k)}{\binom{2k}{k}\log n} \leq 1.1 \quad \text{and} \quad \limsup_{k}\limsup_{n} \frac{\log M(n,k)}{\binom{2k}{k}\log n} \leq 1$$

*holds.*

Recall $N(k) = \max\{|\bigcup_{F\in\mathcal{F}} F| : \mathcal{F}$ is $k$-uniform intersecting with $\tau(\mathcal{F}) = k\}$.

**Proof.** The upper bound follows from Lemma 135 **(ii)** and Theorem 119. The lower bound follows from Construction 118 and the following proposition.

**Proposition 137** *For any positive integers $k$ and $n$ we have $\binom{n}{N(k)} \leq M(n,k)$.*

**Proof of Proposition.** Consider a $k$-uniform intersecting family $\mathcal{F}$ with $\tau(\mathcal{F}) = k$ and $|\cup_{F\in\mathcal{F}} F| = N(k)$. As adding more sets to $\mathcal{F}$ can only increase the size of the union, we may assume that $\mathcal{F}$ is maximal intersecting. Every set $X \in \binom{[n]}{N(k)}$ contains at least one family $\mathcal{F}_X$ isomorphic to $\mathcal{F}$. As $\mathcal{F}_X \neq \mathcal{F}_Y$ whenever $\cup_{F\in\mathcal{F}_X} F = X \neq Y = \cup_{F\in\mathcal{F}_Y} F$, we have at least $\binom{n}{N(k)}$ different maximal intersecting $k$-uniform subfamilies of $\binom{[n]}{k}$. □

For the moreover part, observe that the above proposition implies the lower bound. Also, the upper bound in Theorem 119 for $a = b = k$ is $\sum_{i=1}^{k-1}\binom{i}{\lfloor i/2\rfloor}$. A simple induction gives the first upper bound, while a straightforward analysis of the sequence $\sum_{i=1}^{k-1}\binom{i}{\lfloor i/2\rfloor}/\binom{2k}{k}$ gives the second upper bound. ■

We continue with the number of intersecting families and show that if $n$ is large enough, then almost all intersecting families are trivially intersecting. The number $T(n,k)$ of trivially intersecting families is obviously at least $n2^{\binom{n-1}{k-1}} - \binom{n}{2}2^{\binom{n-2}{k-2}} = (n - o(1))2^{\binom{n-1}{k-1}}$ for any $n$ and $k = k(n)$, if $n > 2k$ and $n$ tends to infinity.

**Theorem 138 (Balogh, Das, Delcourt, Liu and Sharifzadeh [40])** *If $n$ and $k = k(n)$ are integers, such that $n \geq 3k + 8\ln k$ holds, then the number of $k$-uniform intersecting families in $2^{[n]}$ is $(n + o(1))2^{\binom{n-1}{k-1}}$.*

**Proof.** The lower bound follows from the above calculations on $T(n,k)$.

By Theorem 41, the size of a non-trivially intersecting family $\mathcal{F} \subseteq \binom{[n]}{k}$ is at most $1+\binom{n-1}{k-1}-\binom{n-k-1}{k-1}$. Every non-trivially intersecting family is a subfamily of a maximal non-trivially intersecting family. Therefore the number $N(n,k)$ of non-trivially intersecting families is at most $2^{1+\binom{n-1}{k-1}-\binom{n-k-1}{k-1}}M(n,k)$.

$$\frac{N(n,k)}{T(n,k)} \leq M(n,k) \cdot 2^{1-\binom{n-k-1}{k-1}}$$

We have to show that $\log_2 M(n,k) - \binom{n-k-1}{k-1} \to -\infty$. By Lemma 135 **(i)**, we have $\log_2 M(n,k) \leq n\binom{2k-1}{k-1}$, so

$$\log_2 M(n,k) - \binom{n-k-1}{k-1} \leq \left(n - \left(\frac{n-k-1}{2k-1}\right)^{k-1}\right)\binom{2k-1}{k-1}$$

and therefore we need $(\frac{n-k-1}{2k-1})^{k-1} = \omega(n)$. As $n \geq 3k + 8\ln k$ holds, we have $n - k - 1 \geq n/2$ and thus $(\frac{n-k-1}{2k-1})^{k-1} \geq \frac{n^2}{4k^2}(\frac{n-k-1}{2k-1})^{k-3}$. Finally, as there exists a constant $c > 1/2$, such that $1 + x \geq \exp(cx)$, we obtain

$$\left(\frac{n-k-1}{2k-1}\right)^{k-3} \geq \left(1 + \frac{8\ln k}{2k-1}\right)^{k-3} \geq \exp\left(c\frac{8(k-3)\ln k}{2k-1}\right) \geq k^2$$

for large enough $k$. ■

Very recently, Frankl and Kupavskii [212] improved the bound on $n$ and showed that the statement of Theorem 138 holds if $n \geq 2k + 2 + 2\sqrt{k\log k}$. In their paper [40], Balogh et al. also determined the asymptotics of the number of $t$-intersecting families in $\binom{[n]}{k}$. It gives a condition for $n$ that almost all $k$-uniform $t$-intersecting families are trivial. The proof is very similar to that of Theorem 138: they prove an upper bound on the number of maximal $t$-intersecting families using Theorem 123 and then apply Theorem 60 to all maximal $t$-intersecting families.

**Theorem 139 (Balogh, Das, Delcourt, Liu and Sharifzadeh [40])** *If $n$, $k = k(n) \geq 3$, and $t = t(n)$ are integers such that $n \geq (t+1)(k-t+1)+\eta_{k,t}$ holds with*

$$\eta_{k,t} = \begin{cases} 12\ln k & \text{for } t = 2 \text{ and } k-t \geq 3 \\ 1 & \text{for } t \geq 3 \text{ and } k-t \geq 3 \\ 31 & \text{for } t \geq 2 \text{ and } k-t = 2 \\ 18k & \text{for } t \geq 2 \text{ and } k-t = 1, \end{cases}$$

*then the number of $k$-uniform $t$-intersecting families in $2^{[n]}$ is $(\binom{n}{t} + o(1))2^{\binom{n-t}{k-t}}$.*

**Exercise 140 ( [90])** *Prove that $\lambda(n)2^{2^{n-1}-\binom{n}{\lfloor n/2\rfloor}} \leq \Lambda(n)$ holds.*

## 2.8 Cross-intersecting families

In this section we consider the following notion: the families $\mathcal{F}_1, \mathcal{F}_2, \dots, \mathcal{F}_r$ are said to be cross-$t$-intersecting if, for any $1 \leq i < j \leq r$ and $F_i \in \mathcal{F}_i, F_j \in \mathcal{F}_j$, we have $|F_i \cap F_j| \geq t$. We will mainly focus on the case $t = 1$, when the families are simply said to be cross-intersecting. Observe that whenever $\mathcal{F}$ is a $t$-intersecting family of sets, then the families $\mathcal{F}_i = \mathcal{F}$ $(i = 1, 2, \dots, r)$ are cross-$t$-intersecting. Another trivial example of cross-$t$-intersecting families is when the families $\mathcal{F}_2, \mathcal{F}_3, \dots, \mathcal{F}_r$ are empty. We will be interested in theorems on how large cross-$t$-intersecting families can be. There are two natural ways to measure the largeness of $t$-cross-intersecting families: either by the sum $\sum_{i=1}^{r} |\mathcal{F}_i|$ or by the product $\prod_{i=1}^{r} |\mathcal{F}_i|$ of their sizes. We discuss what is known in these two cases in two subsections.

### 2.8.1 The sum of the sizes of cross-intersecting families

We start with a simple case of $t = 1$ with no extra condition on the sizes of sets belonging to the $\mathcal{F}_i$'s. The next exercise generalizes Proposition 8.

**Exercise 141** *Let $\mathcal{F}_1, \mathcal{F}_2, \dots, \mathcal{F}_r \subseteq 2^{[n]}$ be cross-intersecting families. Then $\sum_{i=1}^{r} |\mathcal{F}_i| \leq r2^{n-1}$ holds.*

Let us now turn to the uniform case, when for some $k \leq n/2$ we have $\mathcal{F}_i \subseteq \binom{[n]}{k}$. Remember the two main examples of cross-intersecting families: (i) $\mathcal{F}_1 = \binom{[n]}{k}, \mathcal{F}_2 = \dots = \mathcal{F}_r = \emptyset$ and (ii) all $\mathcal{F}_i$'s being the same intersecting family. In the first case $\sum_{i=1}^{r} |\mathcal{F}_i| = \binom{n}{k}$, while, using the Erdős-Ko-Rado theorem, $\sum_{i=1}^r |\mathcal{F}_i| \le r\binom{n-1}{k-1}$. Which construction gives a larger sum depends on the values of $n, k$ and $r$. The problem of whether one of these examples achieves always the maximum possible sum was settled in the affirmative by Hilton [308]. We will present a simple and short proof by Borg [77]. His proof uses a result on the shadows of intersecting families. To obtain this we will use Theorem 23 and the following simple observation that can be proved similarly to Lemma 2.

**Exercise 142** *Prove the following: if $1 \le s < p \le n$ and $\emptyset \ne \mathcal{A} \subseteq \binom{[n]}{p}$, then $|\Delta_s(\mathcal{A})| \ge \frac{\binom{n}{s}}{\binom{n}{p}}|\mathcal{A}|$ holds, with equality if and only if $\mathcal{A} = \binom{[n]}{p}$.*

**Theorem 143** *Let $k \le n/2$ and $s \le n-k$. Suppose $\mathcal{A} \subseteq \binom{[n]}{k}$ is an intersecting family. Then we have*

$$|\Delta_s(\overline{\mathcal{A}})| \ge \frac{\binom{n-1}{s}}{\binom{n-1}{k-1}}|\mathcal{A}|.$$

*Furthermore, if $s < n-k$, then equality holds if and only if $\mathcal{A} = \{A \in \binom{[n]}{k} : x \in A\}$ for some $x \in [n]$.*

**Proof.** Suppose first $s = n-k$. In that case $\Delta_s(\overline{\mathcal{A}}) = \overline{\mathcal{A}}$, and so $|\Delta_s(\overline{\mathcal{A}})| = |\overline{\mathcal{A}}| = |\mathcal{A}| = \frac{\binom{n-1}{s}}{\binom{n-1}{k-1}}|\mathcal{A}|$.

Suppose next $s < n-k$. By the Erdős-Ko-Rado theorem we have $|\overline{\mathcal{A}}| = |\mathcal{A}| \le \binom{n-1}{k-1} = \binom{n-1}{n-k}$. Recall $I_m^k$ is the $m$ first $k$-sets in the colex ordering. Theorem 23 and Exercise 142 imply

$$|\Delta_s(\overline{\mathcal{A}})| \ge \Delta_s(I_{|\overline{\mathcal{A}}|}^{n-k}) \ge \frac{\binom{n-1}{s}}{\binom{n-1}{n-k}}|\mathcal{A}| = \frac{\binom{n-1}{s}}{\binom{n-1}{k-1}}|\mathcal{A}|$$

and the bound of the theorem is proved.

To see the case of equality, observe that if the second inequality above is satisfied with equality, then Exercise 142 implies that $I_{\overline{\mathcal{A}}}^{n-k} = \binom{[n-1]}{n-k}$, in particular $|\mathcal{A}| = |\overline{\mathcal{A}}| = \binom{n-1}{n-k}$. Then Theorem 27 implies that $\overline{\mathcal{A}} = \binom{S}{n-k}$ for some $(n-1)$-subset $S$ of $[n]$; thus $\mathcal{A} = \{A \in \binom{[n]}{k} : x \in A\}$, with $x$ being the only element of $[n] \setminus S$. ■

With the help of Theorem 143, Borg obtained the following strengthening of the Erdős-Ko-Rado theorem. To state the result, for any family $\mathcal{A}$ of sets we introduce $\mathcal{A}^* = \{A \in \mathcal{A} : \forall A' \in \mathcal{A} \quad A \cap A' \ne \emptyset\}$ and $\mathcal{A}' = \mathcal{A} \setminus \mathcal{A}^*$.

**Corollary 144** *Let $\mathcal{A} \subseteq \binom{[n]}{k}$ and $r \le n/2$. Then*

$$|\mathcal{A}^*| + \frac{k}{n}|\mathcal{A}'| \le \binom{n-1}{k-1}$$

*holds. Furthermore, if $n > 2k$ and $\mathcal{A}^* \neq \emptyset$, then equality holds if and only if $\mathcal{A} = \{A \in \binom{[n]}{k} : x \in A\}$ for some $x \in [n]$.*

**Proof.** As all sets in $\mathcal{A}^*$ intersect every set of $\mathcal{A}$, we must have $\mathcal{A}' \subseteq \binom{[n]}{k} \setminus (\mathcal{A}^* \cup \Delta_k(\overline{\mathcal{A}^*}))$. Applying Theorem 143 with $k = s$, we obtain

$$|\mathcal{A}'| \leq \binom{n}{k} - (|\mathcal{A}^*| + |\Delta_k(\overline{\mathcal{A}^*})|) \leq \binom{n}{k} - (|\mathcal{A}^*| + \frac{n-k}{k}|\mathcal{A}^*|).$$

Rearranging yields the desired inequality. The equality part of the statement follows from that of Theorem 143. ■

Now we are ready to prove Hilton's result.

**Theorem 145 (Hilton [308])** *Let $2 \leq k \leq n/2$ and $\mathcal{F}_1, \mathcal{F}_2, \ldots, \mathcal{F}_r$ be cross-intersecting subfamilies of $\binom{[n]}{k}$ with $\mathcal{F}_1 \neq \emptyset$. Then we have*

$$\sum_{i=1}^{r} |\mathcal{F}_i| \leq \begin{cases} \binom{n}{k} & \text{if } r \leq \frac{n}{k}, \\ r\binom{n-1}{k-1} & \text{if } r \geq \frac{n}{k}. \end{cases} \tag{2.7}$$

*If equality holds, then*

*(i) $\mathcal{F}_1 = \binom{[n]}{k}$, $\mathcal{F}_2, \ldots, \mathcal{F}_r = \emptyset$, provided $r < \frac{n}{k}$,*

*(ii) $\mathcal{F}_1 = \mathcal{F}_2 = \cdots = \mathcal{F}_r$, $\mathcal{F}_i$ is intersecting and $|\mathcal{F}_i| = \binom{n-1}{k-1}$, provided $r > \frac{n}{k}$. Moreover, if $n > 2k$, then $\mathcal{F}_i = \{F \in \binom{[n]}{k} : x \in F\}$ for some $x \in [n]$.*

**Proof.** Let us write $\mathcal{F} = \cup_{i=1}^{r} \mathcal{F}_i$. As every $F \in \mathcal{F}_i^*$ intersects every $F' \in \mathcal{F}_i$ by the definition of $\mathcal{F}_i^*$, and $F$ intersects every $F_j \in \mathcal{F}_j$ by the cross-intersecting property, we have $\mathcal{F}^* = \cup_{i=1}^{r} \mathcal{F}_i^*$ and $\mathcal{F}' = \cup_{i=1}^{k} \mathcal{F}_i'$. Observe that for $i \neq j$ we have $\mathcal{F}_i' \cap \mathcal{F}_j' = \emptyset$, as otherwise $F \in \mathcal{F}_i' \cap \mathcal{F}_j'$ would imply the existence of $F' \in \mathcal{F}_i$ with $F \cap F' = \emptyset$, which would contradict the cross-intersecting property of $\mathcal{F}_i$ and $\mathcal{F}_j$. This implies $|\mathcal{F}'| = \sum_{i=1}^{r} |\mathcal{F}_i'|$. Therefore we obtain

$$\sum_{i=1}^{r} |\mathcal{F}_i| = \sum_{i=1}^{r} (|\mathcal{F}_i^*| + |\mathcal{F}_i'|) \leq |\mathcal{F}'| + r|\mathcal{F}^*| \leq \binom{n}{k} + \left(r - \frac{n}{k}\right)|\mathcal{F}^*|, \tag{2.8}$$

where we applied Corollary 144 to $\mathcal{F}$ to obtain the last inequality.

Suppose first that $r < \frac{n}{k}$ holds. Then (2.8) implies $\sum_{i=1}^{r} |\mathcal{F}_i| \leq \binom{n}{k}$, with equality if and only if $\mathcal{F}^* = \emptyset$ and $\mathcal{F} = \mathcal{F}' = \binom{[n]}{k}$. As $\mathcal{F}_1$ is non-empty, there exists $F \in \mathcal{F}_1$. Therefore, for any $G \in \binom{[n] \setminus F}{k}$ we must have $G \notin \mathcal{F}_j$ with $j > 1$ and therefore $G \in \mathcal{F}_1$. We can apply the same reasoning repeatedly to obtain that all $k$-sets $H \in \binom{[n]}{k}$ belong to $\mathcal{F}_1$. Indeed, if $|F \cup F'| = k+1$, then, as $n \geq 2k+1$, there exists an $k$-set $G$ disjoint from both $F$ and $F'$ yielding $F' \in \mathcal{F}_1$. This proves **(i)**.

Suppose next $r > \frac{n}{k}$. Then, using that Corollary 144 implies $|\mathcal{F}^*| \leq \binom{n-1}{k-1}$, we obtain from (2.8) that

$$\sum_{i=1}^{r} |\mathcal{F}_i| \leq \binom{n}{k} + \left(r - \frac{n}{k}\right)\binom{n-1}{k-1} = r\binom{n-1}{k-1},$$

with equality if and only if $\mathcal{F}_1^* = \cdots = \mathcal{F}_r^* = \mathcal{F}^*$ and $|\mathcal{F}^*| = \binom{n-1}{k-1} = |\mathcal{F}|$. The moreover part of Corollary 144 also implies $\mathcal{F}^* = \{F \in \binom{[n]}{k} : x \in F\}$ for some fixed $x \in [n]$, provided $n > 2k$. This proves **(ii)**. ■

Let us continue with some more results in the case of $r = 2$. In this case Theorem 145 tells us that the pair of cross-intersecting families with largest sum of sizes is $(\binom{[n]}{k}, \emptyset)$. It is quite a natural thought to come up with extra conditions that eliminate the possibility of having the empty family as an extremal example. We describe two such conditions. To obtain the optimal pair in the first case, we will need the following inequality concerning binomial coefficients.

**Exercise 146** *Let $a$, $b$ and $n$ integers with $a + b \leq n$ and $a \leq b$. Then*

$$\binom{n-1}{a-1} + \binom{n-1}{b} \leq \binom{n}{b} - \binom{n-a}{b} + 1$$

*holds.*

The following theorem tells us what the "second best" pair of cross-intersecting families is.

**Theorem 147 (Frankl, Tokushige [221])** *If $\mathcal{F} \subseteq \binom{[n]}{k}, \mathcal{G} \subseteq \binom{[n]}{l}$ are non-empty cross-intersecting with $k \leq l$ and $n \geq k + l$, then*

*(i)* $|\mathcal{F}| + |\mathcal{G}| \leq 1 + \binom{n}{l} - \binom{n-k}{l}$,

*(ii) if* $|\mathcal{F}| \geq \binom{n-1}{n-k} = \binom{n-1}{k-1}$, *then*

$$|\mathcal{F}| + |\mathcal{G}| \leq \begin{cases} \binom{n}{k} - \binom{n-k}{k} + 1 & \text{if } k = l \geq 2 \\ \binom{n-1}{k-1} + \binom{n-1}{l-1} & \text{otherwise} \end{cases} \tag{2.9}$$

*holds.*

**Proof.** We proceed by induction on $n$. The base case $n = k + l$ follows from the fact that the pair of families $\mathcal{F} \subseteq \binom{[k+l]}{k}$, $\mathcal{G} \subseteq \binom{[k+l]}{l}$ is cross-intersecting if and only if for every complement pair of a $k$-set $F$ and an $l$-set $G$ either $F \notin \mathcal{F}$ or $G \notin \mathcal{G}$ holds. So assume $n > k + l$ and both (i) and (ii) hold for $n - 1$. If $\mathcal{F}$ is fixed, then to maximize $|\mathcal{G}|$ one has to pick $\mathcal{G} := \overline{\Delta_l(\overline{\mathcal{F}})}$ where $\overline{\mathcal{A}} = \{[n] \setminus A : A \in \mathcal{A}\}$. Therefore, by Theorem 23, we may assume $\overline{\mathcal{F}} = I^{n-k}_{|\mathcal{F}|}$, the initial segment of the colex ordering. Let us define

$$\mathcal{F}(n) = \{F \backslash \{n\} : n \in F \in \mathcal{F}\} \subseteq \binom{[n-1]}{k-1}, \mathcal{F}'(n) = \{F \in \mathcal{F} : n \notin F\} \subseteq \binom{[n-1]}{k},$$

$$\mathcal{G}(n) = \{G \setminus \{n\} : n \in G \in \mathcal{G}\} \subseteq \binom{[n-1]}{l-1}, \mathcal{G}'(n) = \{G \in \mathcal{G} : n \notin G\} \subseteq \binom{[n-1]}{l}.$$

Observe that $\mathcal{F}(n)$ and $\mathcal{G}'(n)$ are cross-intersecting, and so are $\mathcal{F}'(n)$ and $\mathcal{G}(n)$.

We start by proving **(ii)**. If $|\mathcal{F}| = \binom{n-1}{n-k}$, then the first $|\mathcal{F}|$ sets in the colex order are $\binom{[n-1]}{n-k}$, and so $\mathcal{G} = \{G \in \binom{[n]}{l} : n \in G\}$, and the theorem holds. Thus we may assume $|\mathcal{F}| > \binom{n-1}{n-k}$, $|\mathcal{F}(n)| = \binom{n-1}{n-k}$, and $\mathcal{F}'(n) \neq \emptyset$. Observe that $|\mathcal{F}(n)| = \binom{n-1}{n-k}$ implies $\mathcal{G}'(n) = \emptyset$.

CASE I: $k < l$.

Applying **(i)** to $\mathcal{F}'(n)$ and $\mathcal{G}(n)$ by the induction hypothesis, we obtain $|\mathcal{F}'(n)| + |\mathcal{G}(n)| \leq \binom{n-1}{l-1} - \binom{n-1-k}{l-1} + 1$ and therefore

$$
\begin{aligned}
|\mathcal{F}| + |\mathcal{G}| &= |\mathcal{F}(n)| + |\mathcal{F}'(n)| + |\mathcal{G}(n)| \\
&\leq \binom{n-1}{k-1} + \binom{n-1}{l-1} - \binom{n-1-k}{l-1} + 1 \\
&< \binom{n-1}{k-1} + \binom{n-1}{l-1} \\
&\leq \binom{n}{l} - \binom{n-k}{l} + 1,
\end{aligned}
$$

where to obtain the last inequality we used Exercise 146.

CASE II: $k = l$.

This time the inductive hypothesis implies $|\mathcal{F}(n)| + |\mathcal{G}'(n)| \leq \binom{n-1}{k} - \binom{n-1-(k-1)}{k} + 1$. Therefore, adding $|\mathcal{F}(n)| = \binom{n-1}{k-1}$ to both sides yields

$$
|\mathcal{F}| + |\mathcal{G}| = |\mathcal{F}(n)| + |\mathcal{F}'(n)| + |\mathcal{G}(n)| \leq \binom{n}{k} - \binom{n-k}{k} + 1.
$$

The proof of (i) is very similar to that of (ii). First observe that the bound of (ii) is at most the bound of (i), and since we have already proved (ii), we can assume $|\mathcal{F}| < \binom{n-1}{k-1}$ and therefore $\mathcal{F}'(n) = \emptyset$.

Suppose first $\mathcal{G}'(n) \neq \emptyset$. Then applying induction to $\mathcal{F}(n)$ and $\mathcal{G}'(n)$, we obtain

$$
|\mathcal{F}(n)| + |\mathcal{G}'(n)| \leq \binom{n-1}{l} - \binom{n-1-(k-1)}{l} + 1.
$$

As $\mathcal{G}(n) \leq \binom{n-1}{l-1}$, we conclude to

$$
|\mathcal{F}| + |\mathcal{G}| = |\mathcal{F}(n)| + |\mathcal{G}(n)| + |\mathcal{G}'(n)| \leq \binom{n}{l} - \binom{n-k}{l} + 1.
$$

Suppose next $\mathcal{G}'(n) = \emptyset$. Then $|\mathcal{F}| + |\mathcal{G}| = |\mathcal{F}(n)| + |\mathcal{G}(n)| \leq \binom{n-1}{k-1} + \binom{n-1}{l-1}$ and we are done by Exercise 146. ∎

Let us mention that the case $k = l$ was already proved by Hilton and Milner [309]. The analogous result for cross-$t$-intersecting families has been proved recently.

**Theorem 148 (Frankl, Kupavskii [210])** *If $\mathcal{F}, \mathcal{G} \subseteq \binom{[n]}{k}$ are non-empty cross-$t$-intersecting families and $n > 2k - t$, then $|\mathcal{F}| + |\mathcal{G}| \leq |\mathcal{A}| + 1$, where $\mathcal{A} = \{A \in \binom{[n]}{k} : |A \cap [k]| \geq t\}$.*

In Section 2.1, we considered (among others) bounds on the size of intersecting families $\mathcal{F}$, if we know $\tau(\mathcal{F}) \geq m$ for some integer $m$. A result of this type on cross-intersecting families is the following theorem.

**Theorem 149 (Frankl, Tokushige [222])** *If $\mathcal{F} \subseteq \binom{[n]}{k}, \mathcal{G} \subseteq \binom{[n]}{l}$ are cross-intersecting families with $\tau(\mathcal{F}), \tau(\mathcal{G}) \geq 2$ and $k \leq l$, then $|\mathcal{F}| + |\mathcal{G}| \leq 2 + \binom{n}{l} - 2\binom{n-k}{l} + \binom{n-2k}{l}$ holds, provided $n \geq k + l$. Furthermore, if $n > k + l$ and $l \geq 3$, then equality holds if and only if $\mathcal{F} = \{[k], [k+1, 2k]\}$ and $\mathcal{G} = \{G \in \binom{[n]}{l} : [k] \cap G, [k+1, 2k] \cap G \neq \emptyset\}$.*

It is a natural problem to consider a host family $\mathcal{G}$ of sets, and maximize the sum $\sum_{i=1}^{r} |\mathcal{F}_i|$ over all cross-t-intersecting families $\mathcal{F}_1, \ldots, \mathcal{F}_r$ such that $\mathcal{F}_i \subseteq \mathcal{G}$ for all $i = 1, 2, \ldots, r$. Wang and Zhang obtained [542] a very general condition on $\mathcal{G}$ that implies that one of the basic cross-$t$-intersecting examples is always extremal. Their result covers the case of $\mathcal{G} = 2^{[n]}$ and $\mathcal{G} = \binom{[n]}{k}$. It is also natural to consider downsets (also known as hereditary families) as host families. Chvatal's conjecture [106] states that if $\mathcal{F}$ is an intersecting subfamily of a downset $\mathcal{D}$, then the size of $\mathcal{F}$ is at most the largest degree in $\mathcal{D}$. Borg considered [78, 79] cross-intersecting families in downsets, but we finish this subsection with a result of his that is valid for any host family $\mathcal{H}$.

**Theorem 150 (Borg [81])** *For any family $\mathcal{H}$ of sets and positive integer $t$, let $\ell(\mathcal{H}, t)$ denote the size of the largest $t$-intersecting subfamily of $\mathcal{F}$. There exists a $k_0$ such that if $k \geq k_0$ and $\mathcal{F}_1, \mathcal{F}_2, \ldots, \mathcal{F}_k$ are cross-$t$-intersecting subfamilies of $\mathcal{H}$, then $\sum_{i=1}^{k} |\mathcal{F}_i| \leq k\ell(\mathcal{H}, t)$ holds.*

### 2.8.2 The product of the sizes of cross-intersecting families

In many of the results of the previous section, it is proved that $\sum_{i=1}^{r} |\mathcal{F}_i|$ is maximized by one of the basic examples of cross-$t$-intersecting families. As having the empty-set as one of the cross-intersecting families makes the product $\prod_{i=1}^{r} |\mathcal{F}_i|$ zero, only one of our basic examples remains a candidate for achieving the maximum product of sizes. Obviously, whenever $r$ copies of a largest $t$-intersecting family are proven to obtain the maximum product, such a result strengthens the original $t$-intersection result. The very first result of this type, therefore, strengthens the theorem of Erdős, Ko, and Rado, Theorem 9.

**Theorem 151 (Pyber [471])** *If $\mathcal{F},\mathcal{G} \subseteq \binom{[n]}{k}$ are cross-intersecting with $2k \le n$, then $|\mathcal{F}||\mathcal{G}| \le \binom{n-1}{k-1}^2$ holds.*

**Proof.** The proof we present here is due to Frankl and Kupavskii [211]. Let us recall the colex and the lexicographic ordering of $\binom{[n]}{k}$: in the former we have $F < G$ if and only if $\max\{F \triangle G\} \in G$, while in the latter we have $F < G$ if and only if $\min\{F \triangle G\} \in F$. The initial segments of size $s$ are denoted by $I_s^k$ and $\mathcal{L}_{n,k}(s)$, respectively. We define the *reversed colex order* of $\binom{[n]}{k}$ by $F < G$ if and only if $\min\{F \triangle G\} \in G$. That is, we can obtain the reversed colex order by reversing the *lexicographic* order, or by reversing the base set of the *colex* order. This observation yields that if $\mathcal{F} \subseteq \binom{[n]}{k}$ is an initial segment of the reversed colex order, then $\Delta_s(\mathcal{F})$ is an initial segment of the reversed colex order of $\binom{[n]}{s}$.

We will need the following lemma.

**Lemma 152 (Hilton, [307])** *If $\mathcal{F} \subseteq \binom{[n]}{k}, \mathcal{G} \subseteq \binom{[n]}{l}$ are cross-intersecting, then so are $\mathcal{L}_{n,|\mathcal{F}|}(k)$ and $\mathcal{L}_{n,|\mathcal{G}|}(l)$.*

**Proof of Lemma.** We can assume $k+l \le n$, otherwise $\binom{[n]}{k}$ and $\binom{[n]}{l}$ are cross-intersecting. Observe that $\mathcal{F}$ and $\mathcal{G}$ are cross-intersecting if and only if $\mathcal{F}$ and $\Delta_k(\overline{\mathcal{G}})$ are disjoint. Suppose now, contrary to the statement of the lemma, that $\mathcal{L}_k := \mathcal{L}_{n,|\mathcal{F}|}(k)$ and $\mathcal{L}_l := \mathcal{L}_{n,|\mathcal{G}|}(l)$ are not cross-intersecting; therefore $\mathcal{L}_k$ and $\Delta_k(\overline{\mathcal{L}_l})$ intersect. But $\mathcal{L}_k$ is an initial segment of the lexicographic ordering and $\overline{\mathcal{L}_l}$ (thus $\Delta_k(\overline{\mathcal{L}_l})$, too) is an initial segment of the reversed colex ordering. Therefore $\Delta_k(\overline{\mathcal{L}_l})$ consists of the last sets of the lexicographic ordering. This yields

$$|\mathcal{L}_k| + |\Delta_k(\overline{\mathcal{L}_l})| > \binom{n}{k}. \tag{2.10}$$

By Theorem 23 and the observations about the relationship of the lexicographic and the reversed colex ordering, we obtain that $|\Delta_k(\overline{\mathcal{G}})| \ge |\Delta_k(\overline{\mathcal{L}_l})|$. Together with (2.10) we obtain

$$|\mathcal{F}| + |\Delta_k(\overline{\mathcal{G}})| \ge |\mathcal{L}_k| + |\Delta_k(\overline{\mathcal{L}_l})| > \binom{n}{k},$$

which contradicts the fact that $\mathcal{F}$ and $\mathcal{G}$ are cross-intersecting. □

**Lemma 153** *If $\mathcal{F},\mathcal{G} \subseteq \binom{[n]}{k}$ are cross-intersecting with $2k \le n$ and $\binom{n-2}{k-2} \le |\mathcal{F}| \le |\mathcal{G}|$, then $|\mathcal{F}| + |\mathcal{G}| \le 2\binom{n-1}{k-1}$ holds.*

**Proof of Lemma.** By Lemma 152 we can assume that $\mathcal{F}$ and $\mathcal{G}$ are initial segments of the lexicographic order and $|\mathcal{F}| \le |\mathcal{G}|$. If $|\mathcal{G}| \le \binom{n-1}{k-1}$, then the statement is trivial, so we can assume $\binom{n-1}{k-1} \le |\mathcal{G}|$. This implies $\mathcal{G}$ contains all sets containing 1, so by the cross-intersecting property of $\mathcal{F}$ and $\mathcal{G}$ we obtain that every $F \in \mathcal{F}$ contains 1. As $\binom{n-2}{k-2} \leq |\mathcal{F}|$, all sets containing $\{1,2\}$ belong to both $\mathcal{F}$ and $\mathcal{G}$. This implies that for every set $G \in \mathcal{G}$, we have $G \cap \{1,2\} \neq \emptyset$.

Let us introduce $\mathcal{G}' = \{G \in \mathcal{G} : 1 \notin G\}$ and $\mathcal{H} = \{H \in \binom{[n]}{k} : H \notin \mathcal{F}, 1 \in H\}$. To prove the lemma it is enough to show $|\mathcal{G}'| \leq |\mathcal{H}|$. The above observations imply $G \cap \{1,2\} = \{2\}$ for all $G \in \mathcal{G}'$ and $H \cap \{1,2\} = 1$ for all $H \in \mathcal{H}$. Let us consider the bipartite graph $B = (X_1, X_2, E)$ with $X_i = \{S \in \binom{[n]}{k} : S \cap \{1,2\} = i\}$, where $S_1 \in X_1, S_2 \in X_2$ are joined by an edge if and only if $S_1 \cap S_2 = \emptyset$. The graph $B$ is regular with degree $\binom{n-1-k}{k-1}$ and $\mathcal{H} \subseteq X_1$, $\mathcal{G}' \subseteq X_2$. Observe that the cross-intersecting property of $\mathcal{F}$ and $\mathcal{G}$ implies that if $G \in \mathcal{G}' \subseteq X_2$ is connected to a set $S \in X_2$, then $S \in \mathcal{H}$. That is, the neighborhood $N(\mathcal{G}')$ of $\mathcal{G}'$ is a subfamily of $\mathcal{H}$. As $B$ is regular, we obtain $|\mathcal{G}'| \leq |N(\mathcal{G}')| \leq |\mathcal{H}|$ as required. □

Now we are ready to prove Theorem 151. We can assume $|\mathcal{F}| \leq |\mathcal{G}|$. Observe first that if $|\mathcal{F}| \leq \binom{n-2}{k-2}$, then

$$|\mathcal{F}| \cdot |\mathcal{G}| \leq \binom{n-2}{k-2}\binom{n}{k} < \binom{n-1}{k-1}^2.$$

So we can assume $\binom{n-2}{k-2} < |\mathcal{F}| \leq |\mathcal{G}|$. But then Lemma 153 yields

$$|\mathcal{F}| \cdot |\mathcal{G}| \leq \left(\frac{|\mathcal{F}| + |\mathcal{G}|}{2}\right)^2 \leq \binom{n-1}{k-1}^2.$$

■

Pyber [471] also proved the analogous statement for $\mathcal{F} \subseteq \binom{[n]}{k}, \mathcal{G} \subseteq \binom{[n]}{l}$ but only under the condition $k \leq l$, $n \geq 2l + k - 2$. Matsumoto and Tokushige [411] proved the exact bound on $n$, namely $k \leq l$, $n \geq 2l$. Here is the example that shows that this bound is best possible.

Let $n, k, l$ satisfy the following conditions: $n = 2l - 1$, $k = l - 1$ and $\binom{n}{l}$ is even. Fix an arbitrary family $\mathcal{G} \subseteq \binom{[n]}{l}$ with $|\mathcal{G}| = \frac{1}{2}\binom{n}{l}$ and let $\mathcal{F} := \{F \in \binom{[n]}{k} : [n] \setminus F \notin \mathcal{G}\}$. Then, by definition, $\mathcal{F}$ and $\mathcal{G}$ are cross-intersecting and $|\mathcal{F}| \cdot |\mathcal{G}| = \frac{1}{2}\binom{n}{k}\frac{1}{2}\binom{n}{l} > \binom{n-1}{k-1}\binom{n-1}{l-1}$ holds.

Let us now consider the more general $t$-intersecting case. As opposed to determining the maximum size of a non-uniform intersecting family (Exercise 8), settling the non-uniform $t$-intersection problem (Theorem 48) required a "real proof". Frankl conjectured that the natural generalization of the result of Katona on non-uniform $t$-intersecting families is true. This was verified by the following result.

**Theorem 154 (Matsumoto, Tokushige [412])** *Let $\mathcal{F}, \mathcal{G} \subset 2^{[n]}$ be cross-$t$-intersecting families with $1 \leq t \leq n$. Then*

$$|\mathcal{F}|\cdot|\mathcal{G}|\le\begin{cases}\left(\sum_{i=\frac{n+t}{2}}^{n}\binom{n}{i}\right)^2 & \text{if } n+t \text{ is even,}\\ \left(\binom{n-1}{\frac{n+t-1}{2}}+\sum_{i=\frac{n+t+1}{2}}^{n}\binom{n}{i}\right)^2 & \text{if } n+t \text{ is odd.}\end{cases} \tag{2.11}$$

In the uniform case $\mathcal{F},\mathcal{G}\subseteq\binom{[n]}{k}$, by results of Frankl [186] and Wilson [550], we know that if $n\ge(t+1)(k-t+1)$, then $\mathcal{F}=\mathcal{G}=\{F\in\binom{[n]}{k}:[t]\subseteq F\}$ gives the largest product of sizes in a basic example of cross-$t$-intersecting families. In recent papers [214, 530, 531] it was shown that this pair of families indeed maximizes $|\mathcal{F}|\cdot|\mathcal{G}|$ if $n\ge(t+1)k$.

If the uniformity of $\mathcal{F}$ and $\mathcal{G}$ are different, then first Borg [80] proved that if the base set is large enough, then the largest product of sizes of cross-$t$-intersecting families is given by the following pair. He proved the following bound on the threshold of $n$.

**Theorem 155 ( [83])** *Suppose $1\le t\le k\le l\le n$ and $m$ are all integers and $u$ is a non-negative real such that $u>\frac{6-t}{3}$ and $\min\{n,m\}\ge(t+u+2)(l-t)+k-1$. If $\mathcal{F}\subseteq\binom{[n]}{k},\mathcal{G}\subseteq\binom{[m]}{l}$ are cross-$t$-intersecting families, then $|\mathcal{F}||\mathcal{G}|\le\binom{n-t}{k-t}\binom{m-t}{l-t}$ holds.*

Let us finish this section by mentioning a result when the host family is neither a layer $\binom{[n]}{k}$, nor the complete power set $2^{[n]}$.

**Theorem 156 (Borg [82])** *Let $n_1,n_2,\dots,n_r$ be natural numbers and $k_1,k_2,\dots,k_r$ be integers with $k_i\in[n_i]$. If the families $\mathcal{F}_i\subseteq\binom{[n_i]}{\le k_i}$ are cross-intersecting, then*

$$\prod_{i=1}^{r}|\mathcal{F}_i|\le\prod_{i=1}^{r}\left(\sum_{j=1}^{k_j}\binom{n-1}{j-1}\right)$$

*holds.*

# 3

# Sperner-type theorems

CONTENTS



## 3.1 More-part Sperner families

In this section we consider a generalization of the Sperner and $k$-Sperner properties. In the early 70's, Katona and Kleitman independently proved that the statement of Theorem 1 remains valid, even if the condition of the theorem is relaxed in the following way.

**Theorem 157 (Katona [333], Kleitman [365])** *Suppose $[n]$ is partitioned into $X_1 \cup X_2$, and $\mathcal{F} \subseteq 2^{[n]}$ is a family such that there does not exist distinct sets $F_1, F_2 \in \mathcal{F}$ with*

- $F_1 \subseteq F_2$
- $F_2 \setminus F_1 \subseteq X_1$ *or* $F_2 \setminus F_1 \subseteq X_2$.

*Then* $|\mathcal{F}| \le \binom{n}{\lfloor n/2 \rfloor}$.

We present here Kleitman's proof, that is based on Lemma 158 and Theorem 4. We will need the following definition. A chain $C_1 \subsetneq C_2 \subsetneq \cdots \subsetneq C_l$ in $2^{[n]}$ is said to be *symmetric* if it is obtained from a maximal chain $\mathcal{C}$ by omitting its $\frac{n+1-l}{2}$ smallest and $\frac{n+1-l}{2}$ largest elements. It is equivalent to the conditions that $|C_j \setminus C_{j-1}| = 1$ for all $j = 2, \ldots, l$ and $|C_i| + |C_{l+1-i}| = n$ for all $i = 1, 2, \ldots, l$.

**Lemma 158** *There exists a symmetric chain partition of $2^{[n]}$, i.e. there exist*

$\binom{n}{\lfloor n/2 \rfloor}$ *pairwise disjoint symmetric chains* $\mathcal{C}_1, \mathcal{C}_2, \ldots, \mathcal{C}_{\binom{n}{\lfloor n/2 \rfloor}}$*, such that*

$$2^{[n]} = \bigcup_{i=1}^{\binom{n}{\lfloor n/2 \rfloor}} \mathcal{C}_i$$

*holds. The number of symmetric chains of length* $k$ *in the partition is* $\binom{n}{\frac{n+k-1}{2}} - \binom{n}{\frac{n+k+1}{2}}$ *if* $n+k-1$ *is even, and otherwise 0.*

**Proof.** We proceed by induction on $n$, with the base case $n = 1$ being trivial, as the two subsets of $[1]$ form a symmetric chain. Suppose the statement is true for $n-1$, and let $\mathcal{C}'_1, \mathcal{C}'_2, \ldots, \mathcal{C}'_{\binom{n-1}{\lfloor (n-1)/2 \rfloor}}$ be a symmetric chain partition of $2^{[n-1]}$. For any $j$, let $L_j$ denote the largest set in $\mathcal{C}'_j$, and let us consider the following two symmetric chains

$$\mathcal{C}_{j,1} = \mathcal{C}_j \cup \{L_j \cup \{n\}\}, \qquad \mathcal{C}_{j,2} = \{C \cup \{n\} : C \in \mathcal{C}'_j \setminus \{L_j\}\}.$$

Note that if $n-1$ is even, then some of the $\mathcal{C}'_j$s contain only one set, therefore, for these $j$'s, $\mathcal{C}_{j,2}$ is empty, and we can omit them. Otherwise $\mathcal{C}_{j,1}$ and $\mathcal{C}_{j,2}$ are symmetric, since if the smallest set in $\mathcal{C}'_j$ has size $k$, then the largest one has size $n-1-k$. This implies that the sum of the sizes of the smallest and largest sets is $k + [(n-1-k)+1] = n$ in $\mathcal{C}_{j,1}$, and $k+1+[(n-1-k-1)+1] = n$ in $\mathcal{C}_{j,2}$.

Finally, observe that the $\mathcal{C}_{j,i}$'s partition $2^{[n]}$. Indeed, a set $S \in \mathcal{C}'_j \cap 2^{[n-1]}$ belongs to $\mathcal{C}_{j,1}$. Let us consider a set $S$ containing $n$, then $S' = S \setminus \{n\}$ belongs to a $\mathcal{C}'_j$. If $S'$ is the largest set of $\mathcal{C}'_j$, then $S$ belongs to $\mathcal{C}_{j,1}$, while if $S'$ is not the largest set of $\mathcal{C}'_j$, then $S$ belongs to $\mathcal{C}_{j,2}$.

The statement about the number of chains of length $k$ follows from the fact that every $F \in \binom{[n]}{\frac{n+k-1}{2}}$ belongs to a chain $\mathcal{C}$ of length at least $k$, and if $\mathcal{C}$ contains a set of size $\frac{n+k+1}{2}$, then the length of $\mathcal{C}$ is at least $k+2$. ■

Let us mention that there exist results on partitioning $2^{[n]}$ into copies of different structures. Proving an asymptotic version of a conjecture of Füredi [228], Tomon [533] showed that $2^{[n]}$ is the the disjoint union of $\binom{n}{\lfloor n/2 \rfloor}$ chains, each of size $\Theta(\sqrt{n})$. Results on partitioning $2^{[n]}$ into copies of a poset were obtained in [286, 534].

**Proof of Theorem 157.** Suppose $m = |X_1| \le |X_2|$, and let us consider a symmetric chain partition $\mathcal{C}_1, \mathcal{C}_2, \ldots, \mathcal{C}_{\binom{n}{\lfloor n/2 \rfloor}}$ of $2^{X_1}$, provided by Lemma 158. For any $A \subseteq X_1$ let $\mathcal{F}_A = \{B \subseteq X_2 : A \cup B \in \mathcal{F}\}$. By the property of $\mathcal{F}$, for every $A \subseteq X_1$ the family $\mathcal{F}_A \subseteq 2^{X_2}$ is Sperner, and if $A_1, A_2$ belong to the same $\mathcal{C}_i$, then $\mathcal{F}_{A_1} \cap \mathcal{F}_{A_2} = \emptyset$. Therefore, the family $\mathcal{F}_i = \cup_{A \in \mathcal{C}_i} \mathcal{F}_A$ is $|\mathcal{C}_i|$-Sperner. So applying Theorem 4, we obtain $|\mathcal{F}_i| \le \Sigma(|X_2|, |\mathcal{C}_i|)$. Using the result on the number of $k$-chains in the partition, this yields

$$|\mathcal{F}| = \sum_{i=1}^{\binom{m}{\lfloor m/2\rfloor}} |\mathcal{F}_i| \le \sum_{k=1}^{m+1} \left( \binom{m}{\frac{m+k-1}{2}} - \binom{m}{\frac{m+k+1}{2}} \right) \Sigma(|X_2|, k).$$

We claim that the sum on the right hand side is $\binom{n}{\lfloor n/2\rfloor}$. Consider the family $\mathcal{G} = \binom{[n]}{\lfloor n/2\rfloor}$, and for every $A \subseteq X_1$ let $\mathcal{G}_A = \{B \subseteq X_2 : A \cup B \in \mathcal{G}\}$. Define $\mathcal{G}_i = \cup_{C \in \mathcal{C}_i} \mathcal{G}_C$. Observe that if $\mathcal{C}_i$ has length $k$, then $|\mathcal{G}_i| = \Sigma(|X_2|, k)$. We obtain

$$\binom{n}{\lfloor n/2\rfloor} = |\mathcal{G}| = \sum_{k=1}^{m+1} \left( \binom{m}{\frac{m+k-1}{2}} - \binom{m}{\frac{m+k+1}{2}} \right) \Sigma(|X_2|, k),$$

so the claim follows. ■

Let us now introduce the general notion that we will consider in this section. Let $[n]$ be partitioned into $X_1 \cup X_2 \cup \cdots \cup X_M$. We say that a family $\mathcal{F}$ is *$M$-part $k$-Sperner* with respect to the partition, if $\mathcal{F}$ does not contain a $(k+1)$-chain $F_1 \subsetneq F_2 \subsetneq \cdots \subsetneq F_{k+1}$, such that $F_{k+1} \setminus F_1 \subseteq X_i$ for some $1 \le i \le M$. When $k = 1$, then we say that $\mathcal{F}$ is *$M$-part Sperner*. With this terminology, Theorem 157 is about 2-part Sperner families.

For all the above properties, there are two extremal problems to address: either one considers the maximum size $f(|X_1|, |X_2, \ldots, |X_M|; k)$ of an $M$-part $k$-Sperner family with respect to a fixed $M$-partition $X_1, X_2, \ldots, X_M$ of $X$, or one can try to determine the maximum size of a family $\mathcal{F} \subseteq 2^X$ that is $M$-part $k$-Sperner for *any* $M$-partition of $X$. The latter is the more studied problem, and it can be formulated in a way that has a Ramsey-type flavor (sometimes more part Sperner problems are referred to as Ramsey-Sperner theory, e.g. [229, 237]): what is the minimum number $\phi = \phi(|X|, M; k)$, such that for any family $\mathcal{F} \subseteq 2^X$ with $|\mathcal{F}| > \phi$ and $M$-coloring $c : X \to [M]$, there exists a $(k+1)$-chain $F_1 \subsetneq F_2 \subsetneq \cdots \subsetneq F_{k+1}$, such that the set $F_{k+1} \setminus F_1$ is monochromatic with respect to $c$?

To see the main difference between ordinary and more-part $k$-Sperner families, we introduce the quantity $g(n_1, n_2, \ldots, n_M; k)$ to denote the maximum size of $\cup_{j=1}^k \mathcal{F}_i$, such that all $\mathcal{F}_j$'s are $M$-part Sperner with respect to a partition $X_1 \cup X_2 \cup \cdots \cup X_M$, where $|X_i| = n_i$. Further, let $\gamma(n, M; k)$ be the maximum of $g(n_1, n_2, \ldots, n_M; k)$ over all possible $n_i$'s with $\sum_{i=1}^M n_i = n$. Recall that Erdős's result Theorem 4 states that $\phi(n, 1; k) = \gamma(n, 1; k) = \Sigma(n, k)$, the sum of the $k$ largest binomial coefficients.

A different proof of Theorem 157 was given by P.L. Erdős, Füredi and Katona [170]. Although the relaxation of being Sperner to being 2-part Sperner does not increase the maximum possible size of a family with this property, the number of extremal families does increase. The only maximum size Sperner families are $\binom{[n]}{\lfloor n/2\rfloor}$ and $\binom{[n]}{\lceil n/2\rceil}$, as Theorem 1 shows, while there are many other 2-part Sperner families of size $\binom{n}{\lfloor n/2 \rfloor}$. All maximum families were described first by P.L. Erdős and Katona [171]. To state their result, we need several definitions.

The *type* of a set $F \subset X$ (with respect to a partition $X = X_1 \cup X_2 \cup \cdots \cup X_M$) is the vector $(|F \cap X_1|, \ldots, |F \cap X_M|)$. A family $\mathcal{F} \subseteq 2^X$ is said to be *homogeneous* (with respect to a partition $X = X_1 \cup X_2 \cup \cdots \cup X_M$), if $F \in \mathcal{F}$ implies $G \in \mathcal{F}$ whenever $|F \cap X_i| = |G \cap X_i|$ holds for all $i = 1, 2, \ldots, M$. This means that a homogeneous family $\mathcal{F}$ can be described by the set of types $I(\mathcal{F}) = \{(i_1, i_2, \ldots, i_M) :$ there is $F \in \mathcal{F}$ with $|X_j \cap F| = i_j$ for all $j = 1, 2, \ldots, M\}$. If $\mathcal{F}$ is a homogeneous $M$-part $k$-Sperner family with respect to the partition $X_1, X_2, \ldots, X_M$, and $|X_i| = n_i$ with $n_1 \geq n_2 \geq \cdots \geq n_M$, then $|I(\mathcal{F})| \leq k \prod_{i=2}^{M}(n_i + 1)$. We say that a homogeneous family is *full* if $|I(\mathcal{F})| = k \prod_{i=2}^{M}(n_i + 1)$. In particular, for 2-part Sperner families we have $|I(\mathcal{F})| \leq n_2 + 1$, and for full 2-part Sperner families there exists an injective function $t : [0, n_2] \to [0, n_1]$ such that $I(\mathcal{F}) = \{(t(i), i) : i \in [0, n_2]\}$. The following exercise describes the largest full 2-part Sperner families.

**Exercise 159** *Prove that a full 2-part Sperner family $\mathcal{F} \subseteq 2^{X_1 \cup X_2}$ with $|X_1| = n_1 \geq n_2 = |X_2|$ is of maximum size if and only if the function $t : [0, n_2] \to [0, n_1]$ describing $I(\mathcal{F})$ satisfies the following:*

- *The image of $t$ is an $(n_2 + 1)$-interval around $\lfloor n_1/2 \rfloor$ or $\lceil n_1/2 \rceil$,*
- $\binom{n_2}{i} < \binom{n_2}{j}$ *implies* $\binom{n_1}{t(i)} \leq \binom{n_1}{t(j)}$.

Full families satisfying the conditions of Exercise 159 are called *well-paired.* The result of P.L. Erdős and Katona states that there do not exist non-homogeneous maximum sized 2-part Sperner families.

**Theorem 160 (P.L. Erdős, Katona [171])** *A family $\mathcal{F} \subseteq 2^{[n]}$ is maximum sized 2-part Sperner if and only if $\mathcal{F}$ is homogeneous and well-paired.*

Other proofs of the characterization of all extremal families were obtained by Shahriari [502], and by Aydinian and P.L. Erdős [26].

Concerning general results, let us start with the following asymptotic statement.

**Theorem 161 (Füredi, Griggs, Odlyzko, Shearer, [237])** *(i) For any fixed $M, k$, the limits $\phi(M; k) = \lim_{n\to\infty} \frac{\phi(n,M;k)}{\binom{n}{\lfloor n/2 \rfloor}}$ and $\gamma(M; k) = \lim_{n\to\infty} \frac{\gamma(n,M;k)}{\binom{n}{\lfloor n/2 \rfloor}}$ exist. Moreover, $\gamma(M; k) \leq \phi(M; k)$ and $\gamma(M; k) = k\gamma(M; 1)$ hold.*

*(ii) If $k$ is fixed and $M$ tends to infinity, then we have*

$$(1 + o(1))\gamma(M, k) = \phi(M; k) = (k + o(1))\sqrt{\frac{\pi M}{4 \ln M}}.$$

The following result was obtained independently by P.L. Erdős and Katona [173], and by Griggs, Odlydzko and Shearer [282].

**Theorem 162** *Let $n = n_1 + n_2 + \cdots + n_M$. Then there exists a homogeneous $M$-part $k$-Sperner family $\mathcal{F} \subseteq 2^{X_1 \cup X_2 \cup \cdots \cup X_M}$ with $|X_i| = n_i$, such that $|\mathcal{F}| = f(n_1, n_2, \ldots, n_M; k)$ holds.*

As $\phi(n, M; k)$ is the maximum of $f(n_1, n_2, \ldots, n_M; k)$ over all $M$-tuples of positive integers with $\sum_{i=1}^{M} n_i = n$, it follows that for any $n$, $M$ and $k$, there exists a homogeneous $M$-part $k$-Sperner family $\mathcal{F} \subseteq 2^{[n]}$ of size $\phi(n, M; k)$. We present the proof of Theorem 162 by Griggs, Odlydzko, and Shearer, which uses the idea of Lubell's proof of the LYM-inequality. The proof by P.L. Erdős and Katona is based on the *profile polytope* method that is discussed in more details in Chapter 7.

**Proof.** Let $\mathcal{G} \subseteq 2^{X_1} \times 2^{X_2} \times \cdots \times 2^{X_M}$ be an $M$-part $k$-Sperner family of size $f(n_1, n_2, \ldots, n_M; k)$. Let $\mathcal{C}_i$ be a maximal chain in $X_i$, and let us write $\mathcal{R} = \mathcal{C}_1 \times \mathcal{C}_2 \times \cdots \times \mathcal{C}_M$. Clearly, the number of possible $\mathcal{R}$'s is $\prod_{i=1}^{M} n_i!$. For a set $G \in \mathcal{G} \cap \mathcal{R}$, let $\mathcal{G}_G$ be the family of those sets in $2^{X_1} \times 2^{X_2} \times \cdots \times 2^{X_M}$ that are of the same type as $G$, and finally let $\mathcal{G}_{\mathcal{R}} = \cup_{G \in \mathcal{G} \cap \mathcal{R}} \mathcal{G}_G$. By definition, $\mathcal{G}_{\mathcal{R}}$ is $M$-part $k$-Sperner, independently of the choice of $\mathcal{R}$.

Suppose $G \in \mathcal{G}$ with $|G \cap X_i| = g_i$. Then $G$ belongs to $\prod_{i=1}^{M} g_i!(n_i - g_i)!$ choices of $\mathcal{R}$. For any such $\mathcal{R}$, the family $\mathcal{G}_{\mathcal{R}}$ contains all $\prod_{i=1}^{M} \binom{n_i}{g_i}$ sets of the same type as $G$. This means that $G$ is "responsible" for $\prod_{i=1}^{M} n_i!$ many pairs $(G', \mathcal{R})$ with $G' \in \mathcal{G}_{\mathcal{R}}$. We obtained that $\sum_{\mathcal{R}} |\mathcal{G}_{\mathcal{R}}| \geq |\mathcal{G}| \prod_{i=1}^{M} n_i!$. Therefore there must exist an $\mathcal{R}$ such that $|\mathcal{G}_{\mathcal{R}}| \geq |\mathcal{G}|$ holds. ■

Katona [333] showed the following example yielding $\phi(n, 3; 1) > \binom{n}{\lfloor n/2 \rfloor}$: Let $n$ be odd and partition $[n]$ into $X_1 = [n-2]$, $X_2 = \{n-1\}$, $X_3 = \{n\}$. Define the family $\mathcal{F}$ as

$$\left\{A, A \cup \{n-1, n\} : A \in \binom{[n-2]}{\frac{n-3}{2}}\right\} \cup \left\{B \cup \{n-1\}, B \cup \{n\} : B \in \binom{[n-2]}{\frac{n-1}{2}}\right\}.$$

Observe that $|\mathcal{F}| = 4\binom{n-2}{\frac{n-3}{2}} = 2\binom{n-1}{\frac{n-1}{2}} > \binom{n}{\lfloor n/2 \rfloor}$. Furthermore, $\mathcal{F}$ contains sets of two possible sizes, so if $F, G \in \mathcal{F}$ with $F \subsetneq G$, then $F \in \binom{[n-2]}{\frac{n-3}{2}}$ must hold. If $G = F \cup \{n-1, n\}$, then $G \setminus F$ lies in $X_1$ and $X_2$, while if $G = B \cup \{m\}$ with $F \subsetneq B$ and $m = n-1$ or $m = n$, then $G \setminus F$ lies in either both $X_1$ and $X_2$ or both $X_1$ and $X_3$. So $\mathcal{F}$ is indeed 3-part Sperner.

Another, more general and maybe even simpler construction is due to Aydinian, Czabarka, P.L. Erdős, and Székely [25]. Let $M \geq 3$, let $X_1, X_2, \ldots, X_M$ be pairwise disjoint sets of size $m$ and $X$ be their union. So $n = |X| = |X_1 \cup X_2 \cup \cdots \cup X_M| = Mm$. Let us define the family $\mathcal{F}(M, m)$ as

$$\left\{F \in 2^X : |F| \equiv \lfloor \frac{Mm}{2} \rfloor \pmod{m+1}\right\}.$$

Now $\mathcal{F}(M,m)$ contains $\binom{X}{\lfloor |X|/2 \rfloor}$, and, as $M \geq 3$, $\mathcal{F}(M,m)$ contains some additional sets as well, namely those of size $\lfloor \frac{Mm}{2} \rfloor + m + 1$. Furthermore, $\mathcal{F}(M,m)$ is $M$-part Sperner, as if $F \subsetneq F'$ with $F, F' \in \mathcal{F}(M,m)$, then by definition $|F' \setminus F| \geq m+1$, so it intersects at least two parts $X_i, X_j$.

Katona [338] also showed additional conditions that ensure that a 3-part Sperner family can have size at most $\binom{n}{\lfloor n/2 \rfloor}$. Such conditions were also obtained by Griggs [274] and Griggs and Kleitman [278].

**Theorem 163 (Füredi, Griggs, Odlyzko, Shearer, [237])** *For any $n$ and $k$ we have $\gamma(n,2;k) = \phi(n,2;k)$. Moreover,*

*(i) for $k = 1,2$ these quantities equal $2\binom{n}{\lfloor n/2 \rfloor}$,*

*(ii) for $k \geq 3$ these quantities are strictly smaller than $k\binom{n}{\lfloor n/2 \rfloor}$.*

**Theorem 164 (P.L. Erdős, Katona, [173])** *(i) If at least one of $n_1$ and $n_2$ is odd, then we have*

$$f(n_1, n_2, 1; 1) = 2\binom{n_1 + n_2}{\lfloor \frac{n_1+n_2}{2} \rfloor}.$$

*(ii) If both $n_1$ and $n_2$ are even, then we have*

$$f(n_1, n_2, 1; 1) = 2\binom{n_1 + n_2}{\frac{n_1+n_2}{2}} - \left(\binom{n_1}{\frac{n_1}{2}} - \binom{n_1}{\frac{n_1}{2} - 1}\right)\left(\binom{n_2}{\frac{n_2}{2}} - \binom{n_2}{\frac{n_2}{2} - 1}\right).$$

Asymptotic results on 3-part Sperner families were obtained by Füredi [229] and by Griggs, Odlyzko, and Shearer [282]. The current best bounds on $\phi(3;1)$ are due to Mészáros.

**Theorem 165 (Mészáros [416])** *The following holds for 3-part Sperner families:* $1.05 \leq \phi(3;1) \leq 1.0722$.

**Exercise 166** *Show that the following inequalities hold:*

*(i) $f(n_1, n_2, \ldots, n_M; k) \leq 2f(n_1, n_2, \ldots, n_{i-1}, n_i - 1, n_{i+1}, \ldots, n_M; k)$ (if $n_i = 1$ holds, then we just omit the zero from the argument of the right hand side).*

*(ii) $f(n_1, n_2, 1, \ldots, 1; 1) = f(n_1, n_2, 1; 1) \cdot 2^{M-3}$, if the left hand side is about $M$-part Sperner families with $M-2$ parts of size 1.*

*(iii) $f(n_1, n_2, \ldots, n_M; k) \geq f(n_1, n_2, \ldots, n_{M-2}, n_{M-1} + n_M; k)$.*

**Theorem 167 (Aydinian, Czabarka, P.L. Erdős, Székely, [25])** *For $M \geq 3$ and $n = \sum_{i=1}^{M} n_i$ with $n_1 \geq n_2 \geq \cdots \geq n_M$, we have $f(n_1, n_2, \ldots, n_M; 1) = \binom{n}{\lfloor \frac{n}{2} \rfloor}$ if and only if $M = 3$, $n_3 = 1$ and $n$ is even.*

**Proof.** If $M = 3$, $n_3 = 1$ and $n$ is even, then $n_1 + n_2$ is odd, so by Theorem 164 we have $f(n_1, n_2, n_3; 1) = \binom{n}{n/2}$.

For the other direction, let us assume $f(n_1, n_2, \ldots, n_M; 1) = \binom{n}{\lfloor \frac{n}{2} \rfloor}$.

CASE I: $M = 3$.

If $n_1 \leq \lceil n/2 \rceil - 1$, then the family $\mathcal{F} = \{F \in 2^{[n]} : |F| \equiv \lfloor n/2 \rfloor \pmod{n_1 + 1}\}$ is $M$-part Sperner and $|\mathcal{F}| > \binom{n}{\lfloor \frac{n}{2} \rfloor}$, which contradicts $f(n_1, n_2, \ldots, n_M; 1) = \binom{n}{\lfloor \frac{n}{2} \rfloor}$. So we may assume $n_1 \geq \lceil n/2 \rceil$. By Theorem 164 we can assume $n_3 \geq 2$.

Let $X_1$, $X_2$ and $X_3$ be pairwise disjoint sets with $|X_i| = n_i$ for $i = 1, 2, 3$ and $X = X_1 \cup X_2 \cup X_3$. Let

$$\mathcal{A} = \{A \in \binom{X}{\lfloor n/2 \rfloor} : X_2 \cup X_3 \subseteq A\}$$

and

$$\mathcal{B} = \{B \in 2^X : |B \cap X_1| = \lfloor n/2 \rfloor - n_3 + 1,\ X_2 \cup X_3 \subseteq B\},$$

and let us consider the family

$$\mathcal{F} = \left(\binom{X}{\lfloor n/2 \rfloor} \setminus \mathcal{A}\right) \cup \mathcal{B}.$$

Observe that $\mathcal{F}$ is 3-part Sperner. Indeed, as sets in $\mathcal{F}$ have size either $\lfloor n/2 \rfloor$ or $\lfloor n/2 \rfloor + n_2 + 1$, if we have $F \subsetneq B$ for $F, B \in \mathcal{F}$, then $F \in \binom{X}{\lfloor n/2 \rfloor} \setminus \mathcal{A}$, $B \in \mathcal{B}$ must hold. That means $X_2 \cup X_3 \subseteq B$ and $X_2 \cup X_3 \not\subseteq F$, so $B \setminus F$ intersects $X_2 \cup X_3$, but none of $X_2$ and $X_3$ are large enough to contain $B \setminus F$, a set of size $n_2 + 1$.

To finish the proof of this case, we have to show $|\mathcal{B}| > |\mathcal{A}|$. Clearly, we have $|\mathcal{B}| = \binom{n_1}{\lfloor n/2 \rfloor - n_3 + 1}$ and $|\mathcal{A}| = \binom{n_1}{\lfloor n/2 \rfloor - n_2 - n_3}$. First note

$$\lfloor n/2 \rfloor - n_2 - n_3 = \lfloor \frac{n_1 + n_2 + n_3}{2} \rfloor - n_2 - n_3 \leq \frac{n_1 - n_2 - n_3}{2} < \frac{n_1}{2}.$$

So we have two cases according to whether $\lfloor n/2 \rfloor - n_3 + 1$ is at most or at least $n_1/2$. In the former case, we have $\lfloor n/2 \rfloor - n_2 - n_3 < \lfloor n/2 \rfloor - n_3 + 1 \leq n_1/2$, so clearly $|\mathcal{B}| = \binom{n_1}{\lfloor n/2 \rfloor - n_3 + 1} > \binom{n_1}{\lfloor n/2 \rfloor - n_2 - n_3} = |\mathcal{A}|$, as required. In the latter case, we need to show

$$n_1/2 - \lfloor n/2 \rfloor + n_2 + n_3 > \lfloor n/2 \rfloor - n_3 + 1 - n_1/2.$$

By rearranging, this is equivalent to $n_1 + n_2 + 2n_3 - 1 = n + n_3 - 1 > 2\lfloor n/2 \rfloor$. The only way that this inequality may fail is $n_3 = 1$ and $n$ is even. This finishes the proof of Case I.

CASE II: $M \geq 4$.

Part **(iii)** of Exercise 166 yields

$$\binom{n}{\lfloor n/2 \rfloor} = f(n_1, n_2, \ldots, n_M; 1) \geq f(n_1, n_2, \ldots, n_{M-2}, n_{M-1} + n_M; 1) \geq \binom{n}{\lfloor n/2 \rfloor}.$$

We keep repeating merging the two smallest parts of the partition, until we reach $n_1', n_2', n_3'$ with $f(n_1', n_2', n_3'; 1) = \binom{n}{\lfloor n/2 \rfloor}$. By Case I, we obtain $n_3' = 1$. The only way that this 1 could remain there as smallest part is that in the previous step, we must have merged two singleton parts, so $n_2' = 2$, $n_1' = n-3$, and in the penultimate step we had $f(n-3, 1, 1, 1; 1) = \binom{n}{\lfloor n/2 \rfloor}$. Applying part **(ii)** of Exercise 166 and Theorem 164, we obtain

$$f(n-3, 1, 1, 1; 1) = 2f(n-3, 1, 1; 1) = 4\binom{n-2}{\lfloor \frac{n-2}{2} \rfloor} > \binom{n}{\lfloor n/2 \rfloor}.$$

This contradiction finishes the proof of Case II. ∎

The first general upper bound on $\phi(n, M; k)$ was obtained by Griggs [275]. He showed $\phi(n, M; k) \le k2^{M-2}\binom{n}{\lfloor \frac{n}{2} \rfloor}$. This was improved by Sali [492] to $\phi(n, M; k) \le ck\sqrt{M}\binom{n}{\lfloor \frac{n}{2} \rfloor}$ for some unspecified constant $c$. The current best general bound is the following.

**Theorem 168 (Aydinian, Czabarka, P.L. Erdős, Székely [25])** *For every pair $n, M$ of integers with $n, M \ge 3$, we have*

$$\phi(n, M; k) \le \begin{cases} k\sqrt{\frac{M}{2}}e^{\frac{5}{18n}}\binom{n}{\lfloor \frac{n}{2} \rfloor} & \text{if } n \text{ is even,} \\ k\sqrt{\frac{M(1+1/n)}{2}}e^{\frac{5}{18(n+1)}}\binom{n}{\lfloor \frac{n}{2} \rfloor} & \text{if } n \text{ is odd.} \end{cases} \tag{3.1}$$

Let us finish this section by mentioning further generalizations of the more-part $k$-Sperner property. Katona [336] considered a special case and then Füredi, Griggs, Odlyzko, and Shearer [237] defined in full generality the following strengthened version: $\mathcal{F} \subseteq 2^{X_1 \cup X_2 \cup \cdots \cup X_M}$ has property $Y_k$ if $\mathcal{F}$ does not contain a chain $F_1 \subsetneq F_2 \subsetneq \cdots \subsetneq F_{k+1}$, such that for any $i = 1, 2, \ldots, k$ the set $F_{i+1} \subseteq F_i$ is contained in some $X_j$ ($j$ is not necessarily the same for each value of $i$). If we introduce the quantities $h(n_1, n_2, \ldots, n_M; k)$ and $h(n, M; k)$ analogously, then Katona proved $h(n, 2; k)$ is $\Sigma(n, k)$, the sum of the $k$ largest binomial coefficients of order $n$. Füredi et al. showed that the limit $\eta(M, k) = \lim_{n \to \infty} \frac{h(n,M;k)}{\binom{n}{\lfloor n/2 \rfloor}}$ exists for any fixed $k, M$.

In their survey-like paper, Aydinian, Czabarka, P.L. Erdős and Székely [25] considered the following properties: a family $\mathcal{F}$ is $M$-part $(k_1, k_2, \ldots, k_M)$-Sperner if it does not contain any chain $F_1 \subsetneq F_2 \subsetneq \ldots, \subsetneq F_{k_i+1}$ with $F_{k_i+1} \setminus F_1 \subseteq X_i$ for all $1 \le i \le M$. Finally, a family $\mathcal{F}$ is *$l$-fold $M$-part $k$-Sperner* if it does not contain any chain $F_1 \subsetneq F_2 \subsetneq \ldots, \subsetneq F_{k+1}$ with $F_{k+1} \setminus F_1 \subseteq \cup_{i \in L} X_i$ for some $L \in \binom{[M]}{l}$.

**Exercise 169 (Aydinian, Czabarka, P.L. Erdős, Székely, [25])** *For a family $\mathcal{F} \subseteq 2^{X_1 \cup X_2 \cup \cdots \cup X_M}$ and integers $i_1, i_2, \ldots, i_M$ with $0 \le i_j \le |X_j|$, let $P_{i_1, i_2, \ldots, i_M}(\mathcal{F})$ denote the number of sets $F \in \mathcal{F}$ with $|F \cap X_j| = i_j$ for all*

*$j = 1, 2, \ldots, M$. Suppose $\mathcal{F} \subseteq 2^{X_1 \cup X_2 \cup \cdots \cup X_M}$ is an $M$-part $k$-Sperner family. Then prove that for any $j$ we have*

$$\sum_{i_1, i_2, \ldots, i_M} \frac{P_{i_1, i_2, \ldots, i_M}(\mathcal{F})}{\binom{n_1}{i_1}\binom{n_2}{i_2} \cdots \binom{n_M}{i_M}} \leq k \prod_{i=1, i \neq j}^{M} (n_i + 1).$$

## 3.2 Supersaturation

In this section we consider the supersaturation version of Theorem 1. We know that every family $\mathcal{F} \subseteq 2^{[n]}$ with $|\mathcal{F}| > \binom{n}{\lfloor n/2 \rfloor}$ contains at least one 2-chain. The exact number of 2-chains will be denoted by $c(P_2, \mathcal{F})$. We will be interested in the minimum value of $c(P_2, \mathcal{F})$ over all subfamilies of $2^{[n]}$ of fixed size. We will write $c_2(n, s) = \min_{\mathcal{F} \subseteq 2^{[n]}} \{c(P_2, \mathcal{F}) : |\mathcal{F}| = s\}$. As the largest antichain in $2^{[n]}$ is $\binom{[n]}{\lfloor n/2 \rfloor}$, it is natural to conjecture that, to obtain the least number of 2-chains, one has to consider families that consist of sets of size as close to $n/2$, as possible. We say that a family $\mathcal{F} \subseteq 2^{[n]}$ is *centered*, if $F \in \mathcal{F}$ and $|G - n/2| < |F - n/2|$ imply $G \in \mathcal{F}$. Kleitman [366] proved that for any $s > \binom{n}{\lfloor n/2 \rfloor}$, we have $c_2(n, s) = c(P_2, \mathcal{F})$ for some centered family $\mathcal{F} \subseteq 2^{[n]}$ with $|\mathcal{F}| = s$, and conjectured the same for the number of $k$-chains. (Results concerning this are presented in the supersaturation section of Chapter 7.)

Here we present a proof by Das, Gan, and Sudakov [115], that also characterizes all extremal families. In Sperner's original proof of Theorem 1 (the first proof in Chapter 1), Lemma 2 on bipartite graphs was used. We start with a similar lemma, that will be used in proving that an extremal family must be centered.

**Lemma 170** *Let $G$ be a bipartite graph on $U \cup V$ with minimum degree $\delta_U \geq 1$ in $U$ and maximum degree $d_V$ in $V$. Suppose there is no matching covering $U$. Then there exist non-empty subsets $U_1 \subseteq U$ and $V_1 \subseteq V$ with a perfect matching $M$ between $U_1$ and $V_1$, such that $e(U_1, V) + e(U \setminus U_1, V_1) \leq |U_1| d_V$.*

**Proof.** As there is no matching covering $U$, we can apply Hall's theorem (Theorem 70) to obtain a minimal set $U_0 \subseteq U$ with $|N(U_0)| < |U_0|$. Then $\delta_U \geq 1$ implies $|U_0| \geq 2$. We define $U_1 = U_0 \setminus \{u\}$ for any fixed $u \in U_0$. By the minimality of $U_0$, we have $|N(U_1)| = |U_1|$. Letting $V_1 = N(U_1)$, we can apply Hall's theorem to obtain a perfect matching $M$ between $U_1$ and $V_1$. Indeed, by the minimality of $U_0$, we have $|N(X)| \geq |X|$ for any $X \subseteq U_1 \subset U_0$.

As $V_1 = N(U_1)$ implies $e(U_1, V) = e(U_1, V_1)$, we obtain

$$\begin{aligned} e(U_1, V) + e(U \setminus U_1, V_1) &= e(U_1, V_1) + e(U, V_1) - e(U_1, V_1) = e(U, V_1) \\ &\leq d_V |V_1| = |U_1| d_V. \end{aligned}$$

■

Let us remind the reader, that for a uniform family $\mathcal{F}$, its shadow is denoted by $\Delta(\mathcal{F})$, and for a not necessarily uniform family $\Delta_k(\mathcal{F}) = \{G : |G| = k, \exists F \in \mathcal{F} \text{ with } G \subset F\}$.

**Lemma 171** *If $\mathcal{F} \subseteq 2^{[n]}$ is a family of size $s > \binom{n}{\lfloor n/2 \rfloor}$ that minimizes the number of 2-chains over all families of size $s$, and $F \in \mathcal{F}$ is of maximum size $n/2+m$, then for all sets $G \subseteq F$ with $|G| \geq n/2-m+1$, we have $G \in \mathcal{F}$.*

**Proof.** Suppose $\mathcal{F}$ is a counterexample to the statement. Then we must have $m \geq 1$. Let $l \leq 2m-1$ be the smallest integer such that there exists an $(\frac{n}{2}+m)$-set in $\mathcal{F}$ that contains a subset of size $\frac{n}{2}+m-l$ that does not belong to $\mathcal{F}$. Let us define $\mathcal{A} = \{F \in \mathcal{F} : |F| = \frac{n}{2}+m, \Delta_{\frac{n}{2}+m-l}(F) \not\subset \mathcal{F}\}$ and $\mathcal{B} = \Delta_{\frac{n}{2}+m-l}(\mathcal{F}) \setminus \mathcal{F}$. We will consider the auxiliary bipartite graph $G(\mathcal{A},\mathcal{B})$, where $A \in \mathcal{A}, B \in \mathcal{B}$ are joined by an edge if and only if $B \subset A$. We distinguish two cases according to whether $G$ contains a matching that covers $\mathcal{A}$ or not.

CASE I: There exists a matching $M$ in $G$ that covers $\mathcal{A}$.

Let $M(A)$ denote the neighbor of $A$ in $M$, and let us consider the family $\mathcal{F}' = (\mathcal{F} \setminus \mathcal{A}) \cup \{M(A) : A \in \mathcal{A}\}$. We claim that the number $c(P_2,\mathcal{F}')$ of 2-chains in $\mathcal{F}'$ is smaller than the number $c(P_2,\mathcal{F})$ of 2-chains in $\mathcal{F}$. Let $B = M(A) \in \mathcal{F}' \setminus \mathcal{F}$. If $C \subset B$ is a 2-chain in $\mathcal{F}'$, then so is $C \subset A$ in $\mathcal{F}$, so we only have to deal with 2-chains in which even the smaller set has size at least $\frac{n}{2}+m-l$ (we call them *intermediate chains*).

Suppose first that $l > 1$ holds. Then, when changing from $\mathcal{F}$ to $\mathcal{F}'$, we lose all the 2-chains that contain a set from $\mathcal{A}$ and another set of size larger than $\frac{n}{2}+m-l$. By minimality of $l$, their number is $|\mathcal{A}|\sum_{i=1}^{l-1}\binom{\frac{n}{2}+m}{i}$. The smaller set of every 2-chain that we gain is in $\mathcal{B}$. As no $(\frac{n}{2}+m)$-sets $F \in \mathcal{F}'$ contain any set from $\mathcal{B}$, the number of new 2-chains (in $\mathcal{F}' \setminus \mathcal{F}$) is at most $|\mathcal{A}|\sum_{i=1}^{l-1}\binom{\frac{n}{2}-m+l}{i}$. As $l \leq 2m-1$, the number of new 2-chains is strictly smaller than the number of lost ones.

Assume now $l=1$ holds. Then we cannot gain any intermediate 2-chains, therefore there cannot exist $A \in \mathcal{A}$ and $F \in \Delta(A) \cap \mathcal{F}$, as then we would lose that 2-chain. We obtained $\Delta(\mathcal{A}) \cap \mathcal{F} = \emptyset$, or, equivalently, $\mathcal{B} = \Delta(\mathcal{A})$. Suppose there is a 2-chain $C \subset A$ in $\mathcal{F}$ with $A \in \mathcal{A}$. Then, for an arbitrary $x \in C$, consider a matching $M'$ between $\mathcal{A}$ and $\mathcal{B}$ that covers $\mathcal{A}$, where $A$ is connected to $A \setminus \{x\}$. Such a matching exists by Hall's theorem (Theorem 70), since the degree of a set $A'$ in $G[\mathcal{A} \setminus \{A\}, \mathcal{B} \setminus \{A \setminus \{x\}\}]$ is at least $\frac{n}{2}+m-1$, while the degree of a set $B$ is at most $\frac{n}{2}-m+1$. This means that we lose the 2-chain $C \subset A$ and do not gain any when changing from $\mathcal{F}$ to $\mathcal{F}'$.

Finally, suppose $l=1$, $\mathcal{B} = \Delta(\mathcal{A})$ and no $A \in \mathcal{A}$ is involved in a 2-chain in $\mathcal{F}$. The latter implies that no set in $\Delta(\mathcal{A})$ can form a 2-chain with a set in $\mathcal{F} \setminus \mathcal{A}$. As $|\mathcal{F}| > \binom{n}{\lfloor n/2 \rfloor}$, by Theorem 1, there exists a 2-chain $C \subset D$ in $\mathcal{F}$. Let $B$ be a set in $\mathcal{B} \setminus \{M(A) : A \in \mathcal{A}\}$ and $\mathcal{F}'' = (\mathcal{F} \setminus (\mathcal{A} \cup \{D\})) \cup \{M(A) : A \in \mathcal{A}\} \cup \{B\}$. Then when changing $\mathcal{F}$ to $\mathcal{F}''$, we lose the 2-chain $C \subset D$ and do not gain any, so $c(P_2,\mathcal{F}'') < c(P_2,\mathcal{F})$.

CASE II: There is no matching in $G$ that covers $\mathcal{A}$.

Observe that our auxiliary graph $G$ satisfies the conditions of Lemma 170 with $U = \mathcal{A}$, $V = \mathcal{B}$, $d_V = \binom{\frac{n}{2}-m+l}{l}$. So there is a non-empty subfamily $\mathcal{A}'$ of $\mathcal{A}$ and a matching $M$ covering $\mathcal{A}'$. We again consider $\mathcal{F}' = (\mathcal{F} \setminus \mathcal{A}') \cup \mathcal{B}'$ and observe that again we only have to deal with intermediate 2-chains. Using the minimality of $l$, the same calculation as in CASE I shows that the number of 2-chains $F \subset A'$ lost with $|F'| > \frac{n}{2} + m - l$ is larger than the number of new 2-chains $B' \subset F$ with $|F| < \frac{n}{2} + m$.

So it remains to consider 2-chains from levels $\frac{n}{2} + m$ and $\frac{n}{2} + m - l$. The number of 2-chains gained here is $e(\mathcal{A} \setminus \mathcal{A}', \mathcal{B}')$, while we lose all 2-chains between $\mathcal{A}'$ and $\Delta_{\frac{n}{2}+m-l}(\mathcal{A}') \setminus \mathcal{B}$. Their number is $|\mathcal{A}'|\binom{\frac{n}{2}+m}{l} - e(\mathcal{A}', \mathcal{B})$. We have

$$|\mathcal{A}'|\binom{\frac{n}{2}+m}{l} - e(\mathcal{A}', \mathcal{B}) > |\mathcal{A}'|\binom{\frac{n}{2}+m-l}{l} - e(\mathcal{A}', \mathcal{B}) \geq e(\mathcal{A} \setminus \mathcal{A}', \mathcal{B}'),$$

where the second inequality follows from Lemma 170. This implies $c(P_2, \mathcal{F}') < c(P_2, \mathcal{F})$. ■

Now we are ready to prove the result of Kleitman, together with the characterization of the extremal families obtained by Das, Gan, and Sudakov.

**Theorem 172 (Kleitman [366], Das, Gan, Sudakov [115])** *Let $\mathcal{F} \subseteq 2^{[n]}$ be a family of size $s \geq \binom{n}{\lfloor n/2 \rfloor}$ and let $r \in \frac{1}{2}\mathbb{N}$ be the unique half-integer such that $\sum_{i=n/2-r+1}^{n/2+r-1} \binom{n}{i} < s \leq \sum_{i=n/2-r}^{n/2+r} \binom{n}{i}$. Then $c_2(n, s) = c(P_2, \mathcal{F})$ if and only if the following conditions are satisfied:*

***(i)*** *For every $F \in \mathcal{F}$ we have $|n/2 - |F|| \leq r$.*

***(ii)*** *$|n/2 - |G|| < r$ implies $G \in \mathcal{F}$.*

***(iii)*** *If $s \leq \sum_{i=n/2-r}^{n/2+r-1} \binom{n}{i}$, then $\{F \in \mathcal{F} : |F| = n/2 \pm r\}$ is an antichain.*

***(iv)*** *If $s \geq \sum_{i=n/2-r}^{n/2+r-1} \binom{n}{i}$, then $\{F \in 2^{[n]} \setminus \mathcal{F} : |F| = n/2 \pm r\}$ is an antichain.*

Note that condition **(i)** and **(ii)** are equivalent to $\mathcal{F}$ being centered. Also, families satisfying **(i)-(iii)** or **(i),(ii),(iv)** contain the same number of 2-chains, therefore if one of them is optimal, then all of them are. This implies that it is enough to show that an optimal family satisfies the conditions.

**Proof.** We proceed by induction on $s$, and the base case $s = \binom{n}{\lfloor n/2 \rfloor}$ is covered by the uniqueness part of Theorem 1, as for the families $\binom{[n]}{\lfloor n/2 \rfloor}$ and $\binom{[n]}{\lceil n/2 \rceil}$, properties **(i)-(iv)** are satisfied with $r = 0$ if $n$ is even, and $r = \frac{1}{2}$ if $n$ is odd.

Suppose the statement is true for $s-1$ and let $\mathcal{F}$ be a family with $c_2(n, s) = c(P_2, \mathcal{F})$ and $|\mathcal{F}| = s$. First we prove that **(i)** is satisfied. If not, then we can assume, that there exists a set $F \in \mathcal{F}$ with $|F| = n/2 + t$ and $t > r$ (if necessary, we can change $\mathcal{F}$ to $\overline{\mathcal{F}} = \{[n] \setminus F : F \in \mathcal{F}\}$). Then Lemma 171 implies that $F$ is contained in at least $\sum_{i=1}^{2t-1}\binom{n/2+t}{i} > \sum_{i=1}^{2r}\binom{n/2+r}{i}$ 2-chains, therefore

$$c(P_2,\mathcal{F}) \geq c_2(n,s-1) + \sum_{i=1}^{2t-1}\binom{n/2+t}{i} > c_2(n,s-1) + \sum_{i=1}^{2r}\binom{n/2+r}{i},$$

as $c(P_2,\mathcal{F}\setminus\{F\}) \geq c_2(n,s-1)$. Now let $\mathcal{F}^*$ be a centered family of size $s-1$ with $c_2(n,s-1) = c(P_2,\mathcal{F}^*)$, given by the induction hypothesis. Then there exists $F^* \notin \mathcal{F}^*$ with $|n/2-|F^*|| \leq r$. Let $\mathcal{F}' = \mathcal{F}^* \cup \{F^*\}$. Then, by induction, all set sizes in $\mathcal{F}'$ are between $n/2-r$ and $n/2+r$, therefore the number of 2-chains in $\mathcal{F}'$ containing $F^*$ is at most $\sum_{i=1}^{2r}\binom{n/2+r}{i}$. This implies $c(P_2,\mathcal{F}') \leq c_2(n,s-1) + \sum_{i=1}^{2r}\binom{n/2+r}{i}$, which contradicts the optimality of $\mathcal{F}$.

From now on we assume **(i)** holds. We consider first the case $\sum_{i=n/2-r-1}^{n/2+r-1}\binom{n}{i} < s \leq \sum_{i=n/2-r}^{n/2+r-1}\binom{n}{i}$, and we prove **(ii)** and **(iii)**. The inequality $\sum_{i=n/2-r-1}^{n/2+r-1}\binom{n}{i} < s$ and **(i)** assures that there exists $F \in \mathcal{F}$ with either $|F| = n/2+r$ or $|F| = n/2-r$. By changing from $\mathcal{F}$ to $\overline{\mathcal{F}}$ if necessary, we can assume $|F| = n/2+r$. Then by Lemma 171 $F$ is contained in at least $\sum_{i=1}^{2r-1}\binom{n/2+r}{i}$ 2-chains in $\mathcal{F}$ with the other set of the chain having size between $n/2-r+1$ and $n/2+r-1$, so $c(P_2,\mathcal{F}) \geq c_2(n,s-1) + \sum_{i=1}^{2r-1}\binom{n/2+r}{i}$. Observe that, by induction, for a family $\mathcal{F}$ satisfying **(i)**-**(iii)** we have $c(P_2,\mathcal{F}') = c_2(n,s-1) + \sum_{i=1}^{2r-1}\binom{n/2+r}{i}$, therefore $F$ cannot contain any set $G \in \mathcal{F}$ with $|G| = n/2-r$, and $\mathcal{F}\setminus\{F\}$ must be optimal. These imply **(ii)** and **(iii)**.

Suppose now that $\sum_{i=n/2-r-1}^{n/2+r}\binom{n}{i} \leq s \leq \sum_{i=n/2-r}^{n/2+r}\binom{n}{i}$ holds and we prove **(ii)** and **(iv)**. By **(i)**, we know that $\mathcal{F} \subseteq \mathcal{H} := \cup_{i=n/2-r}^{n/2+r}\binom{[n]}{i}$. Let $\mathcal{G} = \mathcal{H}\setminus\mathcal{F}$ and $c(P_2,(\mathcal{G},\mathcal{H}))$ denote the number of 2-chains that consist of pairs $G,H$ with $G \in \mathcal{G}$, $H \in \mathcal{H}$. Then we have $c(P_2,\mathcal{F}) = c(P_2,\mathcal{H}) - c(P_2,(\mathcal{G},\mathcal{H})) + c(P_2,\mathcal{G})$. Observe that $c(P_2,\mathcal{H})$ is independent of $\mathcal{F}$, while $c(\mathcal{G})$ is minimized when $\mathcal{G}$ is an antichain and thus $c(\mathcal{G}) = 0$. We claim that the quantity $c(P_2,(\mathcal{G},\mathcal{H}))$ is maximized if all sets in $\mathcal{G}$ have size $n/2\pm r$. Indeed, for any $G \in \mathcal{H}$, the number of sets in $\mathcal{H}$ that form a 2-chain with $G$ is $\sum_{i=n/2-r}^{|G|-1}\binom{|G|}{i} + \sum_{i=|G|+1}^{n/2+r}\binom{n-|G|}{i-|G|}$, and this is clearly maximized when $|G| = n/2 \pm r$. As constructions that satisfy **(i)**,**(ii)**,**(iv)** maximize $c(P_2,(\mathcal{G},\mathcal{H}))$ and minimize $c(P_2,\mathcal{G})$, all optimal constructions should possess **(ii)** and **(iv)**. ■

## 3.3 The number of antichains in $2^{[n]}$ (Dedekind's problem)

The problem of finding the number $\phi(n)$ of antichains in $2^{[n]}$ originates from a paper from 1897 by Dedekind [128]. Of course, he posed the problem in a different context. As the collection of downsets of $2^{[n]}$ is in one-to-one correspondence with the set of all antichains through the mapping $\mathcal{D} \mapsto M(\mathcal{D})$ where $M(\mathcal{D})$ is the family of all maximal sets in $\mathcal{D}$, we obtain that $\phi(n)$ equals the number of downsets in $2^{[n]}$. Taking complements or repeating the above argument with maximal sets replaced by minimal sets, $\phi(n)$ is the number of upsets in $2^{[n]}$ and therefore it also equals the number of monotone Boolean functions. Many other reformulations can be given, so determining the asymptotics of $\phi(n)$ was a central problem in many areas of mathematics.

As any subfamily of an antichain $\mathcal{A}$ is also an antichain, we obtain the lower bound $2^{|\mathcal{A}|} \leq \phi(n)$ for any antichain $\mathcal{A} \subseteq 2^{[n]}$. By Theorem 1, we know that the best possible lower bound obtained this way is $2^{\binom{n}{\lfloor n/2 \rfloor}}$.

Dedekind's results were improved by Church [105], Ward [544] and Korobkov [376].

**Theorem 173 (Hansel [302])** *For any integer $n$ we have $\phi(n) \leq 3^{\binom{n}{\lfloor n/2 \rfloor}}$.*

**Proof.** The proof is based on the following lemma about symmetric chain decompositions of $2^{[n]}$.

**Lemma 174** *Let $\mathbf{C}_n = \{\mathcal{C}_1, \mathcal{C}_2, \ldots, \mathcal{C}_{\binom{n}{\lfloor n/2 \rfloor}}\}$ be the symmetric chain partition of $2^{[n]}$ constructed inductively in Lemma 158, and let $C \subsetneq C' \subsetneq C''$ be consecutive sets of some $\mathcal{C}_i$, i.e. $|C| + 2 = |C''|$. Then there exists a chain $\mathcal{C}_j \in \mathbf{C}_n$ with $|\mathcal{C}_j| = |\mathcal{C}_i| - 2$ and a set $D \in \mathcal{C}_j$ such that $C \subsetneq D \subsetneq C''$.*

**Proof of Lemma.** We proceed by induction on $n$, with the base case $n = 1$ being trivial, as $\mathbf{C}_1$ does not contain chains of length more than two. Let us consider the inductive step and remind the reader that in Lemma 158 a chain $C_1 \subsetneq C_2 \subsetneq \cdots \subsetneq C_k$ of $\mathbf{C}_n$ can give birth to two chains in $\mathbf{C}_{n+1}$: $C_1 \subsetneq C_2 \subsetneq \cdots \subsetneq C_k \subsetneq C_k \cup \{n+1\}$ (type 1) and $C_1 \cup \{n+1\} \subsetneq C_2 \cup \{n+1\} \subsetneq \cdots \subsetneq C_{k-1} \cup \{n+1\}$ (type 2). We consider two cases according to the type of the chain that $C, C'$, and $C''$ belong to.

Case I: $C, C', C''$ belong to a chain $\mathcal{C}$ of type 1.

If $n+1 \notin C''$, then we can apply the inductive hypothesis to $\mathcal{C}' = \{C_1, C_2, \ldots, C_k\}$, to obtain a set $D \in \mathcal{C}^* \in \mathbf{C}_n$ with $C \subsetneq D \subsetneq C''$ and $|\mathcal{C}^*| = |\mathcal{C}'| - 2$. Then the set $D$ and the chain $\mathcal{C}^*_+$ of type 2 obtained from $\mathcal{C}^*$ satisfy the statement of the Lemma. If $n+1 \in C''$, then $C = C_{k-1}$ and we can take $D$ to be $C_{k-1} \cup \{n+1\}$ and the chain $\mathcal{C}'$ of type 2 obtained from $\mathcal{C} \setminus \{C''\}$ satisfies the statement of the lemma.

CASE II: $C, C', C''$ belong to a chain $\mathcal{C}$ of type 2.

Then $C \setminus \{n+1\}, C' \setminus \{n+1\}, C'' \setminus \{n+1\}$ belong to a chain $\mathcal{C}' \in \mathbf{C}_n$. Applying induction we find a set $D' \in \mathcal{C}'' \in \mathbf{C}_n$ with $C \setminus \{n+1\} \subsetneq D' \subsetneq C'' \setminus \{n+1\}$ and $|\mathcal{C}''| = |\mathcal{C}'| - 2$. As $\mathcal{C}$ is of type 2, $C'' \setminus \{n+1\}$ is not the maximal element of $\mathcal{C}'$. Therefore, $D'$ cannot be a maximal element of $\mathcal{C}''$ as otherwise we would have $|\mathcal{C}''| < |\mathcal{C}'| - 2$. Hence the set $D = D' \cup \{n+1\}$ and the chain of type 2 obtained from $\mathcal{C}''$ satisfy the statement of the lemma. □

To prove the theorem, we consider $\phi(n)$ as the number of downsets in $2^{[n]}$. We enumerate the chains of $\mathbf{C}_n$ according to their length, in non-decreasing order. We generate a downset $\mathcal{D}$, in such a way that for each $\mathcal{C}_i \in \mathbf{C}_n$, one after the other, we decide which sets of $\mathcal{C}_i$ to include in $\mathcal{D}$. If $C \in \mathcal{C}_i$ is contained in some set $C'$ that was decided to be placed in $\mathcal{D}$ before, then $C$ must be included in $\mathcal{D}$. Similarly, if $C \in \mathcal{C}_i$ contains some set that was decided not to be included in $\mathcal{D}$, then $C$ cannot be added to $\mathcal{D}$. So the sets, that we have to make a decision about, are consecutive sets in $\mathcal{C}_i$. If there were more than two of them, then, by Lemma 174, there would be a set $D$ that belongs to $\mathcal{C}_j$ with $j < i$ and $C^- \subsetneq D \subsetneq C^+$, where $C^-$ denotes the smallest and $C^+$ denotes the largest undecided set of $\mathcal{C}_i$. As $j < i$, we have already decided about $D$, which contradicts the fact that both $C^-$ and $C^+$ are undecided. So for each chain we have to decide about at most two sets, and we cannot include exactly the smaller in $\mathcal{D}$. Therefore we have at most three choices for each chain, which gives the statement of the theorem. ■

Kleitman [363] was the first to determine the asymptotics of $\log \phi(n)$, by showing $\phi(n) \leq 2^{(1+O(\frac{\log n}{\sqrt{n}}))\binom{n}{\lfloor n/2 \rfloor}}$. Kleitman and Markowsky [364] improved the error term by establishing $\phi(n) \leq 2^{(1+O(\frac{\log n}{n}))\binom{n}{\lfloor n/2 \rfloor}}$. This was reproved by Kahn [326] with an explicit constant. He showed $\phi(n) \leq 2^{(1+2\frac{\log_2(n+1)}{n})\binom{n}{\lfloor n/2 \rfloor}}$. The approach of his proof is to bound the number of antichains using entropy.

The asymptotics of $\phi(n)$ was determined by Korshunov [377] in a long and very complicated paper. A simpler but still involved proof was given a couple of years later by Sapozhenko [495]. His proof worked in a more general setting.

**Theorem 175 (Korshunov, [377])** *As $n$ tends to infinity, we have*

$$\frac{\phi(n)}{2^{\binom{n}{\lfloor n/2 \rfloor}}} \sim \begin{cases} e^{\binom{n}{\lfloor n/2-1 \rfloor}(\frac{1}{2^{n/2}} + \frac{n^2}{2^{n+5}} - \frac{n}{2^{n+4}})} & n \text{ is even} \\ 2e^{\binom{n}{\lfloor n/2-1 \rfloor}(\frac{1}{2^{(n+3)n/2}} + \frac{n^2}{2^{n+6}} - \frac{n}{2^{n+3}}) + \binom{n}{\lfloor n/2 \rfloor}(\frac{1}{2^{(n+1)/2}} + \frac{n^2}{2^{n+4}})} & n \text{ is odd.} \end{cases}$$

Next we present a recent but weaker result of Noel, Scott and Sudakov [446]. Its error term is between those of the results of Kleitman and Kleitman and Markowsky. It uses the so-called *container method*, that bounds the number of independent sets in graphs and hypergraphs.

The aim of this method in general is to find a *relatively small* family $\mathcal{C}$ of *not too big* subsets of a (hyper)graph $G$, such that *every* independent set of

$G$ is contained in a set $C \in \mathcal{C}$. If this is done, and we know that the size of the largest independent set is at most $M$, then one can bound the total number of independent sets by $\sum_{C \in \mathcal{C}} \sum_{0 \le i \le M} \binom{|C|}{M}$.

The *comparability graph* $G$ of a poset $P$ has the vertex set of $P$ and elements $p$ and $p'$ are connected by an edge if and only if $p < p'$ or $p' < p$. In this particular case we get an upper bound on the number of independent sets in the comparability graph of $Q_n$, i.e. the antichains, by counting all the subfamilies of the containers. We give further applications of the container method in Chapter 4.

**Lemma 176** *For $k \ge 1$ let $d_1 > \cdots > d_k$ and $m_0 > \cdots > m_k$ be positive integers and let $P$ be a poset such that $|P| = m_0$ and, for $1 \le j \le k$, every subset $S$ of $P$ of cardinality greater than $m_j$ contains at least $|S|d_j$ comparable pairs. Then there exist functions $f_1, \ldots, f_k$ where*

$$f_j : \binom{P}{\le \sum_{r=1}^{j}(m_{r-1}/(2d_r+1))} \to \binom{P}{\le m_j},$$

*such that for every antichain $I \subseteq P$ there exist pairwise disjoint subsets $T_1, T_2, \ldots, T_k$ of $I$ with*

*(i) $|T_j| \le m_{j-1}/(2d_j+1)$ for $1 \le j \le k$,*

*(ii) $T_j \subseteq f_{j-1}(\cup_{r=1}^{j-1} T_r)$ for $2 \le j \le k-1$,*

*(iii) $\cup_{r=1}^{j} T_r \cap f_j(\cup_{r=1}^{j} T_r) = \emptyset$ for $1 \le j \le k$,*

*(iv) $I \subseteq \cup_{r=1}^{k} T_r \cup f_k(\cup_{r=1}^{k} T_r)$.*

**Proof.** The proof is based on an algorithm first introduced by Kleitman and Winston [368]. We fix an ordering $x_1, x_2, \ldots, x_{|P|}$ of the elements of $P$, and then working in the comparability graph $G$ of $P$, at each step we delete a vertex and its entire neighborhood from the ordering. The definition of the sets $T_1, T_2, \ldots, T_k$ and functions $f_1, f_2, \ldots, f_k$ is done during the algorithm. We will build the sets $T_i$ starting with the empty set and adding elements one by one. The input of the algorithm is an antichain $I \subseteq P$ (an independent set in the comparability graph) and the ordering, so once we are done with the description of the algorithm, apart from verifying **(i)**-**(iv)**, we have to check that the $f_i$'s are independent of the choice of $I$, and $f_j(\cup_{r=1}^{j} T_r)$ depends only on $\cup_{r=1}^{j} T_r$ and not on the sets $T_1, \ldots, T_j$.

For a subset $S$ of $P$ and $x \in S$, let $N_S(x)$ denote the neighborhood of $x$ in $G[S]$, i.e. the set of all elements of $S$ that are comparable to $x$, and let $d_S(x)$ denote the degree of $x$ in $G[S]$, i.e. $d_S(x) = |N_S(x)|$. To start the algorithm, let us set $P_0 = P$ and let $u_0$ be the element that, among those with largest $d_{P_0}$-value, comes first in the ordering fixed at the beginning of the proof. Suppose for some $i \ge 1$ we are given an element $u_{i-1} \in P_{i-1}$. Then in the $i$th step either the algorithm terminates or a subset $P_i \subsetneq P_{i-1}$ and an element $u_i \in P_i$ is defined to be used in the next step.

We distinguish two cases according to the value of $d_{P_{i-1}}(u_{i-1})$.

CASE I: $d_{P_{i-1}}(u_{i-1}) \geq 2d_k$.

First set $j$ to be the smallest integer with $d_{P_{i-1}}(u_{i-1}) \geq 2d_j$. If $u_{i-1} \notin I$, then let $P_i = P_{i-1} \setminus \{u_{i-1}\}$. If $u_{i-1} \in I$, then let $P_i = P_{i-1} \setminus (\{u_{i-1}\} \cup N_{P_{i-1}}(u_{i-1}))$ and we add $u_{i-1}$ to $T_j$. In both cases we have $P_i \subsetneq P_{i-1}$. If $P_i \neq \emptyset$, then we let $u_i \in P_i$ be the element that, among those with largest $d_{P_i}$-value, comes first in the ordering. Note that $d_{P_i}(u_i) \leq d_{P_{i-1}}(u_{i-1})$. If $d_{P_i}(u_i) \geq 2d_j$, then we proceed to the next step.

Otherwise if $P_i = \emptyset$, then we set $l = k - j$, while if $d_{P_i}(u_i) < 2d_j$, then we let $l \leq k - j$ be the largest integer with $d_{P_i}(u_i) < 2d_{l+j}$. Then we finalize $T_j$; no more elements will be added to it. At this point $T_{j+1}, T_{j+2}, \dots, T_{j+l}$ are all empty. We define $f_j(\cup_{r=1}^{j} T_r) = f_{j+1}(\cup_{r=1}^{j+1} T_r) = \cdots = f_{j+l}(\cup_{r=1}^{j+l} T_r) = P_i$. Note that $f_j$ cannot change anymore, as $T_r$ is fixed for every $r \leq j$. The functions $f_r$ with $r > j$ are defined because the algorithm might end here; if it does not end, their definition might change later. If $P_i = \emptyset$, then the algorithm terminates, otherwise we proceed to the next step.

CASE II: $d_{P_{i-1}}(u_{i-1}) < 2d_k$.

In this case we terminate the algorithm without any further move.

Note that as $P_i \subsetneq P_{i-1}$ holds for all $i$, and the $u_i$'s are always chosen to have the largest $d_{P_i}$-value, we have that the sequence $d_{P_i}(u_i)$ is monotone decreasing. Because of this and $d_1 > d_2 > \cdots > d_k$, we see that for any $j$, there is a unique step when the definition of $T_j$ and $f_j(\cup_{r=1}^{j} T_r)$ terminates. Furthermore, if $j < j'$, then either $T_{j'} = \emptyset$, or $T_j$ and $f_j(\cup_{r=1}^{j} T_r)$ are defined before $T_{j'}$ and $f_{j'}(\cup_{r=1}^{j'} T_r)$. Then **(iv)** holds by construction, as every element of $I$ was added to some $T_i$. Also **(iii)** and **(ii)** holds, as $f_j(\cup_{r=1}^{j} T_r)$ is defined to consist of all the elements of $P$ that are not in $\cup_{r=1}^{j} T_r$.

By definition, every element of $f_j(\cup_{r=1}^{j} T_r)$ is comparable to less than $2d_j$ many other elements of $f_j(\cup_{r=1}^{j} T_r)$, which implies $c(P_2, f_j(\cup_{r=1}^{j} T_r)) < |f_j(\cup_{r=1}^{j} T_r)|d_j$. By the condition of the Lemma, this yields $|f_j(\cup_{r=1}^{j} T_r)| < m_j$, so the image of $f_j$ is indeed contained in $\binom{P}{\leq m_j}$. For each element that we added to $T_j$ for $j \geq 2$, we deleted at least $2d_j + 1$ elements of $f_{j-1}(\cup_{r=1}^{j-1} T_r)$ and so $|T_j| \leq m_{j-1}/(2d_j + 1)$. Similarly, $|T_j| \leq |P|/(2d_1 + 1) \leq m_0/(2d_1 + 1)$, thus **(i)** holds.

Finally, we need to prove that the functions $f_1, f_2, \dots, f_k$ are well-defined, i.e. $f_j(\cup_{r=1}^{j} T_j)$ does not depend on $I$ and the $T_r$'s, only on $\cup_{r=1}^{j} T_r$. So suppose that for two antichains $I$ and $I'$ the algorithm creates the same $\cup_{r=1}^{j} T_r$. By choosing the smallest possible $j$, we can assume $T_j$ is not empty. Let $u_0, u_1, \dots,$ $P_0, P_1, \dots$ and $u'_0, u'_1, \dots, P'_0, P'_1, \dots$ be the elements and subsets defined during the algorithm when run on $I$ and $I'$, respectively, and let $i$ denote the smallest integer $i$ with $d_{P_i}(u_i) < 2d_j$ (there is such an $i$, when the definition of $T_j$ is finalized). We claim that $u_t = u'_t$ and $P_t = P'_t$ for all $0 \leq t \leq i$. Suppose not and let $t$ be the smallest integer for which this fails. As $u_0$ and $P_0 = P$ does not depend on the input, we have $t \geq 1$. If $P_t = P'_t$, then the next picked element does not depend on $I$ and $I'$, thus we have $u_t = u'_t$. Therefore, we must have $P_t \neq P'_t$ and thus $u'_{t-1} \in I$ and $u_{t-1} \notin I'$ or vice versa. By the definition of $i$ we have $d_{P_{t-1}}(u_{t-1}) \geq 2d_j$, so $u_{t-1} \in I$ implies $u_{t-1} \in \cup_{r=1}^{j} T_r$, while $u'_{t-1} \notin I'$ implies that $u'_{t-1}$ will be deleted when running the algorithm with $I'$ and thus $u'_{t-1}$ will not belong to $\cup_{r=1}^{j} T_r$. This contradicts $u_{t-1} = u'_{t-1}$. This means that $\cup_{r=1}^{j} T_r$ is the same when we run the algorithm for different $I$ and $I'$, thus the elements and subsets that are involved defining $\cup_{r=1}^{j} T_r$ are the same, so $f_j(\cup_{r=1}^{j} T_r)$ will be defined the same. ■

**Corollary 177** *For $k \geq 1$, let $d_1 > \cdots > d_k$ and $m_0 > \cdots > m_k$ be positive integers, and let $P$ be a poset such that $|P| = m_0$, and, for $1 \leq j \leq k$, every subset $S$ of $P$ of cardinality greater than $m_j$ contains at least $|S|d_j$ comparable pairs. Then there is a family $\mathcal{F}$ of subsets of $P$ such that*

***(i)*** $|\mathcal{F}| \leq \prod_{r=1}^{k} \binom{m_{r-1}}{\leq m_{r-1}/(2d_r+1)}$,

***(ii)*** $|A| \leq m_k + \sum_{r=1}^{k} m_{r-1}/(2d_r + 1)$ *for every* $A \in \mathcal{F}$,

***(iii)*** *for every antichain $I$ of $P$, there exists $A \in \mathcal{F}$ with $I \subseteq A$.*

**Proof.** Apply Lemma 176 to obtain the functions $f_1, f_2, \ldots, f_k$. For every antichain $I \subseteq P$ there exist pairwise disjoint sets $T_{1,I}, T_{2,I}, \ldots, T_{k,I}$ assured by Lemma 176 with $I \subseteq \cup_{i=1}^{k} T_{i,I} \cup f(\cup_{i=1}^{k} T_{i,I})$. Let

$$\mathcal{F} := \{\cup_{i=1}^{k} T_{i,I} \cup f(\cup_{i=1}^{k} T_{i,I}) : I \subseteq P \text{ is an antichain}\},$$

then **(ii)** and **(iii)** are satisfied by definition. To see **(i)**, observe first that $T_{1,I}$ can be chosen in at most $\binom{|P|}{\leq |P|/(2d_1+1)}$ ways. Furthermore, if $T_{1,I}, \ldots, T_{j-1,I}$ have already been fixed, then, as $T_{j,I} \subseteq f(\cup_{i=1}^{j-1} T_{i,I})$, there are at most $\binom{m_{j-1}}{\leq m_{j-1}/(2d_j+1)}$ ways to pick $T_{j,I}$. This proves **(i)**. ■

**Theorem 178 (Noel, Scott, Sudakov, [446])** *For the number $\phi(n)$ of antichains in $2^{[n]}$, we have*

$$\phi(n) \leq 2^{\binom{n}{\lfloor n/2 \rfloor}\left(1+O\left(\sqrt{\frac{\log n}{n}}\right)\right)}.$$

**Proof.** We would like to apply Corollary 177 to obtain a a family $\mathcal{F} \subseteq 2^{[n]}$ of containers in the poset $2^{[n]}$ ordered by inclusion. Once this is done, we will bound $\phi(n)$ by the number of subsets of all containers. We set $d_1 = n\sqrt{\log_2 n}$, $d_2 = \sqrt{n \log_2 n}$, $m_0 = 2^n$, $m_1 = \frac{2\binom{n}{\lfloor n/2 \rfloor}}{1-8d_1/n^2}$ and $m_2 = \frac{\binom{n}{\lfloor n/2 \rfloor}}{1-2d_2/n}$. We have to verify that the conditions of Corollary 177 are satisfied. According to Lemma 171, the number $c(P_2, \mathcal{G})$ of comparable pairs over all families of fixed size is attained for some centered families. Therefore, if $|\mathcal{G}| \geq \frac{2\binom{n}{\lfloor n/2 \rfloor}}{1-8d_1/n^2}$, then $\mathcal{G}$ contains at least $(|\mathcal{G}| - 2\binom{n}{\lfloor n/2 \rfloor})\binom{n/2+1}{2}$ comparable pairs. So we have

$$c(P_2, \mathcal{G}) \geq \left(|\mathcal{G}| - 2\binom{n}{\lfloor n/2 \rfloor}\right) \frac{n^2}{8}$$

$$= (|\mathcal{G}| - m_1)\frac{n^2}{8} + \left(m_1 - 2\binom{n}{\lfloor n/2 \rfloor}\right)\frac{n^2}{8} \geq (|\mathcal{G}| - m_1)d_1 + m_1 d_1.$$

Similarly, Lemma 171 implies that if $|\mathcal{G}| \geq m_2$, then $c(P_2, \mathcal{G}) \geq (|\mathcal{G}| - \binom{n}{\lfloor n/2 \rfloor})\frac{n}{2}$. A computation as above yields $c(P_2, \mathcal{G}) \geq |\mathcal{G}|d_2$, so the conditions of Corollary 177 with $k = 2$ and $m_0, m_1, m_2, d_1, d_2$ are indeed satisfied. We thus obtain a family $\mathcal{F}$ of containers of size at most $\binom{m_0}{\leq m_0/d_1}\binom{m_1}{\leq m_1/d_2}$, such that for each $F \in \mathcal{F}$, we have $|F| \leq m_2 + \frac{m_1}{d_2} + \frac{m_0}{d_1}$. As every antichain is contained in at least one $F \in \mathcal{F}$, we obtain that $\phi(n)$ is at most

$$\begin{aligned}
\binom{2^n}{\leq 2^n/d_1}\binom{m_1}{\leq m_1/d_2} 2^{m_2 + \frac{m_1}{d_2} + \frac{m_0}{d_1}} \leq & 2^{2n}\binom{2^n}{2^n/d_1}\binom{m_1}{m_1/d_2} 2^{m_2 + \frac{m_1}{d_2} + \frac{m_0}{d_1}} \\
\leq & 2^{2n}(ed_1)^{2^n/d_1}(ed_2)^{m_1/d_2} 2^{m_2 + \frac{m_1}{d_2} + \frac{m_0}{d_1}} \\
= & 2^{2n + \frac{2^n \log_2(ed_1)}{d_1} + \frac{m_1 \log_2(ed_2)}{d_2} + m_2 + \frac{m_1}{d_2} + \frac{m_0}{d_1}} \\
= & 2^{m_2 + O(\frac{2^n \log_2 n}{d_1}) + O(\frac{m_1 \log_2 n}{d_2})},
\end{aligned}$$

where we used $\binom{n}{\leq a} \leq n\binom{n}{a}$ if $a \leq n/2$ and $\binom{n}{k} \leq (\frac{en}{k})^k$. Finally, using $\binom{n}{\lfloor n/2 \rfloor} = \Theta(\frac{2^n}{\sqrt{n}})$, we obtain that the exponent satisfies

$$\begin{aligned}
& m_2 + O\left(\frac{2^n \log_2 n}{d_1} + \frac{m_1 \log_2 n}{d_2}\right) \\
= & \frac{\binom{n}{\lfloor n/2 \rfloor}}{1 - 2d_2/n} + O\left(\frac{2^n \sqrt{\log_2 n}}{n} + \binom{n}{\lfloor n/2 \rfloor}\sqrt{\log_2 n/n}\right) \\
\leq & \binom{n}{\lfloor n/2 \rfloor} + \frac{4d_2}{n}\binom{n}{\lfloor n/2 \rfloor} + O\left(\binom{n}{\lfloor n/2 \rfloor}\sqrt{\log_2 n/n}\right) \\
= & \binom{n}{\lfloor n/2 \rfloor} + O\left(\binom{n}{\lfloor n/2 \rfloor}\sqrt{\log_2 n/n}\right).
\end{aligned}$$

■

## 3.4 Union-free families and related problems

The starting point of this section is the notion of *union-free families*: $\mathcal{F}$ possesses this property if for any $F, F', G, G' \in \mathcal{F}$, the equality $F \cup F' = G \cup G'$ implies $\{F, F'\} = \{G, G'\}$. Let $f(n)$ be the maximum size of a union-free family $\mathcal{F} \subseteq 2^{[n]}$. The problem of determining $f(n)$ was posed by Erdős and Moser [165]. Being union-free is equivalent to satisfying the following two conditions at the same time:

- For any four distinct sets $F, F', G, G' \in \mathcal{F}$, we have $F \cup F' \neq G \cup G'$. Families satisfying this property are called *weakly union-free*. The maximum size of a weekly union-free family $\mathcal{F} \subseteq 2^{[n]}$ is denoted by $F(n)$.

- For any three distinct sets $F, G, H \in \mathcal{F}$ we have $F \cup H \neq F \cup G$. Families satisfying this property are called *cancellative*. The maximum size of a cancellative family $\mathcal{F} \subseteq 2^{[n]}$ is denoted by $G(n)$. The problem of determining $G(n)$ was raised by Erdős and Katona [339].

For union-free families, Frankl and Füredi [202] proved the bounds $2^{(n-3)/4} \leq f(n) \leq 1+2^{(n+1)/2}$, where the upper bound follows trivially from the fact that all pairwise disjoint unions must be different, so the number of these pairs is at most $2^n$. The current best bounds are as follows.

**Theorem 179 (Coppersmith, Shearer, [109])** *The function $f(n)$ satisfies*

$$2^{(0.31349+o(1))n} \leq f(n) \leq 2^{(0.4998+o(1))n}.$$

Let us consider now weakly union-free families. Frankl and Füredi [202] proved $2^{(n-\log 3)/3}-2 \leq F(n) \leq 2^{(3n+2)/4}$, and Coppersmith and Shearer [109] improved their upper bound and showed $F(n) \leq 2^{(0.5+o(1))n}$.

**Exercise 180** *Prove the lower bound $2^{(n-\log 3)/3} - 2 \leq F(n)$ of Frankl and Füredi. (Hint: consider a random family $\mathcal{F}$ of $2 \cdot 2^{(n-\log 3)/3}$ sets, such that every set of $\mathcal{F}$ contains any $i \in [n]$ with probability $\frac{1-\sqrt{2}}{2}$ independently from any other containment $j \in F'$. Remove a set from any four-tuple of $\mathcal{F}$ contradicting the weakly union-free property.)*

For cancellative families, Erdős and Katona [339] conjectured $G(n) = 3^{(\frac{1}{3}+o(1))n}$. Frankl and Füredi [202] obtained an upper bound $G(n) \leq n1.5^n$. Then Shearer [503] disproved the above conjecture, and finally Tolhuizen [532], based on a uniform construction (presented below) showed $0.288\frac{1.5^n}{\sqrt{n}} \leq G(n)$.

The above properties can be considered for uniform families as well. Let $f_k(n), F_k(n), G_k(n)$ denote the maximum possible size of a union-free, weakly union-free, cancellative family $\mathcal{F} \subseteq \binom{[n]}{k}$.

**Theorem 181 (Frankl, Füredi [205])** *For any positive integer $k \geq 2$, there exist constants $c_k, c'_k$ such that $c_k 2^{\lceil 4k/3 \rceil/2} \leq f_k(n) \leq F_k(n) \leq c'_k 2^{\lceil 4k/3 \rceil/2}$ holds.*

They also obtained the asymptotics of $F_4(n)$ by showing $F_4(n) = (1/24 + o(1))n^3$.

Bollobás [65] proved $G_3(n) = \lfloor \frac{n}{k} \rfloor \cdot \lfloor \frac{n+1}{k} \rfloor \cdot \lfloor \frac{n+2}{k} \rfloor$, with the extremal family being the largest tripartite family (the underlying set is partitioned into three parts $X_1, X_2, X_3$ and $\mathcal{F}$ contains a set $F$ if and only if $|F \cap X_i| = 1$ for all $i = 1, 2, 3$). A stability version of this result was proved by Keevash and Mubayi [351], and later Balogh and Mubayi [46] showed that almost all 3-uniform cancellative families are tripartite. More precisely, let $Canc_3(n)$ denote the number of 3-uniform cancellative families, then for the number $s_3(n)$ of 3-uniform non-tripartite cancellative families we have $s_3(n) = o(Canc_3(n))$. (Actually, this is just a special case of their result.)

Bollobás conjectured that the following generalization of his result should hold: $G_k(n) = \prod_{i=0}^{k-1} \lfloor \frac{n+i}{k} \rfloor$. Sidorenko [507] verified this for $G_4(n)$.

**Proposition 182 (Frankl and Füredi [202])** *(i)* $G_k(n) = 2^{n-k}$ *if* $n \leq 2k$.

*(ii)* $G_k(n) \leq \binom{n}{k} 2^k / \binom{2k}{k}$ *if* $n \geq 2k$.

**Proof.** If $F$ is a set of maximum size in a cancellative family $\mathcal{F}$, then for any other sets $G, G' \in \mathcal{F}$, the condition $F \cup G \neq F \cup G'$ implies $G \cap \overline{F} \neq G' \cap \overline{F}$. Therefore we have $|\mathcal{F}| \leq 1 + 2^{n-|F|}$. If $\mathcal{F} \subseteq \binom{[n]}{k}$, then $G \cap \overline{F} = \emptyset$ is impossible, so we obtain $|\mathcal{F}| \leq 2^{n-k}$ as required.

To prove **(ii)**, we use the Markov inequality (stated at the beginning of Chapter 4). Let $Y$ be a random $2k$-subset of $[n]$ taken uniformly over $\binom{[n]}{2k}$. For any cancellative family $\mathcal{F}$ we consider $\mathcal{F}_Y = \mathcal{F} \cap \binom{Y}{k}$. As $\mathcal{F}_Y \subseteq \mathcal{F}$, we have that $\mathcal{F}_Y$ is cancellative, and therefore **(i)** implies $|\mathcal{F}_Y| \leq 2^{n-k}$. For the expected size of $\mathcal{F}_Y$ we have

$$2^{n-k} \geq \mathbb{E}(|\mathcal{F}_Y|) = \binom{2k}{k} \frac{|\mathcal{F}|}{\binom{n}{2k}}.$$

Rearranging yields **(ii)**. ■

Bollobás's conjecture is false in general, as shown first by a nontrivial construction due to Shearer [503]. Below, we present a construction of Tolhuizen [532], that almost matches the upper bound of Proposition 182. We start with a couple of lemmas.

**Lemma 183** *The number of* $k \times k$ *invertible matrices over* $\mathbb{F}_2$ *is*

$$Inv(k) = \prod_{i=0}^{k} (2^k - 2^i).$$

**Proof.** We pick the columns of the matrix one-by-one. The first column $u_1$ can be any vector but the zero-vector. If $u_1, u_2, \ldots, u_i$ is picked, then $u_{i+1}$ should not belong to $\langle u_1, u_2, \ldots, u_i \rangle$. ■

If $U$ is a $k$-dimensional subspace of $\mathbb{F}_2^n$, then $K \in \binom{[n]}{k}$ is an *information set* of $U$ if for any $u = (u_1, u_2, \ldots, u_n) \in U$ and $v = (v_1, v_2, \ldots, v_n) \in U$, there exists $i \in K$ such that $u_i \neq v_i$.

**Lemma 184** *For any pair $k \leq n$ of integers, there exists a $k$-dimensional subspace $U$ with at least $\gamma\binom{n}{k}$ information sets, where $\gamma = \prod_{i=1}^{\infty}(1-2^{-i}) \approx$* 0.2888.

**Proof.** For any binary $k \times n$ matrix $M$, let us consider the $k$-subspace

$$U(M) = \{\underline{v} \cdot M : \underline{v} \in \mathbb{F}_2^n\}.$$

Observe that $K \in \binom{[n]}{k}$ is an information set of $U(M)$ if and only if the submatrix $M_K$ of $M$, consisting of the columns indexed by $K$, is invertible. So we need to define a binary $k \times n$ matrix with many invertible submatrices.

If we fix $K \in \binom{[n]}{k}$, then the number of $k \times n$ matrices with $M_K$ being invertible is $2^{k(n-k)}Inv(k)$. Indeed, one can pick $M_K$ in $Inv(k)$ many ways, and fill in the rest of the matrix entries arbitrarily. Let us consider the number of pairs $(M, K)$ with $M_K$ being invertible. By the above, this number is $\binom{n}{k}2^{k(n-k)}Inv(k)$. By averaging, there exists a matrix $M$ being in at least

$$\frac{\binom{n}{k}2^{k(n-k)}Inv(k)}{2^{nk}} = \frac{\binom{n}{k}Inv(k)}{2^{k^2}} = \binom{n}{k}\prod_{i=0}^{k-1}(1-2^{i-k}) \geq \gamma\binom{n}{k}$$

pairs. ■

A *coset* $C$ of a subspace $U$ of $\mathbb{F}_2^n$ is a set of the form $\{\underline{u} + \underline{v} : \underline{u} \in U\}$ for some $\underline{v} \in \mathbb{F}_2^n$. A $k$-subset $K \subsetneq [n]$ is an *identifying set* of a coset $C$ if the characteristic vector $\underline{v}_K$ belongs to $C$, and $K$ is an information set of $U$.

**Proposition 185** *If a $k$-subspace $U$ of $\mathbb{F}_2^n$ has at least $T$ information sets, then there exists a coset $C$ of $U$ with at least $\frac{T}{2^{n-k}}$ identifying sets. Moreover, the family consisting of the complements of the identifying sets is cancellative.*

**Proof.** Since $U$ has exactly $2^{n-k}$ pairwise disjoint cosets, one of them should contain the characteristic vector of at least $\frac{T}{2^{n-k}}$ information sets of $U$.

To see the moreover part of the statement, observe that $F \cup G = F \cup G'$ is equivalent to $\overline{F} \cap \overline{G} = \overline{F} \cap \overline{G'}$. So we need to show that there does not exist 3 distinct identifying sets $K, L, L'$ with $K \cap L = K \cap L'$. This follows from the fact that if $K$ is an information set of a subspace $U$, then for any coset $C$ of $U$, any pair of vectors of $C$ differ in at least one coordinate indexed with an element of $K$. That coordinate corresponds to an element of $K$ that belongs to exactly one of $L$ and $L'$. ■

To obtain Tolhuizen's $k$-uniform cancellative family, we can combine Proposition 185 and Lemma 184. First, Lemma 184 provides us an $(n-k)$-subspace $U$ with at least $\gamma\binom{n}{n-k} = \gamma\binom{n}{k}$ information sets, then Proposition 185 turns these sets into a $k$-uniform cancellative family of size at least $\gamma\frac{\binom{n}{k}}{2^{n-(n-k)}} = \gamma\frac{\binom{n}{k}}{2^k}$. So for $2k \leq n$ we can summarize our knowledge on $G_k(n)$ as

$$\gamma\frac{\binom{n}{k}}{2^k} \leq G_k(n) \leq \frac{\binom{n}{k}2^k}{\binom{2k}{k}}.$$

**Exercise 186** *(i) Show that if for disjoint sets $X, Y$, the families $\mathcal{F}_1 \subseteq 2^X$ and $\mathcal{F}_2 \subseteq 2^Y$ are cancellative, then so is $\mathcal{F} = \{F_1 \cup F_2 : F_1 \in \mathcal{F}_1, F_2 \in \mathcal{F}_2\}$.*

***(ii)*** *Show Tolhuizen's lower bound $G(n) \geq 1.5^{(1+o(1))n}$ by taking $k = k(n) = n/3$.*

We call a family $\mathcal{F} \subseteq 2^{[n]}$ *t-cancellative* if for any distinct $t+2$ sets $F_1, F_2, \dots, F_t, H_1, H_2 \in \mathcal{F}$, we have $F_1 \cup F_2 \cup \dots \cup F_t \cup H_1 \neq F_1 \cup F_2 \cup \dots \cup F_t \cup H_2$. Let $G_t(n)$ be the maximum size of a $t$-cancellative family.

Note that part **(i)** of the above exercise implies $G(n+m) \geq G(n) \cdot G(m)$, therefore $\lim_{n\to\infty} G(n)^{1/n}$ exists, but this is not known for $G_t(n)$. Körner and Sinaimeri [375] proved that $0.11 \leq \limsup_{n\to\infty} \frac{1}{n} \log G_2(n) \leq 0.42$. Füredi [235] showed an improved general upper bound $G_t(n) \leq \alpha n^{(t-1)/2} (\frac{t+3}{t+2})^n$ for some absolute constant $\alpha$. This bound can be close to the truth for small values of $t$ and $n$, but he also obtained the following general result.

**Theorem 187 (Füredi [235])** *There exist constants $\beta_1, \beta_2$ and integer $n_0(t)$ such that if $n \geq n_0(t)$, then we have*

$$\frac{\beta_1}{t^2} \leq \frac{\log G_t(n)}{n} \leq \frac{\beta_2 \log t}{t^2}.$$

The paper [235] contains several results on uniform $t$-cancellative families and lots of remarks on the connection of several related properties.

Many of the properties discussed in this section have applications in communication complexity and combinatorial search theory (also known as combinatorial group testing). In Chapter 9, families with similar properties ($r$-cover-free, $r$-single user tracing superimposed, locally thin) are considered.

## 3.5 Union-closed families

The main notion of the last section of this chapter is that of *union-closed families*. We say that a family $\mathcal{F}$ is union-closed if for any pair $F, F'$ of sets in $\mathcal{F}$, the union $F \cup F'$ also belongs to $\mathcal{F}$. Note that union-closed families are to some extent the opposites of Sperner families. Although there exist several research papers on the topic [7, 367], the most well-known conjecture about union-closed families (mostly referred to as Frankl's conjecture) is as follows:

**Conjecture 188** *For any union-closed family $\mathcal{F} \subseteq 2^{[n]}$, there exists an element $x \in [n]$, that belongs to at least half of the sets in $\mathcal{F}$.*

A survey by Bruhn and Schaudt [92] has appeared recently on the different approaches to the conjecture, but there has been progress since its publication.

Let us start enumerating the special cases in which Conjecture 188 has been proved by considering "small cases". There are several parameters that can be small about a union-closed family $\mathcal{F} \subseteq 2^{[n]}$. One of them is the size $n$ of the underlying set, or more importantly the size of $\cup_{F\in\mathcal{F}}F$. Improving on earlier results by Markovic [407] and Morris [421], Bošnjak and Markovic [85] proved that Conjecture 188 holds if $|\cup_{F\in\mathcal{F}} F| \le 11$.

Another possibility for $\mathcal{F}$ to be small is if all sets contained in $\mathcal{F}$ are small. Roberts and Simpson [480] proved that Conjecture 188 is true for all union-closed families with $\max_{F\in\mathcal{F}}|F| \le 46$. Yet another option for smallness is that the number of sets in $\mathcal{F}$ should be small compared to $n$ or $|\cup_{F\in\mathcal{F}} F|$. One has to be careful with this kind of smallness, as if $X$ is disjoint from $\cup_{F\in\mathcal{F}}F$, then $\mathcal{F}' = \{F \cup X | F \in \mathcal{F}\}$ is union-closed with $|\mathcal{F}| = |\mathcal{F}'|$, and the underlying set of $\mathcal{F}'$ can be made arbitrarily large by enlarging $X$. To prevent this, one may consider *separating* families: families $\mathcal{F}$ with the property that for any $x, y \in \cup_{F\in\mathcal{F}}F$ there exists $F \in \mathcal{F}$ with $|F \cap \{x,y\}| = 1$. We note here that we will learn more about separating families in Chapter 9. Maßberg [410] proved that Conjecture 188 holds for any separating union-closed family $\mathcal{F}$ with $|\mathcal{F}| \le 2(n + \frac{n}{\log n - \log\log n})$, where $n$ denotes $|\cup_{F\in\mathcal{F}} F|$. Hu [311] proved that if this bound can be improved to $(2+c)n$ for some constant $c > 0$, this already implies that any union-closed family $\mathcal{F}$ has an element appearing in at least $\frac{c}{2c+2}|\mathcal{F}|$ sets in $\mathcal{F}$. At the moment, the best known lower bound on the maximum degree in any union-closed family $\mathcal{F}$ is $\Omega(\frac{|\mathcal{F}|}{\log|\mathcal{F}|})$ due to Knill [369] and Wójcik [551]. (This is also a consequence of Theorem 189 below.)

On the other end of the spectrum, Conjecture 188 has been verified for families $\mathcal{F} \subseteq 2^{[n]}$ containing a large proportion of subsets of $[n]$. The major tool in this direction is to consider the total size $\sum_{F\in\mathcal{F}}|F|$ of the family (also called the *volume* of $\mathcal{F}$ and denoted by $||\mathcal{F}||$). Observe that if $||\mathcal{F}|| \ge \frac{n|\mathcal{F}|}{2}$, then Conjecture 188 holds for $\mathcal{F}$, as even the average degree is at least $|\mathcal{F}|/2$. Reimer proved the following sharp lower bound on the average set size $\frac{||\mathcal{F}||}{|\mathcal{F}|}$ in a union-closed family.

**Theorem 189 (Reimer [477])** *If $\mathcal{F} \subseteq 2^{[n]}$ is a union-closed family, then the average set size satisfies*

$$\frac{\sum_{F\in\mathcal{F}}|F|}{|\mathcal{F}|} \ge \frac{1}{2}\log|\mathcal{F}|,$$

*and equality holds if and only if $\mathcal{F} = 2^M$ for some $M \subseteq n$.*

**Proof.** Let us recall that the Hamming graph $H_n$ has vertex set $2^{[n]}$ with $F, G$ being joined by an edge if and only if $|F \triangle G| = 1$. For a family $\mathcal{F}$, we define the edge boundary $EB(\mathcal{F})$ of $\mathcal{F}$ to be the set of edges of $H_n$ that have exactly one endpoint in $\mathcal{F}$.

**Lemma 190** *For any upset $\mathcal{U} \subseteq 2^{[n]}$ we have*

$$2\sum_{U\in\mathcal{U}} |U| = n|\mathcal{U}| + |EB(\mathcal{U})|.$$

**Proof of Lemma.**

$$\begin{aligned}2\sum_{U\in\mathcal{U}} |U| &= \sum_{U\in\mathcal{U}}\sum_{U'\in\Delta(U)} 1 + \sum_{U\in\mathcal{U}}\sum_{U'\in\Delta(U),U'\notin\mathcal{U}} 1 + \sum_{U\in\mathcal{U}}\sum_{U'\in\Delta(U),U'\in\mathcal{U}} 1\\ &= \sum_{U\in\mathcal{U}}\sum_{U'\in\Delta(U)} 1 + |EB(\mathcal{U})| + \sum_{U\in\mathcal{U}}\sum_{U''\in\nabla(U)} 1\\ &= \sum_{U\in\mathcal{U}} n + |EB(\mathcal{U})|\\ &= n|\mathcal{U}| + |EB(\mathcal{U})|.\end{aligned}$$

□

Before stating the next lemma, let us recall that for two sets $A, B$ with $A \subseteq B$, the interval $[A, B]$ is the family $\{G : A \subseteq G \subseteq B\}$.

**Lemma 191** *For any union-closed family $\mathcal{F} \subseteq 2^{[n]}$, there exists an upset $\mathcal{U} \subseteq 2^{[n]}$ and a bijection $f : \mathcal{F} \to \mathcal{U}$, such that the following holds:*
*(i) for any $F \in \mathcal{F}$ we have $F \subseteq f(F)$,*
*(ii) for any distinct pair $F, G$ of sets in $\mathcal{F}$ we have $[F, f(F)] \cap [G, f(G)] = \emptyset$,*
*(iii) $\sum_{F\in\mathcal{F}} |f(F) \setminus F| \le n|\mathcal{F}| - |\mathcal{F}| \log |\mathcal{F}|$.*

**Proof of Lemma.** To prove the lemma we will need the following tool. For a family $\mathcal{F} \subseteq 2^{[n]}$ and $i \in [n]$, let us define the *up-shifting* operation $u_i = u_{i,\mathcal{F}}$ by

$$u_{i,\mathcal{F}}(F) = \begin{cases} F \cup \{i\} & \text{if } i \notin F \text{ and } F \cup \{i\} \notin \mathcal{F} \\ F & \text{otherwise,} \end{cases} \tag{3.2}$$

Let us write $u_i(\mathcal{F}) = \{u_{i,\mathcal{F}}(F) : F \in \mathcal{F}\}$. By definition we have $|\mathcal{F}| = |u_i(\mathcal{F})|$ for any $\mathcal{F}$ and $i$. The analogous down-shifting operation was introduced independently by Alon [9] and Frankl [192] (see Chapter 8).

Let $\mathcal{F} \subseteq 2^{[n]}$ be a union-closed family. We define the sequence $\mathcal{F}_0, \mathcal{F}_1, \ldots, \mathcal{F}_n$ of families as follows: let $\mathcal{F}_0 = \mathcal{F}$, and if for some $0 \le j \le n-1$ the family $\mathcal{F}_j$ is defined, then let $\mathcal{F}_{j+1} = u_{j+1}(\mathcal{F}_j)$. We claim that the family $\mathcal{U} = \mathcal{F}_n$ satisfies the statement of the lemma. To define the bijection $f$, we introduce further notation. For any set $F \in \mathcal{F}$, we define the sequence $F_0, F_1, \ldots, F_n$ with $F_0 = F$, and if for some $0 \le j \le n-1$ the set $F_j$ is defined, then let $F_{j+1} = u_{j+1,\mathcal{F}_j}(F_j)$. Observe that, by induction, $F_j \in \mathcal{F}_j$ for all $j = 0, 1, \ldots, n$. Let us define $f(F) = F_n$ for any set $F \in \mathcal{F}$. As $f$ is a composition of bijections, it is indeed a bijection.

**Claim 192** *Suppose $\mathcal{F}_{i-1}$ is obtained in the process described above, $\mathcal{F}_{i-1}$ is union-closed and $F, G \in \mathcal{F}$ with $G_{i-1} = F_{i-1} \cup X$ for some $X \subseteq [n]$. Then $F_i \cup X \in \mathcal{F}_i$, furthermore, if $G_{i-1} \neq F_{i-1}$, then $F_i \cup X \neq F_i$.*

**Proof of Claim.** As $F_{i-1} \cup X = G_{i-1} \in \mathcal{F}_{i-1}$, we have $F_{i-1} \cup X \cup \{i\} \in \mathcal{F}_i$. So $F_i \cup X \notin \mathcal{F}_i$ would imply $i \notin F_{i-1}, F_i, G_{i-1}, X$. In particular, $F_{i-1} = F_i$, so $F_{i-1} \cup \{i\} \in \mathcal{F}_{i-1}$ would hold, and therefore, as $\mathcal{F}_{i-1}$ is union-closed, we would have $F_{i-1} \cup \{i\} \cup G_{i-1} = G_{i-1} \cup \{i\} \in \mathcal{F}_{i-1}$. This would imply $G_i = G_{i-1}$ and thus $F_i \cup X = F_{i-1} \cup X = G_{i-1} = G_i \in \mathcal{F}_i$.

To see the furthermore part, suppose $F_{i-1} \cup X \neq F_{i-1}$ and $F_i \cup X = F_i$ hold. Then $F_i \neq F_{i-1}$ and $F_i = F_{i-1} \cup \{i\}$. But then we have $F_{i-1} \subsetneq F_{i-1} \cup X \subseteq F_i \cup X = F_i = F_{i-1} \cup \{i\}$, which implies $F_{i-1} \cup \{i\} = F_{i-1} \cup X$. We know $F_{i-1} \cup X = G_{i-1} \in \mathcal{F}_{i-1}$, so we should have $F_{i-1} = F_i$. This contradiction finishes the proof of the claim. □

**Claim 193** *For all $i = 0, 1, \ldots, n$, the family $\mathcal{F}_n$ is union-closed.*

**Proof of Claim.** We proceed by induction on $i$, with the base case $i = 0$ being true as $\mathcal{F}$ is union-closed. Let $F_i, F'_i \in \mathcal{F}_i$. Then, as $\mathcal{F}_{i-1}$ is union-closed by induction, we can apply Claim 192 to $F_{i-1}$ and $G_{i-1} := F_{i-1} \cup F'_{i-1}$ to obtain $F_i \cup G_{i-1} \in \mathcal{F}_i$. Now $F_i \cup F'_i$ is either $F_i \cup G_{i-1}$ or $F_i \cup G_{i-1} \cup \{i\}$. But we have shown the former to be in $\mathcal{F}_i$, and for any set $H$, $H \in \mathcal{F}_i$ implies $H \cup \{i\} \in \mathcal{F}_i$. □

**Claim 194** *Let $1 \leq i < j \leq n$ be integers and let $F_i, G_i \in \mathcal{F}_i$ with $G_i = F_i \cup X$ for some $X \subseteq [n]$. Then $F_j \cup X \in \mathcal{F}_j$ holds. Furthermore, if $F_i \cup X \neq F_i$, then $F_j \cup X \neq F_j$.*

**Proof of Claim.** This follows by applying Claim 192 repeatedly. □

**Claim 195** *$\mathcal{U} = \mathcal{F}_n$ is an upset.*

**Proof of Claim.** We need to show that for any $F_n \in \mathcal{F}_n = \mathcal{U}$ and $i \in [n]$, we have $F_n \cup \{i\} \in \mathcal{F}_n$. We know that by definition of $u_i$, we have $F_i \cup \{i\} \in \mathcal{F}_i$. Applying Claim 194 to $F_i$, $X = \{i\}$ and $j = n$, we obtain $F_n \cup \{i\} \in \mathcal{F}_n$. □

Now we are ready to prove Lemma 191. Observe first that **(i)** follows, as $F = F_0 \subseteq F_1 \subseteq \cdots \subseteq F_n = f(F)$.

To prove **(ii)**, let us assume towards a contradiction that there exist $F, G \in \mathcal{F}$ with $Y \in [F, f(F)] \cap [G, f(G)]$. Observe that we may suppose $F \subseteq G$. Indeed, $F, G \subseteq Y$ implies $F \cup G \subseteq Y$, and the union-closed property of $\mathcal{F}$ implies that $F \cup G \in \mathcal{F}$, thus we have $[F, f(F)] \cap [F \cup G, f(F \cup G)] \neq \emptyset$. So if $F, G$ are not in containment, then $F \cup G \neq F$ and $G$ could be replaced with $F \cup G$.

Let us apply Claim 194 to $F$, $X = G$, $i = 0$, and $j = n$. As $F \neq F \cup X = F \cup G = G$, we obtain that $f(F) = F_n \neq F_n \cup X = f(F) \cup G$. This implies $G \not\subseteq f(F)$ and thus $[F, f(F)] \cap [G, f(G)] = \emptyset$ - a contradiction.

To prove **(iii)**, we use the disjointness of the intervals $[F, f(F)]$ from **(ii)** to obtain

$$\sum_{F \in \mathcal{F}} 2^{|f(F) \setminus F|} = \sum_{F \in \mathcal{F}} 2^{|[F, f(F)]|} \leq 2^n.$$

As $2^n$ is a convex function of $n$, we can apply Jensen's inequality to obtain

$$2^{\frac{\sum_{F\in\mathcal{F}}|f(F)\setminus F|}{|\mathcal{F}|}} \leq \frac{\sum_{F\in\mathcal{F}} 2^{|f(F)\setminus F|}}{|\mathcal{F}|} \leq \frac{2^n}{|\mathcal{F}|}.$$

Taking logarithm and rearranging yields

$$\sum_{F\in\mathcal{F}} |f(F)\setminus F| \leq |\mathcal{F}|n - |\mathcal{F}|\log|\mathcal{F}|.$$

□

Having the above lemmas in hand, we are ready to prove Theorem 189. Let $\mathcal{F} \subseteq 2^{[n]}$ be a union-closed family. Applying Lemma 191 we obtain an upset $\mathcal{U}$ and an bijection $f : \mathcal{F} \to \mathcal{U}$. For any $F \in \mathcal{F}$, let us consider the following subset of the edge set of the Hamming graph: $E_F := \{(B, f(F)) \in E(H_n) : F \subseteq B \subseteq f(F)\}$. By definition, we have $|E_F| = |f(F) \setminus F|$, and part **(ii)** of Lemma 191 implies $E_F \cap E_{F'} = \emptyset$ for every distinct $F, F' \in \mathcal{F}$. Another consequence of part **(ii)** of Lemma 191 is that $E_F \subseteq EB(\mathcal{U})$ holds for any $F \in \mathcal{F}$; thus we have $\sum_{F\in\mathcal{F}} |f(F)\setminus F| = \sum_{F\in\mathcal{F}} |E_F| \leq |EB(\mathcal{U})|$. Applying Lemma 190 to $\mathcal{U}$ and using part **(i)** of Lemma 191 we obtain

$$2\sum_{F\in\mathcal{F}} |F| + 2\sum_{F\in\mathcal{F}} |f(F)\setminus F| = 2\sum_{U\in\mathcal{U}} |U| = |\mathcal{F}|n + |EB(\mathcal{U})|.$$

This and $\sum_{F\in\mathcal{F}} |f(F)\setminus F| \leq |EB(\mathcal{U})|$ imply

$$2\sum_{F\in\mathcal{F}} |F| + \sum_{F\in\mathcal{F}} |f(F)\setminus F| \geq |\mathcal{F}|n.$$

By **(iii)** of Lemma 191 we obtain

$$2\sum_{F\in\mathcal{F}} |F| + n|\mathcal{F}| - |\mathcal{F}|\log|\mathcal{F}| \geq |\mathcal{F}|n,$$

and rearranging finishes the proof of the inequality.

Observe that to have equality in the above inequality, all intervals $[F, f(F)]$ must be of the same size, and also they should partition the complete hypercube $2^{[n]}$. As $\mathcal{F}$ is union-closed, it contains a maximal set $M$. Without loss of generality we may assume that $M = [n]$. But then $f(M) = [n]$ must hold, so all intervals $[F, f(F)]$ have size 1, so indeed $\mathcal{F} = 2^M$. ■

Having seen Theorem 189, one might wonder what is the minimum possible volume of a union-closed family $\mathcal{F}$ of size $m$. If we denote the minimum by $f(m)$, then Theorem 189 states $f(m) \geq \frac{m}{2\log m}$, and this is sharp whenever $m = 2^{m'}$ for some integer $m'$. Clearly, if $|\mathcal{F}| = m$, then $|\cup_{F\in\mathcal{F}} F| \geq \log m$ holds. So for any $n, m$ with $2^n \geq m$, one might consider the minimum volume

$f(n,m)$ over all union-closed families $\mathcal{F}$ of size $m$ and $|\cup_{F\in\mathcal{F}} F| \leq n$. It is not obvious by definition if $f(n,m) = f(m)$ holds for all such $n$. Observe that if we let $m_0 = \min\{m' : f(n,m) \geq nm/2\,\forall m \geq m'\}$, then Conjecture 188 holds for all union-closed families $\mathcal{F} \subseteq 2^{[n]}$ with $|\mathcal{F}| \geq m_0$. Czédli [113] proved $m_0 \leq 2^n - \lfloor 2^{n/2} \rfloor$, thus showing that such a reasoning might verify Conjecture 188 for many families. On the other hand, Czédli, Maróti, and Schmidt [114] showed $m_0 \geq 2\lceil 2^n/3 \rceil$, and conjectured this to be the exact value of $m_0$. This was proved by Balla, Bollobás and Eccles [33] in a stronger form: they showed $f(n,m) = f(m) = ||\mathcal{F}(m)||$ for any value of $n$ with $2^n \geq m$, where the family $\mathcal{F}(m)$ is defined as follows:

Recall that the colex order of all finite subsets of $\mathbb{N}$ is defined by

$$F <_{colex} G \Leftrightarrow \max(F \triangle G) \in G.$$

The initial segment of size $m$ of the colex order is denoted by $\mathcal{I}_m$. For any $m < 2^n$, let us write $m$ in the unique form $m = 2^{n-1} + 2^{n-2} + \cdots + 2^k + m'$ with $m' \leq 2^{k-1}$. Let us define $\mathcal{M}(n,m) = \{M \in 2^{[n]} : M \cap [k+1,n] \neq \emptyset\} \cup \{B \cup \{k\} : B \in \mathcal{I}(m')\}$. If $m = 2^n$, then let $\mathcal{M}(n,m) = 2^{[n]}$.

Let us define the *mixed order* of all finite subsets of $\mathbb{N}$ by

$$F <_m G \Leftrightarrow \begin{cases} \max(F) > \max(G), \text{ or} & \\ \max(F) = \max(G) & \text{and} \max(F \triangle G) \in G, \end{cases} \tag{3.3}$$

Note that the mixed order does not have a minimal element among the finite subsets of $\mathbb{N}$; thus all its initial segments are infinite. Let $\mathcal{M}'(n,m)$ be the initial segment of size $m$ of the mixed order *restricted to* $2^{[n]}$.

**Exercise 196** *For any $n$ with $2^n \geq m$, the families $2^{[n]} \setminus \mathcal{M}'(n,m)$ are identical.*

**Exercise 197** *Show that $\mathcal{M}(n,m)$ and $\mathcal{M}'(n,m)$ are identical.*

Finally, for integers $n,m$ with $2^n \geq m$, let us define $\mathcal{F}(m) = 2^{[n]} \setminus \mathcal{M}(n, 2^n - m)$. In the proof of Theorem 199, we will need the following consequence of the shadow theorem (Theorem 23).

**Exercise 198** *For any downset $\mathcal{D}$ we have $||\mathcal{D}|| \leq ||\mathcal{I}(|\mathcal{D}|)||$.*

**Theorem 199 (Balla, Bollobás, Eccles [33])** *Let $m,n \geq 1$ with $2^n \geq m$. For any union-closed family $\mathcal{F} \subseteq 2^{[n]}$ with $|\mathcal{F}| = m$, we have $||\mathcal{F}|| \geq ||\mathcal{F}(m)||$, and equality holds if and only if $\mathcal{F} = \mathcal{F}(m)$.*

**Proof.** We will use Lemma 191 and the notions introduced during its proof. We also define $r_{\mathcal{F}}(F) = |f(F) \setminus F|$.

**Lemma 200** *Suppose that for a union-closed family $\mathcal{F} \subseteq 2^{[n]}$ with $|\mathcal{F}| = m$ the following statements hold:*

***(i)*** *there exists an integer $k$, such that for any $F \in F$, we have $k \leq r_{\mathcal{F}}(F) \leq k+1$,*

***(ii)*** $\cup_{F\in\mathcal{F}}[F, f(F)] = 2^{[n]}$

***(iii)*** $\mathcal{U} = f(\mathcal{F})$ *has minimal volume among all upsets of size $m$.*

*Then $f(n,m) = ||\mathcal{F}||$, i.e. $\mathcal{F}$ minimizes the volume over all union-closed families of size $m$.*

**Proof of Lemma.** Suppose $\mathcal{F}$ satisfies properties **(i)**-**(iii)** and let $\mathcal{F}' \subseteq 2^{[n]}$ be another union-closed family of size $m$. We need to show $||\mathcal{F}'|| \geq ||\mathcal{F}||$. Applying Lemma 191 to $\mathcal{F}'$, we obtain a bijection $f' : \mathcal{F}' \to \mathcal{U}'$ with $\mathcal{U}'$ being an upset in $2^{[n]}$, and we define $r'_{\mathcal{F}'}(F') = |f'(F') \setminus F'|$. Property **(ii)** yields

$$\sum_{F'\in\mathcal{F}'} 2^{r'_{\mathcal{F}'}(F')} \leq \sum_{F\in\mathcal{F}} 2^{r_{\mathcal{F}}(F)} = 2^n.$$

Property **(i)** and the convexity of the function $2^x$ imply $\sum_{F'\in\mathcal{F}'} r'_{\mathcal{F}'}(F') \leq \sum_{F\in\mathcal{F}} r_{\mathcal{F}}(F)$. Property **(iii)** implies $||f'(\mathcal{F}')|| \geq ||f(\mathcal{F})||$. Putting these together we obtain

$$||\mathcal{F}'|| = ||f'(\mathcal{F}')|| - \sum_{F'\in\mathcal{F}'} r'_{\mathcal{F}'}(F') \geq ||f(\mathcal{F})|| - \sum_{F\in\mathcal{F}} r_{\mathcal{F}}(F) = ||\mathcal{F}||.$$

□

**Lemma 201** *The family $\mathcal{F}(m)$ satisfies conditions **(i)**-**(iii)** of Lemma 201.*

**Proof of Lemma.** Observe that any initial segment $\mathcal{I}(t)$ of the colex order is a downset, so its complement with respect to $2^{[n]}$ is an upset for any $n$ with $2^n \geq t$. Therefore, writing $k = \lceil \log m \rceil$, we have $\mathcal{F}(m) = 2^{[n]} \setminus \mathcal{M}(n, 2^n - m) = 2^{[k-1]} \cup \{A \cup \{k\} : A \in \mathcal{V}\}$, where $\mathcal{V}$ is an upset. This means that all $F \in \mathcal{F}(m)$ remains unchanged during the first $k-1$ up-shift operations in the proof of Lemma 191. This implies

$$f(F) = \begin{cases} F \cup [k,n] & \text{if } k \notin F, F \notin \mathcal{V} \\ F \cup [k+1,n] & \text{otherwise,} \end{cases} \tag{3.4}$$

for any $F \in \mathcal{F}(m)$, and therefore we have $n-k \leq r_{\mathcal{F}(m)}(F) \leq n-k+1$. This shows that condition **(i)** of Lemma 200 is satisfied.

To see **(ii)**, let us fix $S \in 2^{[n]}$. If $k \notin S$, then write $S = S_1 \cup S_2$ with $S_1 \subseteq [k-1]$ and $S_2 \subseteq [k+1,n]$. As $S_1 \in \mathcal{F}(m)$ and $f(S_1)$ is either $S_1 \cup [k,n]$ or $S_1 \cup [k+1,n]$, we obtain $S \in [S_1, f(S_1)]$.

Suppose next $k \in S$. Then write $S = S_1 \cup S_2 \cup \{k\}$ with $S_1 \subseteq [k-1]$ and $S_2 \subseteq [k+1,n]$, and observe $S_1 \in \mathcal{F}(m)$. If $S_1 \cup \{k\} \notin \mathcal{F}(m)$, then $f(S_1) = S_1 \cup [k,n]$ and thus $S \in [S_1, f(S_1)]$. If $S_1 \cup \{k\} \in \mathcal{F}(m)$, then $f(S_1 \cup \{k\}) = S_1 \cup [k,n]$ and thus $S \in [S_1 \cup \{k\}, f(S_1 \cup \{k\}]$. This shows that $\mathcal{F}(m)$ satisfies condition **(ii)** of Lemma 200.

To prove that $\mathcal{F}(m)$ satisfies condition **(iii)**, observe that $\mathcal{U} = f(\mathcal{F}(m)) = \{T \in 2^{[n]} : [k,n] \subseteq T\} \cup \{V \cup [k+1,n] : V \in \mathcal{V}\}$, with $\mathcal{V}$ being the complement of the initial segment $\mathcal{I}(m')$ for $2^n - m = 2^{n-1} + \cdots + 2^k + m'$. So if $D$ is the maximal element of $\mathcal{I}(m')$, then

$$2^{[n]} \setminus \mathcal{U} = \{S : S <_{colex} D \cup [k,n]\} = \mathcal{I}(2^n - m)$$

holds. So, by Exercise 198, $2^{[n]} \setminus \mathcal{U}$ maximizes the volume of downsets of size $2^n - m$. Therefore $\mathcal{U} = f(\mathcal{F}(m))$ minimizes the volume of upsets of size $m$. □

Lemma 200 and Lemma 201 finish the proof of the equality $f(n,m) = f(m) = ||\mathcal{F}(m)||$. We omit the proof of the uniqueness of the family $\mathcal{F}(m)$. ■

**Corollary 202** *(i)* $m_0 = 2\lceil 2^n/3 \rceil$.

*(ii) Conjecture 188 holds for union-closed families* $\mathcal{F} \subseteq 2^{[n]}$ *with* $|\mathcal{F}| \geq \frac{2}{3} \cdot 2^n$.

**Proof.** We leave it as an exercise to show that $||\mathcal{F}(m)|| \geq \frac{nm}{2}$ if and only if $m \geq 2\lceil 2^n/3 \rceil$. (This was first proved in [114].) This shows **(i)** and **(ii)** follow the same way as discussed when $m_0$ was introduced. ■

Part **(ii)** of Corollary 202 was extended by Eccles [140] to families with $|\mathcal{F}| \geq (\frac{2}{3} - \frac{1}{104}) \cdot 2^n$. Currently the best known lower bound on $|\mathcal{F}|$ that implies Conjecture 188 is $\frac{1}{2}2^n$ and is due to Karpas [330].

Let us finish this section by mentioning a result concerning the volume of Sperner families. If an antichain $\mathcal{F} \subseteq 2^{[n]}$ contains only sets of the same size, i.e. $\mathcal{F} \subseteq \binom{[n]}{k}$ for some $k$, then the volume of $\mathcal{F}$ is divisible by $k$. Therefore $||\mathcal{F}||$ cannot be equal to any prime number larger than $n$. So if one wants to obtain any possible volume, then one is forced to use at least two levels. Another goal is to obtain a prescribed cardinality at the same time. More precisely, given a Sperner family, we want to find another one with the same cardinality and volume, using few set sizes. Lieby [387] conjectured not only that two possible set sizes suffice, but also that it is possible to use only consecutive integers as set sizes. Such Sperner families are called *flat antichains*. Building on her partial results [388], [389], this was proved by Kisvölcsey [361]. Some special cases of Lieby's conjecture had been verified by Brankovic, Lieby, and Miller [89], Roberts [479] and Maire [401].

**Theorem 203 (Flat antichain theorem, Kisvölcsey [361])** *For any Sperner family* $\mathcal{F} \subseteq 2^{[n]}$*, there exists an integer $k$ and another Sperner family* $\mathcal{S}$ *with* $\mathcal{S} \subseteq \binom{[n]}{k} \cup \binom{[n]}{k+1}$*, such that* $|\mathcal{F}| = |\mathcal{S}|$ *and* $||\mathcal{F}|| = ||\mathcal{S}||$ *hold.*

# 4

# Random versions of Sperner's theorem and the Erdős-Ko-Rado theorem

## CONTENTS



In this chapter we address randomized versions of Theorem 1 and Theorem 9. We will consider random subfamilies of $2^{[n]}$ and $\binom{[n]}{k}$ such that every set belongs to the family with probability $p$ independently from all other sets. More precisely $Q_n(p)$ is the probability space of all families $\mathcal{F} \subseteq 2^{[n]}$ such that

$$\mathbb{P}(Q_n(p) = \mathcal{F}) = p^{|\mathcal{F}|}(1-p)^{2^n - |\mathcal{F}|}$$

holds. Similarly, $Q_{n,k}(p)$ is the probability space of all families $\mathcal{F} \subseteq \binom{[n]}{k}$ such that

$$\mathbb{P}(Q_{n,k}(p) = \mathcal{F}) = p^{|\mathcal{F}|}(1-p)^{\binom{n}{k} - |\mathcal{F}|}$$

holds.

It will be convenient to work with the following graphs. Recall that the *comparability graph* $G_n$ of $2^{[n]}$ has vertex set $2^{[n]}$ and vertices corresponding to sets $A, B$ are connected by an edge if and only if $A \subset B$ or $B \subset A$. The *Kneser graph* $Kn(n,k)$ has vertex set $\binom{[n]}{k}$ and vertices corresponding to sets $A, B$ are connected by an edge if and only if $A \cap B = \emptyset$. Note that an independent set in $G_n$ corresponds to an antichain, and an independent set in $Kn(n,k)$ corresponds to an intersecting family. Finally, for any graph $G$ we denote by $G(p)$ the probability space of all induced subgraphs of $G$ such that every vertex of $G$ belongs to $G(p)$ with probability $p$ independently from all other vertices, i.e.

$$\mathbb{P}(G(p) = G[U]) = p^{|U|}(1-p)^{|V(G)| - |U|}.$$

We will also work with random subgraphs of the complete graph and the Kneser graph. The Erdős-Rényi random graph $G(n,p)$ denotes the probability space of all labeled graphs on $n$ vertices such that every edge appears with probability $p$ independently of all other edges, i.e.

$$\mathbb{P}(G(n,p) = G) = p^{e(G)}(1-p)^{\binom{n}{2}-e(G)}.$$

Similarly, the random subgraph $Kn_p(n,k)$ denotes the probability space of all subgraphs of $Kn(n,k)$ such that for a particular subgraph $G$ we have

$$\mathbb{P}(Kn_p(n,k) = G) = p^{e(G)}(1-p)^{\frac{1}{2}\binom{n}{k}\binom{n-k}{k}-e(G)}.$$

Although $Kn_p(n,k)$, $G(n,p)$ and $Q_n(p)$ are probability spaces, we will often identify them with their instances and make statements about the largest intersecting family in $Kn_p(n,k)$, the largest antichain in $Q_n(p)$, etc.

Finally, let us state some well-known inequalities that we will use (for proofs see e.g. [318]).

**Markov inequality**: if $X$ is a non-negative random variable and $a > 0$ is a real number, then $\mathbb{P}(X > a) \leq \frac{\mathbb{E}(X)}{a}$. In particular, if $X_n{}_{n=1}^{\infty}$ are non-negative integer-valued random variables with $\mathbb{E}(X_n) \to 0$, then $X_n = 0$ w.h.p. Recall that a sequence $\mathcal{E}_n$ of events holds with high probablity (w.h.p) if $\mathbb{P}(\mathcal{E}_n) \to 1$.

**Chebyshev's inequality**: if $X$ is a finite random variable, then

$$\mathbb{P}(|X - \mathbb{E}(X)| \geq k\sigma(X)) \leq \frac{1}{k^2}.$$

In particular, if $\mathbb{E}(X_n) \to \infty$ and $\sigma^2(X_n) = o(\mathbb{E}(X_n))$ (or equivalently $\mathbb{E}(X_n^2) = (1+o(1))\mathbb{E}^2(X_n)$), then $X_n > 0$ w.h.p.

**Chernoff bound**: if $X_1, X_2, \ldots, X_n$ are independent random variables with $\mathbb{P}(X_i = 1) = p_i$ and $\mathbb{P}(X_i = 0) = 1 - p_i$, then for $X = \sum_{i=1}^n X_i$, $\mu = \sum_{i=1}^n p_i$ and $\delta > 0$ we have

- $\mathbb{P}(X > (1+\delta)\mu) \leq \exp(-\frac{\delta^2}{2+\delta}\mu)$,
- $\mathbb{P}(X < (1-\delta)\mu) \leq \exp(-\frac{\delta^2}{2}\mu)$ provided $0 < \delta < 1$.

The inequality $\mathbb{P}(\cup_{i=1}^{\infty}\mathcal{E}_i) \leq \sum_{i=1}^{\infty}\mathbb{P}(\mathcal{E}_i)$ is often referred to as the **union bound**.

## 4.1 The largest antichain in $Q_n(p)$

The problem of determining the size of the largest antichain in $Q_n(p)$ was first considered by Kohakayawa and Kreuter [370], and then by Osthus [449]. Obviously, sets of $Q_n(p)$ of size $\lfloor n/2 \rfloor$ form an antichain, and their number is $(1+o(1))p\binom{n}{\lfloor n/2 \rfloor}$ w.h.p. Osthus proved that w.h.p. there are no larger antichains if $p = \omega(\log n/n)$, and conjectured that this holds if $p = \omega(1/n)$. He showed that his conjecture is sharp in the following stronger form: if, for some fixed positive integer $t$, we have $p = o(n^{-t})$, then there exists an antichain in $Q_n(p)$ of size $(t+1+o(1))p\binom{n}{\lfloor n/2\rfloor}$ w.h.p. Indeed, for the family $\mathcal{F}_t(p)$ of sets of $Q_n(p)$ in its $t+1$ middle levels we have $|\mathcal{F}_t(p)| = (t+1+o(1))p\binom{n}{\lfloor n/2\rfloor}$ w.h.p, while the expected number of pairs of sets from these levels that are in containment is

$$p^2\sum_{j=1}^{t}\binom{n}{\lfloor n/2\rfloor - \lceil t/2\rceil + j}\sum_{i=0}^{j-1}\binom{\lfloor n/2\rfloor - \lceil t/2\rceil + j}{\lfloor n/2\rfloor - \lceil t/2\rceil + i}$$
$$\leq p^2t^2\binom{n}{\lfloor n/2\rfloor}\binom{n}{t} = o\left(p\binom{n}{\lfloor n/2\rfloor}\right).$$

Therefore, the number of such pairs is $o(p\binom{n}{\lfloor n/2\rfloor})$ w.h.p. and so removing those pairs from $\mathcal{F}_t(p)$, we end up with an antichain of size $(t+1+o(1))p\binom{n}{\lfloor n/2\rfloor}$.

The conjecture of Osthus was proved by two groups (Balogh, Mycroft, Treglown [47] and Collares Neto, Morris [443]) independently, both using the container method. (A very recent survey on applications of containers is [44].) Recall that the aim of this method is to find a small family $\mathcal{C}$ of not too big subsets of a (hyper)graph $G$, such that every independent set of $G$ is contained in a set $C \in \mathcal{C}$. If this is done, and we know that size of the largest independent set is at most $M$, then one can bound the total number of independent sets by $\sum_{C\in\mathcal{C}}\sum_{0\leq i\leq M}\binom{|C|}{M}$. Or, as in the proof by Balogh, Mycroft, and Treglown, that we present below, one can use the following estimate

$$\mathbb{P}(\exists I \subseteq G(p) \text{ independent with } |I| \leq M') \leq \sum_{C\in\mathcal{C}}\sum_{J\in\binom{C}{M'}}\mathbb{P}(J \subseteq G(p)).$$

Lemma 206 will provide the containers for the independent set of the comparability graph $G_n$ of $2^{[n]}$.

Before stating the result on the size of the largest antichain in $Q_n(p)$, let us define the *centrality ordering* of $2^{[n]}$. Strictly speaking, it is not one ordering of subsets of $[n]$, but many such orderings exist. All we require is that if $||A| - \lfloor n/2\rfloor| + 1/4$ is smaller than $||B| - \lfloor n/2\rfloor| + 1/4$, then $A$ precedes $B$. In other words, sets in levels that are closer to the middle come before sets in levels further from the middle. The $+1/4$ term is to break ties in case $n$ is even.

**Theorem 204 (Balogh et al. [47], Colares Neto et al. [443])** *For every $\varepsilon > 0$, there exists a constant $C$ such that if $p \geq C/n$, then the size of the largest antichain in $Q_n(p)$ is at most $(1+\varepsilon)p\binom{n}{\lfloor n/2\rfloor}$ w.h.p.*

**Proof.** During the proof we will write $m := \binom{n}{\lfloor n/2\rfloor}$. We work with the comparability graph $G_n$ of $2^{[n]}$. Recall that Theorem 172 states that if $U \subseteq V(G_n)$

is of size $r$, then $e(U) \geq e(M_r)$, where $M_r$ is the initial segment of size $r$ of the centrality ordering of $2^{[n]}$. We will need the following simple consequence.

**Lemma 205** *If $U \subseteq V(G_n)$ with $|U| \geq (t+\varepsilon)m$ for some $0 < \varepsilon < 1/2$ and positive integer $t$, then $e(U) > \varepsilon n^t \frac{|U|}{(2t)^{t+1}}$.*

**Proof of Lemma.** By Theorem 172, it is enough to bound $e(M_r)$ with $|U| = r$. To this end we will use only edges connecting vertices corresponding to sets not in the first $t$ levels of $M_r$ and sets in the first $t$ levels of $M_r$. The number of the former sets is at least $r - tm \geq r(1 - \frac{t}{t+\varepsilon}) \geq \frac{2\varepsilon r}{1+2t}$. Each such set $A$ is connected to at least $\binom{\lceil n/2 \rceil}{t}$ sets of the latter type (we only take into consideration those sets in the level that is the "furthest" from $A$). Therefore

$$e(U) \geq e(M_r) \geq \frac{2\varepsilon r}{1+2t}\binom{\lceil n/2 \rceil}{t} \geq \frac{2\varepsilon r}{1+2t}\left(\frac{n}{2t}\right)^t \geq \frac{\varepsilon n^t r}{(2t)^{t+1}}.$$

□

**Lemma 206** *Let $t$ be a positive integer, $0 < \varepsilon < 1/(2t)^{t+1}$, and $n$ be sufficiently large. Then there exist functions $f : \binom{V(G_n)}{\leq n^{-(t+0.9)}2^n} \rightarrow \binom{V(G_n)}{(t+1+\varepsilon)m}$ and $g : \binom{V(G_n)}{\leq (t+2)m/(\varepsilon^2 n^t)} \rightarrow \binom{V(G_n)}{(t+\varepsilon)m}$ with the following property. For any independent set $I$ in $G_n$, there exist disjoint subsets $S_1, S_2 \subseteq I$ with $|S_1| \leq n^{-(t+0.9)}2^n$ and $|S_1 \cup S_2| \leq (t+2)m/(\varepsilon^2 n^t)$, such that $S_1 \cup S_2$ is disjoint from $g(S_1 \cup S_2)$, $S_2 \subseteq f(S_1)$, and $I \subseteq S_1 \cup S_2 \cup g(S_1 \cup S_2)$.*

**Proof of Lemma.** We fix an arbitrary order of $V(G_n)$ and an independent set $I$. We set $U_0 = V(G_n)$, and, during a process of two phases, we update $U_i$ and add vertices first to $S_1$ and then to $S_2$, both of which are empty at the beginning of the process. At the beginning of each step we select the vertex $u \in U_{i-1}$ with $deg_{G_n[U_{i-1}]}(u)$ maximal, and if there are several such vertices, we take the one with minimal index in the fixed order of $V(G_n)$. The steps in the two phases are as follows:

PHASE I

- If $u \notin I$, then we let $U_i = U_{i-1} \setminus \{u\}$, and proceed to Step $i+1$ in Phase I.
- If $u \in I$ with $deg_{G_n[U_{i-1}]}(u) \geq n^{t+0.9}$, then we add $u$ to $S_1$, let $U_i = U_{i-1} \setminus N_{G_n}[u]$, and proceed to Step $i+1$ in Phase I.
- If $u \in I$ with $deg_{G_n[U_{i-1}]}(u) < n^{t+0.9}$, then we add $u$ to $S_1$, let $U_i = U_{i-1} \setminus N_{G_n}[u]$, define $f(S_1) = U_i$, and proceed to Step $i+1$ in Phase II.

PHASE II

- If $u \notin I$, then we let $U_i = U_{i-1} \setminus \{u\}$, and proceed to Step $i+1$ in Phase II.

- If $u \in I$ with $deg_{G_n[U_{i-1}]}(u) \geq \varepsilon^2 n^t$, then we add $u$ to $S_2$, let $U_i = U_{i-1} \setminus N_{G_n}[u]$, and proceed to Step $i+1$ in Phase II.
- If $u \in I$ with $deg_{G_n[U_{i-1}]}(u) < \varepsilon^2 n^t$, then we add $u$ to $S_2$, let $U_i = U_{i-1} \setminus N_{G_n}[u]$, define $g(S_1 \cup S_2) = U_i$, and stop the process.

Note that by definition, the sets $S_1$ and $S_2$ are disjoint and $S_1 \cup S_2 \subseteq I \subseteq S_1 \cup S_2 \cup g(S_1 \cup S_2)$ holds.

For all vertices $u$ added to $S_1$ in Step $i$ of Phase I (except for the last one), we have $deg_{G_n[U_{i-1}]}(u) \geq n^{t+0.9}$; therefore $|S_1| \leq 1 + 2^n/(1+n^{t+0.9}) \leq 2^n/n^{t+0.9}$. If Phase I ends after Step $i_1$, then, by definition, we know that $deg_{G_n[U_{i_1}]}(v) \leq n^{t+0.9}$ for all $v \in U_{i_1}$; thus $e(U_{i_1}) \leq \frac{1}{2}|U_{i_1}|n^{t+0.9}$. Lemma 205 implies that $U_{i_1} = f(S_1)$ has size at most $(t+1+\varepsilon)m$.

For all vertices $u$ added to $S_2$ in Step $i$ of Phase II (except for the last one), we have $deg_{G_n[U_{i-1}]}(u) \geq \varepsilon^2 n^t$, and their neighbors are never added to $S_1$ (they are not even in $I$); therefore $|S_2| \leq 1 + |f(S_1)|/(1+\varepsilon^2 n^t)$ holds. Hence $|S_1 \cup S_2| \leq 2^n/n^{t+0.9} + (t+1+\varepsilon)m/(\varepsilon^2 n^t) \leq (t+2)m/(\varepsilon^2 n^t)$. If Phase II ends after Step $i_2$, then, by definition we know that $deg_{G_n[U_{i_2}]}(v) \leq \varepsilon^2 n^t$ for all $v \in U_{i_2}$; thus $e(U_{i_2}) \leq \frac{\varepsilon^2}{2}|U_{i_2}|n^t$. Lemma 205 implies that $U_{i_2} = g(S_1 \cup S_2)$ has size at most $(t+\varepsilon)m$.

We can define $f(S)$ and $g(S)$ arbitrarily for sets $S$ that are not obtained as $S_1$ or $S_1 \cup S_2$ for any independent set $I$ in the above process. We are left to prove that if we obtain $S_1$ or $S_1 \cup S_2$ for two different independent sets $I$ and $I'$, then $f(S_1)$ and $g(S_1 \cup S_2)$ do not differ in the two cases. But this follows from the fact that if for $I$ and $I'$ we have $S_1(I) = S_1(I')$, then Phase I is the same for $I$ and $I'$. Similarly, if $S_1(I) \cup S_2(I) = S_1(I') \cup S_2(I')$, then the whole process is the same for $I$ and $I'$ (in particular, $S_1(I) = S_1(I'), S_2(I) = S_2(I')$). These statements can be seen by induction on $i$. If the two processes are identical up to Step $i-1$ (if $i=1$, this trivially holds), then $U_{i-1}$ is the same for both processes, therefore the same vertex $u$ is chosen in Step $i$, and since $S_1(I) = S_1(I')$ (resp. $S_1(I) \cup S_2(I) = S_1(I') \cup S_2(I')$), $u$ is either added to $S_1$ (resp. $S_2$) in both cases or dropped in both cases. □

We prove the following more general statement of which Theorem 204 is the special case $t = 1$: for every $\varepsilon > 0$ and positive integer $t$, there exists a constant $C$ such that if $p \geq \frac{C}{n^t}$, then $Q_n(p)$ does not contain an antichain larger than $(1+\varepsilon)pt\binom{n}{\lfloor n/2 \rfloor}$ w.h.p.

It is enough to prove this statement for $\varepsilon < \frac{1}{(2t)^{t+1}}$. Let us define $C = 10^{10}\varepsilon^{-5}$ and $\varepsilon_1 = \varepsilon/4$. We will show that w.h.p. for every $I$ that is an independent set in $G_n$ of size at least $(1+\varepsilon)ptm$, not all sets corresponding to vertices in $I$ remain in $Q_n(p)$. We will use the union bound, and to bound the number of possibilities we apply Lemma 206 with $\varepsilon_1$ in the role of $\varepsilon$. For every independent set $I$ it gives us $S_1, S_2$ such that

- $S_1 \in \binom{V(G_n)}{\leq n^{-(t+0.9)}2^n}$; therefore the number of possible $S_1$'s is at most

$$\sum_{a \leq n^{-(t+0.9)}2^n} \binom{2^n}{a}.$$

Clearly, we have $\mathbb{P}(S_1 \subseteq Q_n(p)) = p^{|S_1|}$.

- $S_2 \in \binom{V(G_n)}{\leq (t+2)m/(\varepsilon_1^2 n^t)}$ and $S_2 \subseteq f(S_1) \in \binom{V(G_n)}{(t+1+\varepsilon_1)m}$, so for fixed $S_1$ the number of possible $S_2$'s is at most

$$\left|\binom{f(S_1)}{\leq (t+2)m/(\varepsilon_1^2 n^t)}\right| \leq \sum_{b \leq (t+2)m/(\varepsilon_1^2 n^t)} \binom{(t+2)m}{b}.$$

Also, $\mathbb{P}(S_2 \subseteq Q_n(p)) = p^{|S_2|}$.

- for fixed $S_1$ and $S_2$ the corresponding $I$'s are all subsets of $S_1 \cup S_2 \cup g(S_1 \cup S_2)$ and contain $S_1 \cup S_2$. Let $\mathcal{E}_{S_1,S_2}$ be the event that there exists *any* $I$ with $S_1(I) = S_1, S_2(I) = S_2$ and $|I| \geq (1+\varepsilon)ptm$, and $\mathcal{E}_{g(S_1 \cup S_2)}$ be the event that $|Q_n(p) \cap g(S_1 \cup S_2)| \geq (1+\varepsilon)ptm - |S_1 \cup S_2|$ holds.

We bound the probability of the event $\mathcal{E}_{S_1,S_2}$ by the probability of the event $\mathcal{E}_{g(S_1 \cup S_2)}$. Note that

$$(1+\varepsilon)ptm - |S_1 \cup S_2| \geq (1+\varepsilon/2)ptm$$

and

$$|g(S_1 \cup S_2)| \leq (t+\varepsilon_1)m \leq (1+\varepsilon/4)tm.$$

Therefore, $|Q_n(p) \cap g(S_1 \cup S_2)|$ is binomially distributed with $\mathbb{E}(|Q_n(p) \cap g(S_1 \cup S_2)|) \leq (1+\varepsilon/4)pmt$, so by Chernoff's inequality we have

$$\mathbb{P}(\mathcal{E}_{S_1,S_2}) \leq \mathbb{P}[|Q_n(p) \cap g(S_1 \cup S_2)| \geq (1+\varepsilon/2)ptm] \leq e^{-\varepsilon^2 pmt/100}.$$

Note that $S_1$, $S_2$, and $g(S_1 \cup S_2)$ are disjoint, so the above three events are independent; hence the probability, that for fixed $S_1$ and $S_2$ there is a corresponding large antichain, is at most $p^{|S_1|+|S_2|}e^{-\varepsilon^2 pmt/100}$. Summing up for all possible $S_1$ and $S_2$, we obtain that the probability $\Pi$, that there is an antichain in $Q_n(p)$ of size $(1+\varepsilon)ptm$, is at most

$$\sum_{0 \leq a \leq n^{-(t+0.9)}2^n} \sum_{0 \leq b \leq (t+2)m/(\varepsilon_1^2 n^t)} \binom{2^n}{a} p^a \binom{(t+2)m}{b} p^b e^{-\varepsilon^2 pmt/100}.$$

The assumptions $p \geq Cn^{-t}$ and $a \leq n^{-(t+0.9)}2^n$ imply $\frac{\binom{2^n}{a+1}p^{a+1}}{\binom{2^n}{a}p^a} > 1$. Similarly, $p \geq Cn^{-t}$ and $b \leq (t+2)m/(\varepsilon_1^2 n^t)$ imply $\frac{\binom{(t+2)m}{b+1}p^{b+1}}{\binom{(t+2)m}{b}p^b} > 1$, so the largest summand in the above sum belongs to the largest possible values of $a$ and $b$. Therefore the above expression is bounded from above by

$$(n^{-(t+0.9)}2^n + 1)((t+2)m/(\varepsilon_1^2 n^t) + 1)\binom{2^n}{n^{-(t+0.9)}2^n}\binom{(t+2)m}{(t+2)m/(\varepsilon_1^2 n^t)}$$
$$\cdot\, e^{-\varepsilon^2 pmt/100} p^{n^{-(t+0.9)}2^n} p^{(t+2)m/(\varepsilon^2 n^t)}.$$

Note that $pm = \Omega(n^{-(t+1/2)}2^n)$. Therefore

$$(n^{-(t+0.9)}2^n + 1)((t+2)m/(\varepsilon_1^2 n^t) + 1) \le e^{O(n)} \le e^{\varepsilon^2 pmt/(400)}.$$

Also, using $\binom{n}{k} \le (\frac{en}{k})^k$, we have

$$\binom{2^n}{n^{-(t+0.9)}2^n} p^{n^{-(t+0.9)}2^n} \le (en^{t+0.9}p)^{n^{-(t+0.9)}2^n}$$
$$= e^{n^{-(t+0.9)}2^n \ln n} \le e^{\varepsilon^2 pmt/(400)}.$$

Finally, by $C = 10^{10}\varepsilon^{-5}$ and the monotone decreasing property of $\frac{\ln x}{x}$ when $x$ is large enough, we have

$$\binom{(t+2)m}{(t+2)m/(\varepsilon_1^2 n^t)} p^{(t+2)m/(\varepsilon_1^2 n^t)} \le (e\varepsilon_1^2 n^t p)^{(t+2)m/(\varepsilon_1^2 n^t)}$$
$$\le e^{(t+2)mp\frac{\ln(n^t p)}{(\varepsilon^2 n^t p)}} \le e^{\varepsilon^2 pmt/(400)}$$

Therefore, the probability $\Pi$ is at most $e^{-\varepsilon^2 pmt/400} = o(1)$, as required. ■

**Corollary 207** *For every $\varepsilon > 0$ and positive integer $k$, there exists a constant $C$ such that if $p \ge C/n$, then the size of the largest $k$-Sperner family in $Q_n(p)$ is at most $(k+\varepsilon)p\binom{n}{\lfloor n/2 \rfloor}$ w.h.p.*

**Proof.** This follows from the fact that every $k$-Sperner family is the union of $k$ antichains (Lemma 5), and Theorem 204. ■

Let us mention that the proof by Collares Neto and Morris shows directly Corollary 207. They use the following hypergraph container lemma.

**Theorem 208 (Balogh, Morris, Samotij [43], Saxton, Thomason [499])** *For every $k \in \mathbb{N}$ and $c > 0$, there exists a $\delta > 0$, such that the following holds. If $\varepsilon \in (0,1)$ and $\mathcal{H}$ is a $k$-uniform hypergraph on $N$ vertices, such that for every $1 \le l < k$ and $l$-set $L \subseteq V(\mathcal{H})$ we have $|\{H \in E(\mathcal{H}) : L \subseteq H\}| \le c\varepsilon^{l-1}\frac{e(\mathcal{H})}{N}$, then there exists a family $\mathcal{C} \subseteq 2^{V(\mathcal{H})}$ and a function $f : 2^{V(\mathcal{H})} \to \mathcal{C}$ such that:*

*(i) For every independent set $I$ in $\mathcal{H}$ there exists $T \subseteq I$ with $|T| \le k\varepsilon N$ and $I \subseteq T \cup f(T)$,*

*(ii) $|C| \le (1-\delta)N$ for all $C \in \mathcal{C}$.*

Theorem 204 tells us that if $p \geq \frac{C}{n}$ for some constant $C$, then the family of sets of size $\lfloor n/2 \rfloor$ in $Q_n(p)$ is an antichain that is asymptotically largest possible. A natural question to ask if one can eliminate the adjective "asymptotically" for some values of $p$, i.e. if $w(p)$ denotes $\max\{|\binom{[n]}{\lfloor n/2 \rfloor} \cap Q_n(p)|, |\binom{[n]}{\lceil n/2 \rceil} \cap Q_n(p)|\}$, then for what values of $p$ is it true, that the size of the largest antichain in $Q_n(p)$ equals $w(p)$ w.h.p.? Clearly, for this to happen one needs $\binom{[n]}{\lfloor n/2 \rfloor} \cap Q_n(p)$ or $\binom{[n]}{\lceil n/2 \rceil} \cap Q_n(p)$ to be maximal. Let $X$ denote the random variable for the number of sets in $Q_n(p)$ that are not comparable to any set in $\binom{[n]}{\lfloor n/2 \rfloor} \cap Q_n(p)$. Note that $X = \sum_{F \in 2^{[n]} \setminus \binom{n}{\lfloor n/2 \rfloor}} X_F$ where $X_F$ is the indicator random variable of the event that $F \in Q_n(p)$ and $F$ is incomparable to $Q_n(p) \cap \binom{[n]}{\lfloor n/2 \rfloor}$. Then, for the expected value of $X$ we have

$$2\binom{n}{\lfloor n/2 \rfloor - 1} p(1-p)^{n - \lfloor n/2 \rfloor + 1} \leq \mathbb{E}(X),$$

$$\mathbb{E}(X) \leq 2\binom{n}{\lfloor n/2 \rfloor + 1} p(1-p)^{n - \lfloor n/2 \rfloor + 1} + 2^n p(1-p)^{\lfloor n^2/4 \rfloor}.$$

To see the lower bound one only considers summands $X_F$ with $|F| = \lfloor n/2 \rfloor - 1$ or $|F| = \lfloor n/2 \rfloor + 1$. In the upper bound, one dominates $\mathbb{E}(X_F)$ by $p(1-p)^{\lfloor n^2/4 \rfloor}$ as any $F$ is comparable to at least $\lfloor n^2/4 \rfloor$ elements of $\binom{[n]}{\lfloor n/2 \rfloor}$ if $||F| - \lfloor n/2 \rfloor| \geq 2$. Thus, using the fact $\binom{n}{\lfloor n/2 \rfloor} = \Theta(\frac{2^n}{\sqrt{n}})$, we obtain $\mathbb{E}(X) = \Theta(\frac{2^n}{\sqrt{n}} p(1-p)^{n - \lfloor n/2 \rfloor + 1})$. Therefore, $\mathbb{E}(X) \to \infty$ if and only if $p < 3/4$ and $\mathbb{E}(X) = o(1)$ if and only if $p > 3/4$. An easy calculation shows that $\sigma(X) = o(\mathbb{E}(X))$ and therefore, by Chebyshev's inequality, $\binom{[n]}{\lfloor n/2 \rfloor} \cap Q_n(p)$ is not maximal w.h.p if $p < 3/4$.

By Markov's inequality, if $p > 3/4$, then $X = 0$ w.h.p. In probabilistic combinatorics, it happens very often that there is a trivial obstruction for some event $E$ to hold, but once this is gone, $E$ holds w.h.p. Therefore it is natural to formulate the following conjecture (the origin of which we are not absolutely sure of, Hamm and Kahn say in [301] that they heard it from József Balogh).

**Conjecture 209** *If $p > 3/4$, then the size of the largest antichain in $Q_n(p)$ is $w(p)$ w.h.p.*

In [301], Hamm and Kahn claim that using their methods to determine the largest intersecting family in $Q_{2k+1,k}(p)$, they can prove that there exists a positive $\varepsilon$ such that if $p > 1 - \varepsilon$, then the size of the largest antichain in $Q_n(p)$ is $w(p)$ w.h.p.

Let us finish this section by mentioning that after obtaining the appropriate analogue of Lemma 206, Hogenson proved a random version of a theorem of DeBonis and Katona [120]. Let $\bigvee_{r+1}$ be the poset containing $r+2$ elements

$a, b_1, \ldots, b_{r+1}$, with the relations $a < b_i$ for every $i \leq r+1$. The whole Chapter 7 is devoted to families defined by a forbidden subposet. Here we state a random version when $\bigvee_{r+1}$ is forbidden.

**Theorem 210 (Hogenson [310])** *For all $\varepsilon > 0$ and $r \in \mathbb{N}$, there exists a constant $C$ such that, for $t \in \mathbb{N}$, if $p > \frac{C}{n^t}$, then with high probability, the largest $\bigvee_{r+1}$-free family in $Q_n(p)$ has size at most $(1+\varepsilon)p\binom{n}{\lfloor n/2 \rfloor}t$.*

## 4.2 Largest intersecting families in $Q_{n,k}(p)$

The investigation of the largest intersecting family in $Q_{n,k}(p)$ was initiated by Balogh, Bohman and Mubayi [35]. They introduced the following definition.

**Definition 211** *For a family $\mathcal{F} \subseteq \binom{[n]}{k}$ let $i(\mathcal{F})$ denote the size of the largest intersecting subfamily of $\mathcal{F}$. We say that $\mathcal{F}$*

- *satisfies the strong EKR property if all intersecting subfamilies $\mathcal{G} \subseteq \mathcal{F}$ with $|\mathcal{G}| = i(\mathcal{F})$ are trivially intersecting.*
- *satisfies the weak EKR property if there exists a trivially intersecting subfamily $\mathcal{G} \subseteq \mathcal{F}$ with $|\mathcal{G}| = i(\mathcal{F})$.*
- *is not EKR if $|\mathcal{G}| < i(\mathcal{F})$ holds for all trivially intersecting subfamilies $\mathcal{G} \subseteq \mathcal{F}$.*

Let $m = p\binom{n}{k} = \mathbb{E}(|Q_{n,k}(p)|)$ and $\rho = p\binom{n-1}{k-1}$.

**Theorem 212 (Balogh, Bohman, Mubayi [35])** *The following statements hold w.h.p.*

***(i)*** *If $k = o(n^{1/4})$, then $Q_{n,k}(p)$ satisfies the strong EKR property.*

***(ii)*** *If $k = o(n^{1/3})$, then $Q_{n,k}(p)$ satisfies the weak EKR property.*

***iii*** *In the range $n^{1/4} \ll k \ll n^{1/3}$, $Q_{n,k}(p)$ satisfies the strong EKR property if $\rho \ll k^{-1}$ or $n^{-1/4} \ll \rho$, and does not satisfy the strong EKR property if $k^{-1} \ll \rho \ll n^{-1/4}$.*

***(iv)*** *In the range $n^{1/3} \ll k \ll n^{1/2-\varepsilon}$, for any integer $t \geq 3$, if $n^{1/2-1/(2t)} \ll k$ and $n^{(t-3)/2}k^{2-t} \ll \rho \ll n^{-1/t}$, then $Q_{n,k}(p)$ is not EKR.*

***(v)*** *If $k \leq n^{1/2-\varepsilon}$ and $\rho = \Omega(1)$, then $Q_{n,k}(p)$ satisfies the strong EKR property.*

***(vi)*** *Let $(n \ln \ln n)^{1/2} \ll k < n/2$. If*

$$\frac{\ln n}{\binom{n-1}{k}} \ll p \ll \frac{e^{k^2/2n}}{\binom{n}{k}},$$

*then $Q_{n,k}(p)$ is non-trivially intersecting. In particular, $Q_{n,k}(p)$ is not EKR and $i(Q_{n,k}(p)) = (1+o(1))p\binom{n}{k}$.*

The assumption $k \leq n^{1/2-\varepsilon}$ in Theorem 212 **(iv)** and **(v)** is not by chance. The situation changes dramatically depending on whether $k = o(n^{1/2})$ or $k = \omega(n^{1/2})$. This is due to the fact that the probability $q := \frac{\binom{n}{k}-\binom{n-k}{k}}{\binom{n}{k}}$ of the event that two randomly chosen $k$-subsets $A$ and $B$ intersect tends to 0 as long as $k = o(n^{1/2})$, while $q \to 1$ if $k = \omega(n^{1/2})$. The next theorem provides a characterization of all values of $p(n,k)$ for which $Q_{n,k}(p)$ satisfies the strong EKR property if $k \leq \sqrt{Cn \ln n}$ for some fixed positive constant $C < 1/4$. For an integer $t$ let us define $\lambda(t) = \lambda_\rho(t) = \lambda_p(t) = \binom{m}{t} q^{\binom{t}{2}}$. The variable $|Q_{n,k}(p)|$ is binomially distributed and very much concentrated around $m$, therefore $\lambda(t)$ is roughly the expected number of intersecting families of size $t$ in $Q_{n,k}(p)$. Let $\Delta$ denote the maximum degree, i.e. the maximum size of a trivially intersecting family in $Q_{n,k}(p)$. If $\lambda(\Delta)$ is small, then we expect $Q_{n,k}(p)$ to satisfy the strong EKR property. We need one tiny modification due to the fact that every intersecting family of size 2 is trivially intersecting. Let

$$\lambda'_\rho(t) = \begin{cases} 0 & \text{if } t \leq 2 \\ \lambda_\rho(t) & \text{otherwise.} \end{cases}$$

**Theorem 213 (Hamm, Kahn [300])** *For any $C < 1/4$, if $k \leq \sqrt{Cn \ln n}$, then $Q_{n,k}(p)$ satisfies the strong EKR property w.h.p if and only if $\lambda'_\rho(\Delta) = o(1)$ w.h.p.*

The next theorem gives a sufficient condition for $Q_{n,k}(p)$ to satisfy the strong EKR property for a wider range of $k$.

**Theorem 214 (Balogh, Das, Delcourt, Liu, Sharifzadeh [40])** *For a function $k = k_n$ with $3 \leq k \leq n/4$, if*

$$p \geq \frac{9n \log(\frac{ne}{k})\binom{2k}{k}\binom{n}{k}}{\binom{n-k}{k}^2},$$

*then $Q_{n,k}(p)$ satisfies the strong EKR property w.h.p.*

**Proof.** For any $\mathcal{G} \subseteq \binom{[n]}{k}$ let us write $\mathcal{G}_p = \mathcal{G} \cap Q_{n,k}(p)$. Let $\mathcal{F}_0 := \{F \in \binom{[n]}{k} : 1 \in F\}$ and let $\mathcal{F}_1, \mathcal{F}_2, \ldots, \mathcal{F}_M$ be all the maximal non-trivially intersecting families. By Lemma 135, we know that $M \leq \binom{n}{k}^{\binom{2k-1}{k-1}} < 2^{k \log(\frac{ne}{k})\binom{2k-1}{k-1}}$ holds. Also, Theorem 41 implies that $|\mathcal{F}_i| \leq \binom{n-1}{k-1} - \binom{n-k-1}{k-1} + 1$ for all $1 \leq i \leq M$.

We define $D := p\binom{n-k-1}{k-1}/3$ and the following events: $\mathcal{E}_0 = \{|(\mathcal{F}_0)_p| \leq p\binom{n-1}{k-1} - D\}$ and $\mathcal{E}_i = \{|(\mathcal{F}_i)_p| \geq p(\binom{n-1}{k-1} - \binom{n-k-1}{k-1} + 1) + D\}$. If none of these events hold, then $(\mathcal{F}_0)_p$ is larger than all $(\mathcal{F}_i)_p$'s and thus $Q_{n,k}(p)$ satisfies the strong EKR property. Therefore it is enough to show $\sum_{i=0}^{M} \mathbb{P}(\mathcal{E}_i) = o(1)$.

The Chernoff inequality gives

$$\mathbb{P}(\mathcal{E}_0) \le \exp\left(-\frac{D^2}{2p\binom{n-1}{k-1}}\right)$$

and

$$\mathbb{P}(\mathcal{E}_i) \le \exp\left(-\frac{D^2}{2p(\binom{n-1}{k-1} - \binom{n-k-1}{k-1} + 1)}\right) < \exp\left(-\frac{D^2}{2p\binom{n-1}{k-1}}\right).$$

Therefore

$$\begin{aligned}\sum_{i=0}^{M} \mathbb{P}(\mathcal{E}_i) &\le (M+1)\exp\left(-\frac{D^2}{2p\binom{n-1}{k-1}}\right) \\ &\le (2^{k\log(\frac{ne}{k})\binom{2k-1}{k-1}} + 1)\exp\left(-\frac{p\binom{n-k-1}{k-1}^2}{18\binom{n-1}{k-1}}\right).\end{aligned}$$

This tends to 0 as $p \ge \frac{9n\log(\frac{ne}{k})\binom{2k}{k}\binom{n}{k}}{\binom{n-k}{k}^2} \ge \frac{18k\log(\frac{ne}{k})\binom{2k-1}{k-1}\binom{n-1}{k-1}}{\binom{n-k-1}{k-1}^2}$. ■

The most interesting extreme case is when $n = 2k+1$. Let $X$ denote the random variable of the number of pairs $(x, F)$ such that $x \in [n] \setminus F$, where $F \in Q_{n,k}(p)$ and $F$ meets all sets $G \in Q_{n,k}(p)$ with $x \in G$. The event $X = 0$ is equivalent to all stars being maximal with respect to the intersecting property. Clearly, if $n = 2k+1$, then $\mathbb{E}(X) = (2k+1)\binom{2k}{k}p(1-p)^k$. Therefore $\mathbb{E}(X) \to \infty$ as long as $p < 3/4$. It is not very hard to see that in that case no trivially intersecting family is maximal in $Q_{2k+1,k}(p)$, so $p > 3/4$ is a necessary condition for $Q_{2k+1,k}(p)$ to satisfy the strong EKR property w.h.p. The next theorem is a huge step towards proving that this condition is sufficient as well.

**Theorem 215 (Hamm, Kahn [301])** *There exists a positive $\varepsilon$ such that $Q_{2k+1,k}(p)$ satisfies the strong EKR property w.h.p if $p > 1-\varepsilon$.*

Let us turn our attention to the order of magnitude of $i(Q_{n,k}(p))$. Note that Theorem 212 settles the case $k = o(n^{1/2-\varepsilon})$ in a much stronger form.

**Theorem 216 (Balogh, Bohman, Mubayi [35])** *Let $\varepsilon = \varepsilon_n > 0$. If $\log n \ll k < (1-\varepsilon)n/2$ and $p \gg (1/\varepsilon)(\ln n/k)^{1/2}$, then*

$$i(Q_{n,k}(p)) \le (1+o(1))p\binom{n-1}{k-1}$$

*holds w.h.p.*

This was later improved to an almost complete description of the asymptotics of $i(Q_{n,k}(p))$ by the following result.

**Theorem 217 (Gauy, H'an and Oliveira [255])** *Let $p = p_n \in (0,1)$, $k = k_n$ with $2 \le k < n/2$, $N = \binom{n}{k}$ and $D = \binom{n-k}{k}$. For all $0 < \varepsilon < 1$ there exists a constant $C$ such that the following holds w.h.p.*

*(i)* $i(Q_{n,k}(p)) = (1+o(1))pN$, *if* $N^{-1} \ll p \ll D^{-1}$,

*(ii)* $i(Q_{n,k}(p)) \ge (1-\varepsilon)\frac{N}{D}\ln(pD)$, *if* $D^{-1} \ll p \ll (n/k)D^{-1}$ *and* $k \gg \sqrt{n \ln n}$,

*(iii)* $i(Q_{n,k}(p)) \le \frac{CN}{D}\ln(pD)$, *if* $D^{-1} \ll p \le (n/k)^{1-\varepsilon}D^{-1}$,

*(iv)* $i(Q_{n,k}(p)) = (1+o(1))p\frac{k}{n}N = (1+o(1))p\binom{n-1}{k-1}$, *if* $p \ge C(n/k)\ln^2(n/k)D^{-1}$.

## 4.3 Removing edges from $Kn(n,k)$

The Kneser graph $Kn(n,k)$ has vertex set $\binom{[n]}{k}$ with two $k$-sets $F, G$ connected by an edge if and only if they are disjoint. Intersecting subfamilies of $\binom{[n]}{k}$ correspond to independent sets in $Kn(n,k)$. Results of the previous section could be formulated as finding the independence number of the induced subgraph $Kn(n,k)[U]$, where $U$ is the random subset of the vertices of $Kn(n,k)$, such that every vertex $v$ belongs to $U$ with probability $p$, independently of all other vertices.

In this section we will be interested in removing edges of $Kn(n,k)$, instead of removing its vertices. Let $Kn_p(n,k)$ be the random subgraph of $Kn(n,k)$, where every edge is kept with probability $p$ independently of all other edges. As $\alpha(Kn_p(n,k)) \ge \alpha(Kn(n,k)) = \binom{n-1}{k-1}$, the first task is to determine the threshold probability $p_0$ such that if $p \ge p_0$, then the independence number of $Kn(n,k)$ is still $\binom{n-1}{k-1}$. The next theorem establishes the value of $p_0$ if $k$ is relatively small compared to $n$.

**Theorem 218 (Bollobás, Narayanan, and Raigorodski [74])** *For any real $\varepsilon > 0$ and $k = k_n$ with $2 \le k = o(n^{1/3})$ and*

$$p_0 = \frac{(k+1)\ln n - k\ln k}{\binom{n-1}{k-1}},$$

*we have*

$$\lim_{n\to\infty} \mathbb{P}\left(\alpha(Kn_p(n,k)) = \binom{n-1}{k-1}\right) = \begin{cases} 0 & \text{if } p \le (1-\varepsilon)p_0 \\ 1 & \text{if } p \ge (1+\varepsilon)p_0. \end{cases}$$

*Furthermore, when $p \ge (1+\varepsilon)p_0$, the only independent sets of size $\binom{n-1}{k-1}$ are the subsets of vertices corresponding to trivially intersecting families w.h.p.*

Theorem 218 was improved by Balogh, Bollobás and Narayanan to larger values of $k$.

**Theorem 219 (Balogh, Bollobás and Narayanan [38])** *For any $\varepsilon > 0$ there exist constants $c = c(\varepsilon)$ and $c' = c'(\varepsilon)$, such that for any pair $n, k$ of positive integers with $k \leq (1-\varepsilon)n/2$ we have*

$$\lim_{n\to\infty} \mathbb{P}\left(\alpha(Kn_p(n,k) = \binom{n-1}{k-1}\right) = \begin{cases} 0 & \text{if } p \leq \binom{n-1}{k-1}^{-c} \\ 1 & \text{if } p \geq \binom{n-1}{k-1}^{-c'} \end{cases}.$$

*In particular, $\alpha(Kn_{1/2}(n,k)) = \binom{n-1}{k-1}$ holds w.h.p.*

The next theorem determines the asymptotics of the threshold function of the event $\alpha(Kn_p(n,k)) = \binom{n-1}{k-1}$ for $k \leq Cn$ for some constant $C$, and also determines the correct order of magnitude of the threshold function, if $k$ is bounded away from $n/2$, i.e. $k \leq (1-\varepsilon)n/2$ and thus $\frac{2Cn}{n-2k}p_c = \Theta_\varepsilon(p_c)$ for some positive constant $\varepsilon$.

**Theorem 220 (Das, Tran [118])** *For any $\varepsilon > 0$ there exists a constant $C$, such that for any pair $n, k$ of positive integers with $\omega(1) = k$, $\varepsilon = \omega(k^{-1}\ln k)$ and*

$$p_c = \frac{\ln\left(n\binom{n-1}{k}\right)}{\binom{n-k-1}{k-1}},$$

*we have*

$$\lim_{n\to\infty} \mathbb{P}\left(\alpha(Kn_p(n,k) = \binom{n-1}{k-1}\right) = \begin{cases} 0 & \text{if } p \leq (1-\varepsilon)p_c \text{ and } k \leq \frac{n-3}{2} \\ 1 & \text{if } p \geq (1+\varepsilon)p_c \text{ and } k \leq \frac{n}{6C} \end{cases}.$$

*Furthermore, in the latter case trivially intersecting families correspond to all the independent sets of size $\binom{n-1}{k-1}$ w.h.p. Also $\lim_{n\to\infty}\mathbb{P}\alpha(Kn_p(n,k) == \binom{n-1}{k-1})1$ holds for $k \leq (n-3)/2$, if $p \geq \frac{2Cn}{n-2k}p_c$.*

Note that the threshold probabilities in Theorem 218 and Theorem 220 satisfy $p_0 = (1+o(1))p_c$ as long as $k = o(n^{1/2})$. The next result fills the gap of Theorem 220 and determines the order of magnitude of the threshold function for $n \geq 2k+2$.

**Theorem 221 (Devlin, Kahn [132])** *(i) If $n \geq 2k+2$, then the order of magnitude of the threshold function of the event $\alpha(Kn_p(n,k)) = \binom{n-1}{k-1}$ has the same order of magnitude as $p_c$.*

***(ii)*** *There exists a constant $p < 1$ such that $\mathbb{P}(\alpha(Kn_p(2k+1,k)) = \binom{2k}{k-1}) \to 1$ as $k$ tends to infinity.*

In all of the above four theorems, the proofs of statements

$$\lim_{n\to\infty} \mathbb{P}\left(\alpha(Kn_p(n,k) = \binom{n-1}{k-1}\right) = 1$$

for appropriate values of $p$ are quite involved, and exceed the framework of our book. However, the proof of the other implication is very simple. Suppose $2k+2 \le n$, $p \le (1-\varepsilon)p_c$, $k$ tends to infinity and $\varepsilon = \omega(k^{-1}\ln k)$. Let $\mathcal{E}_F$ be the event that a $k$-set $F$ not containing 1 can be added to the independent set $\mathcal{I} = \{H \in \binom{[n]}{k} : 1 \in H\}$ in $Kn_p(n,k)$. Then we have $\mathbb{P}(\mathcal{E}_F) = (1-p)^{\binom{n-k-1}{k-1}}$. Also, the events $\{\mathcal{E}_F : F \in \binom{[n]}{k}, 1 \notin F\}$ are mutually independent; therefore we have

$$\mathbb{P}(\mathcal{I} \text{ cannot be extended in } Kn_p(n,k)) = \left(1-(1-p)^{\binom{n-k-1}{k-1}}\right)^{\binom{n-1}{k}}.$$

As this bound is increasing in $p$, we can assume $p = (1-\varepsilon)p_c = \frac{(1-\varepsilon)\log(n\binom{n-1}{k})}{\binom{n-k-1}{k-1}}$. The assumption $n \ge 2k+2$ implies $p = O(n^{-1}) = o(\varepsilon)$, so using $1-x \ge \exp(-x(1+x))$ we have

$$\begin{aligned}(1-p)^{\binom{n-k-1}{k-1}} &\ge \exp\left(-p(1+p)\binom{n-k-1}{k-1}\right)\\ &= \exp\left(-(1-\varepsilon)(1+p)\log\left(n\binom{n-1}{k}\right)\right)\\ &= \left(n\binom{n-1}{k}\right)^{-(1-\varepsilon)(1+p)} \ge \left(n\binom{n-1}{k}\right)^{-(1-\varepsilon/2)}.\end{aligned}$$

Thus

$$\begin{aligned}\left(1-(1-p)^{\binom{n-k-1}{k-1}}\right)^{\binom{n-1}{k}} &\le \exp\left(-\binom{n-1}{k}(1-p)^{\binom{n-k-1}{k-1}}\right)\\ &\le \exp\left(-n^{-1}\binom{n-1}{k}^{\varepsilon/2}\right) = o(1),\end{aligned}$$

as $\varepsilon = \omega(k^{-1}\ln k)$. We obtained that $\mathcal{I}$ is not a maximal clique w.h.p.; therefore $\lim_{n\to\infty}\mathbb{P}\left(\alpha(Kn_p(n,k) = \binom{n-1}{k-1}\right) = 0$.

Similar calculation works if $n = 2k+1$ and $p < 3/4$.

## 4.4 $G$-intersecting families

In this section we consider the notion of *$G$-intersecting families*, introduced by Bohman, Frieze, Ruszinkó and Thoma [60]. Given a graph $G$ with $V(G) = [n]$, we say that a family $\mathcal{F} \subseteq \binom{[n]}{k}$ is $G$-intersecting if for every $F, F' \in \mathcal{F}$ there exist $x \in F, y \in F'$, such that either $x = y$ (i.e. $F \cap F' \neq \emptyset$) or $(x,y) \in E(G)$. Equivalently, $\mathcal{F}$ is $G$-intersecting if for any $F, F' \in \mathcal{F}$ we have $F \cap N_G[F'] \neq \emptyset$.

Obviously, there is no randomness involved in the definition of $G$-intersecting families. The reasons for which we include this section to this chapter are the following

- just as in the previous section, a family is $G$-intersecting if the corresponding vertices form an independent set in an appropriate subgraph of $Kn(n,k)$,
- there are relatively easy random arguments, showing that families consisting of almost all sets of $\binom{[n]}{k}$ form a $G$-intersecting family if $G$ is sparse and $k \geq cn$, for an appropriate constant $0 < c < 1/2$,
- there are results on the size of $G$-intersecting families with $G$ being the Erdős-Rényi random graph $G(n,p)$.

Observe that if $K$ is a clique in $G$, then $\mathcal{F}_K = \{F \in \binom{[n]}{k} : F \cap K \neq \emptyset\}$ is a $G$-intersecting family; thus denoting by $N(G,k)$ the maximum size of a $G$-intersecting family, we have $N(G,k) \geq \binom{n}{k} - \binom{n-\omega(G)}{k}$. However, such families are not necessarily maximal. Indeed, if $M \subseteq [n] \setminus K$ is a set of vertices with $K \subseteq N_G(M)$, then all $k$-sets containing $M$ could be added to $\mathcal{F}_K$, to obtain a larger $G$-intersecting family. If $K$ is a maximal clique, then any such $M$ should contain at least 2 vertices; hence the number of extra sets is at most $\binom{|N(K)\setminus K|}{2}\binom{n-|K|-2}{k-2}$.

Strengthening a result of [60], Bohman and Martin obtained the following theorem.

**Theorem 222 (Bohman, Martin [61])** *For every $\Delta$ and $\omega$ there exist a constant $C = C_{\Delta,\omega}$ such that if $G$ is graph on $n$ vertices with maximum degree $\Delta$ and clique number $\omega$, $k \leq Cn^{1/2}$ and $\mathcal{F} \subseteq \binom{[n]}{k}$ is $G$-intersecting, then*

$$|\mathcal{F}| \leq \binom{n}{k} - \binom{n-\omega}{k} + \binom{\omega(\Delta-\omega+1)}{2}\binom{n-\omega-2}{k-2}$$

*holds. Furthermore, if $\mathcal{F}$ is a $G$-intersecting family of maximum size, then there exists a maximum size clique $K$ such that $\mathcal{F}$ contains all $k$-sets intersecting $K$.*

**Proof.** Let $\mathcal{F}$ be a $G$-intersecting family and let us define $\mathcal{N} = \{N_G[F] : F \in \mathcal{F}\}$, and from here on we write $N[F]$ for $N_G[F]$. The next lemma shows that if $\mathcal{F}$ is of maximum size, then $\tau(\mathcal{N}) = 1$ must hold.

**Lemma 223** *Let $G$ be a graph with vertex set $[n]$, clique number $\omega$ and maximum degree $\Delta$. If $k \leq \sqrt{\frac{\omega n}{2(\Delta+1)^2}}$, $\mathcal{F} \subseteq \binom{[n]}{k}$ is a $G$-intersecting family, then $\tau(\mathcal{N}) \geq 2$ implies $|\mathcal{F}| < \binom{n}{k} - \binom{n-\omega}{k}$.*

**Proof of Lemma.** Let $\mathcal{F}$ be as in the lemma, and suppose $|\mathcal{F}| \geq \binom{n}{k} - \binom{n-\omega}{k}$. For any $X \subseteq [n]$, we let $\mathcal{F}_X = \{F \in \mathcal{F} : X \subseteq F\}$ and $\mathcal{N}_X = \{N \in \mathcal{N} : X \subseteq N\}$, and we write $\mathcal{F}_x := \mathcal{F}_{\{x\}}$, $\mathcal{N}_x := \mathcal{N}_{\{x\}}$ and $\tau = \tau(\mathcal{N})$. We will sometimes consider $\mathcal{N}_X$ as a multi-set when looking at its cardinality; i.e. we count the number of sets in $\mathcal{F}$ that have $X$ in their neighborhood, and we count $F$ and $F'$ separately, even if $N_G(F) = N_G(F')$.

Note that as $\mathcal{F}$ is $G$-intersecting, we know that for every $F \in \mathcal{F}$ and $N \in \mathcal{N}$ we have $F \cap N \neq \emptyset$. Let $x \in [n]$ arbitrary. As $\tau \geq 2$, there exists $N_1 \in \mathcal{N}$ with $x \notin N_1$. We obtain $\mathcal{F}_x = \cup_{y_1 \in N_1} \mathcal{F}_{\{x,y_1\}}$. Similarly, let $Y_i = \{x, y_1, \ldots, y_{i-1}\}$ be an $i$-set with $i < \tau$; then there exists $N_i \in \mathcal{N}$ with $Y_i \cap N_i = \emptyset$ and thus $\mathcal{F}_{Y_i} = \cup_{y_j \in N_i} \mathcal{F}_{Y_i \cup \{y_j\}}$. As for every $N \in \mathcal{N}$ we have $|N| \leq k(\Delta+1)$, it follows that

$$|\mathcal{F}_x| \leq ((\Delta+1)k)^{\tau-1}\binom{n-\tau}{k-\tau}.$$

On the other hand, by definition, there exists a $u \in [n]$ with

$$\frac{1}{\tau}\left(\binom{n}{k} - \binom{n-\omega}{k}\right) \leq |\mathcal{N}_u|,$$

if the $\mathcal{N}_u$'s are considered as multisets. Therefore, there exists $x \in [n]$ such that

$$\frac{1}{(\Delta+1)\tau}\left(\binom{n}{k} - \binom{n-\omega}{k}\right) \leq |\mathcal{F}_x|$$

holds. We obtained

$$\binom{n}{k} - \binom{n-\omega}{k} \leq \tau(\Delta+1)^{\tau}k^{\tau-1}\binom{n-\tau}{k-\tau}.$$

Note that the right hand side is a decreasing function of $\tau$. Indeed,

$$\frac{\tau(\Delta+1)^{\tau}k^{\tau-1}\binom{n-\tau}{k-\tau}}{(\tau+1)(\Delta+1)^{\tau+1}k^{\tau}\binom{n-\tau-1}{k-\tau-1}} = \frac{\tau(n-\tau)}{(\tau+1)k(\Delta+1)(k-\tau)} \geq \frac{n}{2k^2(\Delta+1)} \geq 1$$

by the assumption on $n$ and $k$.

We obtained

$$\binom{n}{k} - \binom{n-\omega}{k} \leq 2(\Delta+1)^2 k\binom{n-2}{k-2},$$

which is false, as $\binom{n}{k} - \binom{n-\omega}{k} \leq \omega\binom{n-1}{k-1} = \omega\frac{n-1}{k-1}\binom{n-2}{k-2}$ and $tk < \sqrt{\frac{\omega n}{2(\Delta+1)^2}}$. □

By Lemma 223 we may assume that $\mathcal{N}$ is covered by a vertex $u$. By the maximality of $\mathcal{F}$ we know that $|\mathcal{F}_u| = \binom{n-1}{k-1}$. Let us define $K = \{v \in [n] : |\mathcal{F}_v| = \binom{n-1}{k-1}\}$. Note that $K$ is a clique, as if $v_1, v_2 \in K$ are not joined by an edge in $G$, then a $k$-set $F_1$ with $v \in F_1 \subseteq [n] \setminus N[v_2]$ and a $k$-set $F_2$ with $v_2 \in F_2 \subseteq [n] \setminus N[v_1]$ contradicts the $G$-intersecting property of $\mathcal{F}$. As $|N[F]| \leq k(\Delta+1)$ for any $F \in \mathcal{F}$, such sets exist if $n > (\Delta+2)k$.

We next show that $K$ is maximal. Let $w \in [n]$ with $|\mathcal{F}_w| < \binom{n-1}{k-1}$. Then there exists $F \in \mathcal{F}$ with $w \notin N[F]$, as otherwise any $k$-set containing $w$ could be added to $\mathcal{F}$. The $G$-intersecting property implies that every $F'$ must intersect $N[F]$; therefore

$$|\mathcal{F}_w| \leq (\Delta+1)k\binom{n-2}{k-2}.$$

As $\tau = 1$, every $F \in \mathcal{F}$ must meet $N[u]$. Therefore summing the above bound for all vertices $w \in N(u) \setminus K$ we obtain that the number of sets in $\mathcal{F}$ not meeting $K$ is at most

$$\Delta(\Delta+1)k\binom{n-2}{k-2}.$$

If $K$ was not a maximal clique, then we could remove these sets from $\mathcal{F}$ and replace them with $\{H \in \binom{[n]\setminus K}{k} : w_0 \in H\}$, where $w_0$ is a vertex joined to all vertices in $K$. But this contradicts that $\mathcal{F}$ is of maximum size, if $\Delta(\Delta+1)k\binom{n-2}{k-2} < \binom{n-|K|-1}{k-1}$, which does hold, provided $k \leq \sqrt{\frac{n}{\Delta(\Delta+1)}}$.

Note that every $F \in \mathcal{F}$ that is disjoint from $K$, must meet $N[K] \setminus K$ in at least 2 vertices. Indeed, if the intersection is only one vertex $y$, then, by the maximality of $K$, there exists a vertex $x \in K$ that is not adjacent to $y$, which implies $x \notin N[F]$. But then, as $x \in K$ and $n > (\Delta+2)k$, we can find $F' \subseteq [n] \setminus N[F]$ with $x \in F'$. As $F' \in \mathcal{F}$ by definition of $K$, the sets $F$ and $F'$ would contradict the $G$-intersecting property of $\mathcal{F}$. This means we count every $F \in \mathcal{F}$ at least once if we pick two vertices from $N[K] \setminus K$ and then $k-2$ other vertices from $[n] \setminus K$. As $|N[K] \setminus K| \leq \omega(\Delta - \omega + 1)$, this implies

$$|\mathcal{F}| \leq \binom{n}{k} - \binom{n-|K|}{k} + \binom{\omega(\Delta-\omega+1)}{2}\binom{n-|K|-2}{k-2}.$$

Therefore, all we have to show is that $K$ is of size $\omega$. Suppose there is a clique $K'$ in $G$ of size $|K|+1$. Then the family $\{H \in \binom{[n]}{k} : H \cap K' \neq \emptyset\}$ is $G$-intersecting and has size $\binom{n}{k} - \binom{n-|K|-1}{k}$. This contradicts that $\mathcal{F}$ is of maximum size if

$$\binom{n}{k} - \binom{n-|K|}{k} + \binom{\omega(\Delta-\omega+1)}{2}\binom{n-|K|-2}{k-2} < \binom{n}{k} - \binom{n-|K|-1}{k}$$

holds. But $\binom{n-|K|}{k} - \binom{n-|K|-1}{k} = \binom{n-|K|-1}{k-1} = \frac{n-|K|-1}{k-1}\binom{n-|K|-2}{k-2}$, so the above inequality does hold, provided $k = o(n)$. ■

In particular, Theorem 222 implies that $N(C_n, k) = \binom{n}{k} - \binom{n-2}{k} + \binom{n-4}{k-2}$ holds if $k < Cn^{1/2}$. At the other end of the spectrum, for some sparse graphs $N(G,k)$ could almost equal $\binom{n}{k}$ if $k$ is linear in $n$.

**Theorem 224 (Bohman, Frieze, Ruszinkó, Thoma [60])** *(i) If $\delta$ is the minimum degree of $G$ and $k(\delta+1) > cn\log n$ holds, then $N(G,k) \geq (1 - n^{1-c})\binom{n}{k}$.*

*(ii) If $\delta$ is the minimum degree of $G$, $c$ is a constant with $c - (1-c)^{\delta+1} > 0$ and the maximum degree $\Delta = \Delta(G)$ satisfies $\Delta = O(\sqrt{n/\log n})$, then $N(G,k) = (1-o(1))\binom{n}{k}$, provided $k \geq cn$ holds.*

For the cycle $C_n$, Theorem 222 shows that if $k = O(n^{1/2})$, then the "clique construction" is best possible, while Theorem 224 implies that if $k \geq 0.32n$, then almost all $k$-sets can be included in a $C_n$-intersecting family. Bohman et al. conjectured that there is a sharp threshold for $N(C_n, k)$, i.e. there exists a constant $c$ such that for any positive $\varepsilon$, if $k \leq (c-\varepsilon)n$, then $N(C_n, k) = 2\binom{n-1}{k-1} - \binom{n-2}{k} + \binom{n-4}{k-2}$, while if $k \geq (c+\varepsilon)n$, then $N(C_n, k) = (1-o(1))\binom{n}{k}$. They were also interested in whether a similar phenomenon holds for other graphs and what kind of extremal $G$-intersecting families can occur, that are different from the star construction.

Note that not all sparse graphs $G$ with bounded minimum degree have the property, that $N(G, k) = (1-o(1))\binom{n}{k}$, when $k \geq cn$ for some constant $0 < c < 1/2$. Indeed, consider with a fixed $a$ the complete bipartite graph $K_{a,n-a}$ (see Exercise 227). Johnson and Talbot found a graph $G$ (see Exercise 229), for which the extremal $G$-intersecting family does not contain all $k$-sets meeting a fixed clique of $G$. They also showed the sharp threshold phenomenon for the matching $M_n$.

**Theorem 225 (Johnson, Talbot [321])** *Let $n = 2t \geq 1000$, $1 \leq k \leq n$ and $d = 1 - 2^{-1/2}$. Then*

$$N(M_n, k) = \begin{cases} \binom{n}{k} - \binom{n-2}{k} & \text{if } k < dn \\ (1-o(1))\binom{n}{k} & \text{if } k \geq dn(1+\varepsilon_n), \end{cases}$$

*where $\varepsilon_n = 30\sqrt{\log n/n} = o(1)$.*

The second statement follows from the fact that the family

$$\mathcal{F}^+ = \left\{F \in \binom{[n]}{k} : F \text{ meets more than } t \text{ edges of } M_n\right\}$$

is $M_n$-intersecting and $|\mathcal{F}^+| = (1-o(1))\binom{n}{k}$ if $k \geq dn(1+\varepsilon_n)$.

Johnson and Talbot found another sufficient condition for $N(G_n, k) = (1-o(1))\binom{n}{k}$. Their result improves the bound of Theorem 224 for $C_n$ and implies $N(C_n, k) = (1-o(1))\binom{n}{k}$ if $k > \alpha n$, where $\alpha = 0.266...$ is the smallest root of the equation $(1-x)^3(1+x) = 1/2$. They also considered the non-uniform version of the problem and found that if $G_n$ is a sequence of graphs, such that the size $\nu(G_n)$ of the largest matching in $G_n$ is non-decreasing, then the size of the largest $G_n$-intersecting family is $(1-o(1))2^n$ if and only if $\nu(G_n) \to \infty$. They gave upper and lower bounds linear in $2^n$ when $\nu(G_n)$ is fixed.

Finally, let us consider the Erdős-Rényi random graph $G(n,p)$. It is well-known [66] that its clique number is $(2+o(1))\log_{1/p} n$ w.h.p., and therefore $N(G(n,p), k) \geq ((2+o(1))\log_{1/p} n)\binom{n-(2+o(1))\log_{1/p} n}{k-1}$ holds w.h.p. The last result of this section shows that the order of magnitude is correct if $k$ and $p$ are both fixed as $n$ tends to infinity.

**Theorem 226 (Bohman, Frieze, Ruszinkó, Thoma [60])** *Let $0 < p <$*

*1 be a fixed real and $2 \leq k$ a fixed integer. Then*

$$N(G(n,p),k) = \Theta\left(\binom{n}{k-1}\log n\right)$$

*holds w.h.p.*

**Exercise 227** *Determine the maximum size of a $G$-intersecting family with $G = K_{a,n-a}$.*

**Exercise 228** *Prove Theorem 224 (i) by considering the probability, that for a random $k$-set $F$ one has $N_G[F] = [n]$.*

**Exercise 229** *Let $G_n$ be the graph with 3 independent edges and $n-6$ isolated vertices. Show that*

- *the family $\mathcal{F}$ of $k$-sets that meet at least two of the three edges of $G_n$ form a $G_n$-intersecting family, that is larger than the family $\mathcal{F}'$ consisting of all $k$ sets that meet one fixed edge of $H_n$,*
- *$N(G_n,k) \geq \binom{n}{k} - \binom{n-7}{k-1}$, therefore, if $k \leq (1/2-\varepsilon)n$, then $N(G_n,k) \neq (1-o(1))\binom{n}{k}$.*

## 4.5 A random process generating intersecting families

In this section we consider the following random process introduced in [57] that selects a maximal intersecting family in $\binom{[n]}{k}$.

PROCESS GENERATING MAXIMAL INTERSECTING FAMILY.

Let $\mathcal{F}_0 = \emptyset$, and for any $\mathcal{F} \subseteq \binom{[n]}{k}$ let $\mathcal{I}(\mathcal{F}) = \{G \in \binom{[n]}{k} \setminus \mathcal{F} : \forall F \in \mathcal{F}\ G \cap F \neq \emptyset\}$. If $\mathcal{F}_i = \{F_1, F_2, \ldots, F_i\}$ is defined, select $F_{i+1}$ from $\mathcal{I}(\mathcal{F})$ uniformly at random and put $\mathcal{F}_{i+1} = \mathcal{F}_i \cup \{F_{i+1}\}$. The process ends when $\mathcal{I}(\mathcal{F}_j) = \emptyset$ and returns $\mathcal{F}_j$, that is a maximal intersecting family in $\binom{[n]}{k}$.

The following theorem determines for what functions $k = k_n$ this random process generates a trivially intersecting family w.h.p.

**Theorem 230 (Bohman, Cooper, Frieze, Martin, Ruszinkó [57])** *Let $\mathcal{E}_{n,k}$ denote the event that the process returns a trivially intersecting family. Then*

$$\lim_{n\to\infty} \mathbb{P}(\mathcal{E}_{n,k}) = \begin{cases} 1 & \text{if } k = o(n^{1/3}), \\ \frac{1}{1+c^3} & \text{if } \frac{k}{n^{1/3}} \to c, \\ 0 & \text{if } k = \omega(n^{1/3}). \end{cases}$$

**Proof.** We begin with three lemmas that will cover the cases $k = o(n^{1/2})$. Let $A_i$ denote the event that $\mathcal{F}_i = \{F_1, F_2, \ldots, F_i\}$ is a sunflower with a kernel of size one.

**Lemma 231** *Suppose $k = o(n^{1/2})$ holds. Then $\mathbb{P}(A_2) = 1 - o(1)$.*

**Proof of Lemma.** The number $N_j$ of $k$-sets that meet $F_1$ in exactly $j$ vertices is $\binom{k}{j}\binom{n-k}{k-j}$. As $k = o(n^{1/2})$ holds, we have $N_j/N_{j+1} = \frac{(j+1)(n-k+1)}{(k-j+1)^2} \leq 1/2$ for $n$ large enough, and thus $\sum_{j=2}^{k-1} N_j \leq 2N_2 = o(N_1)$. Therefore $\mathbb{P}(A_2) = \frac{N_1}{\sum_{j=1}^{k-1} N_j} = 1 - o(1)$. □

**Lemma 232** *Suppose $k = o(n^{1/2})$ holds. Then*

$$\lim_{n\to\infty} \mathbb{P}(A_3) = \begin{cases} 1 & \text{if } k = o(n^{1/3}), \\ \frac{1}{1+c^3} & \text{if } \frac{k}{n^{1/3}} \to c, \\ 0 & \text{if } k = \omega(n^{1/3}). \end{cases}$$

*Furthermore, if $B_i$ denotes the event that $\cap_{F\in\mathcal{F}_i} F = \emptyset$, then $\mathbb{P}(B_3) = 1 - o(1) - \mathbb{P}(A_3)$ holds.*

**Proof of Lemma.** By Lemma 231, it is enough to compute $\mathbb{P}(A_3|A_2)$. If $F_1 \cap F_2 = \{v\}$, then

- the number $N_1$ of $k$-sets containing $v$, but otherwise disjoint from $F_1 \cup F_2$, is $\binom{n-2k+1}{k-1}$,
- the number $N_2$ of $k$-sets containing $v$ and some other vertex of $F_1 \cup F_2$ is at most $(2k-2)\binom{n-2}{k-2} = \frac{(2k-2)(k-1)}{n-1}\binom{n-1}{k-1} = o(N_1)$ as $k^2 = o(n)$
- the number $N_3$ of $k$-sets not containing $v$, but meeting both $F_1$ and $F_2$, satisfies $(k-1)^2\binom{n-2k+1}{k-2} \leq N_3 \leq (k-1)^2\binom{n-2}{k-2}$. Therefore, just as in the case above, $N_3 = (1+o(1))k^2\binom{n-2}{k-2}$.

Now $\binom{n-1}{k-1} = \frac{n-1}{k-1}\binom{n-2}{k-2}$, so $N_2 + N_3 = (1+o(1))\frac{k^3}{n}\binom{n-1}{k-1}$; therefore

$$\mathbb{P}(A_3|A_2) = \frac{N_1}{N_1 + N_2 + N_3} = (1+o(1))\frac{\binom{n-1}{k-1}}{\binom{n-1}{k-1} + \frac{k^3}{n}\binom{n-1}{k-1}} = (1+o(1))\frac{1}{1+\frac{k^3}{n}}.$$

□

**Lemma 233** *Suppose $k = o(n^{2/5})$. Then $\mathbb{P}(\mathcal{E}_{n,k}|A_3) = 1 - o(1)$.*

**Proof of Lemma.** We first show $\mathbb{P}(A_5|A_3) = \mathbb{P}(A_5|A_4) = \mathbb{P}(A_4|A_3) = 1 - o(1)$. Indeed, as in the proof of Lemma 232, the number $N_{1,3}$ (respectively $N_{1,4}$) of $k$-sets containing the kernel of $\mathcal{F}_3$ (resp. $\mathcal{F}_4$) but no other vertices of $\cup_{i=1}^3 F_i$ (resp. $\cup_{i=1}^4 F_i$) is $\binom{n-3k+2}{k-1}$ ($\binom{n-4k+3}{k-1}$), the number $N_{2,3}$ (resp. $N_{2,4}$) of

$k$-sets containing the kernel and some other vertex of $\cup_{i=1}^3 F_i$ (resp. $\cup_{i=1}^4 F_i$) is at most $3k\binom{n-2}{k-2} = o(N_{1,3})$ (resp. $4k\binom{n-2}{k-2} = o(N_{1,4})$, and the number $N_{3,3}$ (resp. $N_{3,4}$) of $k$-sets not containing the kernel, but meeting all $F_i$'s is at most $k^3\binom{n-3}{k-3} = o(N_{1,3})$ (resp. $k^4\binom{n-4}{k-4} = o(N_{1,4})$) by the assumption $k = o(n^{2/5})$.

Next we claim $\mathbb{P}(B_{3k}|A_5) = o(1)$. Indeed, $\mathbb{P}(B_{i+1}|\overline{B_i}, A_5) \leq \frac{k^5\binom{n-5}{k-5}}{\binom{n-1}{k-1}-i} = O(\frac{k^9}{n^4})$. Therefore,

$$\mathbb{P}(B_{3k}|A_5) = 1 - \prod_{i=5}^{3k-1} \mathbb{P}((\overline{B_i}|\overline{B_{i-1}}, A_5)) \leq 1 - \prod_{i=5}^{3k-1}(1 - O(\frac{k^9}{n^4}) = O\left(\frac{k^{10}}{n^4}\right) = o(1),$$

as $k = o(n^{2/5})$.

Finally, we claim that, conditioned on $\overline{B_{3k}}$ and $A_3$, we have that every $k$-set in $\mathcal{I}(\mathcal{F}_{3k})$ contains the vertex in $\cap_{F\in\mathcal{F}_{3k}} F$ w.h.p. Clearly, conditioning on $\overline{B_{3k}}$, the family $\{F_i \setminus \{v\}\}_{i=1}^{3k}$ is uniformly distributed over all $(3k)$-tuples in $\binom{[n]\setminus\{v\}}{k-1}$. So the probability that our claim is false is at most

$$\binom{n-1}{k}\left(1 - \frac{\binom{n-k-1}{k-1}}{\binom{n-1}{k-1}}\right)^{3k} \leq \left(\frac{en}{k}\right)^k \left(\frac{k^2}{n-2k}\right)^{3k} \leq \left(\frac{2ek^5}{n^2}\right)^k = o(1),$$

as $k = o(n^{2/5})$.

So we have shown that conditioning on $\overline{B_{3k}}$ and $A_5$, that occur w.h.p, the event $\mathcal{E}_{n,k}$ holds. □

Lemma 231 and Lemma 232 show that if $k = \omega(n^{1/3})$ and $k = o(n^{1/2})$, then $B_3$ holds w.h.p., while if $k = O(n^{1/3})$, then Lemma 231, Lemma 232 and Lemma 233 prove the statement of the theorem. We are left to prove that if $k = \Omega(n^{1/2})$, then the probability of $\mathcal{E}_{n,k}$ tends to zero. It is done in the following two lemmas.

**Lemma 234** *If $\omega(n^{1/3}) = k = o(n^{2/3})$, then $B_3$ holds w.h.p.*

**Proof of Lemma.** Let $f(i) = \mathbb{P}(|F_1 \cap F_2| = i) = \frac{\binom{k}{i}\binom{n-k}{k-i}}{\binom{n}{k}-\binom{n-k}{k}-1}$, and $g(i) = \mathbb{P}(\overline{B_3}|\ |F_1 \cap F_2| = i) = \frac{\binom{n}{k}-\binom{n-i}{k}-2}{\binom{n}{k}-2\binom{n-k}{k}+\binom{n-2k+i}{k}}$. Then we have $\mathbb{P}(\overline{B_3}) = \sum_{i=1}^{k-1} f(i)g(i)$.

For $0 \leq s \leq 2k$ we have

$$\begin{aligned}\frac{\binom{n-s}{k}}{\binom{n}{k}} &= \prod_{j=0}^{k-1}\left(1 - \frac{s}{n-j}\right) \\ &= \prod_{j=0}^{k-1} \exp\left(-\frac{s}{n} + O\left(\frac{k^2}{n^2}\right)\right) = \exp\left(-\frac{ks}{n} + O\left(\frac{k^3}{n^2}\right)\right),\end{aligned} \tag{4.1}$$

and applying this we obtain

$$\frac{\binom{k}{i}\binom{n-k}{k-i}}{\binom{n}{k}} \le \frac{k^i}{i!}\frac{\binom{n-k}{k-i}}{\binom{n-k}{k}}\frac{\binom{n-k}{k}}{\binom{n}{k}} \le \frac{k^i}{i!}\frac{k^i}{(n-2k)^i}\exp\left(-\frac{k^2}{n}+O\left(\frac{k^3}{n^2}\right)\right). \quad (4.2)$$

As the last exp term in (4.2) is at most 1, (4.1) and (4.2) imply

$$f(i) \le \frac{k^{2i}}{i!(n-2k)^i}\cdot\frac{1+o(1)}{\exp(\frac{k^2}{n})-1}.$$

Also, $f(i)/f(i+1) = \frac{i+1)(n-2k+i)}{(k-i)^2} \ge \frac{1}{2}\omega$ if $i \ge \frac{k^2\omega}{n}$. Therefore $\sum_{i=\frac{k^2\omega}{n}}^{r-1} f(i) = o(1)$ if $\omega$ tends to infinity. So we can condition on the event $|F_1 \cap F_2| \le \frac{k^2\omega}{n}$ for an arbitrary sequence $\omega = \omega_n$. Let us fix one such that $n^{1/3}\omega = o(k)$.

Substituting (4.1) to the definition of $g(i)$ and using $\frac{k^3}{n^2} = o(1)$ we obtain

$$\begin{aligned} g(i) &\le \frac{1-\exp(-\frac{ik}{n}+O(\frac{k^3}{n^2}))}{1-2\exp(-\frac{k^2}{n}+O(\frac{k^3}{n^2}))+\exp(-\frac{r(2r-i)}{n}+O(\frac{k^3}{n}))} \\ &\le (1+o(1))\frac{\frac{ik}{n}+O(\frac{k^3}{n^2})}{(1-\exp(-\frac{k^2}{n}))^2}. \end{aligned}$$

Plugging back the bounds on $f(i)$ and $g(i)$ and using

$$\frac{1}{(1-\exp(-\frac{k^2}{n}))^l} = \frac{\exp(\frac{lk^2}{n})}{(\exp(\frac{k^2}{n})-1)^l},$$

we obtain that $\mathbb{P}(\overline{B_3})$ is at most

$$\begin{aligned} & o(1)+(1+o(1))\frac{\exp(\frac{2k^2}{n})}{(\exp(\frac{k^2}{n})-1)^3}\sum_{i=1}^{\frac{k^2\omega}{n}}\frac{k^{2i}}{i!(n-2k)^i}\left(\frac{ik}{n}+O(\frac{k^3}{n^2})\right) \\ =& O\left(\frac{\exp(\frac{2k^2}{n})}{(\exp(\frac{k^2}{n})-1)^3}\frac{\omega k^3}{n^2}\exp\left(\frac{k^2}{n-2}\right)\right) \\ =& O\left(\frac{k^3\omega}{n^2}\frac{1}{(1-\exp(\frac{k^2}{n}))^3}\right) \\ =& o(1). \end{aligned}$$

□

**Lemma 235** *If* $k \ge 2n^{1/2}\log n$, *then* $\mathbb{P}(B_i) = 1-o(1)$ *for* $i = \exp(\frac{k^2}{3n})$.

**Proof of Lemma.** The probability that $m$ randomly selected sets (without replacement) from $\binom{[n]}{k}$ do not form an intersecting family is at most

$$\binom{m}{2}\frac{\binom{n-k}{k}}{\binom{n}{k}} \le \frac{m^2}{2}\left(1-\frac{k}{n}\right)^k \le \frac{m^2}{2}\exp\left(-\frac{k^2}{n}\right).$$

In particular, selecting $\exp(\frac{k^2}{3n})$ sets randomly without replacement produces an intersecting family with probability $1 - o(1)$.

Observe that if we condition on the event that they form an intersecting family, selecting $m$ sets has the same probability distribution as $\mathcal{F}_m$. Therefore

$$\mathbb{P}(\overline{B_i}) \leq \exp\left(\frac{-k^2}{3n}\right) + n\left(\frac{\binom{n-1}{k-1}}{\binom{n}{k}}\right)^i = o(1) + k^i n^{1-i} = o(1).$$

□

As Lemma 234 and Lemma 235 cover all cases of $k = \Omega(n^{1/2})$, the proof of the theorem is finished. ■

In a later paper, Bohman et al described the process for the range $n^{1/3}\omega \leq k \leq n^{5/12}/\omega$.

**Theorem 236 (Bohman, Frieze, Martin, Ruszinkó, Smyth [58])** *Let $\omega = \omega_n$ be an arbitrary sequence tending to infinity and let $n^{1/3}\omega \leq k \leq n^{5/12}/\omega$. If $i_d$ denotes the smallest index for which the maximum degree of $\mathcal{F}_i$ is $d$, then the following hold*

- $\lim_{n\to\infty} \mathbb{P}(i_4 \geq ck/n^{1/3}) = e^{-c^3/6}$.
- *$\mathcal{F}_{i_4}$ is simple w.h.p.*
- *$\mathcal{F}_{i_4}$ has a unique vertex of degree 4 and no vertex of degree 3 w.h.p.*
- *The resulting intersecting family consists of $\mathcal{F}_{i_4}$ and all $k$-sets containing $v$ and meeting all sets in $\mathcal{F}_{i_4}$ w.h.p.*

*Furthermore any maximal intersecting family $\mathcal{F} \subseteq \binom{[n]}{k}$ that satisfies the above has size*

$$\sum_{i=0}^{i_4}(-1)^i\binom{i_4-4}{i}\binom{n-1-ki+i(i-1)/2}{k-1} = (1+o(1))\left(\frac{k^2}{n}\right)^{i_4-4}\binom{n-1}{k-1}.$$

**Exercise 237 (Patkós [455])** *Let $\mathcal{E}_{HM}$ denote the event that the process returns a Hilton-Milner type family $\mathcal{F}$, i.e. there exists a vertex $v$ and a $k$-set $F_0$ such that $\mathcal{F} = \{F : v \in F, F \cap F_0 \neq \emptyset\} \cup \{F_0\}$. Show that if $k/n^{1/3} \to c$ for some positive constant $c$, then $\mathbb{P}(\mathcal{E}_{HM}) \to \frac{c^3}{1+c^3} \cdot \frac{1}{1+c^3/3}$.*

**Exercise 238** *Suppose $k/n^{1/3} \to c$ for some positive constant $c$. Let $i_1$ denote the smallest index such that the maximum degree of $\mathcal{F}_i$ is 3, and let $i_2$ denote the smallest index such that the maximum degree of $\mathcal{F}_i$ is at least 3 and there is a unique vertex of maximum degree. For $i_1 \leq i \leq i_2$, let $M_i$ denote the set of vertices of maximum degree. Show that for any $i_1 \leq i \leq i_2 - 1$, we have $M_i \supseteq M_{i+1}$ w.h.p., and the resulting family consists of $\mathcal{F}_{i_2}$ and all $k$-sets containing the unique vertex in $M_{i_2}$ and meeting all sets of $\mathcal{F}_{i_2}$.*

# 5

# *Turán-type problems*

## CONTENTS



Turán in 1941 [535] showed that if a graph does not contain a complete subgraph $K_r$, then the maximum number of edges it can contain is given by the Turán-graph, a complete balanced $r-1$-partite graph. This result (that extends a theorem of Mantel [405] on $K_3$) gave birth to a huge area of graph theory. It is outside the scope of this book; see the monograph [67] by Bollobás for more details. Much less is known about the hypergraph version, but it has received a lot of attention recently. The interested reader may also look at the surveys by de Caen [124], Füredi [230], Sidorenko [511] and Keevash [348].

Families of sets are called hypergraphs in this chapter, as is usual in this area. Members of the family are called hyperedges of the hypergraph $\mathcal{H}$, the underlying set $[n] = V(\mathcal{H})$ is called a vertex set, and its elements are the vertices of the hypergraph. A $k$-uniform hypergraph is briefly called a $k$-graph. (2-graphs are the ordinary graphs then.) We still denote hypergraphs by script capitals, we say hyperedges are members of the hypergraph and denote by $|\mathcal{H}|$ the cardinality of the hypergraph $\mathcal{H}$, i.e. the number of the hyperedges it contains. We denote families of hypergraphs by blackboard bold letters like $\mathbb{F}$.

We say two hypergraphs $\mathcal{F}$ and $\mathcal{G}$ are *isomorphic* if there exists a bijection $f : V(\mathcal{F}) \rightarrow V(\mathcal{G})$ such that $F \in \mathcal{F}$ if and only if $\{f(x) : x \in F\} \in \mathcal{G}$. We say $\mathcal{F}$ is an *induced subhypergraph* of $\mathcal{G}$, if there is a subset $X \subseteq V(G)$ with $|X| = |V(\mathcal{F})|$ such that the hypergraph $\{G \in \mathcal{G} : G \subset X\}$ is isomorphic to $\mathcal{F}$, and a *subhypergraph* of $\mathcal{G}$ if we can delete some hyperedges from $\{G \in \mathcal{G} : G \subset X\}$ to obtain a hypergraph that is isomorphic to $\mathcal{F}$.

Given a family $\mathbb{F}$ of $k$-graphs, the *Turán number* $ex_k(n, \mathbb{F})$ is the maximum cardinality a $k$-graph on $n$ vertices can have, if it does not contain any member of $\mathbb{F}$ as a subhypergraph. If $\mathbb{F} = \{\mathcal{F}\}$, then we write $ex_k(n, \mathcal{F})$ instead of $ex_k(n, \mathbb{F})$.

Many other results mentioned in this book fit into this framework. If the forbidden $k$-graph consists of two disjoint hyperedges, the hypergraph is an intersecting family, and its Turán number is given by the Erdős-Ko-Rado Theorem (Theorem 9). Similarly, if we forbid the $k$-graphs $\mathcal{F}_0, \mathcal{F}_1, \dots, \mathcal{F}_{t-1}$ consisting of two hyperedges intersecting in $0, 1, 2, \dots, t-1$ vertices respectively, then we get a $t$-intersecting family. Any other problem given by a forbidden configuration in a uniform family can be formulated this way. However, there are some questions that naturally occur in this setting, and they are surprisingly difficult.

For $k \leq r$, let $\mathcal{K}_r^k$ be the complete $k$-graph on $r$ vertices (containing all $\binom{r}{k}$ possible edges). Turán's theorem determines exactly $ex_2(n, \mathcal{K}_r^2)$, but for $k > 2$, no Turán number of complete hypergraphs is known, not even its asymptotic value!

For a family $\mathbb{F}$ of $k$-graphs we define its *Turán density*

$$\pi_k(\mathbb{F}) = \lim_{n \rightarrow \infty} \frac{ex_k(n, \mathbb{F})}{\binom{n}{k}}.$$

If $\mathbb{F} = \{\mathcal{F}\}$, then we write $\pi_k(\mathcal{F})$ instead of $\pi_k(\mathbb{F})$.

This function is well-defined, see Exercise 242. Turán [535] conjectured $\pi_3(\mathcal{K}_4^3) = 5/9$.

We call a $k$-graph *$r$-partite* if its vertex set can be partitioned into $r$ subsets such that every hyperedge of $\mathcal{F}$ intersects every part in at most one vertex. It is a *complete $r$-partite hypergraph* if it contains all the $k$-sets that intersect every part in at most one vertex. We denote it by $\mathcal{K}^k(n_1, \dots, n_r)$ if the parts have size $n_1, \dots, n_r$.

For $k \leq r$ let $\mathcal{T}^{(k)}(n, r)$ be the *complete balanced* $r$-partite $k$-graph on $n$ vertices, i.e. $\mathcal{K}^k(n_1, \dots, n_r)$ where for every $i \leq r$ either $n_i = \lfloor n/r \rfloor$ or $n_i = \lceil n/r \rceil$. For $k > r$ let $\mathcal{T}^{(k)}(n, r)$ be defined the same way, but take those hyperedges of size $k$ that intersect every part in $\lfloor k/r \rfloor$ or $\lceil k/r \rceil$ vertices (thus it is not an $r$-partite hypergraph).

**Theorem 239 (Erdős, [151])** *Let $\mathcal{F}$ be a $k$-partite $k$-graph. Then we have $\pi_k(\mathcal{F}) = 0$.*

Note that $\mathcal{T}^{(k)}(n,k)$ shows that if $\mathcal{F}$ is $k$-uniform but not $k$-partite, then $\pi_k(\mathcal{F}) > 0$.

**Proof.** We proceed by induction on $k$ with the base case $k = 1$ being trivial. (Let us mention that the case $k = 2$ of ordinary graphs is proved in a stronger form by the famous Kővári - Sós - Turán Theorem [358].) By adding extra hyperedges to $\mathcal{F}$ if necessary, we can assume $\mathcal{F} = \mathcal{K}^k(t_1, \dots, t_r)$. Let us assume $\mathcal{H}$ is an $\mathcal{F}$-free $k$-uniform hypergraph with $|\mathcal{H}| = cn^k$. Let us consider all the link hypergraphs $\mathcal{H}_i = \{H \setminus \{i\} : i \in H \in \mathcal{H}\}$. As $\mathcal{H}$ is $k$-uniform, we obtain $\sum_{i=1}^n |\mathcal{H}_i| = ckn^k$.

If a copy of $\mathcal{K}^{k-1}(t_2, \dots, t_r)$ was contained in $t_1$ different link hypergraphs, they together would form a copy of $\mathcal{F}$ contradicting the $\mathcal{F}$-free property of $\mathcal{H}$. Thus, by the induction hypothesis, the intersection of any $t_1$ link hypergraphs contains $o(n^{k-1})$ hyperedges. For a $(k-1)$-edge $S$ let $d(S)$ be the number of link hypergraphs containing it. Then $t_1$ link hypergraphs can be chosen $\binom{n}{t_1}$ ways, and every $(k-1)$-edge $S$ is counted $\binom{d(S)}{t_1}$ times this way (which is 0 if $d(S) < t_1$). This implies

$$\sum_{S \subseteq \binom{[n]}{k-1}} \binom{d(S)}{t_1} = o(n^{t_1+k-1}). \tag{5.1}$$

On the other hand we have

$$\sum_{S \subseteq \binom{[n]}{k-1}} d(S) = \sum_{i=1}^{n} |\mathcal{H}_i| = ckn^k. \tag{5.2}$$

Therefore, there exists a constant $c'$ such that $\sum_S d(S) \geq c'n^k$ if we sum over only those $(k-1)$-sets $S$ with $d(S) \geq t_1$. Furthermore, as $d(S) \leq n - k + 1$, the number of such $(k-1)$-sets is at least $c''n^{k-1}$. The function $\binom{x}{t_1}$ is convex if $x \geq t_1$, so by Jensen's inequality we obtain

$$\sum_{S \subseteq \binom{[n]}{k-1}} \binom{d(S)}{t_1} \geq c''n^{k-1} \binom{\sum_{S \subseteq \binom{[n]}{k-1}} d(S)}{t_1} = c''n^{k-1} \binom{c'n}{t_1} = \Omega(n^{t_1+k-1})$$

contradicting (5.1). This finishes the proof. ■

In the rest of this chapter we will give bounds on the Turán number or Turán density of several specific hypergraphs, and briefly describe some methods. Before that we state two theorems that give general bounds. The following theorem of Keevash improves a result of Sidorenko [509].

**Theorem 240 (Keevash, [345])** *Let $\mathcal{F}$ be a $k$-graph with $f$ edges.*

*(i) If $k = 3$ and $f \geq 4$, then $\pi_3(\mathcal{F}) \leq \frac{1}{2}\sqrt{f^2 - 2f - 3} - f + 3$.*

*(ii) For a fixed $k \geq 3$ and $f \to \infty$ we have $\pi_k(\mathcal{F}) < \frac{f-2}{f-1} - \frac{1+o(1)}{2k!^{2/k} f^{3-2/k}}$.*

Lu and Székely [395] applied the Lovász local lemma [164] to obtain the following theorem.

**Theorem 241 (Lu, Székely, [395])** *Let $\mathcal{F}$ be a $k$-graph such that every hyperedge intersects at most $d$ other hyperedges. Then $\pi_k(\mathcal{F}) \leq 1 - \frac{1}{e(d+1)}$, where $e$ is the base of natural logarithm.*

**Exercise 242 (Katona, Nemetz, Simonovits, [331])** *Prove that for any $k$-uniform hypergraph $\mathcal{H}$ the sequence $\frac{ex_k(n,\mathcal{H})}{\binom{n}{k}}$ is monotone decreasing, in particular $\pi_k(\mathcal{H}) = \lim_{n\to\infty} \frac{ex_k(n,\mathcal{H})}{\binom{n}{k}}$ exists.*

## 5.1 Complete forbidden hypergraphs and local sparsity

As we have mentioned, $\pi_k(\mathcal{K}_r^k)$ is unknown for $r > k > 2$; thus we can only show some bounds. Let us mention first that $\mathcal{G}$ is $\mathcal{K}_r^k$-free if and only if the complement hypergraph $\overline{\mathcal{G}} = \binom{[n]}{k} \setminus \mathcal{G}$ satisfies the following property: any $r$-subset of $[n]$ contains a hyperedge of $\overline{\mathcal{G}}$. In that case one tries to minimize the number of hyperedges in $\overline{\mathcal{G}}$. In fact, most of the literature uses this formulation, and this minimum is often called the Turán number. As we deal with the more general case of forbidding other hypergraphs later, we will continue to use the functions $ex_k(n, \mathcal{K}_r^k)$ and $\pi_k(\mathcal{K}_r^k)$.

Let us start with some general bounds.

**Theorem 243 (de Caen, [123])** $ex_k(n, \mathcal{K}_r^k) \leq \binom{n}{k}(1 - \frac{n-k+1}{(n-r+1)\binom{r-1}{k-1}})$.

**Proof.** Let us consider an arbitrary $k$-graph $\mathcal{H}$ and let $N_l$ denote the number of copies of $\mathcal{K}_l^k$ in $\mathcal{G}$. Note that if $l < k$ then $N_l = \binom{n}{l}$.

**Claim 244** *If $N_{l-1} \neq 0$, then we have*

$$N_{l+1} \geq \frac{l^2 N_l}{(l-k+1)(l+1)} \left( \frac{N_l}{N_{l-1}} - \frac{(k-1)(n-l)+l}{l^2} \right).$$

**Proof of Claim.** Let $A_1, \dots, A_{N_{l-1}}$ be the vertex sets of the copies of $\mathcal{K}_{l-1}^k$ in $\mathcal{H}$ and let $a_i$ be the number of copies of $\mathcal{K}_l^k$ containing $A_i$. Similarly let $B_1, \dots, B_{N_l}$ be the vertex sets of the copies of $\mathcal{K}_l^k$ in $\mathcal{H}$ and let $b_i$ be the number of copies of $\mathcal{K}_{l+1}^k$ containing $B_i$.

Let us count the number $P$ of pairs $(S,T)$ where $S$ and $T$ are both $l$-sets, $S$ spans $\mathcal{K}_l^k$ in $\mathcal{H}$ while $T$ does not, and $|S \cap T| = k - 1$. On the one hand for each $B_i$ there are $n-l-b_i$ ways to choose another vertex $x \notin B_i$ such that $B_i \cup \{x\}$ does not span $\mathcal{K}_{l+1}^k$ in $\mathcal{H}$. It means that there is an $X \subseteq B_i$ with $|X| = k-1$ such that $X \cup \{x\} \notin \mathcal{H}$. Then for every $x' \in B_i \setminus X$ if $B_i$ plays the role of $S$ then $B_i \cup \{x\} \setminus \{x'\}$ can play the role of $T$. This implies

$$P \geq \sum_{i=1}^{N_l}(n-l-b_i)(l-k-1) = (l-k-1)((n-l)N_l - (l+1)N_{l+1}), \quad (5.3)$$

as $\sum_{i=1}^{N_l} b_i = (l+1)N_{l+1}$.

On the other hand for each $A_i$ we have $a_i$ ways to add another vertex to obtain a copy of $\mathcal{K}_l^k$ and $n-l+1-a_i$ ways to obtain a set $T$ that does not span $\mathcal{K}_l^k$. This implies

$$P = \sum_{i=1}^{N_{l-1}} a_i(n-l+1-a_i) \leq (n-l+1)lN_l - \frac{l^2N_l^2}{N_{l-1}} \quad (5.4)$$

as $\sum_{i=1}^{N_{l-1}} a_i = lN_l$. Combining (5.3) with (5.4) and rearranging finishes the proof. □

Let us define

$$F(n,l,k) = \frac{\binom{l-1}{k-1}(n-k+1) - (n-l+1)}{k\binom{l-1}{k-1}}\binom{n}{k-1}.$$

**Claim 245** *For $l \geq k+1$ we have $N_l \geq N_{l-1}\frac{k^2\binom{l}{k}}{l^2\binom{l-1}{k-1}}(N_k - F(n,l,k))$*

**Proof of Claim.** We proceed by induction on $l$. The base case $l = k+1$ follows from Claim 244 using $N_{k-1} = \binom{n}{k-1}$. Let $l > k+1$. By Claim 244 and the inductive assumption we have

$$\begin{aligned} N_{l+1} &\geq N_l\frac{l^2}{(l-k+1)(l+1)}\left(\frac{N_l}{N_{l-1}} - \frac{(k-1)(n-l)+l}{l^2}\right) \\ &\geq N_l\frac{l^2}{(l-k+1)(l+1)}\left(\frac{k^2}{l^2}\frac{\binom{l}{k}}{\binom{n}{k-1}}(N_k - F(n,l,k)) - \frac{(k-1)(n-l)+l}{l^2}\right) \\ &= N_l\frac{k^2\binom{l}{k}}{(l-k+1)(l+1)\binom{n}{k-1}}\left[N_k - \left(F(n,l,k) - \frac{(k-1)(n-l)+l}{k^2\binom{l}{k}}\binom{n}{k-1}\right)\right] \\ &= N_l\frac{k^2\binom{l+1}{k}}{(l+1)^2\binom{n}{k-1}}(N_k - F(n,l+1,k)). \end{aligned}$$

□

The above claim implies that $N_l > 0$ whenever $|\mathcal{H}| = N_k > F(n,l,k)$, which means $\mathcal{H}$ contains a copy of $\mathcal{K}_r^k$ if $|\mathcal{H}| > F(n,r,k)$, which implies $ex_k(n,\mathcal{K}_r^k) \leq F(n,r,k)$. Simple rearranging finishes the proof. ■

**Corollary 246** $\pi_k(\mathcal{K}_r^k) \leq 1 - \frac{1}{\binom{r-1}{k-1}}$.

The best general lower bound is the following.

**Theorem 247 (Sidorenko, [506])** $\pi_k(\mathcal{K}_r^k) \geq 1 - \frac{k-1}{r-1}^{k-1}$.

A natural case to focus on is when $r = k+1$. The above theorem of de Caen gives the upper bound $(k-1)/k$, and there are only small improvements in this special case. Chung and Lu [102] showed $\pi_k(\mathcal{K}_{k+1}^k) \leq \frac{k-1}{k} - \frac{1}{k(k+3)} + O(1/k^3)$. Lu and Zhao [396] gave some other improvements in the lower order terms if $k$ is even.

The following theorem of Sidorenko [512] gives the best lower bound for large $k$.

**Theorem 248** $\pi_k(\mathcal{K}_{k+1}^k) \geq 1 - \frac{(1+o(1))\log k}{2k}$.

The bound $\pi_k(\mathcal{K}_{k+1}^k) \geq 1 - (\frac{1+2\log k}{k})$, given by Kim and Roush [359] is better for smaller $k$, just as the bound $\pi_{2k}(\mathcal{K}_{2k+1}^{2k}) \geq 3/4 - 2^{-2k}$ by de Caen, Kreher and Wiseman [126] for even-uniform hypergraphs.

Let us now consider some specific complete hypergraphs. As we have mentioned, Turán conjectured $\pi_3(\mathcal{K}_4^3) = 5/9$ based on the following construction.

**Construction 249** *Let us partition* $[n]$ *into three sets of roughly equal size* $V_0, V_1, V_2$*, and take the 3-edges that intersect each part, and the 3-edges that intersect (for some* $i$*)* $V_i$ *in two elements and* $V_{i+1}$ *in one element, where* $V_3 = V_0$.

Later Kostochka [381] found exponentially many non-uniform constructions achieving the same size. Most likely this is one of the main reasons why this problem proved to be so difficult to tackle. The best known upper bound is $\pi_3(\mathcal{K}_4^3) \leq 0.561666$ due to Razborov [475]. Subsection 5.5.2 describes his flag algebra method that he used to obtain this result. He improved earlier bounds by de Caen [124], Giraud (unpublished) and Chung and Lu [102].

Let us consider now $\mathcal{K}_5^4$. Giraud [269] gave a construction showing $\pi_4(\mathcal{K}_5^4) \geq 11/16$. Markström [408] obtained $\pi_4(\mathcal{K}_5^4) \leq 1753/2380$. Note that Exercise 242 shows that an upper bound on $ex_k(n_0, \mathcal{F})$ gives an upper bound on $ex_k(n, \mathcal{F})$ for every $n \geq n_0$; thus it gives an upper bound on $\pi_k(\mathcal{F})$. Markström obtained his result by determining $ex_k(n, \mathcal{K}_5^4)$ for $n \leq 16$ using extensive computer search.

Similarly, for $\pi_3(\mathcal{K}_5^3)$ the best known upper bound $9/11$ is a consequence of determining $ex_k(n, \mathcal{K}_5^4)$ for $n \leq 13$ by Boyer, Kreher, Radziszowski, Zhou and Sidorenko [86]. Turán conjectured $\pi_3(\mathcal{K}_5^3) = 3/4$, and if true, $\mathcal{T}^{(3)}(n, 2)$ would be an extremal example.

Brown, Erdős and Sós [91] considered forbidding all the $k$-graphs on $v$ vertices which have $e$ hyperedges (they studied this for 3-graphs in [518]).

Let us denote their family by $\mathbb{F}^k(v,e)$. One could say that an $\mathbb{F}^k(v,e)$-free hypergraph is locally sparse, as there are less than $e$ edges spanned by any $v$ vertices. This is a generalization of the problems studied above, as $\{\mathcal{K}_r^k\} = \mathbb{F}^k(r, \binom{r}{k})$.

**Theorem 250 (Brown, Erdős, Sós, [91])** *For $e > 1$ and $v > k$ we have $ex_k(n, \mathbb{F}^k(v,e)) = \Omega(n^{\frac{ke-v}{e-1}})$.*

**Proof.** Let $m = cn^{\frac{ke-v}{e-1}}$, and let $\mathbb{G}$ be the family of all the $k$-graphs of size $m$. We call a $v$-set *bad* for a $\mathcal{G} \in \mathbb{G}$ if it spans a $k$-graph in $\mathbb{F}^k(v,e)$ in $\mathcal{G}$. We count the pairs $(V, \mathcal{G})$, where $V$ is a bad set for $\mathcal{G}$ and it spans $\mathcal{F} \in \mathbb{F}^k(v,e)$ in $\mathcal{G}$. We can pick the $v$-set $V$ first, then $e$ sets of size $k$ in it, and finally $m - e$ other $k$-sets. Thus the total number is at most

$$\binom{n}{v}\binom{\binom{v}{k}}{e}\binom{\binom{n}{k}-e}{m-e} \leq c_1 n^v \binom{\binom{n}{k}}{m}\left(\frac{m}{\binom{n}{k}}\right)^e \leq c_2 \binom{\binom{n}{k}}{m}\frac{m}{2\binom{v}{k}}$$

for appropriate constants $c_1 = c_1(k,v,e)$ and $c_2 = c_2(k,v,e)$. This implies there is a $\mathcal{G} \in \mathbb{G}$ that contains at most $\frac{m}{2\binom{v}{k}}$ bad sets. If we delete every $k$-edge from $\mathcal{G}$ that appears in any bad set for $\mathcal{G}$, we still have at least $m/2$ hyperedges but no subhypergraph in $\mathbb{F}^k(v,e)$, finishing the proof. ■

**Corollary 251** *For $e > 1$ and $1 \leq r < k$ we have $ex_k(n, \mathbb{F}^k(e(k-r)+r, e)) = \Theta(n^r)$.*

Observe that $v = e(k-r)+r$ is chosen in such a way that any $r$-set is contained in at most $e-1$ hyperedges. This implies the upper bound, while the lower bound here follows from Theorem 250. Brown, Erdős and Sós asked whether $ex_3(n, \mathbb{F}^3(6,3)) = o(n^2)$. This was answered by the celebrated 6-3 theorem of Ruzsa and Szemerédi [491].

**Theorem 252 (Ruzsa, Szemerédi, [491])** $n^{2-o(1)} < ex_3(n, \mathbb{F}^3(6,3)) < o(n^2)$.

Erdős, Frankl and Rödl generalized Theorem 252 to show the following.

**Theorem 253 (Erdős, Frankl, Rödl [157])** *For any $k$ we have $n^{2-o(1)} < ex_k(n, \mathbb{F}^k(3(k-1),3)) < o(n^2)$.*

Corollary 251 motivates the question: what happens when we have $v > e(k-r)+r$. Sárközy and Selkow [497] (extending their earlier result [496]) proved the following.

**Theorem 254 (Sárközy, Selkow, [497])** $ex_k(n, \mathbb{F}^k(e(k-r)+r+\lfloor \log_2 e \rfloor, e)) = o(n^r)$.

*If $r \geq 3$ then $ex_k(n, \mathbb{F}^k(4(k-r)+r+1, 4)) = o(n^r)$.*

Alon and Shapira showed the following.

**Theorem 255 (Alon, Shapira, [16])** *For* $2 \leq r < k$ *we have* $n^{r-o(1)} < ex_k(n, \mathbb{F}^k(3(k-r)+r+1, 3)) = o(n^r)$.

Balachandran and Bhattacharya [32] observed that this problem is related to the so-called combinatorial batch codes. We omit the definition and note that (for some parameters $k$, $r$ and $n$) they are equivalent to $k$-graphs not containing $\mathbb{G}^k(r) = \bigcup_{e \leq r} \mathbb{F}^k(e,e)$. They proved the following.

**Theorem 256 (Balachandran, Bhattacharya, [32])** *Let* $r \geq 7$, $3 \leq k \leq r - 1 - \lceil \log r \rceil$ *and* $n$ *large enough. Then* $ex_k(n, \mathbb{G}^k(r)) \leq n^{k - \frac{1}{2^{k-1}}}$. *On the other hand if* $r \geq 6$ *and* $r - \lceil \log r \rceil \leq k \leq r-1$, *then* $ex_k(n, \mathbb{G}^k(r)) = \Theta(n^k)$.

Bujtás and Tuza [94] studied the more general version $\mathbb{G}^k(q,r) = \bigcup_{e \leq r} \mathbb{F}^k(e, e+q)$. Note that $q$ can be negative here, but if $q < -k$, then $e = 1 - q$ gives a $k$-graph with $e \geq k$ vertices and one hyperedge, which means $ex_k(n, \mathbb{G}^k(q,r)) = 0$ for $n$ large enough. They proved the following.

**Theorem 257 (Bujtás, Tuza, [94])** *Let* $k \geq 2$, $q \geq -k$ *and* $n \geq r \geq 2q + 2k$. *Then* $ex_k(n, \mathbb{G}^k(q,r)) = O(n^{k-1+\frac{1}{\lfloor \frac{r}{q+k} \rfloor}})$.

**Theorem 258 (Bujtás, Tuza, [94])** *Let* $k \geq 2$, $q \geq -k$ *and* $r \geq q+k$. *Then* $ex_k(n, \mathbb{G}^k(q,r)) = \Omega(n^{k-1+\frac{q+k-1}{r-1}})$

They also connected $\mathbb{G}$ and $\mathbb{F}$ in the following way.

**Theorem 259** *Let* $k \geq 2$ *and* $n \geq r \geq q+k \geq 2$. *Then* $ex_k(n, \mathbb{F}^k(r-q, r)) - ex_k(n, \mathbb{G}^k(q,r)) \leq (r-1)\binom{n-1}{k-1}$.

This implies that the two above families have the same Turán density.

Another generalization of forbidding complete hypegraphs is to describe exactly how many hyperedges can $v$ vertices can span. Note that this is an induced version of hypergraph Turán problems, as we are allowed to forbid here hypergraphs which contain one hyperedge on $v$ vertices, without forbidding those containing two hyperedges on $v$ vertices.

Frankl and Füredi [201] studied 3-graphs where no 4 vertices span exactly 1, 3 or 4 hyperedges. In addition to determining their maximum size they completely characterized such hypergraphs.

As a weakening of Turán's conjecture about $\mathcal{K}_4^3$, Razborov [475] considered 3-graphs where no 4 vertices span exactly 1 or 4 hyperedges, and proved their Turán density is at most $5/9$. As Construction 249 possesses this property, Razborov's bound is sharp. Pikhurko [466] showed that if $n$ is large enough, then Construction 249 is the unique extremal hypergraph.

Lu and Zhao [396] considered $k$-graphs where any $k+1$ vertices span exactly 0 or $k$ hyperedges. Answering a question of de Caen [124], they showed that if $n$ is large enough, such a hypergraph is either the empty hypergraph or a complete star $\{E \in \binom{[n]}{k} : v \in E\}$ for some fixed vertex $v \in [n]$.

## 5.2 Graph-based forbidden hypergraphs

In this section we consider a special kind of forbidden configurations: those that are defined by extending a graph to a hypergraph. Even for the simplest graphs there are many different ways to do this. Depending on these ways, one can use graph theoretic results and methods in this kind of problems, which make them easier to handle. On the other hand it means that some of these methods are outside the scope of the present work; hence we omit many details that are otherwise both important and interesting.

We use written capitals to denote graphs, as opposed to hypergraphs.

Two ways to extend graphs to hypergraphs, namely expansions and Berge hypergraphs have been studied for several specific graphs and general graphs as well. Some other ways have been studied, even defined only for some specific classes of graphs, or more classes of graphs using different names and different notation.

Here we try to unify our notation. With the exception for expansions, we write a letter referring to the way the graph is extended, the uniformity of the hypergraph, and the name of the graph, like $\mathcal{S}^k K_r$. The first letter is script capital if we obtain a hypergraph from the graph, and blackboard bold if we obtain a family of hypergraphs, like $\mathbb{B}^k C_l$. In case of expansions, there is a well-known notation $F^+$, but we change it to $F^{+k}$ for the $k$-uniform expansion. This way it is always clear what the hypergraph exactly is.

Note that many hypergraphs that can be defined as a graph-based hypergraph were originally defined differently and studied for a different reason. Here we group them according to how they can be obtained from a graph.

Some of these forbidden configurations have been studied with the additional assumption that the hypergraph not containing them is *linear* which means that any two hyperedges intersect in at most one vertex (note that the word *simple* is also used for this notion in the literature, very confusingly. We call a hypergraph simple if there are no multiple edges. For graphs the two notions coincide). As linear hypergraphs are more similar to graphs, it makes sense to consider this additional assumption. Note that it already implies there are at most $\binom{n}{2}$ hyperedges. We denote the maximum number of hyperedges in a linear $k$-graph that does not contain $\mathcal{H}$ by $ex_k^L(n, \mathcal{H})$.

In most of these cases one can start with an $r$-graph instead of a graph and extend it to a $k$-graph similarly. As even graph-based hypergraphs are a new topic with large unexplored areas, the more general topic of hypergraph-based hypergraphs is not very well-studied. Still we describe some results of that type in Section 5.3.

### 5.2.1 Expansions

For a graph $F$ its $k$-uniform *expansion* (or briefly $k$-expansion) $F^{+k}$ is the hypergraph obtained by adding $k-2$ new elements to each edge. These new elements are distinct from each other and the vertices of the original graph, thus $F^{+k}$ has $(k-2)|E(F)|+|V(F)|$ vertices and $|E(F)|$ hyperedges. For a recent and very thorough survey on expansions see [435].

Many extremal results can be stated this way. For example, if $F$ consists of two disjoint edges, we get once again the Erdős-Ko-Rado Theorem (Theorem 9). If $F$ is a matching, we get Erdős's matching conjecture (Subsection 2.3.1).

If $F$ is a star $S_r$, the forbidden configuration consists of $r$ sets of size $k$ intersecting in the same vertex (and nothing else). This problem was first studied in 1976 by Erdős and Sós [517]. A trivial lower bound on $ex_k(S_r^{+k})$ is the maximum cardinality of a 2-intersecting family, given by Theorem 50. It was shown by Frankl [184] that for $r=2$ and large $n$ this is sharp, i.e. $ex_k(n,S_2^{+k})=\binom{n-2}{k-2}$ if $n$ is large enough. Duke and Erdős showed [138] $ex_k(n,S_r^{+k})=\theta(n^{k-2})$. For $k=3$, improving and generalizing earlier results by Frankl [187] and Chung [104], Chung and Frankl proved the following.

**Theorem 260 (Chung, Frankl, [103])** *For $\geq 3$ we have*

$$ex_3(n,S_r^{+3})=\begin{cases} r(r-1)n+2\binom{r}{3} & \text{if } n>r(r-1)(5r+2)/2 \text{ and } r \text{ is odd} \\ \frac{r(2r-3)n}{2}-\frac{2r^3-9r+6}{2} & \text{if } n\geq 2r^3-9r+7 \text{ and } r \text{ is even.} \end{cases} \tag{5.5}$$

The case $k=4$, $r=2$ was completely solved by Keevash, Mubayi and Wilson.

**Theorem 261 (Keevash, Mubayi, Wilson, [354])**

$$ex_4(n,S_2^{+4})=\begin{cases} \binom{n}{4} & \text{if } n=4,5,6 \\ 15 & \text{if } n=7 \\ 17 & \text{if } n=8 \\ \binom{n-2}{2} & \text{if } n\geq 9. \end{cases} \tag{5.6}$$

The Turán number of $K_3^{+k}$ was first asked by Erdős [155], who conjectured $ex_k(n,K_3^{+k})=\binom{n-1}{k-1}$ if $n\geq 3k/2$. Mubayi and Verstraëte [433] proved the conjecture, improving results by Chvátal [106], Bermond and Frankl [54], Frankl [183, 190], Frankl and Füredi [206] and Csákány and Kahn [111]. In fact they proved the following more general theorem.

**Theorem 262 (Mubayi, Verstraëte, [433])** *Let $k\geq d+1\geq 3$ and $n\geq$*

*$(d+1)k/d$. Suppose that a $k$-graph $\mathcal{H}$ does not contain a non-trivially intersecting family of size $d+1$. Then its cardinality is at most $\binom{n-1}{k-1}$ and equality holds only if $\mathcal{H}$ consists of all the hyperedges containing a fixed vertex.*

Note that $d = 2$ gives the result about $ex_k(n, K_3^{+k})$.

Observe that if $F$ is neither a star nor a triangle, then an intersecting $k$-uniform hypergraph does not contain $F^{+k}$. Indeed, two edges of $F$ that do not share a vertex would give us two disjoint hyperedges. This together with Theorem 9 gives us $ex_k(n, F^{+k}) \geq \binom{n-1}{k-1}$ (more precisely the construction of a maximal trivially intersecting family gives this bound). If $F$ contains $s+1$ independent edges, then the simple constructions $\mathcal{A}(k,s)$ and $\mathcal{A}(n,k,s)$ from Subsection 2.3.1 (avoiding $s$ pairwise disjoint $k$-edges) give lower bounds for $ex_k(n, F^{+k})$.

Mubayi [427] determined $\pi_k(K_r^{+k})$ and conjectured that the complete balanced $(r-1)$-partite $k$-graph is the (unique) maximum if $n$ is large enough. It was later proved by Pikhurko.

**Theorem 263 (Pikhurko, [467])** *For any $r \geq k \geq 3$ if $n$ is large enough, we have $ex_k(n, K_{r+1}^{+k}) = |\mathcal{T}^{(k)}(n,r)|$ and $\mathcal{T}^{(k)}(n,r)$ is the unique extremal hypergraph.*

The proof of this theorem illustrates the so-called stability method for finding exact results. After proving that a hypergraph $\mathcal{H}$ is asymptotically the largest for an extremal problem, a stability result is often almost immediate. Then any larger hypergraph would also be close to $\mathcal{H}$. Analyzing the difference between $\mathcal{H}$ and an arbitrary hypergraph $\mathcal{H}'$ that is close to it sometimes shows $\mathcal{H}'$ cannot be larger.

**Lemma 264** *For any $r \geq k \geq 3$ and $\varepsilon > 0$, there is $\delta > 0$ such that if $\mathcal{H}$ is a $k$-uniform $K_{r+1}^{+k}$-free hypergraph with at least $|\mathcal{T}^{(k)}(n,r)| - \delta\binom{n}{k}$ edges on $n$ vertices, where $n$ is large enough, then we can remove or add at most $\varepsilon\binom{n}{k}$ edges to get $\mathcal{T}^{(k)}(n,r)$.*

The proof of Lemma 264 uses the following theorem of Mubayi [427]. Let $\mathbb{K}_r^k$ denote the family of all $k$-graphs with at most $\binom{r}{2}$ edges that contain a set $R$ of $r$ vertices, called *core* such that each pair of vertices in $R$ is contained by a hyperedge. In particular $K_r^{+k} \in \mathbb{K}_r^k$.

**Theorem 265 (Mubayi, [427])** *Let $n \geq r \geq k \geq 3$. Then $ex_k(n, \mathbb{K}_{r+1}^k) = |\mathcal{T}^{(k)}(n,r)|$ and $\mathcal{T}^{(k)}(n,r)$ is the unique extremal hypergraph. Moreover, for any $\varepsilon' > 0$ there exists $\delta' > 0$ such that if $n$ is large enough, every $\mathbb{K}_{r+1}^k$-free hypergraph on $n$ vertices and least $(\pi_k(\mathbb{K}_r^k) - \delta')\binom{n}{k}$ hyperedges can be turned into $\mathcal{T}^{(k)}(n,r)$ by adding and/or removing at most $\varepsilon'\binom{n}{k}$ hyperedges.*

**Proof of Lemma 264**. Let $\varepsilon > 0$ and choose $0 < \delta \leq \varepsilon$ such that $2\delta$ is at most the $\delta'$ given by Theorem 265 for $\varepsilon' = \varepsilon/2$. Let $\mathcal{H}$ be a $k$-uniform $K_{r+1}^{+k}$-free hypergraph with at least $|\mathcal{T}^{(k)}(n,r)| - \delta\binom{n}{k}$ hyperedges.

We call a pair of vertices a *blue pair* if there are at most

$$m = \left(\binom{r+1}{2}(k-2) + r + 1\right)\binom{n}{k-3}$$

hyperedges in $\mathcal{H}$ containing both of them. Let $\mathcal{H}'$ be the hypergraph obtained from $\mathcal{H}$ by deleting all the hyperedges containing a blue pair, at most $m\binom{n}{2}$ hyperedges. As $n$ is large enough, we have $m\binom{n}{2} \leq \delta\binom{n}{k}/2$

Now we show that $\mathcal{H}'$ is $\mathbb{K}_{r+1}^{k}$-free. Assume otherwise and let $R$ be the core. Then we can find a copy of $K_{r+1}^{+k}$ greedily. Let us order the pair of vertices in $R$ arbitrarily, and in this order we always pick a hyperedge containing the pair such that it intersects $R$ in those two vertices, and does not intersect outside $R$ the hyperedges chosen earlier. When we arrive at the $i$th pair $(x,y)$, there are $i < \binom{r+1}{2}$ hyperedges chosen earlier; thus the new hyperedge has to avoid the other $r-2$ vertices of $R$ and at most $(k-2)i \leq (k-2)(\binom{r+1}{2} - 1)$ other vertices of the earlier hyperedges. The number of hyperedges that contain $x, y$ and another of those $r - 2 + (k-2)(\binom{r+1}{2} - 1)$ vertices is at most $(r - 2 + (k-2)(\binom{r+1}{2} - 1))\binom{n}{k-3} < m$, and the pair $(x,y)$ is not blue, thus we can always find the hyperedge we want.

$\mathcal{H}'$ contains at least $|\mathcal{T}^{(k)}(n,r)| - 2\delta\binom{n}{k}$ hyperedges, thus we can add and/or remove $\varepsilon\binom{n}{k}/2$ hyperedges to obtain $\mathcal{T}^{(k)}(n,r)$, by the choice of $\delta$. All together we added and/or removed at most $(\frac{\delta}{2} + \frac{\varepsilon}{2})\binom{n}{k} \leq \varepsilon\binom{n}{k}$ hyperedges to obtain $\mathcal{T}^{(k)}(n,r)$.

■

**Proof of Theorem 263**. We will choose constants $c_1, \dots, c_5$ such that each of them is sufficiently small depending on the previous constants. Let $n$ be large enough and let us consider the largest $k$-uniform $K_{r+1}^{+k}$-free hypergraph $\mathcal{H}$ on $n$ vertices. It must have at least $|\mathcal{T}^{(k)}(n,r)|$ edges. Let $V_1 \cup \cdots \cup V_r$ be a partition of $[n]$ such that $f = \sum_{H \in \mathcal{H}} |\{i \leq r : H \cap V_i \neq \emptyset\}|$ is maximum possible. Let $\mathcal{T}$ be the complete $r$-partite $k$-graph with parts $V_1, \dots, V_r$. Clearly $|\mathcal{T}| \leq |\mathcal{T}^{(k)}(n,r)|$ and $f \geq k|\mathcal{T} \cap \mathcal{H}|$.

Lemma 264 states that the difference between $\mathcal{H}$ and $\mathcal{T}^{(k)}(n,r)$ is at most $c_5\binom{n}{k}$ edges. Note that as $\mathcal{H}$ has more or equal edges to $\mathcal{T}^{(k)}(n,r)$, we can pick $\delta$ to be arbitrarily small. This gives $f \geq k(|\mathcal{H}| - c_5\binom{n}{k})$, as $\mathcal{T}^{(k)}(n,r)$ gives a partition on $[n]$, and for that partition $f$ would be at least $k(|\mathcal{H}| - c_5\binom{n}{k})$. As $f$ is chosen to be maximum possible, we have the desired inequality.

On the other hand $f \leq k(|\mathcal{H}| - |\mathcal{H} \setminus \mathcal{T}|) + (k-1)|\mathcal{H} \setminus \mathcal{T}|$. These imply $|\mathcal{H} \setminus \mathcal{T}| \leq c_5 k\binom{n}{k}$, which in turn implies $|\mathcal{T} \setminus \mathcal{T}^{(k)}(n,r)| \leq (k+1)c_5\binom{n}{k}$. Informally, $\mathcal{T}$ is also close to $\mathcal{T}^{(k)}(n,r)$. We claim that each part $V_i$ has to be large enough, for example $|V_i| \geq n/2r$. Indeed, if not, then even if all other part sizes are as close to each other as possible, then still $|\mathcal{T}|$ would be some constant times $\binom{n}{k}$ smaller than $|\mathcal{T}_{n,r}^{(k)}|$, which is a contradiction if $c_5$ is small enough. We also know that $|\mathcal{T}| \leq |\mathcal{T}^{(k)}(n,r)|$.

Let us call the hyperedges of $\mathcal{H} \setminus \mathcal{T}$ the *bad hyperedges*, and the hyperedges of $\mathcal{T} \setminus \mathcal{H}$ the *missing hyperedges.* Then the number of bad hyperedges is at least the number of missing hyperedges.

We call a pair of vertices a *blue pair* again if there are at most

$$m = \left(\binom{r+1}{2}(k-2) + r + 1\right)\binom{n}{k-3}$$

hyperedges in $\mathcal{H}$ containing them together, and *red pair* otherwise.

**Claim 266** *For any (blue or red) pair $e = (x, y)$ that is covered by a hyperedge of $\mathcal{H}$ and every set $S$ of $r+1$ vertices that contains $e$, there is a blue pair not equal to $e$ in $S$.*

**Proof of Claim.** Let us assume otherwise, i.e. every pair in $\binom{S}{2} \setminus \{e\}$ is red. Let us consider a hyperedge $H \in \mathcal{H}$ containing $x$ and $y$, and then greedily take a hyperedge in $\mathcal{H}$ for every other pair of vertices to build a $k$-graph $\mathcal{F}$. We make sure that the new hyperedge we pick for a pair $e'$ uses $k-2$ new vertices not contained in any of the earlier hyperedges of $\mathcal{F}$.

This is doable, as if $H_1, H_2, \ldots H_j$ are the hyperedges picked for earlier pairs, then $j$ is at most $\binom{r+1}{2} - 1$, so $\cup_{i=1}^{j} H_i$ contain at most $r + 1 + (k - 2)(\binom{r+1}{2} - 1)$ vertices; thus there are fewer than $m$ hyperedges that contain $e'$ and intersect $(S \cup \cup_{i=1}^{j} H_i) \setminus e'$. As $e'$ is red, there are at least $m$ hyperedges of $\mathcal{H}$ containing it; thus we can pick a new hyperedge with the desired property. At the end of this procedure we obtain a copy of $K_{r+1}^{+k}$ that is a subhypergraph of $\mathcal{H}$, a contradiction. □

Observe that there are at most $c_4 n^2$ blue pairs, as every blue pair is contained in $\Omega(n^{k-2})$ missing hyperedges, and a missing hyperedge contains at most $\binom{k}{2} = O(1)$ blue pairs, while there are at most $c_5 k\binom{n}{k}$ missing hyperedges. (Note that as we introduce $c_4$ when establishing an upper bound on the number of blue pairs, we can do that while satisfying $c_4 \geq c_5$.)

Let us call a pair of vertices *bad*, if they belong to the same part $V_i$ and are covered by a hyperedge in $\mathcal{H}$ (which is then a bad hyperedge). There exists a bad pair, otherwise we are done. Let us assume that the vertices of this bad pair are $x_0, x_1 \in V_1$. If we pick a vertex $x_i$ from $V_i$ for every $2 \leq i \leq r$, then there must be another blue pair $x_0', x_1'$ among them by Claim 266. For a fixed blue pair $x_0', x_1'$ with $\{x_0', x_1'\} \cap \{x_0, x_1\} = \emptyset$, there are less than $n^{r-3}$ ways to pick the other vertices $x_i \in V_i$. As there are at most $c_4 n^2$ blue pairs, the number of ways we can pick the $x_i$'s to obtain blue pairs disjoint from $\{x_0, x_1\}$ is at most $c_4 n^{r-1}$. This is at most half the ways we can choose $x_2, \ldots, x_r$, thus at least half of the choices $x_2, \ldots, x_r$ give us a blue pair that intersects $\{x_0, x_1\}$.

Observe that as all $V_i$'s have size at least $n/2r$, the number of choices for $x_2, \ldots, x_r$ is at least $(\frac{n}{2r})^{r-1}$. A blue pair $x_0, x_1'$ occurs in at most $n^{r-2}$

choices, so the number of blue pairs that intersect $\{x_0, x_1\}$ is at least $c_1 n$. Observe that $c_1$ is independent of $c_4$ and $c_5$, so if $c_4, c_5$ are sufficiently small, then $c_1 \geq c_4$.

Let $A$ be the set of those vertices that are incident to at least $c_1 n/2$ blue pairs. Then by the above every bad pair intersects $A$. The linearly many blue pairs intersecting $A$ give us

$$\frac{|A|c_1}{2} \cdot \frac{1}{2}\left(\frac{n}{2r}\right)^{k-1} \geq c_2 n^{k-1}$$

missing hyperedges; thus there are at least that many bad hyperedges. Let $\mathcal{B}$ consist of the pairs $(H, e)$ where $e = (x, y)$ is a bad pair, and $H \in \mathcal{H}$ with $x, y \in H$. Every bad hyperedge contains a bad pair; thus we have $|\mathcal{B}| \geq |A|c_2 n^{k-1}$. On the other hand if $e = (x, y)$ is a bad pair, then by the above, one of $x$ and $y$, say $x$ is in $A$. If we fix $x$ and $H$, there are at most $k-1$ ways to pick $y$ from $H$. Thus there is a vertex $x \in A$, say $x \in V_1$, that belongs to at least $\frac{|\mathcal{B}|}{(k-1)|A|}$ bad hyperedges, each intersecting $V_1$ in another vertex. Let $Y$ be the set of these vertices; then we have $|Y| \geq c_3 n$ with $c_3$ depending only on $c_1, c_2$.

For $2 \leq j \leq r$, let $Z_j \subset V_j$ be the set of vertices in $V_j$ that create a red pair with $x$. We consider two cases. Suppose first $|Z_j| \geq c_3 n$ for every $j$. For $y \in Y$, choose a hyperedge $H \in \mathcal{H}$ that contains both $x$ and $y$, and an $(r+1)$-set $R = \{x, y, z_2, \ldots, z_r\}$ where $z_j \in Z_j \setminus H$. Since the pairs $(xz_j)$ are red, there must be a blue pair among the other $r$ vertices by Claim 266. There are at least $(c_3 n - k)^r$ choices for $R$, and a blue pair is chosen at most $n^{r-2}$ times; thus we have at least $(c_3 n - k)^r / n^{r-2}$ blue pairs, which is larger than $c_4 n^2$ if $c_3$ is large enough compared to $c_4$. This contradiction finishes the proof of the first case.

Suppose now that, for example, $|Z_2| < c_3 n$. Then there are at most $c_3 n^{k-1} + O(n^{k-2})$ hyperedges in $\mathcal{H}$ that contain $x$ and intersect $V_2$. Let us move $x$ from $V_1$ to $V_2$. Then $f$ decreases by the number of hyperedges that contain $x$ and already intersected $V_2$; thus by at most $c_3 n^{k-1} + O(n^{k-2})$. On the other hand $f$ increases by the number of hyperedges that contain $x$, intersect $V_1$ in another vertex, and do not intersect $V_2$. By the definition of $x$, there are at least $\frac{|\mathcal{B}|}{(k-1)|A|}$ hyperedges that satisfy the first two properties, and at most $c_3 n^{k-1} + O(n^{k-2})$ of them intersect $V_2$. As $c_3$ is small enough compared to $c_1$, we obtain that $f$ increases, which contradicts the definition of $f$. ■

Mubayi and Verstraëte showed that the asymptotic version of Theorem 263 can be extended to graphs with chromatic number $r$.

**Theorem 267 (Mubayi, Verstraëte, [435])** *For any $r \geq k \geq 3$ and graph $F$ with chromatic number $r+1$ we have $ex_k(n, F^{+k}) = (1 + o(1))|\mathcal{T}^{(k)}(n, r)|$.*

Expansions of cycles and paths were studied earlier. They are also called *linear cycles* (paths), as they are all linear hypergraphs. Note that the length (number of vertices) of a $k$-uniform linear cycle is always divisible by $k-1$. Another name for them is *loose cycle*, but sometimes this name refers to a larger family of hypergraphs, where one of the consecutive pairs of hyperedges can intersect in more than one vertex (this way a cycle can have any length). The name loose cycle is also used for hypergraphs, where all of the consecutive pairs of hyperedges can intersect in more than one vertex, but non-consecutive ones must be disjoint. These are more often called *minimal cycles*, and we will return to them in Subsection 5.2.3.

Expansions of cycles and paths were studied by Frankl and Füredi [206], Csákány and Kahn [111], Füredi [234], Füredi, Jiang and Seiver [244], Füredi and Jiang [239], Jackowska, Polcyn and Ruciński [317] and Kostochka, Mubayi and Verstraëte [378]. The following theorem summarizes their results.

**Theorem 268** *Let $k \geq 3$, $l \geq 4$ and $t = \lceil (l-1)/2 \rceil$. For $n$ large enough we have*

$$ex_k(n, P_l^{+k}) = \binom{n}{k} - \binom{n-t}{k} + \begin{cases} 0 & \text{if } t \text{ is odd} \\ \binom{n-t-2}{k-2} & \text{if } t \text{ is even.} \end{cases}$$

*The same result holds for $C_l^{+k}$, except for the case $l = 4$, $k = 3$, where $ex_3(n, C_4^{+3}) = \binom{n}{2} + \max\{n-3, 4\lfloor \frac{n-1}{4} \rfloor\}$.*

Jackowska, Polcyn and Ruciński considered a very strong stability version of a special case of this theorem. Let the *first order Turán number* of a $k$-graph $\mathcal{F}$ be the ordinary Turán number, and the hypergraphs achieving this are called *1-extremal.* Then the *ith order Turán number* of $\mathcal{F}$ is the maximum number of $k$-edges in an $\mathcal{F}$-free $k$-graph $\mathcal{H}$ that is not a subhypergraph of a $j$-extremal one for $j < i$ (if such a hypergraph $\mathcal{H}$ exists), and the *i-extremal* hypergraphs are those achieving this maximum. In a series of papers [469,470, 487] they determined up to the fifth order of the Turán numbers of $P_4^{+3}$.

Bushaw and Kettle [98] determined the Turán number $ex_k(n, F^{+k})$ for expansions of graphs consisting of disjoint copies of paths. Gu, Li and Shi [290] did the same for graphs consisting of disjoint copies of cycles.

The result of Bushaw and Kettle deals with forests consisting of path components. The asymptotics of the Turán number is known for the expansion of every forest. Before stating that result, we need a definition due to Frankl and Füredi [206]. A *crosscut* is a set of vertices containing exactly one vertex from every hyperedge. The minimum size of a crosscut in a hypergraph $\mathcal{H}$ is denoted by $\sigma(\mathcal{H})$. The following theorem was proved by Füredi for $k \geq 4$ and by Kostochka, Mubayi and Verstraëte for $k = 3$.

**Theorem 269 (Füredi [236], Kostochka, Mubayi, Verstraëte, [380])** *Let $k \geq 3$ and $F$ be a forest. Then*

$$ex_k(n, F^{+k}) = (1 + o(1))(\sigma(F^{+k}) - 1)\binom{n}{r-1}.$$

Crosscuts are useful for other graphs as well. Kostochka, Mubayi and Verstraëte found the asymptotics of $ex_3(n, F^{+3})$ for every graph $F$ with a crosscut of size two.

**Theorem 270 (Kostochka, Mubayi, Verstraëte, [380])** *For any graph $F$ with $\sigma(F) = 2$ we have $ex_3(n, F^{+3}) = (1 + o(1))\binom{n}{2}$.*

If $\sigma(F) \geq 3$, then for any fixed pair $x, y$ of vertices the hypergraph consisting of all triples containing exactly one of $x$ and $y$ is $F^{+3}$-free. Thus we have the following.

**Corollary 271** *For any graph $F$ either $ex_3(n, F^{+3}) \leq (\frac{1}{2} + o(1))n^2$ or $ex_3(n, F^{+3}) \geq (1 + o(1))n^2$.*

Finally, for complete bipartite graphs Kostochka, Mubayi and Verstraëte proved the following.

**Theorem 272 (Kostochka, Mubayi, Verstraëte, [379])** *For $t \geq s \geq 3$ we have $ex_3(n, K_{s,t}^{+3}) = O(n^{3-1/s})$ and for $t > (s-1)!$ we have $ex_3(n, K_{s,t}^{+3}) = \Theta(n^{3-1/s})$.*

As expansions are linear hypergraphs, it makes sense to study (the size of) the largest linear hypergraphs that do not contain $F^{+k}$. Note that the results are very different: instead of the trivial lower bound $\binom{n-2}{k-2}$ (and $\binom{n-1}{k-1}$ for non-stars), we have the trivial upper bound $\binom{n}{2}$. We are only aware of results about cycles. Let us note first that forbidding $C_3^{+3}$ in a linear hypergraph is almost equivalent to saying that there are no 3 hyperedges on 6 vertices. It is easy to see that the celebrated Theorem 252 due to Ruzsa and Szemerédi implies $n^{2-o(1)} < ex_3(n, C_3^{+3}) < o(n^2)$.

Collier-Cartaino, Graber and Jiang proved the following.

**Theorem 273 (Collier-Cartaino, Graber, Jiang, [108])** *For all $k, l \geq 3$ we have*

$$ex_k^L(n, C_l^{+k}) = O(n^{1+\frac{1}{\lfloor l/2 \rfloor}}),$$

*and*

$$ex_k^L(n, C_l^{+k}) = \Omega(n^{1+\frac{1}{l-1}}).$$

Ergemlidze, Győri and Methuku improved the lower bound for 3-graphs and odd cycles.

**Theorem 274 (Ergemlidze, Győri, Methuku, [175])** *For $l = 2, 3, 4, 6$ we have $ex_3^L(n, C_{2l+1}^{+3}) = \Theta(n^{1+1/l})$. Moreover, for $l \geq 2$ we have $ex_3^L(n, C_{2l+1}^{+3}) = \Omega(n^{1+\frac{2}{3l-4+\delta}})$, where $\delta = 0$ if $l$ is odd and $\delta = 1$ if $l$ is even.*

### 5.2.2 Berge hypergraphs

Berge [50] defined cycles in hypergraphs in the following way. A cycle of length $l$ consists of $l$ vertices $v_1, \dots, v_l$ and $l$ hyperedges $e_1, \dots, e_l$ such that $v_i, v_{i+1} \in e_i$ for $i < l$ and $v_l v_1 \in e_l$. Note that it is important that we consider here $l$ distinct vertices and hyperedges. Also note that there are several different $k$-graphs that satisfy this property. We denote their family by $\mathbb{B}^k C_l$. Berge did not study extremal properties of hypergraph cycles. The first such result deals with hypergraphs that are (among other properties) $\mathbb{B}^3 C_2$-free. $C_2$ is a cycle of length 2, i.e. two copies of the same edge. As we do not deal with multisets or multiple edges, it makes no sense to additionally forbid this subgraph. However, a $\mathbb{B}^k C_2$ consists of two hyperedges that intersect in at least two vertices. Forbidding them is equivalent to assuming the hypergraph is linear.

Lazebnik and Verstraëte proved the following.

**Theorem 275 (Lazebnik, Verstraëte, [386])**

$$ex_3(n, \mathbb{B}^3 C_2 \cup \mathbb{B}^3 C_3 \cup \mathbb{B}^3 C_4) = \frac{1}{6} n^{3/2} + o(n^{3/2}).$$

The systematic study of the Turán number of Berge cycles started with $\mathbb{B}^k C_3$ by Győri [293], and continued with $\mathbb{B}^3 C_5$ by Bollobás and Győri [69]. Győri and Lemons studied them in a series of papers.

**Theorem 276 (Győri, Lemons, [295–298])** *$ex_k(n, \mathbb{B}^k C_{2l}) = O(n^{1+1/l})$.*
*If $k \geq 3$, then $ex_k(n, \mathbb{B}^k C_{2l+1}) = O(n^{1+1/l})$.*

The factors depending on $l$ were improved by Jiang and Ma [319], by Füredi and Özkahya [248], and by Gerbner, Methuku and Vizer [263].

Győri, G.Y. Katona and Lemons [294] extended the definition to paths in a natural way: omit an edge, say $e_l$ from the definition. We denote by $\mathbb{B}^k P_l$ the family of $k$-uniform Berge paths of length $l$. They proved the following theorem for all values of $k$ and $l$ except for $l = k + 1$. This remaining case was solved by Davoodi, Győri, Methuku and Tompkins.

**Theorem 277 (Győri et al. [294], Davoodi et al. [119])** *If $l > k$, then $ex_k(n, \mathbb{B}^k P_l) = \frac{n}{l}\binom{l}{k}$.*
*If $l \leq k$, then we have $ex_k(n, \mathbb{B}^k P_l) = \frac{n(l-1)}{k+1}$.*

Gerbner and Palmer [264] generalized this notion to any graph. A hypergraph $\mathcal{H}$ is a *Berge* copy of a graph $G$, if $V(G) \subseteq V(\mathcal{H})$ and there is a bijection $f$ from the hyperedges of $\mathcal{H}$ to the edges of $G$, such that $f(H) \subseteq H$ for every hyperedge $H \in \mathcal{H}$. We denote by $\mathbb{B}^k F$ the family of $k$-uniform Berge copies of $F$. Obviously $\mathbb{B}^2 F$ contains only $F$. They proved the following.

**Theorem 278 (Gerbner, Palmer, [264])** *If $k \geq |V(F)|$, then $ex_k(n, \mathbb{B}^k F) = O(n^2)$.*

*If $k \geq s+t$, then $ex_k(n, \mathbb{B}^k K_{s,t}) = O(n^{2-1/s})$.*

Notice that we have mostly presented only upper bounds so far. One reason is that some of the above results show that $ex_k(n, \mathbb{B}^k F) = O(ex_2(n, F))$. This is sharp in many cases. However, this does not give us the final answer if we do not know the order of magnitude of $ex_2(n, F)$, as is the case with cycles and $K_{s,t}$ for most values of $s$ and $t$. (The celebrated result of Erdős, Stone, and Simonovits states that $ex_2(n, F) = (1 - \frac{1}{\chi(F)-1} + o(1))\binom{n}{2}$ for any graph $F$; thus $ex_2(n, F) = \Theta(n^2)$ if and only if $F$ is not bipartite.) This makes Theorem 275 especially interesting. It determines the correct asymptotics of the Turán number when forbidding Berge cycles of length at most 4, although the analogous graph problem is still unsolved.

Gerbner and Palmer [265] connected Berge hypergraphs to a different graph parameter. Let $ex(n, K_k, F)$ be the maximum number of copies of $K_k$ in an $F$-free graph on $n$ vertices.

**Theorem 279 (Gerbner, Palmer, [265])** *For any $F$ and $r$ we have $ex(n, K_k, F) \leq ex(n, \mathbb{B}^k F) \leq ex(n, K_k, F) + ex_2(n, F)$.*

**Proof.** Let us consider a graph $G$ that is $F$-free on $n$ vertices and contains $ex(n, K_k, F)$ copies of $K_k$. Let $\mathcal{H}$ be the $k$-uniform hypergraph that contains these copies of $K_k$ as hyperedges. Then it is $\mathbb{B}^k F$-free and has cardinality $ex(n, K_k, F)$, proving the lower bound.

For the upper bound, let us consider a $\mathbb{B}^k F$-free hypergraph, and order its hyperedges arbitrarily. We go through the hyperedges, and greedily pick a new 2-edge from each of them if we can. More precisely, let the hyperedges be $H_1, \ldots, H_m$. At the first step we pick a subedge $f(H_1)$ of size 2 of $H_1$. Then in step $i$ for every $i = 2, \ldots, m$ we check if there is a subedge of size 2 of $H_i$ that is not $f(H_j)$ for $j < i$. If there is, then an arbitrary such subedge is chosen to be $f(H_i)$. If there is no such edge, no $f(H_i)$ is chosen. At the end the graph $G$ consisting of the edges $f(H_i)$ is $F$-free. For every $i$ either there is an edge $f(H_i)$, or every subedge of size 2 of $H_i$ is in $G$; thus $H_i$ corresponds to a copy of $K_k$ in $G$. This proves the upper bound. ∎

The systematic study of $ex(n, K_k, F)$ and in general the maximum number of copies of a graph $H$ in $F$-free graphs has been recently started by Alon and Shikhelman [17]. For every $k > 2$ when determining $\pi_k(\mathbb{B}^k F)$, $ex_2(n, F)$ is a lower order error term; thus it would be enough to find the maximum number of cliques in $F$-free graphs. Even when determining the order of magnitude or the asymptotics of $ex_k(n, \mathbb{B}^k F)$, for many graphs $ex_2(n, F)$ is negligible (for example if $ex(n, K_k, F)$ is super-quadratic, in particular if $F$ has chromatic number at least $k+1$). For other graphs, not many sharp bounds are known. Improving a result by Palmer, Tait, Timmons and Wagner [451], Gerbner, Methuku and Vizer determined the asymptotics for $\mathbb{B}^3 K_{2,t}$ if $t \geq 7$. Note that they proved a more general theorem giving upper bounds for every $r$ and every forbidden Berge hypergraph, in particular improving the known bounds for $\mathbb{B}^3C_{2l}$.

**Theorem 280 (Gerbner, Methuku, Vizer, [263])** *Let $t \geq 7$. Then*

$$ex_3(n, \mathbb{B}^3K_{2,t}) = \frac{1}{6}(t-1)^{3/2}n^{3/2}(1+o(1)).$$

**Proof.** Let us consider a 3-graph $\mathcal{H}$ that is $\mathbb{B}^3K_{2,t}$-free. We call a pair of vertices $u, v$ a *blue edge* if they are contained together in exactly one hyperedge in $\mathcal{H}$ and a *red edge* if they are contained together in more than one hyperedges. We call a hyperedge in $\mathcal{H}$ *blue* if it contains a blue edge, and *red* otherwise. Let us denote the set of blue edges by $S$, and the number of blue hyperedges by $s$. Let $S'$ be a subset of $S$ maximal with respect to the property that every blue hyperedge contains at most one edge of $S'$. Observe that the size of $S'$ equals the number of blue hyperedges as the intersection of two blue hyperedges cannot be a blue edge by definition. Now we consider the graph $G$ that consists of all the red edges and the blue edges in $S'$.

Let us assume there is a copy of $K_{2,t}$ in $G$. We build an auxiliary bipartite graph $B$. One of its classes $B_1$ consists of the edges of that copy of $K_{2,t}$, and the other class $B_2$ consists of the hyperedges of $\mathcal{H}$ that contain at least one of them. We connect a vertex of $B_2$ to a vertex of $B_1$ if the hyperedge contains the edge. Note that every hyperedge can contain at most 2 edges from a copy of $K_{2,t}$ (since $K_{2,t}$ is triangle-free); thus vertices of $B_2$ have degree at most 2.

Notice that if there is a matching in $B$ covering $B_1$, then the other endpoints of that matching would form a $\mathbb{B}^3K_{2,t}$, contradicting the $\mathbb{B}^3K_{2,t}$-free property of $\mathcal{H}$. Thus, by Hall's theorem (Theorem 70), there is a subset $X \subset B_1$ with $|N(X)| < |X|$. We call such sets of edges *bad.*

**Claim 281** *There is a blue edge in every copy of $K_{2,t}$ in $G$.*

**Proof.** If there is a red copy of $K_{2,t}$ in $G$, we can find a bad set $X \subset B_1$ such that every element in it is red; thus the corresponding vertices have degree at least 2 in $B$. On the other hand since every vertex of $N(X)$ is in $B_2$, they have degree at most 2 in $B$. This implies $|N(X)| \geq |X|$, contradicting the assumption that $X$ is a bad set. □

Our plan is to delete a red edge from every bad set in $G$, eliminating every copy of $K_{2,t}$. Note that a red edge might be contained in several blue hyperedges; those remain blue when we delete the red edge.

**Claim 282** *Every bad set $X$ contains a red edge that is contained in at most one red hyperedge.*

**Proof.** Let us assume there are $x$ red and $y$ blue hyperedges in $N(X)$. Then, by the definition of $S'$, we have at most $y$ blue, thus at least $x+1$ red edges in $X$. There are at most $2x$ edges in $B_2$ incident to the $x$ red hyperedges, so there must be a red edge in $X$ that is incident to at most one of those. □

Now we delete such an edge in a bad set from $G$, and recolor the incident red hyperedge to *green.* If there are other copies of $K_{2,t}$, we repeat this. Note that it is possible that some of the incident red hyperedges have already turned green. However, the number of green hyperedges is obviously at most the number of deleted edges.

Let us now color the deleted edges green, and recolor the remaining (non-green) red edges and hyperedges to *purple* to avoid confusion. Thus $G$ contains blue, green and purple edges, while $\mathcal{H}$ contains blue, green and purple hyperedges. A blue hyperedge contains at least one blue edge of $G$, a green hyperedge contains at least one green edge of $G$, while a purple hyperedge contains exactly 3 purple edges of $G$. On the other hand, a blue edge is contained in exactly one hyperedge, which is blue. A green edge is contained in at least one green hyperedge.

Let $G_1$ be the graph consisting of the blue and purple edges, and let $G_2$ be the graph consisting of the green and purple edges. Clearly $G_1$ is $K_{2,t}$-free because we get $G_1$ by eliminating every copy of $K_{2,t}$ from $G$. $G_2$ is also $K_{2,t}$-free, as we recolored only red edges to green or purple, so the edges in $G_2$ were all originally red. Hence it cannot contain a copy of $K_{2,t}$ by Claim 281.

**Claim 283** *If a $K_{2,t}$-free graph $G$ contains $x$ edges, then it contains at most $\frac{(t-1)x}{3}$ triangles.*

**Proof.** Observe that the neighborhood of every vertex is $S_t$-free. An $S_t$-free graph on $d(v)$ vertices contains at most $(t-1)d(v)/2$ edges. It means $v$ is in at most that many triangles, so summing up for every vertex $v$, every triangle is counted 3 times. On the other hand as $\sum_{v\in V(G)} d(v) = 2x$, we have

$$\sum_{v\in V(G)} (t-1)d(v)/2 = (t-1)x,$$

showing that the number of triangles in $G$ is at most $\frac{(t-1)x}{3}$. □

Now we continue the proof of Theorem 280. Let $x$ be the number of purple edges. Then, by Claim 283, the number of copies of $K_3$ consisting of purple edges is at most $\frac{(t-1)x}{3}$, but then the number of purple hyperedges is also at most this number, as for every purple hyperedge, the set of edges contained in it on the same three vertices form a purple triangle.

Let $y := ex_2(n, K_{2,t})$. Then the number of blue edges is at most $y-x$ as $G_1$ is $K_{2,t}$-free, and similarly the number of green edges is at most $y-x$ as $G_2$ is $K_{2,t}$-free. This implies that the number of blue hyperedges is at most $y-x$ and the number of green hyperedges is also at most $y-x$ by Claim 282 (indeed, any deleted red edge that was colored green, was contained in at most one red hyperedge, so a green edge is in at most one green hyperedge). Therefore the total number of hyperedges is at most

$$\frac{(t-1)x}{3} + 2(y-x) \leq \max\{(t-1)/3, 2\}(x+y-x).$$

This finishes the proof of the upper bound as Füredi showed [231] $y = ex_2(n, K_{2,t}) = (1+o(1))\frac{\sqrt{t-1}}{2}n^{3/2}$.

The lower bound follows from Theorem 279 and a theorem of Alon and Shikhelman [17] stating $ex(n, K_3, K_{2,t}) = \frac{1}{6}(t-1)^{3/2}n^{3/2}(1+o(1))$. ■

Gerbner, Methuku and Palmer [261] proved the following strengthening of Theorem 279. Note that an essentially equivalent statement was obtained by Füredi, Luo and Kostochka [245].

**Theorem 284** *Let $\mathcal{H}$ be a $k$-uniform $\mathbb{B}^k F$-free hypergraph. Then we can construct an $F$-free graph $G$ whose edges can be partitioned into two parts $E_1$ and $E_2$ such that $|\mathcal{H}|$ is at most the number of edges in $E_1$ plus the number of copies of $K_k$ in $E_2$.*

**Proof.** Let $B$ be an auxiliary bipartite graph with classes $A$ and $A'$, where $A$ is the set of hyperedges of $\mathcal{H}$ and $A'$ is the set of all pairs of vertices contained in some hyperedge of $\mathcal{H}$. A vertex $a \in A$ is joined to $a' \in A'$ if both vertices corresponding to $a'$ are contained in the hyperedge corresponding to $a$. A matching in $B$ induces a set of vertices in $A'$; these vertices correspond to a set of pairs on the vertex set of $\mathcal{H}$, i.e., they form a graph on the vertex set of $\mathcal{H}$.

Let $M$ be a maximum size matching in $B$ and let $G$ be the graph corresponding to the endpoints of $M$ in $A'$. Note that $G$ is $F$-free, as each edge of $G$ is associated with a hyperedge in $\mathcal{H}$ that contains it. If $M$ saturates $A$, then $G$ satisfies the statement of the theorem with $E(G) = E_1$ and $E_2 = \emptyset$. Therefore, we can assume $M$ does not saturate $A$. We call a path in $B$ *alternating* if it alternates between edges in $M$ and edges not in $M$ (beginning with an edge of $M$).

Let $A_1 \subset A$ and $A'_1 \subset A'$ be the vertices of $B$ that are not in $M$. As $M$ is maximum, there are no edges between $A_1$ and $A'_1$. Let $A_2 \subset A$ be vertices of $M$ in $A$ that are connected by an alternating path to a vertex in $A'_1$. Let $A'_2 \subset A'$ be vertices matched to $A_2$ by $M$. Suppose there exists an edge $aa'$ where $a \in A \setminus A_2$ and $a' \in A'_2$. Then there is an alternating path from $a$ to a vertex in $A_1$ by definition. Adding the edge $aa'$ to this alternating path gives an alternating path with both start and end edges not in $M$, which implies $M$ is not maximal, a contradiction. Therefore, every edge incident to $A'_2$ is incident to $A_2$.

Similarly, let $A'_3 \subset A'$ be vertices of $M$ in $A'$ that are connected by an alternating path to a vertex in $A_1$. Let $A_3 \subset A$ be the vertices matched to $A'_3$ by $M$. For any edge $aa'$ of $M$, if there is an alternating path from $a$ to a vertex in $A'_1$, then there is no alternating path from $a'$ to a vertex in $A_1$, otherwise $M$ can be increased. Therefore, $A_2$ and $A_3$ are disjoint as are $A'_2$ and $A'_3$.

Finally, let $A_4$ and $A'_4$ be the remaining vertices in $A$ and $A'$, respectively. Now $G$ is the graph spanned by the pairs represented by vertices in $A'_2 \cup A'_3 \cup A'_4$. The vertices $A'_2 \cup A'_3 \cup A'_4$ are all endvertices of edges in $M$, so $G$ is $F$-free. Let $E_1$ be the set of edges of $G$ represented by vertices of $A'_2$ and let $E_2$ be the set of edges of $G$ represented by vertices of $A'_3 \cup A'_4$. The number of hyperedges in $\mathcal{H}$ is

$$|\mathcal{H}| = |A_1| + |A_2| + |A_3| + |A_4| = |A'_2| + |A_1| + |A_3| + |A_4|.$$

The vertices in $A_1 \cup A_3 \cup A_4$ are only adjacent to vertices in $A'_3 \cup A'_4$. Thus we have that $|A_1| + |A_3| + |A_4|$ is at most the number of cliques formed by the edges of $E_2$. ■

Let us combine the above theorem with Claim 283. Let $x = |E_2|$ and $y = ex_2(n, K_{2,t})$, then $|E_1| \le y - x$. As $E_2$ is $K_{2,t}$-free, it contains at most $(t-1)x/3$ triangles by Claim 283. Thus Theorem 284 gives the bound $ex_3(n, \mathbb{B}^3 K_{2,t}) \le (t-1)x/3 + y - x$. This is at most $(t-1)y/3$ if $t \ge 4$, which implies that Theorem 280 holds for $t \ge 4$.

Consider now Berge complete graphs. Recall that Theorem 263 states that for $r \ge k \ge 3$ and $n$ large enough, the complete balanced $r$-partite $k$-graph contains the largest number of $k$-edges among $K_{r+1}^{+k}$-free $k$-graphs. Recall also that the expansion of a graph is a Berge-copy of it, thus the upper bound holds for $ex_k(n, \mathbb{B}^k K_{r+1})$ as well. Since an $r$-partite hypergraph does not contain $\mathbb{B}^k K_{r+1}$, a sharp result follows for large enough $n$. The threshold in the 3-uniform case was improved by Maherani and Shahsiah [400], who showed that the bound $ex_3(n, \mathbb{B}^3 K_{r+1}) \le |\mathcal{T}^{(3)}(n, r)|$ holds in the Berge case for every $n$ in case $r \ge 13$. Using Theorem 284, Gerbner, Methuku and Palmer [261] moved the threshold down. In particular, if $r > k$, then for every $n$ we have $ex_k(n, \mathbb{B}^k K_{r+1}) = |\mathcal{T}^{(k)}(n, r)|$.

Recall that forbidding a $\mathbb{B}^k C_2$ is equivalent to assuming the hypergraph is linear. Gerbner and Palmer considered this version and proved the following.

**Proposition 285 (Gerbner, Palmer, [264])**

$$ex_3(n, \mathbb{B}^k C_2 \cup \mathbb{B}^k C_3 \cup \cdots \cup \mathbb{B}^k C_l) \le \frac{ex_2(n, C_4, C_5 \ldots, C_l)}{3}.$$

**Proof.** Let us consider a hypergraph $\mathcal{H}$ on $n$ vertices that does not have a Berge cycle of length at most $l$, and replace each hyperedge by the three edges contained in it. By the linearity of $\mathcal{H}$ the resulting graph $G$ has three times more edges, hence it is enough to show that it does not contain any cycles of length between 4 and $l$. Assume otherwise and let $C$ be one of the shortest such cycles. If all the edges of $C$ come from different hyperedges, then we get a Berge cycle of the same length, a contradiction. Hence at least two edges of $C$ come from the same hyperedge $H$. We can replace them with the third edge $e$ from that hyperedge to get a shorter cycle $C'$. That is possible only if $C'$ is a triangle. However, all the triangles in $G$ correspond to a hyperedge in $\mathcal{H}$. As $e$ is in $C'$, that hyperedge must be $H$, but $C'$ does not contain the other two edges from $H$, a contradiction. ■

Note that by choosing $l = 4$ and using the well-known result $ex_2(n, C_4) = (1/2 + o(1))n^{3/2}$ [231], we get the upper bound in Theorem 275.

Timmons [529] studied $ex(n, \mathbb{B}^k C_2 \cup \mathbb{B}^k K_{2,t})$. His results were improved by Gerbner, Methuku and Vizer [263], who proved the following.

**Theorem 286 (Gerbner, Methuku, Vizer, [263])** *For all $k, t \geq 2$, we have*

$$ex_k(n, \mathbb{B}^k C_2 \cup \mathbb{B}^k K_{2,t}) \leq \frac{\sqrt{t-1}}{k(k-1)} n^{3/2} + O(n).$$

**Theorem 287 (Gerbner, Methuku, Vizer, [263])** *There is an absolute constant $c$ such that for any $t \geq 2$, we have,*

$$ex_3(n, \mathbb{B}^3 C_2 \cup \mathbb{B}^3 K_{2,t}) \geq \left(1 - \frac{c}{\sqrt{t-1}} \ln^{3/2}(t-1)\right) \frac{\sqrt{t-1}}{6} n^{3/2}.$$

Note that Theorem 286 implies the following result of Ergemlidze, Győri and Methuku [175], which is a strengthening of Theorem 275.

**Theorem 288 (Ergemlidze, Győri, Methuku, [175])**

$$ex_3(n, \mathbb{B}^3 C_2 \cup \mathbb{B}^3 C_4) = (\frac{1}{6} + o(1))n^{3/2}.$$

Ergemlidze, Győri and Methuku [175] also proved the following.

**Theorem 289 (Ergemlidze, Győri, Methuku, [175])**

$$ex_3(n, \mathbb{B}^3 C_2 \cup \mathbb{B}^3 C_5) = \frac{1}{3\sqrt{3}} n^{3/2} + O(\sqrt{n}).$$

Let us note that $F^{+k} \in \mathbb{B}^k F$; thus $ex_k(n, \mathbb{B}^k F) \leq ex_k(n, F^{+k})$. Hence the upper bounds from Subsection 5.2.1 all hold here. On the other hand Theorem 275 gives a lower bound for linear $C_4^{+3}$-free hypergraphs.

Finally we note that the above theorems show that as $k$ increases starting from two, $ex_k(n, \mathbb{B}^k F)$ can increase above $n^2$, as there are more $k$-sets, but after a while it stops increasing and goes back to $O(n^2)$. Grósz, Methuku and Tompkins [285] examined this phenomenon and showed the decrease does not stop here. For any $F$ if $k$ is large enough, we have $ex_k(n, \mathbb{B}^k F) = o(n^2)$. They examined the threshold, and showed that $ex_k(n, \mathbb{B}^k K_3) = o(n^2)$ if and only if $k \geq 5$. Note that by increasing $k$, the Turán number cannot decrease by much. Indeed, one can observe $\mathbb{B}^k K_3 \subseteq \mathbb{F}^k(3k-3, 3)$; thus Theorem 253 implies $ex_k(n, \mathbb{B}^k K_3) \geq n^{2-o(1)}$ for all $k$.

**Exercise 290** *Modify the proof of Theorem 279 to show that* $ex_k(n, \mathbb{B}^k C_2 \cup \mathbb{B}^k F) \le ex_2(n, F)$.

### 5.2.3 Minimal Berge hypergraphs

Recall that a hypergraph $\mathcal{H}$ is a Berge copy of a graph $F$ is there is a bijection $f$ from the hyperedges of $\mathcal{H}$ to the edges of $G$, such that $f(H) \subseteq H$ for every hyperedge $H$ of $\mathcal{H}$. We say that it is a *minimal* Berge copy, if the hyperedges $H_1$ and $H_2$ only intersect if their images $f(H_1)$ and $f(H_2)$ intersect. Let us denote the family of minimal Berge copies of $F$ by $\mathbb{M}^k F$. We are only aware of results concerning minimal Berge cycles and paths. They are often called *weak cycles* or simply *minimal cycles* and we will also usually omit the word Berge in this subsection.

Here we show a simple construction that avoids $\mathbb{M}^k F$. Let $\nu(F)$ be the largest number of independent edges in $F$. Then a hypergraph without $\nu(F)$ pairwise disjoint hyperedges cannot be a minimal copy of $F$. Also note that $F^{+k} \in \mathbb{M}^k F$; thus upper bounds from Subsection 5.2.1 hold here as well. On the other hand all hypergraphs in $\mathbb{M}^k F$ are Berge copies of $F$; thus lower bounds from Subsection 5.2.2 also hold here.

Mubayi and Verstraëte [434] studied minimal paths and cycles. The exact Turán numbers were found by Füredi, Jiang and Seiver [244] for paths, and by Füredi and Jiang [239] for cycles.

**Theorem 291 (Füredi, Jiang, Seiver [239, 244])** *For* $k \ge 4$, $l \ge 1$ *and* $n$ *large enough we have* $ex_k(n, \mathbb{M}^k P_{2l+1}) = ex_k(n, \mathbb{M}^k C_{2l+1}) = \binom{n}{k} - \binom{n-2}{k-l}$ *and* $ex_k(n, \mathbb{M}^k P_{2l+2}) = ex_k(n, \mathbb{M}^k C_{2l+2}) = \binom{n}{k} - \binom{n-2}{k-l} + 1$. *The same holds for* $\mathbb{M}^3 P_{2l+1}$.

Note that this theorem does not say anything about $\mathbb{M}^3 C_{2l+1}$. The best bounds we are aware of are the simple lower bound mentioned in the beginning of this subsection and the upper bound that follows from Theorem 268.

### 5.2.4 Inflated graphs

The *inflated* version of a graph $F$ is a $2k$-graph $\mathcal{I}^{2k} F$ which we get when we replace each vertex by $k$ new vertices; thus each edge becomes a $2k$-set. This operation is often called blowing up. Turán problems regarding (mostly) $\mathcal{I}^{2k} K_r$ have been studied earlier; they were called *expanded* graphs, or even *expansions.* To avoid confusion, we renamed them.

**Theorem 292 (Sidorenko [509, 510])** $\pi_{2k}(\mathcal{I}^{2k} K_r) \le \frac{r-2}{r-1}$. *On the other hand if* $r = 2^p + 1$, *then we have* $\pi_{2k}(\mathcal{I}^{2k} K_r) \ge \frac{r-2}{r-1}$.

**Proof.** Let us start with the upper bound. Assume we are given an $\mathcal{I}^{2k} K_r$-free $2k$-graph $\mathcal{H}$ on the vertex set $V$ of size $n$. We construct an auxiliary graph $G$

in the following way. Let $\binom{V}{k}$ be the vertex set, and two $k$-sets be connected, if their union is a hyperedge of $\mathcal{H}$. As $\mathcal{H}$ is $\mathcal{I}^{2k}K_r$-free, $G$ must be $K_r$-free; thus it has at most $ex_2(\binom{n}{k}, K_r) = (1+o(1))\pi(K_r)\binom{\binom{n}{k}}{2}$ edges. By definition, every hyperedge of $\mathcal{H}$ creates $\binom{2k}{k}/2$ edges of $G$ and distinct hyperedges create disjoint sets of edges. Therefore $\mathcal{H}$ has at most

$$\frac{(1+o(1))\pi_2(K_r)\binom{\binom{n}{k}}{2}}{\binom{2k}{k}/2} = (1+o(1))\pi_2(K_r)\binom{n}{2k}$$

hyperedges. This implies $\pi_{2k}(\mathcal{I}^{2k}K_r) \leq \pi_2(K_r) = \frac{r-2}{r-1}$.

Let us continue with the lower bound and partition a vertex set $V$ of size $n$ into $r-1$ parts $V_i$ of size either $\lfloor n/(r-1)\rfloor$ or $\lceil n/(r-1)\rceil$, where $i$ is an element of the vector space $\mathbf{F}_2^p$. For an element $x \in V_i$ let $\phi(x) = i$. Now we define a hypergraph $\mathcal{H}$. Let a $2k$-set $H$ be a hyperedge in $\mathcal{H}$ if and only if $\sum_{x\in H}\phi(x) \neq 0$.

Now a copy of $\mathcal{I}^{2k}K_r$ would consist of $r$ $k$-sets, such that the union of any two form a hyperedge. Then there would be two of these $k$-sets $A$ and $B$ with $\sum_{x\in A}\phi(x) = \sum_{x\in B}\phi(x)$, as there are at most $r-1$ different values in $\mathbf{F}_2^p$. But then $A \cup B$ is not a hyperedge, a contradiction; thus $\mathcal{H}$ is $\mathcal{I}^{2k}K_r$-free.

Let us consider a random $2k$-set $H$, and an arbitrary element $x$ of it. No matter what the other elements are, $H$ is a hyperedge in $\mathcal{H}$ if and only if $x$ does not belong to the only part that makes the sum 0. It obviously happens with probability $\frac{r-2}{r-1} + o(1)$; thus $\mathcal{H}$ indeed has the required density. ■

Keevash and Sudakov [356] showed that for other values the upper bound is not necessarily sharp. More precisely they showed that if $r$ is not $2^p+1$ for any integer $p$, then $\pi_4(\mathcal{I}^4K_r) \leq \frac{r-2}{r-1} - 10^{-33}r^{-70}$.

The case of the triangle was studied by Frankl [195], who proved $\pi_{2k}(\mathcal{I}^{2k}K_3) = 1/2$. The construction giving the lower bound is the following. We partition $[n]$ into two not necessarily equal parts and consider the family of all $2k$-sets that intersect both parts in an odd number of vertices. Frankl conjectured that this is the best one can do. Let $b_{2k}(n)$ denote the largest family one can get this way. Note that the optimum is not given by a balanced partition. Keevash and Sudakov proved Frankl's conjecture.

**Theorem 293 (Keevash, Sudakov, [356])** *Let $n$ be large enough. Then $ex_{2k}(n, \mathcal{I}^{2k}K_3) = b_{2k}(n)$.*

Keevash [348] gave a construction showing $\pi_4(\mathcal{I}^4K_4) \geq 9/14$.

If $F = (A, B, E)$ is bipartite, one can replace each vertex in part $A$ by $a$ copies of it, and each vertex in part $B$ by $b$ copies of it. This way each edge becomes an $(a+b)$-set. We denote the resulting graph by $\mathcal{I}^{(a,b)}F(A,B)$. Füredi, Kostochka, Mubayi, Jiang and Verstraëte [243] called it the $(a,b)$-blowup of $F$. Note that we usually cannot omit $A$ and $B$, as the graph $F$

can be partitioned other ways, and that would give us a different hypergraph. In particular usually $\mathcal{I}^{(a,b)}F(A,B) \neq \mathcal{I}^{(a,b)}F(B,A)$. However, for even paths they are equal.

Let us consider the graph $P_3$ and let $A$ consist of the two endpoints, while $B$ consist of the remaining vertex. Then $\mathcal{I}^{(a,b)}F(A,B)$ consist of two hyperedges sharing exactly $b$ vertices. One can think of it as an $L$-intersecting problem, where $L = [k-1]\setminus\{b\}$. The maximum cardinality of $\mathcal{I}^{(a,b)}F(A,B)$-free families (they are also called *b-avoiding*) was asked by Erdős [156], and mostly solved by Frankl and Füredi.

**Theorem 294 (Frankl, Füredi, [203])** *(i) For any $a, b$ we have*

$$ex_{a+b}(n, \mathcal{I}^{(a,b)}P_3(A,B)) = \Theta(n^{\max\{a,b-1\}}).$$

*(ii) If $b-1 > a$ and $n$ is large enough, then we have*

$$ex_{a+b}(n, \mathcal{I}^{(a,b)}P_3(A,B)) = \binom{n-a-1}{b-1}.$$

*(iii) If $b-1 \leq a$ and $k-a$ has a prime divisor $q$ with $q > l$, then we have*

$$ex_{a+b}(n, \mathcal{I}^{(a,b)}P_3(A,B)) = (1+o(1))\binom{n}{a}\binom{a+2b-1}{b-1}/\binom{a+2b-1}{a}.$$

The $\mathcal{I}^{(1,k-1)}P_3(A,B)$-free property means that every $(k-1)$-set is contained in at most one hyperedge. This implies $ex_k(n, \mathcal{I}^{(1,k-1)}P_3(A,B)) \leq \frac{1}{k}\binom{n}{k-1}$, with equality if every $(k-1)$-set is contained in exactly one hyperedge. This means the hypergraph is a *design*, and the celebrated theorem of Keevash [349] states that it exists if $n$ is large enough and some obvious divisibility conditions are satisfied; thus in that case we have an exact result (for other values of $n$ this bound is asymptotically tight). Füredi and Özkahya [247] showed that for $n$ large enough $ex_{a+b}(n, \mathcal{I}^{(a,b)}P_4(A,B)) = \binom{n-1}{a+b-1}$.

Füredi, Jiang, Kostochka, Mubayi, and Verstraëte [243] considered paths and trees. Once again let $A$ contain at least one of the endpoints of a path; this property determines both $A$ and $B$. They showed $ex_{a+b}(n, \mathcal{I}^{(a,b)}P_{2l}(A,B)) = (l-1+o(1))\binom{n}{a+b-1}$ if $2l \geq 6$. They proved the same for $P_{2l+1}$ if $a > b$. The other cases remain open.

If $A$ has size $t$, a simple $\mathcal{I}^{(a,b)}F(A,B)$-free hypergraph is if we take a set of $t-1$ vertices and all the hyperedges of size $a+b$ intersecting that set. Let $\mathcal{H}^{a+b}(t-1)$ be this hypergraph; then $|\mathcal{H}^{a+b}(t-1)| = \binom{n}{a+b} - \binom{n-t+1}{a+b} = (t-1)\binom{n}{a+b-1} + o(n^{a+b-1})$.

**Theorem 295 (Füredi et al. [243])** *Let $b < a$ and $T(A,B,E)$ be a tree with $|A| = t$, $|B| = s$; then $ex_{a+b}(n, \mathcal{I}^{(a,b)}T(A,B)) \leq (t-1)\binom{n}{a+b-1} + o(n^{a+b-1})$. If $t \leq s$, this is asymptotically sharp. If $t \leq s$, $b \geq 2$ and there is a leaf in $B$, then $ex_{a+b}(n, \mathcal{I}^{(a,b)}T(A,B)) = \binom{n}{a+b} - \binom{n-t+1}{a+b}$.*

Mubayi and Verstraëte [432] considered $\mathcal{I}^{1,k-1}K_{s,t}$. Note that $\mathcal{I}^{1,k-1}K_{s,t} \neq \mathcal{I}^{1,k-1}K_{t,s}$, but their results apply to both cases. They proved the following.

**Theorem 296 (Mubayi and Verstraëte [432])** *Let $t \geq s \geq 2$. Then $ex_3(n, \mathcal{I}^{1,2}K_{s,t}) = O(n^{3-1/s})$. On the other hand for every $s$ and $t$ we have $ex_3(n, \mathcal{I}^{1,2}K_{s,t}) = \Omega(\frac{ex_2(n,K_{s,t})^2}{n})$.*

**Proof.** For the lower bound we consider a $K_{s,t}$-free bipartite graph $G$ with parts $A = \{a_1, \dots, a_n\}$ and $B = \{b_1, \dots, b_n\}$. It is well-known that we can find such a graph with $ex_2(2n, K_{s,t})/2$ edges. Indeed, take a $K_{s,t}$-free graph $G$ with $ex_2(2n, K_{s,t})$ edges. Consider a random bipartition of the vertex set to two equal sizes. The expected number of edges between the two parts is exactly half of the number of edges, so there exists a bipartite subgraph of $G$ with the required properties.

Let $A' = \{a'_1, \dots, a'_n\}$ be a set of $n$ new vertices. Let the vertex set of $\mathcal{H}$ be $A \cup B \cup A'$ and the hyperedges be $\{a_i, b_j, a'_l\}$ for which $a_i b_j a_l$ is a path in $G$. Then the number of hyperedges in $\mathcal{H}$ is

$$2\sum_{v\in B}\binom{deg(v)}{2} \geq 2n\binom{e(G)/n}{2} = e(G)^2/n - e(G).$$

Note that $\mathcal{H}$ is 3-partite. It implies that if it contains a copy of $\mathcal{I}^{1,2}K_{s,t}$, then the set $S$ of the $s$ unchanged vertices of the original $K_{s,t}$ belongs to one of the parts of $\mathcal{H}$. For the $t$ other vertices of the original $K_{s,t}$, each is replaced by two vertices, one of which belongs to another part, and the other belongs to the third part. Thus if $S \subset A$, it forms a copy of $K_{s,t}$ in $G$ together with the $t$ vertices in $B$. A similar argument works if $S \subset B$. If $S = \{a'_1, \dots, a'_s\} \subset A'$, then $\{a_1, \dots, a_s\}$ form a copy of $K_{s,t}$ together with the $t$ vertices in $B$. In all cases, we found a copy of $K_{s,t}$ in $G$, a contradiction.

For the upper bound we use the following lemma. (Prove it by taking a random tripartition and considering the expected number of hyperedges that form a transversal of this partition.)

**Lemma 297 (Erdős, Kleitman [161])** *Let $\mathcal{G}$ be a 3-graph on $3n$ vertices. Then there is a 3-partite subhypergraph of $\mathcal{G}$, with parts of size $n$, and with at least $2|\mathcal{G}|/9$ hyperedges.*

With that in hand, it is enough to prove the statement for an arbitrary $\mathcal{I}^{1,2}K_{s,t}$-free 3-partite 3-graph $\mathcal{G}$ with parts $A, B, C$. Let us partition the elements of $A \times B$ into $n$ matchings $M_1, \dots, M_n$ and let $\mathcal{G}_i$ be the set of hyperedges containing a pair from $M_i$. By the pigeonhole principle one of them, say $\mathcal{G}_1$ contains at least $|\mathcal{G}|/n$ hyperedges. Let $G$ be the graph with vertex set $B \cup C$ where $bc$ is an edge if and only if it is contained in a 3-edge of $\mathcal{G}_1$.

We claim that $G$ cannot contain a copy of $K_{s,t}$ such that $t$ vertices are in $B$. Indeed, otherwise we could add to those vertices their neighbor in $M_1$, and we would obtain a copy of $\mathcal{I}^{1,2}K_{s,t}$. This means, using the well-known result of Kővári, Sós and Turán [358], that there are $O(n^{2-1/s})$ edges in $G$. Observe that every hyperedge of $\mathcal{G}_1$ contains exactly one edge in $G$, and every edge of $G$ is contained in exactly one edge of $\mathcal{G}_1$; thus we have $|\mathcal{G}|/n \leq |\mathcal{G}_1| = e(G) = O(n^{2-1/s})$.

■

Note that the above theorem gives the correct order of magnitude if $s = 2$. But in that case Mubayi and Verstraëte [432] have even stronger results. For the general case, they conjecture that $ex_3(n, \mathcal{I}^{1,2}K_{s,t}) = \Theta(n^{3-2/s})$. They managed to improve the upper bound for $s = 3$.

**Theorem 298 (Mubayi, Verstraëte, [432])** *(i)* $ex_3(n, \mathcal{I}^{1,2}K_{3,t}) = O(n^{13/5})$. *(ii) Let* $t \geq 2$ *and* $n \geq 3t$. *Then* $ex_3(n, \mathcal{I}^{1,2}K_{2,t}) <$

$$\begin{cases} 3\binom{n}{2} + 6n & \text{if } t = 2 \\ t^4\binom{n}{2} & \text{if} t > 2. \end{cases}$$

*Moreover, for infinitely many* $n$ *we have* $ex_3(n, \mathcal{I}^{1,2}K_{2,t}) \geq \frac{2t-1}{3}\binom{n}{2}$.

### 5.2.5 Suspensions

For a graph $G$ its $k$-uniform *suspension* is the hypergraph we obtain by taking $k-2$ new vertices and adding them to every edge. We denote it by $\mathcal{S}^kG$. Sidorenko [508] considered this notion for trees, and called this operation *enlargement.* In the survey of Keevash [348] they are called *extended $k$-trees.* However, the same operation is called suspension there, when applied to the complete graph. Moreover, it is also called *cone* there. Sós, Erdős and Brown [518] also call it cone, and they call $\mathcal{S}^3C_l$ a *wheel.* Bollobás, Leader and Malvenuto [72] studied $\mathcal{S}^kK_r$ and called it *daisy.*

**Proposition 299** *For any graph* $F$ *we have* $\pi_k(\mathcal{S}^kF) \leq \pi_{k-1}(\mathcal{S}^{k-1}F)$.

**Proof.** Let $\varepsilon > 0$, $n$ is large enough and consider an $\mathcal{S}^kF$-free $k$-graph $\mathcal{H}$ on $n$ vertices and an arbitrary vertex $v$ of it. The link of $v$, i.e. the $(k-1)$-graph $\{H \setminus \{v\} : v \in H \in \mathcal{H}\}$ is obviously $\mathcal{S}^{k-1}F$-free, thus contains at most $(\pi(\mathcal{S}^{k-1}F) + \varepsilon)\binom{n-1}{k-1}$ hyperedges. This is true for every vertex, which gives at most $(\pi_{k-1}(\mathcal{S}^{k-1}F) + \varepsilon)\binom{n-1}{k-1}n/k$ hyperedges, as each hyperedge would be counted $k$ times. This means $|\mathcal{H}| \leq (\pi_{k-1}(\mathcal{S}^{k-1}F) + \varepsilon)\binom{n}{k}$, which implies the statement.

■

Sidorenko considered $\mathcal{S}^kT$ for trees, but he applied the extension operation (that we define in Subsection 5.3.1) to them, and determined Turán densities for the resulting hypergraphs.

Regarding $\mathcal{S}^k K_r$, Keevash [348] mentions a couple of unpublished observations due to Bukh and Alon, and a paper by Bollobás, Leader and Malvenuto [72] states several conjectures. The main conjecture appears to be that for any $r$ the density $\pi_k(\mathcal{S}^k K_r)$ goes to 0 if $k$ goes to infinity. Note that it is trivial for $r = 3$, as $\mathcal{S}^k K_3$ is the unique $k$-graph consisting of three hyperedges on $k+1$ vertices, and it easily implies $\pi_k(\mathcal{S}^k K_3) \leq 2/(k+1)$. Shi [505] proved $1/2 \leq \pi_3(\mathcal{S}^3 K_4) \leq 2/3$.

Note that $\mathcal{S}^3 K_3$ is often denoted by $K_4^-$, when it is obvious that only 3-graphs are considered, as it is obtained by deleting an edge from the complete 3-graph on 4 vertices. It has been studied a lot. The best known bounds are $2/7 \leq \pi_3(\mathcal{S}^3 K_3) \leq 0.2871$. The construction is due to Frankl and Füredi [201]. They start with the following 3-graph $\{1,2,3\}, \{2,3,4\}, \{3,4,5\}, \{4,5,1\}, \{5,1,2\}, \{1,3,6\}, \{2,4,6\}, \{3,5,6\}, \{4,1,6\}, \{5,2,6\}\}$ and blow it up, i.e. replace each of the six vertices by $\lfloor n/6 \rfloor$ or $\lceil n/6 \rceil$ copies and each hyperedge by the corresponding complete 3-partite 3-graph. Then iteratively substitute the same construction inside each of the six parts.

The upper bound is due to Baber and Talbot [30]. They improved earlier results by de Caen [123], Matthias [413], Mubayi [426], Talbot [523], Mubayi and Talbot [431], Markström and Talbot [409] and Razborov [475]. Let us mention that we present a proof of the upper bound 1/3 in Subsection 5.5.2 to illustrate the flag algebra method.

Gunderson and Semeraro [291] showed that if $n-1 = 4r+3$ and is a prime power, then $ex_4(n, \mathcal{S}^4 K_3) = \frac{n}{16}\binom{n}{3}$.

If $G$ is a complete bipartite graph, then $\mathcal{S}^k G$ is a complete multipartite $k$-graph where $k-2$ of the parts are of size 1. Mubayi considered such hypergraphs and proved the following.

**Theorem 300 (Mubayi, [425])** ***(i)*** *$ex_k(n, \mathcal{S}^k K_{2,t}) = (1+o(1))\frac{\sqrt{t}}{k!} n^{k-1/2}$.*

***(ii)*** *$ex_k(n, \mathcal{S}^k K_{3,3}) = (1+o(1))\frac{n^{k-1/3}}{k!}$.*

***(iii)*** *Let $s > 3$, $t \geq (s-1)!+1$. Then we have $ex_k(n, \mathcal{S}^k K_{s,t}) = \Theta(n^{k-1/s})$.*

**Proof.** The upper bound is a slight extension of Proposition 299 for each statement. We prove it by induction on $k$. The base steps $k = 2$ for these statements (i.e. bounds on the graph Turán number of $K_{2,t}$ and $K_{s,t}$) were proved in [233] and [358].

For a complete bipartite graph $F$ we consider an $\mathcal{S}^k F$-free $k$-graph $\mathcal{H}$ on $n$ vertices and an arbitrary vertex $v$ of $\mathcal{H}$. The link of $v$, i.e. the $(k-1)$-graph $\{H \setminus \{v\} : v \in H \in \mathcal{H}\}$ is obviously $\mathcal{S}^{k-1}F$-free, thus contains at most $ex_{k-1}(n, \mathcal{S}^{k-1}F)$ hyperedges. This is true for every vertex, which gives at most $ex_{k-1}(n, \mathcal{S}^{k-1}F)n/k$ hyperedges, as each hyperedge is counted $k$ times. Plugging in the bound of the induction hypothesis finishes the proof of the upper bound.

The lower bounds are given by simple extensions of the constructions for the graph case. Let us prove **(i)** first. It is well-known that we can find a prime power $q$ such that $(q-1)/(t-1)$ is an integer, and $(q-1)^2/(t-1) = (1+o(1))n$ (see for example [314]). It means we will be able to assume $(q-1)^2/(t-1) = n$ without changing the asymptotics.

Recall $\mathbb{F}_q$ is the $q$-element finite field and let $H$ be a $(t-1)$-element subgroup of $\mathbb{F}_q^\times$. Consider the equivalence relation on $(\mathbb{F}_q^\times) \times (\mathbb{F}_q^\times)$ defined by $(a,b) \sim (x,y)$ if and only if there is an $\alpha \in H$ such that $a = \alpha x$ and $b = \alpha y$. The class represented by $(a,b)$ is denoted by $\langle a,b\rangle$. These equivalence classes form the underlying set $V$. A set of $k$ vertices $\langle a_1,b_1\rangle, \langle a_2,b_2\rangle, \dots, \langle a_k,b_k\rangle$ form a hyperedge in $\mathcal{G}$ if and only if $\prod_{i=1}^k a_i + \prod_{i=1}^k b_i \in H$. Observe that $\mathcal{G}$ is well-defined as if $(a_i,b_i) \sim (a'_i,b'_i)$ for $i = 1,2,\dots,k$, then there exist $\alpha_i \in H$ $i = 1,2,\dots,k$ with $a'_i = \alpha_i a_i, b'_i = \alpha_i b_i$, then $\prod_{i=1}^k a'_i + \prod_{i=1}^k b'_i = \pi_{i=1}^k \alpha_i(\prod_{i=1}^k a_i + \prod_{i=1}^k b_i)$ and $\prod_{i=1}^k \alpha_i \in H$ as $H$ is a subgroup of $\mathbb{F}_q^\times$, so $\prod_{i=1}^k a_i + \prod_{i=1}^k b_i \in H$ if and only if $\prod_{i=1}^k a'_i + \prod_{i=1}^k b'_i \in H$.

The number of hyperedges is at least

$$\binom{n}{k-1} \cdot \frac{q-k}{k} = \frac{q^{2k-1}}{t^{k-1}k!} + O(q^{2k-2}) = \left(\frac{\sqrt{t}}{k!}\right) n^{k-1/2}$$

as for every $(k-1)$-set $U$ of vertices the degree $|\{v \in V : U \cup \{v\} \in \mathcal{G}\}|$ of $U$ is at least $q-k$.

What is left to show is that $\mathcal{G}$ is $\mathcal{S}^k K_{2,t}$-free.

Let $(a,b),(a',b') \in \mathbb{F}_q^\times$ with $(a,b) \not\sim (a',b')$. Then for every $\alpha,\beta \in H$ the equation system

$$ax + by = \alpha$$

$$a'x + b'y = \beta$$

has at most one solution $(x,y)$. Hence if we have $ax+by, a'x+b'y \in H$, there are at most $t-1$ non-equivalent solutions $(x,y)$.

Assume there is a copy of $\mathcal{S}^k K_{2,t}$ in $\mathcal{G}$, where $\langle u_1,v_1\rangle$ and $\langle u_2,v_2\rangle$ form the part of size 2 in the original $K_{2,t}$, $\langle x_i,y_i\rangle$ $(1 \le i \le t)$ form the part of size $t$ in the original $K_{2,t}$ and $\langle a_j,b_j\rangle$ $(1 \le j \le k-2)$ are the additional vertices contained in every hyperedge. Let

$$p_1 = \left(\prod_{j=1}^{k-2} a_j\right) u_1,\ p_2 = \left(\prod_{j=1}^{k-2} a_j\right) u_2,\ q_1 = \left(\prod_{j=1}^{k-2} b_j\right) v_1,\ q_2 = \left(\prod_{j=1}^{k-2} b_j\right) v_2.$$

Observe that $(p_1,p_2) \not\sim (q_1,q_2)$ as otherwise $(u_1,v_1) \sim (u_2,v_2)$. The $2t$ hyperedges give $p_1x_i + q_1y_i \in H$ and $p_2x_i + q_2y_i \in H$ for $1 \le i \le t$ with $t$ non-equivalent solutions. But we have already seen that such a system of equations can have at most $t-1$ non-equivalent solutions, a contradiction that finishes the proof of **(i)**.

For **(ii)** and **(iii)** we consider *norm hypergraphs*. Let the vertex set be $\mathbb{F}_{q^{s-1}} \times \mathbb{F}_q^\times$. For $X \in \mathbb{F}_{q^{s-1}}$ let $N(X) = X^{1+q+\dots+q^{s-1}}$ be the norm of $X$ over $\mathbb{F}_q$. Note that we have $(N(X))^q = N(X)$; thus $N(X) \in \mathbb{F}_q$. The vertices $(A_i,a_i)$ $(1 \le i \le k)$ form a hyperedge in $\mathcal{G}$ if and only if $N(\sum_{i=1}^k A_i) = \prod_{i=1}^k a_i$.

For every choice of $k-1$ vertices there are exactly $q^{s-1}-k$ hyperedges containing them; thus a calculation similar to the one above shows $\mathcal{G}$ contains the desired number of hyperedges. To show that $\mathcal{G}$ is $\mathcal{S}^kK_{s,t}$-free, we need the following lemma by Alon, Rónyai and Szabó.

**Lemma 301 (Alon, Rónyai, Szabó, [15])** *If* $(D_1,d_1),(D_2,d_2),\ldots,(D_s,d_s)$ *are distinct vertices of* $\mathcal{G}$, *then the system of $s$ equations*

$$N(D_j+X)=d_jx \quad (1\le j\le s)$$

*has at most $(s-1)!$ solutions* $(X,x)\in\mathbb{F}_{q^{s-1}}\times\mathbb{F}_q^{\times}$.

Suppose there is a copy of $\mathcal{S}^kK_{s,t}$ in $\mathcal{G}$, and let $(A_i,a_i)$ $(1\le i\le s)$ form the part of size $s$ in the original $K_{s,t}$, $(B_j,b_j)$ $(1\le j\le t)$ form the part of size $t$ and $(C_l,c_l)$ $(1\le l\le k-2)$ the additional vertices. We define

$$P_i=A_j+\sum_{l=1}^{k-2}C_l \text{ and } p_i=a_j+\sum_{l=1}^{k-2}c_l.$$

Again we have $(P_i,p_i)\ne(P_{i'},p_{i'})$ for $i\ne i'$ as otherwise we have $(B_i,b_i)=(B_{i'},b_{i'})$. The $st$ hyperedges forming $\mathcal{S}^kK_{s,t}$ yield $N(P_i+B_j)=p_ib_j$ for $1\le i\le s$ and $1\le j\le t$. This means $t$ solutions $(B_j,b_j)$ for a system of $s$ equations, contradicting Lemma 301. ■

Let us mention a variant of suspensions. The $k$-uniform *$r$-suspension* of a graph $G$ is obtained by taking $r$ pairwise disjoint $(k-2)$-sets, and adding each of them to every edge of $G$ (creating $r|E(G)|$ hyperedges on $r(k-2)+|V(G)|$ vertices). Let us denote the resulting $k$-graph by $\mathcal{S}^k_rG$. Simonovits [514, 515] called $\mathcal{S}^3_rC_l$ an *$r$-pyramid*. However, he forbade the infinite family $\mathbb{P}_r$ of all the $r$-pyramids $\mathcal{S}^3_rC_l$ (for all $l$). He proved $ex_3(n,\mathbb{P}_r)=O(n^{3-1/r})$, and $ex_3(n,\mathbb{P}_3)=\Theta(n^{8/3})$, generalizing results of Sós, Erdős and Brown [518].

Let us note that $\mathcal{S}^k_rM_t=\mathcal{I}^{1,k-1}K_{r,t}$, and we mentioned these in the previous subsection.

**Exercise 302** *Show that* $ex_k(n,\mathcal{S}^kG)=O(n^{k-2}ex_2(n,G))$.

### 5.2.6 Other graph-based hypergraphs

In this subsection we consider hypergraphs that are related to some specific graphs, but not in a way that could be generalized to arbitrary graphs.

A *$t$-tight path* $\mathbb{T}^kP_l(t)$ in a $k$-graph is a $\mathbb{B}^kP_l$ such that consecutive edges intersect in at least $t$ vertices. Note that 1-tight paths are the same as Berge paths. Győri, Katona and Lemons studied them and showed the following.

**Theorem 303 (Győri, Katona, Lemons, [294])** *Let $k\ge 2$ and $1\le t<k$. Then* $ex_k(n,\mathbb{T}^kP_l(t))=(1+o(1))\frac{\binom{n}{t}\binom{l}{k}}{\binom{l}{t}}$.

The proof uses the following theorem of Rödl.

**Theorem 304 (Rödl, [481])** *There is an $l$-graph $\mathcal{H}$ with $|\mathcal{H}| = (1-o(1))\frac{\binom{n}{t}}{\binom{l}{t}}$ such that every $t$-set is contained in at most one hyperedge.*

Note that this is sharp, as the hyperedges each cover $\binom{l}{t}$ $t$-sets, and those must be all different. This means there have to be at least $|\mathcal{H}|\binom{l}{t}$ different $l$-sets. By the celebrated result of Keevash [349], for infinitely many $n$ the factor $1 - o(1)$ is not needed.

**Proof of Theorem 303.** For the lower bound, let $\mathcal{H}$ be the $l$-graph given by Theorem 304. Let us replace each of its hyperedges by all of their $k$-subsets, and let $\mathcal{H}'$ denote the $k$-graph obtained this way, i.e. $\mathcal{H}' = \Delta_k(\mathcal{H})$ the $k$-shadow of $\mathcal{H}$. As for any $H_1, H_2 \in \mathcal{H}$ we have $\Delta_t(H_1) \cap \Delta_t(H_2) = \emptyset$; therefore the number of hyperedges in $\mathcal{H}'$ is $(1 + o(1))\frac{\binom{n}{t}\binom{l}{k}}{\binom{l}{t}}$.

We claim that $\mathcal{H}'$ is $\mathbb{T}^k P_l(t)$-free. Indeed, if $\mathcal{H}'$ contained a copy $\mathcal{F}$ of $\mathbb{T}^k P_l(t)$, then it is on $l+1$ vertices; thus it contains vertices from different hyperedges of $\mathcal{H}$. There must be two consecutive hyperedges in the path containing vertices from different hyperedges $H_1$ and $H_2$, one of them must be fully in $H_1$, the other must be fully in $H_2$, and they intersect in $t$ vertices by the $t$-tightness property. Thus $|H_1 \cap H_2| \geq t$, which contradicts the assumption of Theorem 304.

For the upper bound observe first that the case $t = 1$ is solved by Theorem 277. For larger $t$ let us consider a $k$-graph $\mathcal{G}$ with more than $\frac{\binom{n}{t}\binom{l}{k}}{\binom{l}{t}}$ hyperedges. It contains a vertex $x_1$ of degree more than

$$\frac{k}{n}\frac{\binom{n}{t}\binom{l}{k}}{\binom{l}{t}}.$$

Consider its link hypergraph $\mathcal{G}_1 = \{H : H \cup \{x_1\} \in \mathcal{G}\}$; it has a vertex of degree more than

$$\frac{k(k-1)}{n(n-1)}\frac{\binom{n}{t}\binom{l}{k}}{\binom{l}{t}} = \frac{\binom{n-1}{t-1}\binom{l-1}{k-1}}{\binom{l-1}{t-1}}.$$

Continuing this we can find a sequence of vertices $x_1, x_2, \ldots x_{t-1}$ and their link hypergraphs $\mathcal{G}_1, \mathcal{G}_1, \ldots, \mathcal{G}_{t-1}$ such that $|\mathcal{G}_i| \geq \frac{\binom{n-i}{t-i}\binom{l-i}{k-i}}{\binom{l-i}{t-i}}$. Then we can find a 1-tight path in $\mathcal{G}_{t-1}$ by Theorem 277. All its hyperedges can be extended to hyperedges of $\mathcal{G}$ with $x_1, x_2, \ldots, x_{t-1}$, resulting in a $t$-tight path. ■

A $k$-uniform *tight path* $\mathcal{T}^k P_l$ of length $l$ consists of $l - k + 1$ vertices $v_1, \ldots, v_{l-k+1}$ and $l$ hyperedges $H_1, \ldots, H_l$ such that for each $1 \leq i \leq l$ $H_i = \{v_i, \ldots, v_{i-k+1}\}$. Note that even a $(k-1)$-tight path is not necessarily tight. Győri, Katona and Lemons proved the following.

**Theorem 305 (Győri, Katona, Lemons, [294])** $(1 + o(1))\frac{l-1}{k}\binom{n}{k-1} \leq ex_k(n, \mathcal{T}^k P_l) \leq (l-1)\binom{n}{k-1}$.

Füredi, Jiang, Kostochka, Mubayi and Verstraëte gave the following non-trivial upper bound.

**Theorem 306 (Füredi, Jiang, Kostochka, Mubayi, Verstraëte, [242])** $ex_k(n, \mathcal{T}^k P_l) \leq \frac{(l-1)(k-1)}{k}\binom{n}{k-1}$.

**Exercise 307 (Patkós [456])** *Prove by induction on $l$ that for any pair $k, l$ of integers with $l \leq k$ we have*

$$ex_k(n, \mathcal{T}^k P_l) \leq \sum_{j=2}^{l} \frac{j-1}{k-j+2}\binom{n}{k-1}.$$

One can similarly define the *tight cycle* $\mathcal{T}^k C_l$ of length $l$ that consists of $l$ vertices $v_1, \ldots, v_l$ and $l$ hyperedges $H_1, \ldots, H_l$ such that for each $1 \leq i \leq l$ $H_i = \{v_i, \ldots, v_{i-k+1}\}$, where the addition is taken modulo $l$. Mubayi and Rödl [430] considered $\mathcal{T}^3 C_5 = \{\{1,2,3\},\{2,3,4\},\{3,4,5\},\{4,5,1\},\{5,1,2\}\}$. They proved $0.464 \leq \pi(\mathcal{T}^k C_l) \leq 2 - \sqrt{2}$. Razborov [475] improved the upper bound to 0.4683.

Note that extremal properties of tight cycles were mostly studied in the case the cycle has length $n$ (Hamiltonian cycle) or close to $n$, see for example [8]. One exception is the maximum number $f_k(n)$ of hyperedges that a $k$-uniform hypergraph on $n$ vertices can have if it does not contain *any* tight cycles. Disproving a conjecture of Sós and Verstraëte, Huang and Ma [313] showed that there exists a constant $c_k$ such that $f_k(n) \geq (1 + c_k)\binom{n-1}{k-1}$ holds. It is still unknown if $f_k(n) = \Theta(n^{k-1})$ holds.

A $k$-graph $\mathcal{H}$ is a *hypertree* if its hyperedges can be ordered as $H_1, \ldots, H_m$ such that for any $1 \leq i \leq m$ there is $i' < i$ with

$$H_i \cap (\cup_{j=1}^{i-1} H_j) \subseteq H_{i'}.$$

It is called a *tight* hypertree if this intersection has size $k-1$, i.e. there is exactly one new element in $H_i$. Both definitions generalize ordinary trees. The following conjecture of Kalai appeared in [206].

**Conjecture 308 (Kalai)** *Let $\mathcal{F}$ be a $k$-uniform tight hypertree with $v$ vertices. Then we have $ex_k(n, \mathcal{F}) \leq \frac{v-k}{k}\binom{n}{k-1}$.*

A hypertree is *star-shaped* if every $i' = 1$ in the definition of the hypertree, i.e. there is a hyperedge $H_1$ that intersects every other hyperedge in $k-1$ vertices. Frankl and Füredi [206] proved the above conjecture for star-shaped hypertrees.

As Exercise 310 shows, the order of magnitude in this conjecture is correct, for every hypertree. Füredi and Jiang [240] considered this problem and proved exact and asymptotic results for some hypertrees. Recently, Füredi, Jiang, Kostochka, Mubayi, and Verstaëte proved [241] Kalai's conjecture for tight hypertrees with at most 4 hyperedges, and showed the upper bound $(\frac{t-1}{k} + O_{r,c}(1))\binom{n}{k-1}$ for tight hypertrees with $t$ edges and at most $c$ vertices of degree 1.

A hypergraph $\mathcal{H}$ is a *generalized 4-cycle* if it consists of four hyperedges $A, B, C, D$ such that $A \cup C = B \cup D$ and $A \cap C = B \cap D = \emptyset$. Let $\mathbb{G}^k$ be their family. Note that $\mathbb{G}^k \subseteq \mathbb{M}^k C_4$ and the unique 3-uniform generalized 4-cycle is $\mathcal{I}^{1,2}C_4$. Füredi [227] proved

$$\binom{n-1}{k-1} + \lfloor\frac{n-1}{k}\rfloor \leq ex_k(n, \mathbb{G}^k) < \frac{7}{2}\binom{n}{k-1}.$$

The upper bound was improved by Mubayi and Verstraëte [432] and further improved by Pikhurko and Verstraëte [468] to $(1+\frac{2}{\sqrt{r}}+o(1))\binom{n}{k-1}$. They also proved $ex_3(n, \mathbb{G}^3) \leq \frac{13}{9}\binom{n}{2}$.

Let $\mathbb{G}_k$ be the family of $k$-graphs that consists of three hyperedges $A, B, C$ such that $|A \cap B| = k - 1$, and $C$ contains $A \triangle B$. Note that if a $k$-graph does not contain any member of $\mathbb{G}_k$, then it is cancellative. The *generalized triangle* $\mathcal{G}_k$ is the member of $\mathbb{G}_k$ where $C$ contains $k-2$ new vertices not in $A \cup B$. Frankl and Füredi proved [202] $ex_3(n, \mathcal{G}_3) = ex_3(n, \mathbb{G}_3)$ for $n > 3000$. Keevash and Mubayi [351] showed it for $n > 32$. Sidorenko [507] showed $ex_4(n, \mathbf{G}_4)$ is given by $\mathcal{T}^{(4)}(n, 4)$.

Frankl and Füredi [208] conjectured $ex_k(n, \mathcal{G}_k) = ex_k(n, \mathbb{G}_k)$ holds for larger $k$, and proved $ex_k(n, \mathbb{G}_k) \leq ex_k(n, \mathcal{G}_k) \leq (1 + O(1/n))ex_k(n, \mathbb{G}_k)$ (see Exercise 311 for a weaker result). They also showed the following.

**Theorem 309 (Frankl, Füredi, [208])** $ex_5(n, \mathbb{G}_5) \leq \frac{6}{11^4}n^5$ *with equality holding if and only if* $11|n$.

$ex_6(n, \mathbb{G}_6) \leq \frac{11}{12^5}n^6$ *with equality holding if and only if* $12|n$.

Pikhurko [465] proved $ex_4(n, \mathcal{G}_4) = ex_4(n, \mathbb{G}_4)$ for $n$ large enough. Norin and Yepremyan [447] showed $ex_5(n, \mathcal{G}_5) = ex_5(n, \mathbb{G}_5)$ and $ex_6(n, \mathcal{G}_6) = ex_6(n, \mathbb{G}_6)$.

Let us mention one more simple way to define hypergraphs from graphs besides those studied in this section. Let $F$ be a graph, then let $\mathbb{T}^k F$ be the family of hypergraphs $\mathcal{H}$ such that there exists a set $A \subset [n]$ with $|A| = |V(F)|$ such that $\mathcal{H}|_A$ is isomorphic to $F$, i.e. $F$ appears as a trace. Here any other forbidden configuration could be used instead of a graph, even non-uniform ones. In fact, there are several such results, but we consider them in Section 8.3 in the chapter about traces.

**Exercise 310** *Show that if $\mathcal{F}$ is a $k$-uniform hypertree, then $ex_k(n, \mathcal{F}) = \Theta(n^{r-1})$.*

**Exercise 311** *Use supersaturation to show $ex_k(n, \mathcal{G}_k) - ex_k(n, \mathbb{G}_k) = o(n^k)$.*

## 5.3 Hypergraph-based forbidden hypergraphs

Some problems we have already mentioned in the previous section can be described better when we extend hypergraphs rather than graphs. For example, inflated $P_3$ led us to the problem of determining the maximum size of a $k$-graph without hyperedges sharing exactly $t$ vertices. One could describe the forbidden hypergraph in a simpler way: let $\mathcal{F}$ be the $t+1$-graph consisting of 2 hyperedges sharing $t$ vertices. Then the forbidden hypergraph is the expansion of $\mathcal{F}$.

Let us give a more precise definition. For a hypergraph $\mathcal{H}$, its $k$-expansion $\mathcal{H}^{+k}$ is the hypergraph obtained by adding $k - |H|$ new elements to each hyperedge $H$, such that the sets of these new elements are disjoint from each other and from the vertex set of the original graph.

Keller and Lifschitz [357] considered expansions of hypergraphs. If $k$ grows, the part that defines the forbidden hypergraph remains the same. Thus one can expect that the extremal hypergraph is also mostly defined by a small part.

Let $J \subset [n]$. We say that a hypergraph $\mathcal{H}$ is a *$J$-junta* if for every hyperedge $H$ and another $k$-set $F$, $H \cap J = F \cap J$ implies that $F$ is also a hyperedge of $\mathcal{H}$. Informally, it means that for any $k$-set its intersection with $J$ decides if it is a hyperedge of $\mathcal{H}$ or not. We say that $\mathcal{H}$ is a *$j$-junta* if it is a $J$-junta for some $|J| = j$. We are interested in the case when $j$ is a small constant.

**Theorem 312 (Keller, Lifschitz, [357])** *For any fixed hypergraph $\mathcal{F}$ there exist constants $c$ and $j$ such that the following holds. Let $c < k < n/c$. Suppose $\mathcal{H}$ is an $\mathcal{F}^{+k}$-free $k$-graph. Then there exists an $\mathcal{F}^{+k}$-free $j'$-junta $\mathcal{H}'$ with $j' \leq j$ such that $|\mathcal{H} \setminus \mathcal{H}'| \leq \max\{e^{-k/c}, ck/n\}|\mathcal{H}'|$.*

In particular, $|\mathcal{H}|$ and $|\mathcal{H}'|$ have the same order of magnitude, and if $k$ tends to infinity, but grows sublinearly in $n$, they have the same asymptotics. Note that an important feature of this theorem is that it is enough to assume $n > ck$. This means that in its corollaries the threshold for $n$ being large enough is linear in $k$.

A *$d$-dimensional simplex* (or briefly $d$-simplex) consists of $d+1$ sets, such that their intersection is empty, but any $d$ sets among them have non-empty intersection. Chvátal [106] conjectured that the largest hypergraph without a $d$-simplex has $\binom{n-1}{k-1}$ hyperedges, provided $d < k < dn/(d+1)$, and the unique extremal hypergraph is where every hyperedge contains a fixed vertex. This conjecture has attracted the attention of many researchers. Frankl [183] showed it holds for $k \ge (d-1)n/d$. Frankl and Füredi [206] showed that it holds for $n$ large enough. Mubayi and Verstraëte [433], as we have mentioned in Subsection 5.2.1, proved it for $d = 2$. Keevash and Mubayi [352] proved it for $cn \le k \le n/2 - O_d(1)$, where $c$ depends only on $d$.

Note that a simplex itself is not an expansion of any hypergraph. However, let us consider the special $d$-simplex $\mathcal{F}$ that consists of all the $d$-subsets of a $d+1$-element set. Then $\mathcal{F}^{+k}$ is a $d$-simplex for every $k > d$. Applying Theorem 312 to $\mathcal{F}$ gives the same upper bound asymptotically. Using this, Keller and Lifschitz proved Chvátal's conjecture.

**Theorem 313 (Keller, Lifschitz, [357])** *For any $d$ there exists an $n_0 = n_0(d)$ such that if $n \ge n_0$, $d < k < dn/(d+1)$ and $\mathcal{F}$ denotes the family of $k$-graphs that are $d$-simplices, then $ex_k(n, \mathcal{F}) = \binom{n-1}{k-1}$ and the unique extremal hypergraph is where every hyperedge contains a fixed vertex.*

One can easily extend the definition of Berge hypergraphs similarly. Let $\mathcal{F}$ be a hypergraph with hyperedges of size at most $k$. We say that a $k$-graph $\mathcal{H}$ is a $k$-uniform Berge copy of $\mathcal{F}$ if there is a bijection $f$ from the hyperedges of $\mathcal{H}$ to the hyperedges of $\mathcal{F}$ such that $f(H) \subset H$. Let us denote their family by $\mathbb{B}^k\mathcal{F}$.

If $F$ is a graph and $\mathcal{H} \in \mathbb{B}^l F$, then $\mathbb{B}^k\mathcal{H} \subseteq \mathbb{B}^k F$ for $l \le k$. This implies $ex_k(n, \mathbb{B}^k\mathcal{H}) \ge ex_k(n, \mathbb{B}^k F)$. Győri and Lemons [297] proved Theorem 276 using a so-called reduction lemma, that states that if $F$ is a cycle and $\mathcal{H} \in \mathbb{B}^k F$, then we also have $ex_k(n, \mathbb{B}^k\mathcal{H}) = O(ex_k(n, \mathbb{B}^k F))$. Let us remark that their proof extends to the case when $F$ has treewidth at most two.

Anstee and Salazar [21] considered Berge problems in the matrix setting, where characteristic vectors of the hyperedges form the columns of a matrix. Similarly to Section 8.3, repeated columns are equivalent to multiple edges. Also, as it is more natural in that setting, both the forbidden hypergraph $\mathcal{F}$ and the $\mathcal{F}$-free hypergraph can be non-uniform.

The *t-blowup* of a $k$-graph $\mathcal{F}$ is defined in the following way. We replace each vertex by $t$ copies, and we replace each hyperedge $F \in \mathcal{F}$ by $t^k$ hyperedges: for each vertex of that hyperedge, we choose one of its $t$ copies arbitrarily, and all the resulting hyperedges are in $\mathcal{F}(t)$. In general we say that $\mathcal{H}$ is a blowup of $\mathcal{F}$ if we can obtain it by replacing each vertex of $\mathcal{F}$ by some (potentially different for each vertex) number of copies of it. Thus a hyperedge is replaced by a complete $k$-partite $k$-graph. Theorem 239 shows that for a complete $k$-partite $k$-graph $\mathcal{F}$ we have $\pi_k(\mathcal{F}) = 0 = \pi_k(\mathcal{E})$, where $\mathcal{E}$ is the $k$-graph consisting of a single hyperedge. This statement easily extends to the following.

**Theorem 314** *For any $k$-graph $\mathcal{F}$ we have $\pi_k(\mathcal{F}(t)) = \pi_k(\mathcal{F})$.*

**Proof.** Here we follow the elegant proof by Keevash [348] that uses the supersaturation phenomenon described in Theorem 335. Let $n$ be large enough, $a$ be a fixed constant, and let us consider a $k$-graph $\mathcal{H}$ with more than $(\pi_k(\mathcal{F})+a)\binom{n}{k}$ hyperedges. Then, by Theorem 335, there exists $b$ such that $\mathcal{H}$ contains at least $b\binom{n}{v}$ copies of $\mathcal{F}$, where $v=|V(\mathcal{F})|$. Let us build an auxiliary hypergraph $\mathcal{H}'$ with vertex set $V(\mathcal{H})$ where every hyperedge consists of the vertex set of a copy of $\mathcal{F}$ in $\mathcal{H}$. This auxiliary hypergraph has density greater than zero, thus by Theorem 239 it contains a copy $\mathcal{T}$ of $\mathcal{T}^{(v)}(m,v)$ for any fixed $m$ (once again, $n$ is large enough).

Let $A_1,\dots,A_v$ be the classes of $\mathcal{T}$ and $a_1,\dots,a_v$ be the elements of $V(\mathcal{F})$. Let us consider a hyperedge $T\in\mathcal{T}$. It is a copy of $\mathcal{F}$, which gives a bijection $\phi$ from the vertex set of $\mathcal{F}$ to $T$. On the other hand, each element of $T$ belongs to a different class $A_j$. This gives us a permutation on $[v]$, which maps $i$ to the index of the class containing $\phi(a_i)$. Let that permutation be the color of that hyperedge, then standard Ramsey theory results imply that for large enough $m$ there is a monochromatic copy of $\mathcal{T}^{(v)}(t,v)$, which gives us a copy of $\mathcal{F}(t)$. ■

### 5.3.1 Extensions and Lagrangians of hypergraphs

For a hypergraph $\mathcal{F}$, its *extension* $\mathcal{E}^k\mathcal{F}(t)$ is defined in the following way. First we add $t$ new vertices. For two new vertices, or an old vertex and a new vertex, or for a pair of old vertices that are not contained in an edge of $\mathcal{F}$, we add a hyperedge containing them and $k-2$ new vertices (disjoint from everything). Also, we keep the hyperedges of $\mathcal{F}$. For example if $\mathcal{F}$ is the hypergraph with no vertices, then $\mathcal{E}^k\mathcal{F}(t)=K_t^{+k}$.

Let us consider another way to extend hypergraphs. For a $k$-graph $\mathcal{F}$ let $\mathbb{H}^k\mathcal{F}(t)$ be the family of $k$-graphs $\mathcal{H}$ that contain a set $S$ of $t+|V(\mathcal{F})|$ vertices such that the subhypergraph induced on $S$ contains $\mathcal{F}$, and every pair in $S$ is covered in $\mathcal{H}$. Obviously $\mathcal{E}^k\mathcal{F}(t)\in\mathbb{H}^k\mathcal{F}(t)$.

Let $\mathcal{F}$ be the hypergraph with no vertices. Then $\mathbb{H}^k\mathcal{F}(t)\subseteq\mathbb{K}_r^k$ which was defined in Subsection 5.2.1. It is easy to see then that Theorem 265 determines the exact Turán number of $\mathbb{H}^k\mathcal{F}(t)$. Brandt, Irwing and Jiang [88] defined the general operation that we denote by $\mathbb{H}^k$, in order to generalize Mubayi's argument and determine exact results on extensions. In order to state their main result, we need to introduce Lagrangians. This theory was developed by Sidorenko [507] and independently by Frankl and Füredi [207], generalizing earlier work by Motzkin and Straus [424] and by Zykov [556] on graphs.

Here we follow the presentation of Brandt, Irwing and Jiang [88], which is based on that of Keevash [348]. We say that a hypergraph $\mathcal{H}$ *covers pairs* if any two vertices are contained in a hyperedge together. If $\mathcal{H}$ does not cover pairs, i.e. there are two vertices $u$ and $v$ not contained in any hyperedge of $\mathcal{H}$ together, the *symmetrization* of $u$ to $v$ is the operation of removing all the hyperedges containing $u$ and adding the hyperedges $\{H \cup \{u\} \setminus \{v\} : v \in H \in \mathcal{H}\}$.

**Proposition 315** *Let $p \geq k+1$. Let $\mathcal{F}$ be a $k$-graph on at most $p$ vertices and $\mathcal{H}$ be a $k$-graph that is $\mathbb{H}^k\mathcal{F}$-free. Let $u$ and $v$ be two vertices not covered by any hyperedge of $\mathcal{H}$, and let $\mathcal{H}'$ be the $k$-graph obtained from $\mathcal{H}$ by symmetrizing $u$ to $v$. Then $\mathcal{H}'$ is also $\mathbb{H}^k\mathcal{F}$-free.*

We say that two vertices are *equivalent* if this operation does not change anything. We take two non-equivalent vertices $u$ and $v$ with $d(u) \geq d(v)$ and symmetrize every vertex in the equivalence class of $v$ to $u$. It is easy to see that we can do this for those vertices at the same time and the number of equivalence classes decreases, thus this process ends. Let $\mathcal{H}^*$ be the resulting hypergraph.

**Proposition 316** *(i)* $|\mathcal{H}| \leq |\mathcal{H}^*|$.

*(ii) Let $S$ consist of a vertex from each equivalence class in $\mathcal{H}^*$, and $\mathcal{H}^*_S$ be the subhypergraph induced by $S$. Then $\mathcal{H}^*_S$ covers pairs and $\mathcal{H}^*$ is a blowup of $\mathcal{H}^*_S$.*

We define the *Lagrange polynomial* of $\mathcal{F}$ on the variables $x = (x_1, \ldots, x_n)$ as

$$p_{\mathcal{F}}(x) = k! \sum_{F \in \mathcal{F}} \prod_{i \in F} x_i.$$

Note that the definition of this polynomial also appears in the literature without the factor of $k!$.

We define the *Lagrangian* of $\mathcal{F}$ to be

$$\lambda(\mathcal{F}) = \max\{p_{\mathcal{F}}(x) : x_i \geq 0 \text{ for all } i = 1, \ldots n, \sum_{i=1}^{n} x_i = 1\},$$

and the *Lagrangian density* as $\pi^\lambda_k(\mathcal{F}) = \max\{\lambda(\mathcal{H}) : \mathcal{H} \not\supset \mathcal{F}\}$.

**Proposition 317** *Let $\mathcal{F}$ be a $k$-graph, $\mathcal{G}$ be an $\mathcal{F}$-free $k$-graph, and $\mathcal{H}$ a blowup of $\mathcal{G}$ on $n$ vertices. Then $|\mathcal{H}| \leq \pi^\lambda_k(\mathcal{F})n^k/k!$.*

**Proof.** Let $[s]$ be the set of vertices of $\mathcal{G}$, and let $V_1, \ldots, V_s$ be the corresponding partition of the vertex set of $\mathcal{H}$. Since $\mathcal{H}$ is a blowup of $\mathcal{G}$, we have

$$|\mathcal{H}| = \sum_{G \in \mathcal{G}} \prod_{i \in G} |V_i|.$$

Let $x_i = |V_i|/n$ for every $i \leq s$. Then we have

$$\sum_{G \in \mathcal{G}} \prod_{i \in G} |V_i| = n^k \sum_{G \in \mathcal{G}} \prod_{i \in G} x_i = \frac{n^k}{k!} p_{\mathcal{G}}(x) \leq \frac{n^k}{k!} \pi^\lambda_k(\mathcal{F}).$$

■

Note that $\lambda(\mathcal{H})$ is also called the *blowup density* of $\mathcal{H}$. Keevash [348] observed the following.

**Proposition 318** *If $\mathcal{H}$ is a $k$-graph that covers pairs, then $\pi_k^\lambda(\mathcal{H}) = \pi_k(\mathcal{H})$.*

The following theorem was proved by Keevash as a generalization of a theorem of Sidorenko [508].

**Theorem 319 (Keevash, [348])** *Let $\mathcal{F}$ be a $k$-graph that covers pairs, and $t \geq |V(\mathcal{F})|$. If $\pi_k(\mathcal{F}) \leq \prod_{i=1}^{k-1}(1-i/t)$, then $\pi_k(\mathcal{E}^k\mathcal{F}(t)) = \prod_{i=1}^{k-1}(1-i/t)$.*

One can use this theorem to determine the Turán density of $\mathcal{E}^k\mathcal{F}(t)$, but one can similarly determine the Turán number of $\mathbb{H}^k\mathcal{F}(t)$, then prove a stability version of that, and use that to determine the exact Turán number of $\mathcal{E}^k\mathcal{F}(t)$. This is what we have already seen in the proof of Theorem 263 by Pikhurko, and Brandt, Irwing and Jiang generalized this.

**Theorem 320 (Brandt, Irwing, Jiang, [88])** *Let $m$, $k$ be positive integers. Let $\mathcal{F}$ be a $k$-graph that either has at most $m$ vertices or has $m+1$ vertices, one of which has degree 1. Suppose either $\pi_k^\lambda(\mathcal{F}) < \frac{m!}{(m-r)!m^r}$, or $\pi_k^\lambda(\mathcal{F}) = \frac{m!}{(m-r)!m^r}$ and $\mathbb{H}^k\mathcal{F}(t)$ is stable in the following sense. For any $\varepsilon > 0$ there is $\delta > 0$ such that if $n$ is large enough, and $\mathcal{H}$ is a $\mathcal{F}$-free $k$-graph on $n$ vertices with at least $(\pi_k(\mathcal{F}) - \delta)\binom{n}{k}$ hyperedges, then $\mathcal{H}$ can be made $m$-partite by deleting at most $\varepsilon n$ vertices. Then for $n$ large enough we have $ex_k(\mathcal{E}^k\mathcal{F}(t+1)) = |\mathcal{T}^{(k)}(n,t)|$.*

Here we state some corollaries.

A *generalized fan* is a $k$-graph consisting of $k$ edges that share a vertex $x$ but otherwise disjoint, and another one that intersects each of them in a vertex different from $x$. The following result is also implied by Theorem 320.

**Theorem 321 (Mubayi, Pikhurko, [429])** *Let $\mathcal{F}_k$ be the generalized fan and $n$ be large enough. Then $ex_k(n,\mathcal{F}_k) = |\mathcal{T}^{(k)}(n,k)|$.*

The next two theorems follow from Theorem 320, but they were also independently proved by Norin and Yepremyan [448].

Sidorenko [508] determined the Turán density for extensions of suspensions of trees that satisfy the Erdős-Sós conjecture. This conjecture [150] states that the Turán number of any tree on $t$ vertices is $(t-2)n/2$. It is still open, although Ajtai, Komlós, Simonovits and Szemerédi claimed a solution for $t$ large enough, and it is known to hold for several families of trees.

**Theorem 322 (Norin, Yepremyan, [448])** *For every $k$, if $n$ is large enough and $T$ is a tree on $r$ vertices that satisfies the Erdős-Sós conjecture, then $ex_k(n,\mathcal{E}^k\mathcal{S}^kT(0)) = |\mathcal{T}^{(k)}(n,r+k-3)|$, and $\mathcal{T}^{(k)}(n,r+k-3)$ is the unique extremal hypergraph.*

**Theorem 323 (Norin, Yepremyan, [448])** *Let $\mathcal{F}$ be a $k$-graph that covers pairs, $t-1 \geq |V(F)|$, and $n$ large enough. If $\pi_k(\mathcal{F}) \leq \prod_{i=1}^{k-1}(1-i/t)$, then $ex_k(n, \mathcal{E}^k\mathcal{F}(t)) = |\mathcal{T}^{(k)}(n,t)|$, and $\mathcal{T}^{(k)}(n,t)$ is the unique extremal hypergraph.*

Let $\mathcal{M}_t^k$ be the $k$-graph consisting of $t$ independent edges. The Erdős-Ko-Rado theorem determines $ex_k(n, \mathcal{M}_2^k)$. Hefetz and Keevash [305] determined $ex_3(n, \mathcal{E}^3\mathcal{M}_2^3)$ by determining the maximum Lagrangian of intersecting 3-graphs. Wu, Peng and Chen [553] determined $ex_4(n, \mathcal{E}^4\mathcal{M}_2^4)$, and Watts, Norin and Yepremyan [548] determined $ex_k(n, \mathcal{E}^k\mathcal{M}_2^k)$ for $n$ large enough. See also the thesis of Yepremyan [555].

Jiang, Peng and Wu [320] generalized the result of Hefetz and Keevash to a different direction and showed the following.

**Theorem 324 (Jiang, Peng, Wu, [320])** *Let $t \geq 2$. Then $ex_3(n, \mathcal{E}^3\mathcal{M}_t^3) = |\mathcal{T}^{(3)}(n, 3t-1)|$ for $n$ large enough. Moreover, $\mathcal{T}^{(3)}(n, 3t-1)$ is the unique largest $\mathcal{E}^3\mathcal{M}_t^3$-free hypergraph.*

They also determined the Turán numbers of the extensions of the expansions of stars $\mathcal{E}^3S_t^{+3}$ and $\mathcal{E}^4S_t^{+4}$.

Finally, let us note that Lagrangians are more than just a tool in Turán theory. They make their own area in extremal graph and hypergraph theory, when instead of maximizing the number of edges, one tries to maximize of the Lagrangian of hypergraphs avoiding a given configuration. Here we just mention the intriguing conjecture of Frankl and Füredi [208].

**Conjecture 325** *For any positive integers $n, m, k$ with $m \leq \binom{n}{k}$ the maximum Lagrangian of a $k$-graph on $n$ vertices with $m$ hyperedges is given by the first $m$ $k$-edges in the colex ordering.*

## 5.4 Other forbidden hypergraphs

In this section we describe some further results, where the forbidden hypergraphs do not fit into the earlier categories. We mostly focus on exact results, by which we mean that the exact value of $ex_k(n, \mathcal{F})$ is known for $n$ large enough. For a long time, hardly any exact results were found when only one hypergraph was forbidden, and in 2011 Keevash could list them all in his survey [348]. Here we only state some of them (in addition to those stated earlier).

The Fano plane $PG_2(2)$ is the projective plane over a field of two elements. In hypergraph language it is the 3-graph on [7] consisting of seven hyperedges

$\{1,2,3\}, \{3,4,5\}, \{5,6,1\}, \{1,4,7\}, \{3,6,7\}, \{2,5,7\}, \{2,4,6\}$. A simple construction avoiding the Fano plane can be obtained by partitioning $[n]$ into two parts of almost equal size and taking all triples that intersect both parts. Sós [517] conjectured that this is the maximum possible size of a $PG_2(2)$-free 3-graph. De Caen and Füredi [125] showed that it is asymptotically correct, i.e $\pi_3(PG_2(2)) = 3/4$. Then Füredi and Simonovits, and independently Keevash and Sudakov proved the following exact result.

**Theorem 326 (Füredi, Simonovits [253], Keevash, Sudakov [355])** *Let $n$ be large enough. Then $ex_3(n, PG_2(2)) = \binom{n}{3} - \binom{\lceil n/2 \rceil}{3} - \binom{\lfloor n/2 \rfloor}{3}$.*

Let us consider now $PG_m(q)$, the projective geometry of dimension $m$ over the field of $q$ elements. The vertex set of $PG_m(q)$ consists of all 1-dimensional subspaces of an $(m+1)$-dimensional vector space $V$ over $\mathbb{F}_q$, and the edge set of $PG_m(q)$ consists of all 2-dimensional subspaces of $V$. Note that $PG_m(q)$ is a $(q+1)$-uniform hypergraph. This more general version was first studied by Cioaba [107], who proved $27/32 \leq \pi(PG_3(2)) \leq 27/28$. This was improved and generalized by Keevash.

**Theorem 327 (Keevash [346])**

$$\prod_{i=1}^{k}\left(1 - \frac{i}{\sum_{j=1}^{m} g^j}\right) \leq \pi_{q+1}(PG_m(q)) \leq 1 - \frac{1}{\binom{q^m}{q}},$$

$$\frac{(2^{m+1}-3)(2^{m+1}-4)}{(2^{m+1}-2)^2} \leq \pi_3(PG_m(2)) \leq \begin{cases} 1 - \frac{3}{2^{2m}-1} & \text{if } m \text{ is odd} \\ 1 - \frac{6}{(2^m-1)(2^{m+1}+1)} & \text{if } n \text{ is even,} \end{cases}$$

$$3\sqrt{3} + 2\sqrt{2(9-5\sqrt{3})} \leq \pi_3(PG_3(2)) \leq 13/14.$$

The *k-book* with $p$ pages $\mathcal{B}_p^k$ consists of $p \leq k$ hyperedges of size $k$ containing a common $k-1$-set $T$, and one more $k$-edge that is disjoint from $T$ and contains the other vertices of the other $p$ hyperedges. Füredi, Pikhurko and Simonovits [251] studied the 4-book with 3 pages, i.e. the hypergraph $\mathcal{B}_3^4 = \{\{1,2,3,4\}, \{1,2,3,5\}, \{1,2,3,6\}, \{4,5,6,7\}\}$.

**Theorem 328** *If $n$ is large enough, then $ex_4(n, \mathcal{B}_3^4) = \binom{\lfloor n/2 \rfloor}{2}\binom{\lceil n/2 \rceil}{2}$.*

Let us note that the 3-book with 2 pages is the generalized triangle we examined in Section 5.2.6.

Let us consider now $k$-books with $k$ pages. A $k$-graph $\mathcal{H}$ does not contain this if and only if for any $k-1$-set $T$, the neighborhood of $T$ $\{v \in [n] : T \cup \{v\} \in \mathcal{H})\}$ does not contain any hyperedges of $\mathcal{H}$. Füredi, Pikhurko and Simonovits considered $\mathcal{B}_3^3$ and showed the following.

**Theorem 329 (Füredi, Pikhurko, Simonovits [250])** *If $n$ is large enough, we have $ex_3(n, \mathcal{B}_3^3) = \max_{\alpha \in [n]}(n-\alpha)\binom{\alpha}{2}$.*

Note that the unique extremal construction is obtained by taking a partition into two parts and taking the hyperedges that intersect one of the parts in 1 vertex and the other part in 2 vertices. It is the largest when one of the parts contains about $n/3$ vertices.

Füredi, Mubayi and Pikhurko [246] conjectured something similar is true for $\mathcal{B}_k^k$: the unique extremal construction is obtained by taking a partition into two parts and taking the hyperedges that intersect one of the parts in an odd number of vertices that is less than $k$. Note that it is the same construction that gives the maximum number of hyperedges if we forbid the inflated triangle $\mathcal{I}^{2k}K_3$. They proved this conjecture for $k=4$, but Bohman, Frieze, Mubayi and Pikhurko disproved it for $k>7$. In fact they showed the following.

**Theorem 330 (Bohman, Frieze, Mubayi, Pikhurko, [59])** *As $k \to \infty$, we have*

$$1 - \frac{2\log k}{k} + (1+o(1))\frac{\log\log k}{k} \le \pi_k(\mathcal{B}_k^k) \le 1 - \frac{2\log k}{k} + (5+o(1))\frac{\log\log k}{k}.$$

*Furthermore, for $k \ge 7$ we have $\pi_k(\mathcal{B}_k^k) > 1/2$.*

They also showed $40/81 \le \pi_k(\mathcal{B}_5^5) \le 0.534$.

Balogh, Bohman, Bollobás and Zhao [34] considered the following hypergraphs. Let us consider two disjoint sets $S$ and $T$ with $|S| = i$ and $|T| = k-1$. Then $\mathcal{B}_i^{(k)}$ consists of the $k$-edges that are contained in $S \cup T$ and contain either $S$ or $T$. Note that $\mathcal{B}_k^{(k)} = \mathcal{B}_k^k$, and $\mathcal{B}_2^{(k)} = \mathcal{K}_{k+1}^k$. They proved the following.

**Theorem 331 (Balogh, Bohman, Bollobás, Zhao, [34])** *Fix $i \ge 1$. As $k \to \infty$, we have*

$$1 - \Theta(\frac{\log k}{k^i}) \le \pi_k(\mathcal{B}_{i+1}^{(k)}), \pi_k(\mathcal{B}_{k-i}^{(k)}) \le 1 - \frac{1}{k^i}.$$

They also defined the following variant. The vertex set of $\hat{\mathcal{B}}_i^{(k)}$ is $S \cup T \cup \{v\}$ with $v$ being a new vertex and the hyperedges of $\hat{\mathcal{B}}_i^{(k)}$ are $T \cup \{v\}$ and the $k$-edges $S \cup T'$, where $T' \in \binom{T}{k-i}$. In other words, we replace the hyperedges containing $T$ with $T \cup \{v\}$.

**Theorem 332 (Balogh, Bohman, Bollobás, Zhao, [34])** *Fix $i \ge 1$. As $k \to \infty$, we have $\pi_k(\hat{\mathcal{B}}_i^{(k)}) = 1 - \Theta(\frac{\log k}{k^i})$.*

Finally, they defined $\mathcal{B}^{(k)}$, where the vertex set is $S \cup T$ with $S \cap T = \emptyset$, $|S| = |T| = k-1$, and the edge set consists of all $k$-edges containing $S$ and a single $k$-edge containing $T$.

**Theorem 333 (Balogh, Bohman, Bollobás, Zhao, [34])** *As $k \to \infty$, we have $\pi_k(\hat{\mathcal{B}}_i^{(k)}) = 1 - \Theta(\frac{\log k}{k})$.*

Goldwasser and Hansen [270] considered the hypergraph $\mathcal{F}(3,3)$ with edge set $\{\{1,2,3\},\{1,4,5\},\{1,4,6\},\{1,5,6\},\{2,4,5\},\{2,4,6\},\{2,5,6\},\{3,4,5\},\{3,4,6\},\{3,5,6\}\}$. They proved that if $n \neq 5$, then the unique optimal construction is $T(n,3)$. It was also independently proved by Keevash and Mubayi [353].

Let $\mathcal{F}_{3,2} = \{\{1,2,3\},\{1,4,5\},\{2,4,5\},\{3,4,5\}\}$. Füredi, Pikhurko and Simonovits [249] determined its Turán density, proving a conjecture of Mubayi and Rödl [430], who showed $4/9 \le \pi_3(\mathcal{F}_{3,2}) \le 1/2$.

**Theorem 334 (Füredi, Pikhurko, Simonovits, [249])** $\pi_3(\mathcal{F}_{3,2}) = 4/9$.

## 5.5 Some methods

In this section we get a quick glance into a couple of important topics and methods. They would all deserve their own sections and much more attention, but due to space constraints we only briefly summarize them, stating only a couple of theorems.

### 5.5.1 Supersaturation

Erdős and Simonovits [167] noticed that whenever we have more than $ex_k(n,\mathcal{F})$ hyperedges in a $k$-graph, we find not only one, but many copies of $\mathcal{F}$. This phenomenon is called supersaturation. Note that we have seen and will see similar phenomena for other forbidden configurations in other chapters.

**Theorem 335 (Erdős, Simonovits, [167])** *For any $k$-graph $\mathcal{F}$ and $\varepsilon > 0$ there exists $\eta > 0$ such that if $n$ is large enough and a $k$-graph $\mathcal{H}$ on $n$ vertices has more than $(\pi_k(\mathcal{F}) + \varepsilon)\binom{n}{k}$ hyperedges, then it contains at least $\delta\binom{n}{|V(\mathcal{F})|}$ copies of $\mathcal{F}$.*

**Proof.** The proof uses the following lemma proved by Erdős.

**Lemma 336 (Erdős, [154])** *For every fixed integer $m$ and constants $c$ and $\varepsilon$, there exists another constant $\eta$ such that whenever a $k$-uniform hypergraph $\mathcal{H}$ on $n$ vertices contains at least $(c+\varepsilon)\binom{n}{k}$ edges, then at least $\eta\binom{n}{m}$ induced subhypergraphs $\mathcal{H}[M]$ with $|M| = m$ contain at least $(c+\varepsilon/2)\binom{m}{k}$ edges.*

**Proof of Lemma.** Counting pairs $(M,H)$ with $H\in\mathcal{H}$, $M\in\binom{V(\mathcal{H})}{m}$, and $H\subset M$, we obtain

$$\sum_{M\in\binom{V(\mathcal{H})}{m}}|\mathcal{H}[M]|\geq(c+\varepsilon)\binom{n}{k}\binom{n-k}{m-k}.$$

If the statement of the lemma did not hold, then the left hand side of the above inequality could be upper bounded by $(1-\eta)\binom{n}{m}(c+\varepsilon/2)\binom{m}{k}+\eta\binom{n}{m}\binom{m}{k}$. For $\eta$ sufficiently small this is at most $\binom{n}{m}(c+3\varepsilon/4)\binom{m}{k}$. This contradiction proves the lemma. □

Choose $m$ large enough such that $ex_k(m,\mathcal{F})\leq\pi_k(\mathcal{F})(c+\varepsilon)\binom{m}{k}$. Applying Lemma 336 with $c=\pi_k(\mathcal{F})$ we obtain at least $\eta\binom{n}{m}$ many $m$-sets $M$ with $|\mathcal{F}[M]|>\pi_k(\mathcal{F})(c+\varepsilon/2)\binom{m}{k}$ and thus for each such $M$ the induced hypergraph $\mathcal{H}[M]$ contains a copy of $\mathcal{F}$. Every copy is contained in at most $\binom{n-|V(\mathcal{F})|}{m-|V(\mathcal{F})|}$ induced hypergraphs, therefore the number of distinct copies of $\mathcal{F}$ in $\mathcal{H}$ is at least

$$\frac{\eta\binom{n}{m}}{\binom{n-|V(\mathcal{F})|}{m-|V(\mathcal{F})|}}\geq\delta\binom{n}{m}/n^{m-|V(\mathcal{F})|}\geq\delta\binom{n}{|V(\mathcal{F})|}$$

for small enugh $\delta$. ■

The interested reader can find more precise results of this type for several specific hypergraphs in [428].

### 5.5.2 Flag algebras

In this subsection, we give a short summary of Razborov's flag algebra method that he introduced in [474] and [475].

Razborov's papers are quite a task to read as his method is very much involved and complicated in its full generality. One of the first papers that gave a combinatorial overview is [30] by Baber and Talbot. Later, nice introductions were given in Keevash's survey [348], and Baber's and Falgas-Ravry's PhD theses [29, 176]. Two other texts that concentrate only on graphs are [127] and [288].

Let us fix a family $\mathbb{F}$ of forbidden $k$-graphs and let $\mathbb{H}_{\mathbb{F}}(n)$ denote the set of $\mathbb{F}$-free $k$-graphs on $n$ vertices. For small values of $n$ one can find all members of $\mathbb{H}_{\mathbb{F}}(n)$ (possibly with the use of computer search). Clearly, if $\mathcal{G}\in\mathbb{H}_{\mathbb{F}}(n)$ and $l\leq n$, then for any $l$-subset $L$ of the vertices we have $\mathcal{G}[L]\in\mathbb{H}_{\mathbb{F}}(l)$.

Let us denote for $\mathcal{H}\in\mathbb{H}_{\mathbb{F}}(l),\mathcal{G}\in\mathbb{H}_{\mathbb{F}}(n)$ by $p(\mathcal{H},\mathcal{G})$ the probability that for a random $l$-subset $L$ of $V(\mathcal{G})$ the induced subgraph $\mathcal{G}(L)$ is isomorphic to $\mathcal{H}$. Then by averaging we obtain

$$d(\mathcal{G})=\sum_{\mathcal{H}\in\mathbb{H}_{\mathbb{F}}(l)}d(\mathcal{H})p(\mathcal{H},\mathcal{G}),\tag{5.7}$$

where $d(\mathcal{G}) = \frac{|\mathcal{G}|}{\binom{n}{k}}$ is the *density* of $\mathcal{G}$. In particular, this implies

$$d(\mathcal{G}) \leq \max_{\mathcal{H} \in \mathbb{H}_{\mathbb{F}}(l)} d(\mathcal{H}), \tag{5.8}$$

which is equivalent to the statement of Exercise 242. However, equality could hold only if all induced subhypergraphs of $\mathcal{H}$ would have the highest density in $\mathcal{H} \in \mathbb{H}_{\mathbb{F}}(l)$. In other words, when moving from (5.7) to (5.8) we disregard the structural properties of extremal $\mathbb{F}$-free $k$-graphs, like how two of these can or cannot overlap. Razborov's method tries to account for these overlapping possibilities by obtaining linear inequalities of the probabilities $p(\mathcal{H}, \mathcal{G})$. We will need further definitions.

- A *flag* $F$ is a pair $(\mathcal{G}_F, \theta)$ with $\mathcal{G}_F$ being a $k$-graph and $\theta$ an injection from $[s]$ to $V(\mathcal{G}_F)$ for some integer $s$. The *order* of a flag (denoted by $|F|$) is $|V(\mathcal{G}_F)|$.

- If $\theta$ is a bijection, then we call the flag a *type*.

- Let $\sigma = (\mathcal{G}, \eta)$ be a type. We call $F = (\mathcal{G}_F, \theta)$ a *$\sigma$-flag*, if $Im(\theta) = Im(\eta)$ and the mapping $i : Im(\theta) \to V(\mathcal{G})$ defined by $i(v) = \eta(\theta^{-1}(v))$ is an isomorphism between $\mathcal{G}_F(Im(\theta))$ and $\mathcal{G}$. This means the induced labeled subhypergraph of $\mathcal{G}_F$ given by $\theta$ is $\sigma$.

- A flag $F$ is $\mathbb{F}$-*admissible* (or simply admissible if $\mathbb{F}$ is clear from context) if $\mathcal{G}_F$ is $\mathbb{F}$-free.

- The next definitions will let us consider overlapping subhypergraphs of the $\mathbb{F}$-free hypergraph $\mathcal{G}$. Let $\sigma = (\mathcal{G}_\sigma, \eta)$ be a type. If $|V(\mathcal{G}_\sigma)| \leq m \leq \frac{l+|V(\mathcal{G}_\sigma)|}{2}$, then there is room for two $k$-subgraphs on $m$ vertices of an $l$-vertex $k$-graph to overlap in $|V(\mathcal{G}_\sigma)|$ vertices. Let $\mathbf{F}_m^\sigma$ denote the set of all admissible $\sigma$-flags of order $m$ up to isomorphism. For a hypergraph $\mathcal{H}$ let $\Theta_{\mathcal{H}}$ be the set of all injective functions from $[|\mathcal{G}|]$ to $V(\mathcal{H})$ (note that we fixed $\mathcal{G}$ at the beginning of this paragraph). For a flag $F \in \mathbf{F}_m^\sigma$ and function $\theta \in \Theta_{\mathcal{H}}$ let $p(F, \theta, \mathcal{G})$ denote the conditional probability that a randomly picked $m$ subset $U$ of $V(\mathcal{G})$ induces a $\sigma$-flag $(\mathcal{G}[U], \theta)$ that is isomorphic to $F$, conditioned on the event that $Im(\theta) \subseteq U$.

- Therefore the conditional probability that for given flags $F_1, F_2 \in \mathbf{F}_m^\sigma$ and $\theta \in \Theta_{\mathcal{H}}$ two independently picked random $m$-sets $U_1, U_2$ induce $\sigma$-flags that are isomorphic to $F_1, F_2$ is $p(F_1, \theta, G) \cdot p(F_2, \theta, G)$, provided $Im(\theta) \subseteq U_1 \cap U_2$. This sampling of $\sigma$-flags can be seen as picking the non-$\sigma$ parts of $V(\mathcal{G}_{F_1})$ and $V(\mathcal{G}_{F_2})$ with replacement.

- The next definition stands for the case when we can sample the non-$\sigma$ parts of $V(\mathcal{G}_{F_1})$ and $V(\mathcal{G}_{F_2})$ without replacement. We pick randomly an $m$-set $U_1$ conditioned on $Im(\theta) \subseteq U_1$ and then another $m$-set $U_2$ conditioned on $U_1 \cap U_2 = Im(\theta)$. We define $p(F_1, F_2, \theta, G)$ to be the probability that $(\mathcal{G}[U_1], \theta)$ and $(\mathcal{G}[U_2], \theta)$ are isomorphic to $F_1$ and $F_2$, respectively.

The next simple exercise states that if $|V(\mathcal{G})|$ is large enough, then the above probabilities differ only in a negligible quantity.

**Exercise 337** *Let $\sigma$ be fixed. For any $F_1, F_2 \in \mathbf{F}_m^\sigma$ and $\theta \in \Theta_\mathcal{G}$, if $|V(\mathcal{G})|$ tends to infinity, then the following equation holds*

$$p(F_1, \theta, \mathcal{G})p(F_2, \theta, \mathcal{G}) = p(F_1, F_2, \theta, \mathcal{G}) + o(1).$$

As the $o(1)$ term in Exercise 337 does not depend on $\theta$, by averaging, we obtain

$$\mathbb{E}_{\theta \in \Theta_\mathcal{G}}[p(F_1, \theta, \mathcal{G})p(F_2, \theta, \mathcal{G})] = \mathbb{E}_{\theta \in \Theta_\mathcal{G}}[p(F_1, F_2, \theta, \mathcal{G})] + o(1). \tag{5.9}$$

As $F_1, F_2 \in \mathbf{F}_m^\sigma$ and $l$ is large enough, the right hand side of the above equation can be calculated by considering only $l$-vertex subhypergraphs of $\mathcal{G}$. Therefore we obtain

$$\mathbb{E}_{\theta \in \Theta_\mathcal{G}}[p(F_1, F_2, \theta, \mathcal{G})] = \sum_{\mathcal{H} \in \mathbb{H}_\mathbb{F}} \mathbb{E}_{\theta \in \Theta_\mathcal{H}}[p(F_1, F_2, \theta, \mathcal{H})]p(\mathcal{H}, \mathcal{G}). \tag{5.10}$$

As the probabilities $p(\mathcal{H}, \mathcal{G})$ have appeared, we just have to gather the appropriate coefficients to obtain the desired linear inequalities. Let $Q = (q_{ij})$ be an $|\mathbf{F}_m^\sigma| \times |\mathbf{F}_m^\sigma|$ positive semidefinite matrix and for $\theta \in \Theta_\mathcal{G}$ let us define the vector $\mathbf{p}_\theta$ of length $|\mathbb{F}_m^\sigma|$ by $\mathbf{p}_\theta = (p(F, \theta, \mathcal{G}) : F \in \mathbb{F}_m^\sigma)$. As $Q$ is positive semidefinite we obtain

$$0 \le \mathbb{E}_{\theta \in \Theta_\mathcal{G}}[\mathbf{p}_\theta^T Q \mathbf{p}_\theta] = \sum_{F_i, F_j \in \mathbf{F}_m^\sigma} \sum_{\mathcal{H} \in \mathbb{H}_{\mathbb{F}(l)}} q_{ij} \mathbb{E}_{\theta \in \Theta_\mathcal{H}} p(F_i, F_j, \theta, \mathcal{H}) p(\mathcal{H}, \mathcal{G}) + o(1), \tag{5.11}$$

where we used (5.9), (5.10) and linearity of expectation. Thus for every positive semidefinite matrix $Q$, type $\sigma$ and integer $m > |V(\mathcal{G}_\sigma)|$, we obtained a linear inequality of the $p(\mathcal{H}, \mathcal{G})$'s, where the coefficient of $p(\mathcal{H}, \mathcal{G})$ is

$$c_\mathcal{H}(\sigma, m, Q) = \sum_{F_i, F_j \in \mathbf{F}_m^\sigma} q_{ij} \mathbb{E}_{\theta \in \Theta_\mathcal{H}} p(F_i, F_j, \theta, \mathcal{H}). \tag{5.12}$$

Then one can consider such linear inequalities for several choices of types, integers, and matrices. Let us fix $(\sigma_1, m_1, Q_1), \dots, (\sigma_t, m_t, Q_t)$ with $\sigma_i$ being a type, $m_i$ an integer satisfying $m_i \le \frac{l + |V(\mathcal{G}_{\sigma_i})|}{2}$ and $Q_i$ an $|\mathbf{F}_{m_i}^{\sigma_i}| \times |\mathbf{F}_{m_i}^{\sigma_i}|$ positive semidefinite matrix. Then for any $\mathcal{H} \in \mathbb{H}_\mathbb{F}(l)$ we can introduce

$$c_\mathcal{H} = \sum_{i=1}^{t} c_\mathcal{H}(\sigma_i, m_i, Q_i),$$

which, in view of (5.12), is independent of $\mathcal{G}$. Summing (5.11) for $i = 1, 2, \dots, t$ we obtain

$$\sum_{\mathcal{H} \in \mathbb{H}_\mathbb{F}(l)} c_\mathcal{H} p(\mathcal{H}, \mathcal{G}) + o(1) \ge 0.$$

Adding this to (5.7) yields

$$d(\mathcal{G}) \leq \sum_{\mathcal{H} \in \mathbb{H}_{\mathbb{F}}(l)} (d(\mathcal{H}) + c_{\mathcal{H}}) p(\mathcal{H}, \mathcal{G}) + o(1)$$

and therefore

$$\pi_k(\mathbb{F}) \leq \max_{\mathcal{H} \in \mathbb{H}_{\mathbb{F}_l}} \{d(\mathcal{H}) + c_{\mathcal{H}}\}. \tag{5.13}$$

As $c_{\mathcal{H}}$ can be negative, the bound (5.13) can indeed improve (5.8). It is far from being obvious how to pick the types, the $m_i$'s, and given those the $Q_i$'s to minimize (5.13). This is beyond the scope of this brief introduction to flag algebras. Instead, we illustrate the method with a simple example and then list some of the results concerning Turán-type problems that were obtained via flag algebras.

Let us recall that $\mathcal{S}^3K_3$ is the unique 3-uniform hypergraph on 4 vertices with 3 edges. We show how to apply the flag algebra method to obtain de Caen's old result [123] $\pi(\mathcal{S}^3K_3) \leq 1/3$. To do so, we will use one fixed type $\sigma$, an integer $m$, another one $l$ and then look for an appropriate positive semidefinite matrix $Q$. Let $\sigma = (G_\sigma, \eta)$ be defined by $\mathcal{G}_\sigma$ being the empty hypergraph with vertex set $\{2\}$ and $\eta(x) = x$ for $x = 1, 2$. We take $m = 3$; therefore there are only two $\sigma$-flags $F_0 = (\mathcal{G}_{F_0}, \theta_0)$ and $F_1 = (\mathcal{G}_{F_1}, \theta_1)$, as there can be at most one hyperedge in $\mathcal{G}_F$ and, assuming $V(\mathcal{G}_F) = \{1, 2, 3\}$, $\theta_0 = \theta_1$ must be the identity function on $\{1, 2, 3\}$. $F_i$ denotes the flag with $\mathcal{G}_{F_i}$ having $i$ edge. We fix $l = 4$ and instead of $\mathbb{H}_{\{\mathcal{S}^3K_3\}}(4)$ we will simply write $\mathbb{H}(4)$ to denote the set of $\mathcal{S}^3K_3$-free hypergraphs with vertex set $\{1, 2, 3, 4\}$. Then $\mathbb{H}(4) = \{\mathcal{H}_0, \mathcal{H}_1, \mathcal{H}_2\}$, where $\mathcal{H}_i$ contains $i$ edges.

We have to calculate $\mathbb{E}_{\theta \in \Theta_{\mathcal{H}}}[p(F_i, F_j, \theta, \mathcal{H})]$ for every $\mathcal{H} \in \mathbb{H}(4)$, and ordered pair $(F_i, F_j)$ with $F_i, F_j \in \{F_0, F_1\}$. To pick $\theta \in \Theta_{\mathcal{H}}$, we have 12 choices (the value of $\theta(1)$ and $\theta(2)$), independently of $\mathcal{H}$, and as $V_0$ and $V_1$ are 3-sets with $V_0 \cap V_1 = Im(\theta)$, we must pick $x \in [4] \setminus Im(\theta)$ to be the third element of $V_0$ and $y \in [4] \setminus (\{x\} \cup Im(\theta))$ to be the third element of $V_1$. So this random choice can be described by a permutation $(u, v, x, y)$ of $[4]$ with $\theta(1) = u$, $\theta(2) = v$, $V_0 = \{u, v, x\}$, $V_1 = \{u, v, y\}$. Hence if $\mathcal{H}_1$ contains the hyperedge $\{1, 2, 3\}$, then the permutations satisfying $(\mathcal{H}_1[V_0], \theta) = F_0$ and $(\mathcal{H}_1[V_1], \theta) = F_1$ are those where $V_0 \neq \{1, 2, 3\}$ and $V_1 = \{1, 2, 3\}$. The following six permutations have this property:

$$(1, 2, 4, 3), (3, 1, 4, 2), (2, 3, 4, 1), (2, 1, 4, 3), (3, 2, 4, 1), (1, 3, 4, 2).$$

Therefore we have $\mathbb{E}_{\theta \in \Theta_{\mathcal{H}_1}}[p(F_0, F_1, \theta, \mathcal{H}_1)] = 6/24 = 1/4$. Similarly, one can obtain

- $\mathbb{E}_{\theta \in \Theta_{\mathcal{H}_0}}[p(F_0, F_0, \theta, \mathcal{H}_0)] = 1$, $\mathbb{E}_{\theta \in \Theta_{\mathcal{H}_1}}[p(F_0, F_0, \theta, \mathcal{H}_1)] = 1/2$ and $\mathbb{E}_{\theta \in \Theta_{\mathcal{H}_2}}[p(F_0, F_0, \theta, \mathcal{H}_2)] = 1/6$,
- $\mathbb{E}_{\theta \in \Theta_{\mathcal{H}_0}}[p(F_0, F_1, \theta, \mathcal{H}_0)] = 0$, $\mathbb{E}_{\theta \in \Theta_{\mathcal{H}_2}}[p(F_0, F_1, \theta, \mathcal{H}_2)] = 1/3$,

- $\mathbb{E}_{\theta\in\Theta_{\mathcal{H}_0}}[p(F_1,F_1,\theta,\mathcal{H}_0)]=0$, $\mathbb{E}_{\theta\in\Theta_{\mathcal{H}_1}}[p(F_1,F_1,\theta,\mathcal{H}_1)]=0$ and $\mathbb{E}_{\theta\in\Theta_{\mathcal{H}_2}}[p(F_1,F_1,\theta,\mathcal{H}_2)]=1/6$.

As we have only two $\sigma$-flags of order 3, $Q$ must be a $2\times 2$ positive semidefinite matrix

$$\begin{pmatrix} q_{00} & q_{01} \\ q_{10} & q_{11} \end{pmatrix}.$$

The above values of $\mathbb{E}_{\theta\in\Theta_{\mathcal{H}}}[p(F_i,F_j,\theta,\mathcal{H})]$ and (5.12) yield:

$$c_{\mathcal{H}_0}=q_{00},\quad c_{\mathcal{H}_1}=\frac{1}{2}q_{00}+\frac{1}{2}q_{01}\quad c_{\mathcal{H}_2}=\frac{1}{6}q_{00}+\frac{2}{3}q_{01}+\frac{1}{6}q_{11}.$$

As the densities of $\mathcal{H}_0,\mathcal{H}_1,\mathcal{H}_2$ are $0,1/4+1/2$ respectively, (5.13) implies

$$\pi_3(\mathcal{S}^3K_3)\le\max_Q\left\{q_{00};\frac{1}{2}q_{00}+\frac{1}{2}q_{01}+\frac{1}{4};\frac{1}{6}q_{00}+\frac{2}{3}q_{01}+\frac{1}{6}q_{11}+\frac{1}{2}\right\}.$$

Now any choice of a $2\times 2$ positive semidefinite matrix gives an upper bound. If one picks

$$Q=\begin{pmatrix} \frac{1}{3} & -\frac{2}{3} \\ -\frac{2}{3} & \frac{4}{3} \end{pmatrix};$$

then we obtain the desired upper bound of $1/3$.

As we have mentioned in Subsection 5.2.5, the best known upper bound is 0.2871, due to Baber and Talbot [30]. They used flag algebras to obtain it, choosing $l=7$ and four types. It obviously makes the calculations much more complicated. Instead of carrying them out similarly, they used a computer program called Densitybounder. Later Emil R. Vaughan developed Flagmatic to implement the flag algebra method. It takes a family $\mathbb{F}$ of graphs or 3-graphs and $m$ as input, determines all $\mathbb{F}$-admissible hypergraphs and generates a set of types and flags to use. Flagmatic is publicly available at http://www.maths.qmul.ac.uk/ ev/flagmatic/.

**Theorem 338 (Baber, Talbot [31])** ***(i)*** *Let $\mathcal{H}_1$ be the hypergraph with vertex set $[6]$ and edge set $\{\{1,2,3\},\{1,2,4\},\{3,4,5\},\{1,5,6\}\}$. Then $\pi_3(\mathcal{H}_1)=2/9$ holds.*

***(ii)*** *Let $\mathcal{H}_2$ and $\mathcal{H}_3$ be hypergraphs with vertex set $[6]$ and edge sets*

$$\{\{1,2,3\},\{1,2,4\},\{1,3,4\},\{2,3,5\},\{2,4,5\},\{1,5,6\}\}$$

*and*

$$\{\{1,2,3\},\{1,2,4\},\{1,3,5\},\{3,4,5\},\{1,4,6\},\{2,5,6\}\},$$

*respectively. Then $\pi_3(\mathcal{H}_2)=\pi_3(\mathcal{H}_3)=4/9$ holds.*

Baber and Talbot in [31] also gave 12 examples of 3-graphs with Turán density 5/9 and 3 examples of 3-graphs with Turán-density 3/4. They also determined the Turán-densities of several finite families of 3-graphs.

Falgas-Ravry and Vaughan [179] obtained $\pi_3(\mathbb{F})$ for some finite sets of 3-graphs. In all cases, $\mathbb{F}$ contained $\mathcal{F}_{3,2} = \{\{1,2,3\},\{1,4,5\},\{2,4,5\},\{3,4,5\}\}$ (recall that $\pi_3(\mathcal{F}_{3,2}) = 4/9$ by Theorem 334). Note that $\mathcal{G}$ is $\mathcal{F}_{3,2}$-free if and only if for any pair $x, y$ of vertices of $\mathcal{G}$ the neighborhood $\{z \in V(\mathcal{G}) : \{x,y,z\} \in \mathcal{G}\}$ is an independent set in $\mathcal{G}$. In [178], Falgas-Ravry and Vaughan determined the following parameter for many pairs $\mathcal{H}$, $\mathcal{G}$: $\pi_k(\mathcal{H},\mathcal{G}) = \lim_{n\to\infty} \frac{ex_k(n,\mathcal{H},\mathcal{G})}{\binom{n}{|V(\mathcal{G})|}}$, where $ex_k(n,\mathcal{H},\mathcal{G})$ denotes the maximum number of *induced* copies of $\mathcal{G}$ that an $\mathcal{H}$-free $n$-vertex hypergraph can have. (The existence of the limit follows just as Exercise 242.)

### 5.5.3 Hypergraph regularity

The original regularity lemma of Szemerédi [522] is about graphs. We do not state it, as we are going to state the hypergraph version. Briefly summarizing, it says that any large enough graph can be decomposed into $r$ parts such that almost all pairs of parts are regular in a sense. It proved to be very useful in many areas of graph theory, including Turán theory. For detailed introduction focusing only on graphs, the reader may consult the surveys [372, 373].

Hypergraph regularity theory was developed by several different groups of researchers. Among the most important papers are the ones by Rödl and Skokan [485], Rödl and Schacht [484], and by Gowers [271]. There are many different versions of the hypergraph regularity lemma; see also the papers by Tao [525, 526] for more variants. Here we only state the *weak hypergraph regularity lemma*, which we have chosen for its (relative) simplicity. It is a straightforward extension of the original regularity lemma. We follow the presentation of Kohayakawa, Nagle, Rödl and Schacht [371].

Let $V_1, \ldots, V_k$ be pairwise disjoint subsets of the vertex set of a $k$-graph $\mathcal{F}$. Then $e_{\mathcal{F}}(V_1, \ldots, V_k)$ is the number of $k$-edges in $\mathcal{F}$ intersecting each $V_i$ in exactly one vertex. Their *density* is $d(V_1, \ldots, V_k) = \frac{e_{\mathcal{F}}(V_1,\ldots,V_k)}{|V_1|\cdots|V_k|}$. We say that the $k$-tuple $(V_1, \ldots, V_k)$ is $(\varepsilon, d)$*-regular* if $|d(W_1, \ldots, W_k) - d| \le \varepsilon$ for all subsets $W_i \subset V_i$ with $|W_i| \ge \varepsilon|V_i|$ for every $i \le k$. We say a $k$-tuple is $\varepsilon$*-regular* if it is $(\varepsilon, d)$-regular for some $d \ge 0$.

**Theorem 339** *For all integers $k \ge 2$ and $t_0 \ge 1$, and every $\varepsilon > 0$, there exist $T_0 = T_0(k, t_0, \epsilon)$ and $n_0 = n_0(k, t_0, \varepsilon)$ such that for every $k$-uniform hypergraph on $n \ge n_0$ vertices, there exists a partition of the vertex set into parts $V_0, V_1, \ldots, V_t$ so that the following hold.*

- $t_0 \le t \le T_0$,
- $|V_0| \le \varepsilon n$ *and* $|V_1| = \cdots = |V_t|$, *and*
- *for all but at most* $\varepsilon\binom{t}{k}$ *sets* $\{i_1, \ldots, i_k\} \subset [t]$, *the $k$-tuple* $(V_{i_1}, \ldots, V_{i_k})$ *is $\varepsilon$-regular.*

An important consequence of regularity is the counting lemma that estimates the number of certain subgraphs. Its hypergraph versions were proved by Nagle, Rödl and Schacht [439] and Rödl and Schacht [483], building on the work of Frankl and Rödl [219] and Nagle and Rödl [438]. Kohayakawa, Nagle, Rödl and Schacht [371] proved a version for linear hypergraphs. Again, we will only state one version. Before that, we need some definitions.

Recall that for a $k$-graph $\mathcal{F}$, $\mathcal{K}_r^k(\mathcal{F})$ denotes the $r$-graph consisting of the underlying sets of copies of $\mathcal{K}_r^k$ in $\mathcal{F}$. Let $V = V_1 \cup \cdots \cup V_l$ with $|V_i| = n$ for every $i \le l$. An $(n, l, j)$-cylinder is an $l$-partite $j$-graph with vertex set $V$ and parts $V_i$. Let $\mathcal{G}_j$ be an $(n, l, j)$-cylinder for every $j \le k$ (with parts $V_i$). Observe that $\mathcal{G}_1$ consists of at most $ln$ singletons, and assume it consists of all $ln$ singletons in $V$. Then $\mathbb{G} = \{\mathcal{G}_1, \ldots, \mathcal{G}_k\}$ is an *$(n, l, k)$-complex* if $\cup_{j=1}^k \mathcal{G}_j$ is a downset. Equivalently, every hyperedge in $\mathcal{G}_j$ corresponds to a clique of order $j$ in $\mathcal{G}_{j-1}$, i.e. $\mathcal{G}_j \subseteq \mathcal{K}_j^k(\mathcal{G}_{j-1})$.

Let $\mathcal{G}_j$ be an $(n, l, j)$-cylinder and $\mathbb{H} = \{\mathcal{H}_1, \ldots, \mathcal{H}_r\}$ be a family of $(n, l, j-1)$-cylinders. Then the *density of $\mathcal{G}_j$ with respect to $\mathbb{H}$* is

$$d(\mathcal{G}_j \mid \mathbb{H}) = \begin{cases} \frac{|\mathcal{G}_j \cap \bigcup_{s\in[r]} \mathcal{K}_j^{j-1}(\mathcal{H}_s)|}{|\bigcup_{s\in[r]} \mathcal{K}_j^{j-1}(\mathcal{H}_s)|} & \text{if } |\bigcup_{s\in[r]} \mathcal{K}_j^{j-1}(\mathcal{H}_s)| > 0 \\ 0 & \text{otherwise.} \end{cases}$$

Let $\delta_j$ and $d_j$ be positive reals, $\mathcal{G}_j$ be an $(n, j, j)$-cylinder and $\mathcal{G}_{j-1}$ be an $(n, j, j-1)$-cylinder with $\mathcal{G}_j \subset \mathcal{K}_j^k(\mathcal{G}_{j-1})$. We say $\mathcal{G}_j$ is *$(\delta_j, d_j, r)$-regular with respect to $\mathcal{G}_{j-1}$* if whenever $\mathbb{H} = \{\mathcal{H}_1, \ldots, \mathcal{H}_r\}$ is a family of $(n, j, j-1)$-cylinders with $\mathcal{H}_s \subseteq \mathcal{G}_{j-1}$ for every $s \le r$ satisfying

$$|\bigcup_{s\in[r]} \mathcal{K}_j^{j-1}(\mathcal{H}_s)| \ge \delta_j |\mathcal{K}_j^{j-1}(\mathcal{G}_{j-1})|,$$

then $d_j - \delta_j \le d(\mathcal{G}_j \mid \mathbb{H}) \le d_j + \delta_j$. This definition can be extended to $(n, l, j)$-cylinders. Let $\mathcal{G}_j$ be an $(n, l, j)$-cylinder and $\mathcal{G}_{j-1}$ be an $(n, l, j-1)$-cylinder. We say $\mathcal{G}_j$ is *$(\delta_j, d_j, r)$-regular with respect to $\mathcal{G}_{j-1}$* if whenever $V' = V_{i_1} \cup \ldots V_{i_j}$ is the union of $j$ parts, $\mathcal{G}_j|_{V'}$ is $(\delta_j, d_j, r)$-regular with respect to $\mathcal{G}_{j-1}|_{V'}$.

Let $\delta = (\delta_2, \ldots, \delta_k)$ and $\mathbf{d} = (d_2, \ldots, d_k)$. An $(n, l, k)$-complex $\mathbb{G} = \{\mathcal{G}_1, \ldots, \mathcal{G}_k\}$ is $(\delta, \mathbf{d}, r)$-regular if $\mathcal{G}_j$ is *$(\delta_j, d_j, r)$-regular* with respect to $\mathcal{G}_{j-1}$ for $3 \le j \le k$, and $\mathcal{G}_2$ is $(\delta_2, d_2, 1)$-regular with respect to $\mathcal{G}_1$.

**Theorem 340 (Counting lemma, Nagle, Rödl, Schacht [439])** *For all integers $2 \le l \le k$, the following is true:*

$$\forall \gamma > 0\ \forall d_k > 0\ \exists \delta_k > 0\ \forall d_{k-1} > 0\ \exists \delta_{k-1} > 0 \ldots \forall d_2 > 0\ \exists \delta_2 > 0$$

*and there are integers $r$ and $n_0$ so that, with $\delta = (\delta_2, \ldots, \delta_k)$, $\mathbf{d} = (d_2, \ldots, d_k)$ and $n \ge n_0$, whenever $\mathbb{G} = \{\mathcal{G}_1, \ldots, \mathcal{G}_k\}$ is a $(\delta, \mathbf{d}, r)$-regular $(n, l, k)$-complex, then*

$$(1-\gamma) \prod_{j=2}^k d_j^{\binom{l}{j}} n^l \le |\mathcal{K}_l^k(\mathcal{G}_k)| \le (1+\gamma) \prod_{j=2}^k d_j^{\binom{l}{j}} n^l.$$

We also state a version of the removal lemma by Rödl, Skokan [486] and by Rödl, Nagle, Skokan, Schacht and Kohayakawa [482].

**Theorem 341** *For integers $l \geq k$ and $\varepsilon > 0$ there exists $\delta > 0$ such that the following holds. Suppose $n$ is large enough and $\mathcal{F}$ is a $k$-graph on $l$ vertices, while $\mathcal{H}$ is a $k$-graph on $n$ vertices. If $\mathcal{H}$ contains at most $\delta\binom{n}{l}$ copies of $\mathcal{F}$, then one can delete at most $\varepsilon\binom{n}{k}$ hyperedges from $\mathcal{H}$ such that the resulting hypergraph is $\mathcal{F}$-free.*

### 5.5.4 Spectral methods

A huge area barely outside the scope of this book is studying eigenvalues of the adjacency hypermatrices of hypergraphs. They are closely connected to every area of the theory of hypergraphs, including extremal theory. Here we will present a theorem of Nikiforov [444] that shows a particular connection.

A *hypermatrix* or *$k$-dimensional matrix* or *$k$-matrix* is an $n_1 \times n_2 \times \cdots \times n_k$ size ordered array. A hypermatrix $M$ has entries $M(\underline{i})$, where $\underline{i} = (i_1, i_2, \ldots, i_k)$, with $1 \leq i_j \leq n_j$ for all $1 \leq j \leq k$. The *adjacency hypermatrix* $M_{\mathcal{H}}$ of a $k$-graph $\mathcal{H}$ is defined by

$$M_{\mathcal{H}}(\underline{i}) = \begin{cases} 1 & \text{if } \{i_1, \ldots, i_k\} \in \mathcal{H} \\ 0 & \text{otherwise.} \end{cases}$$

Consider a $k$-graph $\mathcal{H}$ and a vector $\mathbf{x} = (x_1, \ldots, x_n) \in \mathbb{R}^n$; the *polyform* of $\mathcal{H}$ is

$$P_{\mathcal{H}}(x) = k! \sum_{\{i_1,\ldots,i_k\} \in \mathcal{H}} x_{i_1} x_{i_2} \ldots x_{i_k}.$$

Note that the *largest eigenvalue* of $M_{\mathcal{H}}$ in $l^{\alpha}$-norm is

$$\lambda^{(\alpha)}(\mathcal{H}) = \max_{||\mathbf{x}||_{\alpha}=1} P_{\mathcal{H}}(x) = \max_{|x_1|^{\alpha}+|x_2|^{\alpha}+\cdots+|x_n|^{\alpha}} k! \sum_{\{i_1,\ldots,i_k\} \in \mathcal{H}} x_{i_1} x_{i_2} \ldots x_{i_k}.$$

Observe that for $\alpha = 1$ it is the Lagrangian of $\mathcal{H}$.

Let us now consider induced Turán problems. For a family of $k$-graphs $\mathbb{F}$ let $ex_k^I(n, \mathbb{F})$ be the maximum cardinality a $k$-graph on $n$ vertices can have, if it does not contain any member of $\mathbb{F}$ as an induced subhypergraph. Let

$$\pi_k^I(\mathbb{F}) = \lim_{n \to \infty} \frac{ex_k^I(n, \mathbb{F})}{\binom{n}{k}}.$$

Now we can state the main theorem of Nikiforov [444].

**Theorem 342 (Nikiforov, [444])** *For every family* $\mathbb{F}$ *of* $k$*-graphs and* $\alpha \geq 1$ *the limit*

$$\lim_{n\to\infty} n^{k/\alpha-k} \max\{\lambda^{(\alpha)}(\mathcal{H}) : \mathcal{H} \textit{ is induced } \mathbb{F}\textit{-free on } n \textit{ vertices}\}$$

*exists and if* $\alpha > 1$*, it is equal to* $\pi_k^I(\mathbb{F})$.

See [445] and [350] for more analytic methods for hypergraph problems.

## 5.6 Non-uniform Turán problems

Turán problems on non-uniform hypergraphs have been studied much less. The straightforward definition does not make much sense: If we forbid a particular non-uniform hypergraph $\mathcal{F}$, then it has a largest hyperedge of size $k$. If $n$ is large enough, almost all the possible sets have size larger than $k$, thus they could be added to any $\mathcal{F}$-free family.

Mubayi and Zhao [436] were perhaps the first to use the term non-uniform Turán problems. They avoided the above problem by minimizing the number of the hyperedges in the complements of hypergraphs avoiding a given hypergraph. More precisely, motivated by a computer science problem, they considered hypergraphs where every $k$-set contains at least $t$ hyperedges and every $k-1$-set contains at least $t-1$ hyperedges. Note that the uniform version of this is a variant of the local sparsity condition from Section 5.1. They determined the order of magnitude for every fixed $k$ and $t$ and studied related problems.

Another way to avoid the problem of having too large $\mathcal{F}$-free families is to forbid a large family of hypergraphs the size of which grows with $n$. We have already mentioned that Anstee and Salazar [21] studied Berge-type problems in the non-uniform case. Let us recall that a hypergraph $\mathcal{H}$ is a *Berge* copy of a graph $G$, if there is a bijection $f$ from the hyperedges of $\mathcal{H}$ to the edges of $G$, such that $f(H) \subseteq H$ for every hyperedge $H \in \mathcal{H}$. Note that this is the definition from Subsection 5.2.2 and there is no uniformity mentioned here.

In this area there is surprisingly small difference between the uniform and the non-uniform version. Some of the results in Subsection 5.2.2 were originally stated in the more general, non-uniform way. Note that when we forbid all Berge-copies of a graph $F$ in $k$-uniform hypergraphs, we forbid only a constant number (depending on $k$ and $F$) of $k$-graphs. In the non-uniform version the number of forbidden hypergraphs grows with $n$. We denote by $\mathbb{B}F$ the family of Berge copies of $F$.

If a $\mathbb{B}F$-free hypergraph has large hyperedges, replacing one with any of its subsets cannot create a copy of $\mathbb{B}F$. This is a reason why having larger hyperedges does not help. In fact, instead of the number of hyperedges, one often studies the *total size* or *volume*$\sum_{H\in\mathcal{H}} |H|$ of $\mathbb{B}F$-free hypergraphs $\mathcal{H}$. Even then the order of magnitude is typically the same. For example one of the earliest results is a theorem of Győri [293] which states that if $n$ is large enough and $\mathcal{H}$ is a $\mathbb{B}C_3$-free hypergraph, then we have $\sum_{H\in\mathcal{H}}(|H|-2) \le n^2/8$, and this is sharp. In fact he dealt with multihypergraphs. In this setting it makes more sense to consider $|H|-2$ instead of $|H|$, as the hypergraph consisting of multiple copies of a particular 2-edge does not contain $\mathbb{B}C_3$.

Gerbner and Palmer [265] proved a version of Theorem 279 for non-uniform hypergraphs.

**Theorem 343** *For any graph $F$ the Turán number $ex(n, \mathbb{B}F)$ is equal to the maximum number of complete subgraphs of an $F$-free graph on $n$ vertices.*

**Proof.** Let us consider the $F$-free graph with the largest number of complete subgraphs on $n$ vertices and replace its cliques by hyperedges. We obtain a $\mathbb{B}F$-free hypergraph. On the other hand, let us consider a $F$-free hypergraph, and order its hyperedges in such a way that we start with the hyperedges of size 2 (otherwise arbitrarily). In this order, for every hyperedge we pick one of its subedges (i.e. a subhyperedge of size 2) that has not been picked earlier. If there is no such subedge, then the hyperedge corresponds to a clique of the same size, so we pick that clique. This way for every hyperedge we picked a (distinct) complete subgraph of the graph of the picked subedges, and that graph is $F$-free by the definition of Berge copies.

■

Note that the proof shows that a stronger, weighted version of the statement could be formulated, as a clique is always replaced by a hyperedge of the same cardinality, while a hyperedge is replaced by a clique the cardinality of which is either the same or two.

Johnston and Lu [323] chose a third approach to avoid the problems of large hyperedges. They defined the set of *edge-types* of $\mathcal{F}$, denoted by $R(\mathcal{F})$, that is the set of hyperedge sizes in $\mathcal{F}$. We say that $\mathcal{F}$ is an $R$-graph if it has edge-type set $R$. We are interested only in $\mathcal{F}$-free hypergraphs with the same edge-types. Then the Turán number $ex_R(n, \mathcal{F})$ of $\mathcal{F}$ is the maximum number of hyperedges that an $\mathcal{F}$-free hypergraph on $n$ vertices with edge-type $R$ can have. One could define its *Turán density* $\pi_R(\mathcal{F})$ the straightforward way to be

$$\lim_{n\to\infty} \frac{ex_R(n, \mathcal{F})}{\sum_{i\in R(\mathcal{F})} \binom{n}{i}},$$

but it would have the same problem we have mentioned earlier: the large hyperedges would dominate. Indeed, if $k$ is the largest element of $R(\mathcal{F})$, then hyperedges of any other sizes would give 0 to this number. Instead, Johnston and Lu used the Lubell function $\lambda(\mathcal{H}, n) = \sum_{H\in\mathcal{H}} \frac{1}{\binom{n}{|H|}}$ and defined

$$\pi_R(\mathcal{F}) = \lim_{n\to\infty} \max\{\lambda(\mathcal{H}_n, n) : \mathcal{H}_n \text{ is } \mathcal{F}\text{-free on } n \text{ vertices with edge-type } R(\mathcal{F})\}.$$

Averaging shows that $\pi_R(\mathcal{F})$ exists. One can define these notions for families of hypergraphs analogously.

**Proposition 344** *(i)* $|R(\mathcal{H})| - 1 \le \pi_{R(\mathcal{H})}(\mathcal{H}) \le |R(\mathcal{H})|$.
*(ii) If $\mathcal{F} \subset \mathcal{H}$, then $\pi_{R(\mathcal{F})}(\mathcal{F}) \le \pi_{R(H)}(\mathcal{H}) - |R(\mathcal{H})| + |R(\mathcal{F})|$.*

**Proof.** Consider a hypergraph that avoids sets of size $r$ for some $r \in R$. It is obviously $\mathcal{H}$-free. If it contains all the hyperedges of the other sizes, then its Lubell function is $|R(\mathcal{H})| - 1$, giving the lower bound of **(i)**. The upper bound of **(i)** follows from the fact that we consider only hypergraphs with edge-type $R(\mathcal{H})$. The proof of **(ii)** is similar; we omit the details.

■

One can extend the supersaturation result Theorem 335 to non-uniform hypergraphs.

**Theorem 345 (Johnston, Lu, [323])** *For any $R$-graph $\mathcal{F}$ and $\varepsilon > 0$ there is $\delta > 0$ such that if $n$ is large enough and an $R$-graph $\mathcal{G}$ on $n$ vertices has $\lambda(\mathcal{G}) > \pi_R(\mathcal{F}) + \varepsilon$, then it contains at least $\delta\binom{n}{|V(\mathcal{F})|}$ copies of $\mathcal{F}$.*

**Proof.** By the definition of $\pi_R$ there is an $m$ such that for all $n \ge m$ we have $\lambda(\mathcal{H}_n) \le \pi_R(\mathcal{F}) + \varepsilon/2$ if $\mathcal{H}_n$ is an $\mathcal{F}$-free $R$-graph on $n$ vertices. An averaging argument shows that for any $R$-graph $\mathcal{G}$ on $n \ge m$ vertices we have

$$\lambda(\mathcal{G}, n) = \frac{1}{\binom{n}{m}} \sum_{M \in \binom{[n]}{M}} \lambda(\mathcal{G}[M], m).$$

This implies that if $\lambda(\mathcal{G}, n) > \pi_R(\mathcal{F}) + \varepsilon$, then there are at least $\frac{\varepsilon}{2|R|}\binom{n}{m}$ $m$-sets $M$ such that the Lubell function of the subhypergraph induced on $M$ is larger than $\pi_R(\mathcal{F}) + \varepsilon/2$. Indeed, otherwise we have $\lambda(\mathcal{G}, n) \le (\pi_R(\mathcal{F}) + \varepsilon/2)\binom{n}{m} + \frac{\varepsilon}{2|R|}\binom{n}{m}|R| = (\pi_R(\mathcal{F}) + \varepsilon)\binom{n}{m} < \binom{n}{m}\pi_R(\mathcal{G})$, a contradiction.

Each of those $\frac{\varepsilon}{2|R|}\binom{n}{m}$ $m$-sets contains at least one copy of $\mathcal{F}$ by the definition of $m$. An $m$-set is counted $\binom{n-|V(\mathcal{F})|}{m-|V(\mathcal{F})|}$ times, which gives the desired result with $\delta = \frac{\varepsilon}{2|R|m^v}$.

■

One can similarly extend several other results to the non-uniform case.

**Exercise 346** *Define the blowup of non-uniform hypergraphs, and show that the Turán density of the blowup is the same as that of the original hypergraph.*

Johnston and Lu determined the Turán density for all hypergraphs with edge-type set $\{1, 2\}$.

Johnston and Lu [322] and independently Peng, Peng, Tang and Zhao [462] defined and examined Lagrangians in the non-uniform case. Since then a lot of research has been carried out in this direction, see for example [100, 289, 552].

# 6

# *Saturation problems*

## CONTENTS



Most of the theorems discussed in this book address problems on the maximum possible size of a family that possesses some prescribed property $\mathcal{P}$. In this chapter we present a natural dual problem and will look for the minimum possible size of certain set families. Almost all properties discussed in the book are monotone, i.e. if a family $\mathcal{F}$ possesses $\mathcal{P}$, then so do all subfamilies of $\mathcal{F}$. In this case, the empty family has property $\mathcal{P}$, so asking for the minimal possible size would not be too interesting. Therefore we will be interested in *$\mathcal{P}$-saturated* families: families that have property $\mathcal{P}$, but adding any new set to the family would result in not having property $\mathcal{P}$ anymore. The largest of these families is the same as the largest family that does not have property $\mathcal{P}$, but this time we are interested in the smallest of these families.

More formally, given a host family $\mathcal{G}$, a monotone property $\mathcal{P}$ is a collection of subfamilies of $\mathcal{G}$ such that $\mathcal{F}' \subset \mathcal{F} \in \mathcal{P}$ implies $\mathcal{F}' \in \mathcal{P}$. In this situation we say that $\mathcal{F}$ is $\mathcal{P}$-saturated if $\mathcal{F} \in \mathcal{P}$ holds, but for any $G \in \mathcal{G} \setminus \mathcal{F}$ we have $\mathcal{F} \cup \{G\} \notin \mathcal{P}$. We will be interested in the minimum size $sat(\mathcal{P}, \mathcal{G})$ that a $\mathcal{P}$-saturated family can have. In most cases the host family will be either $2^{[n]}$ or $\binom{[n]}{k}$.

Let us consider the two basic properties addressed in extremal finite set theory from this point of view. Let $\mathbb{I}_n$ denote the set of intersecting subfamilies of $2^{[n]}$ and $\mathbb{I}_{n,k}$ denote the set of intersecting subfamilies of $\binom{[n]}{k}$. As Exercise 15 shows, *every* maximal intersecting family $\mathcal{F} \subseteq 2^{[n]}$ has size $2^{n-1}$; hence $sat(\mathbb{I}_n, 2^{[n]}) = 2^{n-1}$ holds. For the uniform case, note that if $T$ is transversal/cover of a maximal intersecting family $\mathcal{F} \subseteq \binom{[n]}{k}$, then for any $k$-set $K$ with $T \subseteq K \subseteq [n]$ we have $K \in \mathcal{F}$. Therefore $|\mathcal{F}| \geq \binom{n-\tau(\mathcal{F})}{k-\tau(\mathcal{F})}$ holds. Let us consider now a $k$-uniform intersecting family $\mathcal{F}$ with covering number $k$. For every $k$-set $G$ that could be added to $\mathcal{F}$ such that it remains intersecting, we must have $G \subseteq \cup_{F \in \mathcal{F}} F$. This shows that if we find a $k$-uniform intersecting family with covering number $k$ on an underlying set $X$, it can be extended to a subfamily of $\binom{X}{k}$, that is saturated on any underlying set containing $X$. It implies $sat(\mathbb{I}_{n,k})$ does not increase with $n$ (after some threshold). Many results in Section 2.6 can be transformed into theorems on saturating intersecting families.

One generalization of the intersecting property we addressed in Section 2.3 is that of not containing $s$ pairwise disjoint sets. We say that $\mathcal{F} \subseteq 2^{[n]}$ is $s$-matching saturated if $\mathcal{F}$ does not contain $s$ pairwise disjoint sets, but for any $G \in 2^{[n]} \setminus \mathcal{F}$ the family $\mathcal{F} \cup \{G\}$ does. Erdős and Kleitman [162] conjectured that if $\mathcal{F} \subseteq 2^{[n]}$ is $s$-matching saturated, then $|\mathcal{F}| \le (1-2^{-(s-1)})2^n$ holds. The conjecture, if true, is sharp as shown by the family $\{F \in 2^{[n]} : F \cap [s-1] \neq \emptyset\}$ (but there exist many other $s$-matching saturated families of this size). For a long time only the following lower bound was known on the size of $s$-matching saturated families.

**Exercise 347** *Show that if $\mathcal{F} \subseteq 2^{[n]}$ is s-matching saturated, then $|\mathcal{F}| \ge 2^{n-1}$. (Hint: show that $\{\overline{G} : G \in 2^{[n]} \setminus \mathcal{F}\}$ is intersecting and apply Exercise 15.)*

Very recently Bucić, Letzter, Sudakov, and Tran [93] obtained a lower bound that does depend on $s$.

**Theorem 348 ( [93])** *If $\mathcal{F} \subseteq 2^{[n]}$ is s-matching saturated, then we have*

$$|\mathcal{F}| \ge (1 - 1/s)2^n.$$

The other basic property in extremal finite set theory is that of being an antichain. If $\mathbb{S}_n$ denotes the set of Sperner families $\mathcal{F} \subseteq 2^{[n]}$, then clearly one has $sat(\mathbb{S}_n, 2^{[n]}) = 1$ as $\{\emptyset\}$ and $\{[n]\}$ are both maximal antichains. In Section 6.2, we will consider the problem of determining $sat(\mathbb{S}_{n,k}, 2^{[n]})$, where $\mathbb{S}_{n,k}$ denotes the set of $k$-Sperner families $\mathcal{F} \subseteq 2^{[n]}$ and we will address some related problems as well.

Another area where saturation problems occur very naturally is that of Turán type problems. In the first Section of this chapter, we will gather some basic results in this topic. Just as in Chapter 5, we will use the hypergraph terminology as all research papers of the area do so. We will only focus on uniform problems and, as throughout the book, will not consider the graph case $k = 2$ separately. The interested reader might find the survey of Faudree, Faudree, and Schmitt [180] a good starting point for the latter. For a $k$-uniform hypergraph $\mathcal{H}$, we will write $sat(n, \mathcal{H})$ for the minimum size of a saturating $\mathcal{H}$-free $k$-graph and $sat(\mathcal{G}, \mathcal{H})$ for the minimum size of a saturating $\mathcal{H}$-graph $k$-subgraph of $\mathcal{G}$.

## 6.1 Saturated hypergraphs and weak saturation

Bollobás considered the first instances of the $\mathcal{H}$-free saturation problem and proved the following.

**Theorem 349 (Bollobás [62])** *If $\mathcal{F}$ is a $k$-uniform saturated $\mathcal{K}_t^k$-free hypergraph on $n$ vertices, then $|\mathcal{F}| \geq \binom{n}{k} - \binom{n-t+k}{k}$ holds.*

He used his inequality Theorem 6 to obtain the result. In another paper [63] he introduced the following weak version of saturation. Let $\mathcal{H}$ be a fixed $k$-uniform hypergraph. We say that a $k$-uniform hypergraph $\mathcal{F}$ with vertex set $[n]$ is *weakly saturated $\mathcal{H}$-free* if $\mathcal{F}$ is $\mathcal{H}$-free and $\binom{V}{k} \setminus \mathcal{F}$ can be enumerated $E_1, E_2, \dots, E_{\binom{n}{k}-|\mathcal{F}|}$ in such a way that if we add these sets to $\mathcal{F}$ in this order, then for any $1 \leq j \leq \binom{n}{k} - |\mathcal{F}|$ the addition of $E_j$ creates a new copy of $\mathcal{H}$. The minimum size of a weakly saturated $\mathcal{H}$-free hypergraph $\mathcal{F}$ on $n$ vertices will be denoted by $wsat(n, \mathcal{H})$ and $wsat(\mathcal{G}, \mathcal{H})$ denotes the minimum size of a weakly saturated $\mathcal{H}$-free subhypergraph $\mathcal{F}$ of $\mathcal{G}$. Since for a saturated $\mathcal{H}$-free hypergraph $\mathcal{F}$ there exists a copy of $\mathcal{H}$ in $\mathcal{F} \cup \{E\}$ containing $E$ for any $E \in \binom{V}{k} \setminus \mathcal{E}$, any saturated $\mathcal{H}$-free hypergraphs are weakly saturated, and therefore $wsat(n, \mathcal{H}) \leq sat(n, \mathcal{H})$ holds. Bollobás [67] conjectured that his sharp lower bound on $sat(n, \mathcal{K}_k^t)$ also holds for $wsat(n, \mathcal{K}_k^t)$. Once one obtains the bound on skew-ISPS's (Theorem 120), Bollobás's proof of the lower bound goes through for the weak saturation version.

**Theorem 350 (Frankl [191], Kalai [327])** *If $\mathcal{F}$ is a $k$-uniform weakly saturated $\mathcal{K}_t^k$-free hypergraph on $n$ vertices, then $|\mathcal{F}| \geq \binom{n}{k} - \binom{n-t+k}{k}$ holds.*

**Proof.** Let $\mathcal{F}$ be a $k$-uniform weakly saturated $\mathcal{K}_t^k$-free hypergraph on $[n]$. Then, by definition, we can enumerate $\binom{[n]}{k} \setminus \mathcal{F}$ as $G_1, G_2, \dots, G_l$ with $l = \binom{n}{k} - |\mathcal{F}|$ such that for every $1 \leq j \leq l$ there exists a $t$-subset $T_j \subseteq [n]$ with $G_j \subseteq T_j$ and $\binom{T_j}{k} \subseteq \mathcal{F} \cup \{G_1, G_2, \dots, G_j\}$. Let us define $Z_j = [n] \setminus T_j$. Then we claim that the pairs $\{(Z_j, G_j)\}_{j=1}^l$ form a skew-ISPS. Indeed, $G_j \subset T_j$ implies $G_j \cap Z_j = \emptyset$ and $\binom{T_j}{k} \subseteq \mathcal{F} \cup \{G_1, G_2, \dots, G_j\}$ implies that for any $j' > j$ we have $G_{j'} \notin \binom{T_j}{k}$ and therefore $Z_j \cap G_{j'} \neq \emptyset$. Therefore Theorem 120 yields $l \leq \binom{n-t+k}{k}$, which finishes the proof of the theorem. ■

Tuza conjectured [537, 539] that for any fixed $k$-uniform hypergraph $\mathcal{H}$ we have $sat(n, \mathcal{H}) = O(n^{k-1})$ (as Theorem 350 shows, this is sharp). Pikhurko verified this in the following form.

**Theorem 351 (Pikhurko [463])** *Let $\mathbb{H}$ be a finite family of $k$-uniform hypergraphs with $s = \max\{\alpha(\mathcal{H}) : H \in \mathbb{H}\}$ and $s' = \min\{|V(\mathcal{H})| : \mathcal{H} \in \mathbb{H}\}$. Then*

$$sat(n, \mathbb{H}) < (s' - s + 2^{k-1}(s-1))\binom{n}{k-1}$$

*holds.*

**Proof.** To prove the theorem, we need to construct an $\mathbb{H}$-saturated $k$-uniform hypergraph $\mathcal{F}$ with $|\mathcal{F}| < (s' - s + 2^{k-1}(s-1))\binom{n}{k-1}$. We will need the following definition: the *tail* $\mathcal{T}_B$ of an $i$-subset $B$ of $[n]$ with $i \leq k$ is $\{T \in \binom{[n]}{k} :$

$T \cap [\max B] = B\}$, i.e. the family of those $k$-subsets of $[n]$ that have $B$ as initial segment. In particular, $\mathcal{T}_B = \{B\}$ for any $B \in \binom{[n]}{k}$. The following algorithm builds $\mathcal{F}$ starting from the empty hypergraph such that in every step it either adds a complete tail to the edge set of $\mathcal{F}$ or leaves it unchanged (for that particular step).

**for** $x = 1$ to $n$ **do**
  **for** $i = 0$ to $k-1$ **do** Choose an enumeration $A_1, A_2, \dots$ of $\binom{[x-1]}{i}$
    **for** $l = 1$ to $\binom{x-1}{i}$ **do**
      **if** $\mathcal{F} \cup \mathcal{T}_{A_l \cup \{x\}}$ is $\mathbb{H}$-free **then** let $\mathcal{F} := \mathcal{F} \cup \mathcal{T}_{A_l \cup \{x\}}$
      **else** $\mathcal{F} = \mathcal{F}$.
      **end if**
    **end for**
  **end for**
**end for**

**Claim 352** *The hypergraph $\mathcal{F}$ generated by the above algorithm is $\mathbb{H}$-saturated.*

**Proof of Claim.** $\mathcal{F}$ is clearly $\mathbb{H}$-free as we add sets to it only if they do not create copies of hypergraphs in $\mathbb{H}$. To see that $\mathcal{F}$ is saturated let $G$ be any set from $\binom{[n]}{k} \setminus \mathcal{F}$. Let us consider the algorithm when the value of $x$ is the largest element of $G$, when $i = k-1$ and when $l$ is the index of $G \setminus \{x\}$ in the enumeration of $\binom{[x]}{k-1}$. Then at that moment $\mathcal{T}_{A_l \cup \{x\}} = \{G\}$; therefore the only reason for which $G$ is not added to $\mathcal{F}$ is that it would have created some forbidden copy of a hypergraph in $\mathbb{H}$. □

We still have to bound the size of $\mathcal{F}$. To this end let $\mathcal{F}_y$ denote the subhypergraph of those sets in $\mathcal{F}$ that were added not later than when the value of $x$ was $y$. Let $\mathcal{G}_i$ be the hypergraph of the $i$-subsets $B$ of $[n]$ with the property that the tail $\mathcal{T}_B$ of $B$ was added to $\mathcal{F}$ when $B$ was considered by the algorithm.

**Claim 353** $|\mathcal{G}_1| \le s' - 1$ *and* $|\mathcal{G}_i| \le (s-1)\binom{n}{i-1}$ *for all* $i = 2, 3, \dots, k$.

**Proof of Claim.** Let us consider first $\mathcal{G}_1 = \{\{i_1\}, \{i_2\}, \dots, \{i_m\}\}$ with $i_1 < i_2 < \dots < i_m$ and let $S$ denote the set $\{i_1, i_2, \dots, i_m\}$. Then $\binom{S}{k} \subseteq \mathcal{F}$. Indeed, if $K \in \binom{S}{k}$ and $i_j$ is the smallest element of $K$, then $K \in \mathcal{T}_{\{i_j\}} \subseteq \mathcal{F}$. As, by definition, there exists an $\mathcal{H} \in \mathbb{H}$ with $|V(\mathcal{H})| = s'$, the $\mathbb{H}$-free property of $\mathcal{F}$ implies that we must have $|S| < s'$.

For $i = 2, 3, \dots, k$, suppose $|\mathcal{G}_i| > (s-1)\binom{n}{i-1}$ holds. Then for some $v_1 < v_2 < \dots < v_{i-1}$ the $(i-1)$-set $V = \{v_1, v_2, \dots, v_{i-1}\}$ is the initial segment of at least $s$ $i$-sets $G_j = V \cup \{z_j\} \in \mathcal{G}_i$ with $v_{i-1} < z_1 < z_2 < \dots < z_s$.

Note that $G_1 \in \mathcal{G}_i$ means that all $F \in \mathcal{F}$ of which $G_1$ is an initial segment are added to $\mathcal{F}$ at the moment when $A = V$ and $x = z_1$. This implies that $V \notin \mathcal{G}_{i-1}$, as otherwise these sets would already be present in $\mathcal{F}$. As $V \notin \mathcal{G}_{i-1}$, at the moment when $A = V \setminus \{v_{i-1}\}$ and $x = v_{i-1}$, the hypergraph $\mathcal{F}' = \mathcal{F}_{v_{i-1}} \cup \mathcal{T}_V$ contains a copy $\mathcal{H}^*$ of a forbidden hypergraph $\mathcal{H} \in \mathbb{H}$. Note that by definition $U = \{v_{i-1}+1, v_{i-1}+2, \ldots, n\}$ is an independent set in $\mathcal{F}'$; therefore $|U \cap V(\mathcal{H}^*)| \leq s$ holds.

Let us observe that if for a permutation $\sigma \in S_n$ we have $\sigma(y) = y$ for all $1 \leq y \leq z$, then $\sigma$ is an automorphism of $\mathcal{F}_z$, because for all $B \subseteq [z]$ we have $\sigma(\mathcal{T}_B) = \mathcal{T}_B$. Applying this with $z = v_{i-1}$ we can assume that $U \cap V(\mathcal{H}^*) \subseteq \{z_1, z_2, \ldots, z_s\}$ holds.

Now let $H \in \mathcal{H}^*$ be arbitrary. Then either we have $H \in \mathcal{F}_{v_{i-1}} \subset \mathcal{F}$ or $H \cap \{z_1, z_2, \ldots, z_s\} \neq \emptyset$. In the latter case, let $z_j = \min H \cap \{z_1, z_2, \ldots, z_s\}$. Then the assumption $U \cap V(\mathcal{H}^*) \subseteq \{z_1, z_2, \ldots, z_s\}$ implies $H \in \mathcal{T}_{G_j}$. As $G_j \in \mathcal{G}_i$, we have $\mathcal{T}_{G_j} \subseteq \mathcal{F}$. In either case, we obtained $H \in \mathcal{F}$; thus the forbidden copy $\mathcal{H}^*$ is in $\mathcal{F}$, which is clearly a contradiction. □

Using Claim 353 we obtain

$$
\begin{aligned}
|\mathcal{F}| &\leq \sum_{i=1}^{k} \binom{n-i}{k-i} |\mathcal{G}_i| < (s'-1)\binom{n}{k-1} + (s-1)\sum_{i=2}^{k} \binom{n-i+1}{k-i}\binom{n}{i-1} \\
&= (s'-s)\binom{n}{k-1} + (s-1)\sum_{i=1}^{k} \binom{n-i+1}{k-i}\binom{n}{i-1} \\
&= (s'-s)\binom{n}{k-1} + (s-1)\sum_{i=1}^{k} \binom{n}{k-1}\binom{k-1}{i-1} \\
&= (s'-s+2^{k-1}(s-1))\binom{n}{k-1}.
\end{aligned}
$$

■

Erdős, Füredi and Tuza considered saturation of $\mathbb{F}^k(v,e)$, the family of $k$-graphs on $v$ vertices with $e$ hyperedges. They showed the following.

**Theorem 354 (Erdős, Füredi, Tuza, [158])** *For $n \geq 5$, we have*

$$sat(n, \mathbb{F}^3(6,3)) = \lfloor (n-1)/2 \rfloor.$$

**Proof.** For the upper bound consider the following 3-graph. We take a vertex and $\lfloor (n-1)/2 \rfloor$ hyperedges that contain it, and otherwise are disjoint.

For the lower bound we apply induction on $n$; the statement is trivial for $n = 5$ and $n = 6$. Consider a $\mathbb{F}^3(6,3)$-saturated hypergraph $\mathcal{H}$ on $n$ vertices and observe first that if two hyperedges of $\mathcal{H}$ intersect in two vertices, then none of them can intersect any other hyperedges, as those three hyperedges would form a forbidden 3-graph in $\mathbb{F}^3(6,3)$. Observe that by removing these two hyperedges and the four vertices they contain from $\mathcal{H}$, we get a $\mathbb{F}^3(6,3)$-saturated hypergraph on $n-4$ vertices; thus altogether we have at most $2 + \lfloor (n-5)/2 \rfloor$ hyperedges in $\mathcal{H}$.

Hence we can assume any two hyperedges in $\mathcal{H}$ share at most one vertex. If $\mathcal{H}$ has at least three connected components, then any hyperedge intersecting each of those components in one vertex can be added to $\mathcal{H}$ without creating a copy of $\mathbb{F}^3(6,3)$, a contradiction. Hence there are at most two connected components. As every 3-edge can add at most two vertices to its connected component (apart from the first edge), a component of size $m$ must contain at least $1+\lceil\frac{m-3}{2}\rceil$ 3-edges. Applying this we obtain that the number of 3-edges in $\mathcal{H}$ is at least $1+\frac{n-3}{2}=\lfloor(n-1)/2\rfloor$ if $\mathcal{H}$ is connected and $1+\lceil\frac{m-3}{2}\rceil+1+\lceil\frac{n-m-3}{2}\rceil$ if $\mathcal{H}$ consists of two components. ■

They also obtained the following results.

**Theorem 355 (Erdős, Füredi, Tuza, [158])** *(i) For $n>k\ge 2$ we have*

$$sat(n,\mathbb{F}^k(k+1,k))=(1/2-o(1))\binom{n}{k-1}.$$

*(ii) For $n\ge 4$ we have $sat(n,\mathbb{F}^3(4,3))=\lfloor(n-1)^2/4\rfloor$.*
*(iii) For $n>k\ge 2$ we have $wsat(n,\mathbb{F}^k(k+1,3))=n-k+1$.*

Pikhurko [464] determined the weak saturation number for a large class of hypergraphs. Let us consider two vectors $r=(r_1,\dots,r_t)$ and $s=(s_1,\dots,s_t)$ with $r_i\le s_i$ for every $i\le t$. Let $k=\sum_{i=1}^t r_i$ and $S_i$ be disjoint sets of size $s_i$. Let $P(s,r)$ be the $k$-graph on the vertex set $\cup_{i=1}^t S_i$ where a $k$-set $F$ is a hyperedge if and only if $|F\cap(\cup_{i=1}^{t'}S_i)|\ge\sum_{i=1}^{t'}r_i$ for every $t'\le t$. Several hypergraphs we have already studied belong to this class. Among others, the following generalization of statement **(iii)** of Theorem 355 is implied.

**Theorem 356 (Pikhurko [464])** *For $n>k\ge l-1\ge 3$ we have $wsat(n,\mathbb{F}^k(k+1,l))=\binom{n-k+l-2}{l-2}$.*

Let us mention without further details that weak saturation is related to bootstrap percolation. That motivated a few papers. Alon [10] showed that if $\mathcal{H}_m$ denotes the complete $k$-partite $k$-graph $\mathcal{H}$ with all parts of size exactly $m$, then we have $wsat(\mathcal{H}_n,\mathcal{H}_p)=n^k-(n-p+1)^k$. Balogh, Bollobás, Morris and Riordan [37] extended this result, and finally Moshkowitz and Shapira [423] determined the weak saturation number of every complete $k$-partite $k$-graph.

English, Graber, Kirkpatrick, Methuku and Sullivan [146] considered saturation for Berge-hypergraphs. They proved for example that $sat(n,\mathbb{B}^kK_3)=\lceil(n-1)/(k-1)\rceil$. They conjectured that the saturation number of any Berge-hypergraph is $O(n)$ for any $k$. This was proved for $k=3,4,5$ by English, Gerbner, Methuku and Tait [145].

## 6.2 Saturating $k$-Sperner families and related problems

In this section we deal with the saturation problem for $k$-Sperner families and for brevity we will write $sat(n,k)$ for $sat(\mathbb{S}_{n,k}, 2^{[n]})$. The problem of determining $sat(n,k)$ was first considered by Gerbner, Keszegh, Lemons, Palmer, Pálvölgyi, and Patkós [257], and most of their bounds were improved later by Morrison, Noel, and Scott [422], who also considered the following variant: a family $\mathcal{F} \subseteq 2^{[n]}$ is *oversaturating* $k$-Sperner if for any $G \in 2^{[n]} \setminus \mathcal{F}$, the family $\mathcal{F} \cup \{G\}$ contains a $(k+1)$-chain $\mathcal{C}$ with $G \in \mathcal{C}$. In other words, an oversaturating $k$-Sperner family $\mathcal{F}$ is not necessarily $k$-Sperner, but adding any new set to $\mathcal{F}$ increases the number of $(k+1)$-chains in $\mathcal{F}$. In particular, any saturating $k$-Sperner family is oversaturating. Let us denote by $osat(n,k)$ the minimum size of an oversaturating $k$-Sperner family $\mathcal{F} \subseteq 2^{[n]}$; then $osat(n,k) \le sat(n,k)$ holds for all $n$ and $k$. We include almost all results from the papers [257] and [422] - some with proof, some as exercises. The starting point is the following construction that shows that $sat(n,k) \le 2^{k-1}$ holds independently of $n$.

**Exercise 357** *Let us write $X = X_{n,k} = [n] \setminus [k-2]$ and define $\mathcal{F}_{n,k} = 2^{[k-2]} \cup \{Y \cup X : Y \in 2^{[k-2]}\}$. Verify that $\mathcal{F}_{n,k}$ is a saturating $k$-Sperner family.*

The following simple proposition states the only known general lower bound both for $sat(n,k)$ and $osat(n,k)$. Readers are advised to think about the proof before reading.

**Proposition 358 ( [257])** *For any $k$ and $n$ we have $2^{k/2-1} \le osat(n,k) \le sat(n,k)$.*

**Proof.** Let $\mathcal{F} \subseteq 2^{[n]}$ be an oversaturating $k$-Sperner family and consider any set $G \in 2^{[n]} \setminus \mathcal{F}$. By definition, there exist $k$ distinct sets $F_1, ..., F_k \in \mathcal{F}$ such that $F_1 \subsetneq ... \subsetneq F_i \subsetneq G \subsetneq F_{i+1} \subsetneq ... \subsetneq F_k$ holds. Thus $G$ is a set from the interval $I_{F_i,F_{i+1}} = \{S : F_i \subsetneq S \subsetneq F_{i+1}\}$ which has size at most $2^{n-k+2}$ as $|F_{i+1} \setminus F_i| \le n-k+2$ holds by the existence of the other $F_j$'s. We obtain that $2^{[n]}$ can be covered by intervals of size at most $2^{n-k+2}$ and thus we have

$$\frac{|\mathcal{F}|^2}{2} 2^{n-k+2} \ge \binom{|\mathcal{F}|}{2} 2^{n-k+2} \ge 2^n.$$

Now the proposition follows by rearranging. ∎

Let us start investigating the structure of saturating $k$-Sperner families. We will need the following definitions. We say that $(\mathcal{F}_i)_{i=1}^t$ is a *layered sequence of families* if any $F \in \mathcal{F}_{i+1}$ strictly contains a set of $\mathcal{F}_i$. A set $H \subseteq X$ is *homogeneous* for a family $\mathcal{F} \subseteq 2^X$ if $|H| \geq 2$ and it is maximal with respect to the property

$$\forall F \in \mathcal{F} \quad (H \cap F = \emptyset \text{ or } H \subseteq F).$$

If a set $H$ is homogeneous for $\mathcal{F}$, then $\mathcal{F}$ can be partitioned into $\mathcal{F}^{small} \cup \mathcal{F}^{large}$, where $\mathcal{F}^{small} = \{F \in \mathcal{F} : F \cap H = \emptyset\}$ and $\mathcal{F}^{large} = \{F \in \mathcal{F} : H \subseteq F\}$. The next couple of lemmas and exercises establish some useful properties of saturating $k$-Sperner families and related notions.

**Lemma 359** *If $\mathcal{A} \subseteq 2^X$ is a saturating antichain with homogeneous set $H$ and $S \in 2^X \setminus \mathcal{A}$, then either $S$ contains a set from $\mathcal{A}^{small}$ or is contained in a set from $\mathcal{A}^{large}$.*

**Proof.** Suppose $S$ is a set from $2^X \setminus \mathcal{A}$ that does not have the required property. Then as $\mathcal{A}$ is a saturating antichain, either there exists $A$ with $S \subsetneq A \in \mathcal{A}^{small}$ or there exists a set $B$ with $B \subsetneq S$ and $B \in \mathcal{A}^{large}$.

Suppose first that there exists $A$ with $S \subsetneq A \in \mathcal{A}^{small}$. Let us fix $x \in H$, $a \in A \setminus S$ and consider $T = (A \setminus \{a\}) \cup \{x\}$. As $H$ is homogeneous for $\mathcal{A}$ and $T \cap H = \{x\} \subsetneq H$, subsets of $T$ cannot be in $\mathcal{A}^{large}$ and supersets of $T$ cannot be in $\mathcal{A}^{small}$, in particular we have $T \notin \mathcal{A}$. As $\mathcal{A}$ is saturating and $x \in T$, we either have a set $A' \in \mathcal{A}^{small}$ with $A' \subsetneq T$ or a set $A'' \in \mathcal{A}^{large}$ with $T \subsetneq A''$. The former case is impossible as $x \notin A'$ and thus $A' \subsetneq A$ which contradicts $\mathcal{A}$ being an antichain. In the latter case $S \subsetneq T \subsetneq A'' \in \mathcal{A}^{large}$ as claimed by the lemma.

Now the case $B$ with $B \subsetneq S$ and $B \in \mathcal{A}^{large}$ follows by applying the same argument to $\overline{\mathcal{A}} = \{X \setminus A : A \in \mathcal{A}\}$, since $H$ is homogeneous for $\overline{\mathcal{A}}$ and $\overline{\mathcal{A}}^{large} = \{X \setminus A : A \in \mathcal{A}^{small}\}$, $\overline{\mathcal{A}}^{small} = \{X \setminus A : A \in \mathcal{A}^{large}\}$. ■

**Lemma 360** *If $(\mathcal{A}_i)_{i=1}^k$ is a layered sequence of pairwise disjoint saturating antichains in $2^X$, then $\mathcal{F} = \cup_{i=1}^k \mathcal{A}_i$ is a saturating $k$-Sperner family in $X$.*

**Proof.** Clearly $\mathcal{F}$ is $k$-Sperner as it is the union of $k$ antichains. For a set $T \in 2^X \setminus \mathcal{F}$, let $t$ be the largest integer with $\mathcal{A}_t$ containing a set $A_t$ with $A_t \subsetneq T$. As $(\mathcal{A}_i)_{i=1}^k$ is layered, we find a chain $A_1 \subsetneq A_2 \subsetneq \cdots \subsetneq A_t \subsetneq T$ with $A_i \in \mathcal{A}_i$. Note that as $\mathcal{A}_{t+1}$ is a saturating antichain there is a set $A_{t+1} \in \mathcal{A}_{t+1}$ in containment with $T$, so by definition of $t$ we must have $T \subsetneq A_{t+1}$. Finally, we will extend the $(t+2)$-chain $A_1 \subsetneq A_2 \subsetneq \cdots \subsetneq A_t \subsetneq T \subsetneq A_{t+1}$ to a $(k+1)$-chain. Once $A_j \in \mathcal{A}_j$ with $j \geq t+1$ is found, then by the saturating antichain property of $\mathcal{A}_{j+1}$, there is a set $A_{j+1} \in \mathcal{A}_{j+1}$ that is in containment with $A_j$. If $A_j \subsetneq A_{j+1}$, then we are done extending the chain. If $A_{j+1} \subsetneq A_j$, then as $(\mathcal{A}_i)_{i=1}^k$ is layered, there exists $A'_{j+1} \supsetneq A_j \supsetneq A_{j+1}$ contradicting that $\mathcal{A}_{j+1}$ is an antichain. This shows that for any set $T \in 2^X \setminus \mathcal{F}$, there exists a $(k+1)$-chain in $\mathcal{F} \cup \{T\}$. ■

**Lemma 361** *Let $(\mathcal{A}_i)_{i=1}^k$ be a sequence of pairwise disjoint saturating antichains in $2^X$, each of which has a homogeneous set. Then $(\mathcal{A}_i)_{i=1}^k$ is layered if and only if $(\mathcal{A}_i^{small})_{i=1}^k$ is layered.*

**Proof.** Suppose first $(\mathcal{A}_i)_{i=1}^k$ is layered and let $A \in \mathcal{A}_{i+1}^{small}$. If $A$ does not contain a set from $\mathcal{A}_i^{small}$, then, by the layered property, it must contain some $B \in \mathcal{A}_i^{large}$. As $\mathcal{A}_i$ is an antichain, $A$ cannot be contained in any set of $\mathcal{A}_i$, in particular in any set of $\mathcal{A}_i^{large}$. Thus by Lemma 359 $A$ must contain a set $A_i \in \mathcal{A}_i^{small}$.

Suppose next that $(\mathcal{A}_i^{small})_{i=1}^k$ is layered. We have to show that any $A \in \mathcal{A}_{i+1}^{large}$ contains a set from $\mathcal{A}_i$. If not, then as $\mathcal{A}_i$ is a saturating antichain, there must exist an $A' \in \mathcal{A}_i$ with $A \subsetneq A'$. Applying Lemma 359 to $A'$ and $\mathcal{A}_{i+1}$ we either obtain a set $S \in \mathcal{A}_{i+1}^{large}$ with $A' \subsetneq S$ or a set $T \in \mathcal{A}_{i+1}^{small}$ with $T \subsetneq A'$. The former is impossible as $A \subsetneq A' \subsetneq S$ contradicts that $\mathcal{A}_{i+1}$ is an antichain. In the latter case, we also get a contradiction as $(\mathcal{A}_i^{small})_{i=1}^k$ is layered; therefore $T$ contains a set $A'' \in \mathcal{A}_i^{small}$ and thus $A'' \subsetneq T \subsetneq A'$ with $A'', A' \in \mathcal{A}_i$. ■

**Exercise 362** ***(i)*** *For any saturating $k$-Sperner family $\mathcal{F} \subseteq 2^X$ there exists at most one homogeneous set $H$.*

***(ii)*** *If $\mathcal{F} \subseteq 2^X$ and $|X| > 2^{|\mathcal{F}|}$, then there is a set $H \subseteq X$ that is homogeneous for $\mathcal{F}$.*

**Exercise 363** *Show that for any fixed $k$ if $n, m$ are large enough, then $sat(n,k) = sat(m,k)$ and $osat(n,k) = osat(m,k)$ hold.*

According to Exercise 363 we can introduce $sat(k) = \lim_{n \to \infty} sat(n,k)$ and $osat(k) = \lim_{n \to \infty} osat(n,k)$. The upper bound provided by the construction of Exercise 357 is roughly the square of the lower bound of Proposition 358. The next exercise shows that $\lim_{k \to \infty} \sqrt[k]{sat(k)}$ exists.

**Exercise 364** *Show that for any pair $k, l$ of positive integers we have $sat(k + l) \leq 4sat(k)sat(l)$. As a consequence prove that there exists a constant $c$ with $1/2 \leq c \leq 1$ such that $sat(k) = 2^{(c+o(1))k}$ holds.*

As the simple Lemma 5 shows, every $k$-Sperner family can be decomposed into $k$ antichains, $\mathcal{F} = \cup_{j=1}^k \mathcal{F}_j$, where $\mathcal{F}_j$ is the family of minimal elements of $\mathcal{F} \setminus \cup_{i=1}^{j-1} \mathcal{F}_i$ (we called this its canonical partition). One might wonder whether this remains valid for saturating $k$-Sperner families and saturating antichains. It is not true in general that a saturating $k$-Sperner family can be decomposed into $k$ saturating antichains. (Find a counterexample!) The next exercise gives a sufficient condition when this is still true.

**Exercise 365** *Let $\mathcal{F} \subsetneq 2^X$ be a saturating $k$-Sperner family with homogeneous set $H$. Show that all $\mathcal{F}_j$'s in the canonical partition of $\mathcal{F}$ are saturating antichains.*

We now start turning our attention to upper bounds on $sat(n,k)$ and $osat(n,k)$. To this end we will need constructions. The next lemma tells us how one can combine two appropriate saturating $k$-Sperner families to obtain a saturating $k'$-Sperner family.

**Lemma 366 (Morrison, Noel, Scott, [422])** *Let $X_1, X_2$ be disjoint sets and $\mathcal{F}_1 \subseteq 2^{X_1}$ be saturating $k_1$-Sperner with $\emptyset, X_1 \in \mathcal{F}_1$ and let $\mathcal{F}_2 \subseteq 2^{X_2}$ be saturating $k_2$-Sperner with $\emptyset, X_2 \in \mathcal{F}_2$. Assume furthermore that $H_1 \subset X_1$ is homogeneous for $\mathcal{F}_1$ and $H_2 \subset X_2$ is homogeneous for $\mathcal{F}_2$; then*

$$\mathcal{F} = \{A \cup B : A \in \mathcal{F}_1^{small}, B \in \mathcal{F}_2^{small}\} \cup \{A \cup B : A \in \mathcal{F}_1^{large}, B \in \mathcal{F}_2^{large}\}$$

*is saturating $(k_1 + k_2 - 2)$-Sperner with $|\mathcal{F}| = |\mathcal{F}_1^{small}||\mathcal{F}_2^{small}| + |\mathcal{F}_1^{large}||\mathcal{F}_2^{large}|$, $H = H_1 \cup H_2$ is homogeneous for $\mathcal{F}$ and $\emptyset, X_1 \cup X_2 \in \mathcal{F}$.*

**Proof.** By definition, $\emptyset \in \mathcal{F}$, $X_1 \cup X_2 \in \mathcal{F}$, $H_1 \cup H_2$ is homogeneous for $\mathcal{F}$, and $|\mathcal{F}| = |\mathcal{F}_1^{small}||\mathcal{F}_2^{small}| + |\mathcal{F}_1^{large}||\mathcal{F}_2^{large}|$ holds. We still need to show that $\mathcal{F}$ is saturating $(k_1 + k_2 - 2)$-Sperner.

First, we prove that $\mathcal{F}$ is $(k_1 + k_2 - 2)$-Sperner. Let $F_1 \subsetneq F_2 \subsetneq \cdots \subsetneq F_r$ be a chain with $F_i \in \mathcal{F}$. We can assume that $F_1 = \emptyset$ and $F_r = X_1 \cup X_2$. For $i = 1,2$ let us define $I_i = \{j : F_j \cap X_i \subsetneq F_{j+1} \cap X_i\}$. As the $F_j$'s form a chain, we have $I_1 \cup I_2 = \{1, 2, \ldots, r-1\}$. Also, by definition of $\mathcal{F}$ we have $F_j \cap X_i \in \mathcal{F}_i$, so the $k_i$-Sperner property of $\mathcal{F}_i$ implies $|I_i| \leq k_i - 1$. Let $t$ denote the maximum index with $F_t \cap X_1 \in \mathcal{F}_1^{small}$. As $F_1 = \emptyset$, $t$ exists, and as $F_r = X_1 \cup X_2$, we have $t \leq r-1$. Clearly, $t \in I_1$. But by definition of $\mathcal{F}$ we must have $F_t \cap X_2 \in \mathcal{F}_2^{small}$ and $F_{t+1} \cap X_2 \in \mathcal{F}_2^{large}$; therefore $t$ also belongs to $I_2$. We obtained

$$r - 1 = |I_1 \cup I_2| = |I_1| + |I_2| - |I_1 \cap I_2| \leq (k_1 - 1) + (k_2 - 1) - 1.$$

Rearranging yields that $\mathcal{F}$ is indeed $(k_1 + k_2 - 2)$-Sperner.

Now let $S$ be a set in $2^{X_1 \cup X_2} \setminus \mathcal{F}$. We need to show that there exists a $(k_1 + k_2 - 1)$-chain in $\mathcal{F} \cup \{S\}$. To do so, let us fix elements $x_1 \in H_1, x_2 \in H_2$ and define $T = (S \setminus (H_1 \cup H_2)) \cup \{x_1, x_2\}$, $T_i = T \cap X_i$ for $i = 1,2$. As $H_i$ is homogeneous for $\mathcal{F}_i$, we have $T_i \notin \mathcal{F}_i$ for $i = 1,2$. This implies that there exist sets $\emptyset = F_1^1, F_2^1, \ldots, F_{k_1}^1 = X_1$, $\emptyset = F_1^2, F_2^2, \ldots, F_{k_2}^2 = X_2$ and indices $1 \leq t_1 < k_1$, $1 \leq t_2 < k_2$ such that

$$F_1^1 \subsetneq \cdots \subsetneq F_{t_1}^1 \subsetneq T_1 \subsetneq F_{t_1+1}^1 \subsetneq \cdots \subsetneq F_{k_1}^1$$

and

$$F_1^2 \subsetneq \cdots \subsetneq F_{t_2}^2 \subsetneq T_2 \subsetneq F_{t_2+1}^2 \subsetneq \cdots \subsetneq F_{k_2}^2$$

hold. As $H_i$ is homogeneous for $\mathcal{F}_i$, we must have $F_j^i \in \mathcal{F}_i^{small}$ if and only if $j \leq t_i$. Therefore $F_{j_1}^1 \cup F_{j_2}^2 \in \mathcal{F}$ if and only if either $j_1 \leq t_1, j_2 \leq t_2$ or $j_1 > t_1, j_2 > t_2$. Note also that $S$ and $T$ differ only in elements of $H_1 \cup H_2$, so

$F^1_{t_1} \cup F^2_{t_2} \subsetneq T \subsetneq F^1_{t_1+1} \cup F^2_{t_2+1}$ and $S \notin \mathcal{F}$ imply $F^1_{t_1} \cup F^2_{t_2} \subsetneq S \subsetneq F^1_{t_1+1} \cup F^2_{t_2+1}$. Therefore $\mathcal{F} \cup \{S\}$ does indeed contain the following chain of length $k_1+k_2-1$:

$$F^1_1 \cup F^2_1 \subsetneq F^1_2 \cup F^2_1 \subsetneq \cdots \subsetneq F^1_{t_1} \cup F^2_1 \subsetneq F^1_{t_1} \cup F^2_2 \subsetneq \cdots \subsetneq F^1_{t_1} \cup F^2_{t_2} \subsetneq S \subsetneq$$

$$F^1_{t_1+1} \cup F^2_{t_2+1} \subsetneq F^1_{t_1+2} \cup F^2_{t_2+1} \subsetneq \cdots \subsetneq F^1_{k_1} \cup F^2_{t_1+1} \subsetneq F^1_{k_1} \cup F^2_{t_2+2} \subsetneq \cdots \subsetneq F^1_{k_1} \cup F^2_{k_2}.$$

■

To be able to use Lemma 366, we need a starting point. Let us remind the reader that our aim is to give better bounds than Exercise 357, so we have to come up with a construction of a saturating $k$-Sperner family $\mathcal{F} \subseteq 2^{[n]}$ with $|\mathcal{F}| < 2^{k-1}$. This was already done in [257], but that construction does not satisfy the other assumptions of Lemma 366. The following proposition provides us a saturating 6-Sperner family which is somewhat similar to that in [257], but possesses the required homogeneous set.

**Proposition 367 (Morrison, Noel, Scott, [422])** *If $X$ is a set of size at least 8, then there exists a saturating 6-Sperner family $\mathcal{F} \subseteq 2^X$ of size 30 with a homogeneous set such that $|\mathcal{F}^{large}| = |\mathcal{F}^{small}| = 15$.*

**Proof.** We define $\mathcal{F}$ via its canonical partition $\cup_{i=1}^6 \mathcal{F}_i$. For notation we fix $x_1, x_2, y_1, y_2, w, z \in X$ and set $H = X \setminus \{x_1, x_2, y_1, y_2, w, z\}$ ($|X| \geq 8$ is needed to ensure $|H| \geq 2$ and therefore it satisfies this requirement of a homogeneous set). We let $\mathcal{F}_1 = \{\emptyset\}$, $\mathcal{F}_6 = \{X\}$ and

$$\mathcal{F}_2 = \{\{x_1\}, \{x_2\}, \{y_1\}, \{z\}, H \cup \{y_2, z\}\}, \mathcal{F}_5 = \{X \setminus F : F \in \mathcal{F}_2\}.$$

We define first $\mathcal{F}_3^{small}, \mathcal{F}_4^{small}$ by

$$\mathcal{F}_3^{small} = \{\{x_i, y_j\} : 1 \leq i, j \leq 2\} \cup \{\{w, z\}\},$$

$$\mathcal{F}_4^{small} = \{\{x_1, y_1, z\}, \{x_1, y_1, w\}, \{x_2, y_2, z\}, \{x_2, y_2, w\}\}.$$

For $i = 3, 4$ the families $\mathcal{F}_i^{large}$ consist of the maximal independent sets of $\mathcal{F}_i^{small}$, i.e. those sets that do not contain any $F \in \mathcal{F}_i^{small}$ and are maximal with respect to this property. The maximality condition implies that $\mathcal{F}_i^{large}$ is an antichain and that any set of $\mathcal{F}_i^{large}$ contains $H$. Therefore $\mathcal{F}_i = \mathcal{F}_i^{small} \cup \mathcal{F}_i^{large}$ is an antichain for $i = 3, 4$. Moreover they are saturating as every set in $2^X$ either contains a set from $\mathcal{F}_i^{small}$ or is contained in a set of $\mathcal{F}_i^{large}$. It can be seen that $(\mathcal{F}_i^{small})_{i=1}^6$ is layered; therefore, by Lemma 361, so is $(\mathcal{F}_i)_{i=1}^6$. Lemma 360 implies that $\mathcal{F}$ is saturating $k$-Sperner. To count the sets in $\mathcal{F}$ observe that

$$\mathcal{F}_3^{large} = \{H \cup \{x_1, x_2, w\}, H \cup \{x_1, x_2, z\}, H \cup \{y_1, y_2, w\}, H \cup \{y_1, y_2, w\}\},$$

$$\mathcal{F}_4^{large} = \{H \cup \{x_1, x_2, y_1, y_2\}, H \cup \{x_1, x_2, w, z\}, H \cup \{y_1, y_2, w, z\},$$
$$H \cup \{x_1, y_2, w, z\}, H \cup \{x_2, y_1, w, z\}\}.$$

■

Now we are ready to combine Lemma 366 and Proposition 367 to obtain an improved upper bound on $sat(n,k)$.

**Theorem 368 (Morrison, Noel, Scott, [422])** *There exists $\varepsilon > 0$ such that $sat(n,k) \leq 2^{(1-\varepsilon)k}$ holds for all positive integers $k \geq 6$ and large enough $n$.*

**Proof.** First we prove the result for integers of the form $k = 4j+2$. If $n \geq 8j$, then $[n]$ can be partitioned into $j$ sets $X_1, \ldots, X_j$ each of size at least 8. We obtain saturating 6-Sperner families $\mathcal{F}_j \subseteq 2^{X_j}$ provided by Proposition 367 and then apply Lemma 366 repeatedly to obtain a saturating $k$-Sperner family $\mathcal{F}$ with $|\mathcal{F}^{small}| = |\mathcal{F}^{large}| = 15^j$ and $|\mathcal{F}| = 2 \cdot 15^j = \frac{1}{2}(\frac{15}{16})^j 2^k$.

For other values of $k$ observe that $sat(n,k) \leq 2sat(n-1,k-1)$ as if $\mathcal{G} \subseteq 2^{[n-1]}$ is saturating $(k-1)$-Sperner; then $\mathcal{G} \cup \{G \cup \{n\} : G \in \mathcal{G}\} \subseteq 2^{[n]}$ is saturating $k$-Sperner. (Check it!) Using this for $k = 4j+2+s$ $(s = 1,2,3)$ we can obtain a saturating $k$-Sperner family of size $2^{s+1}15^j$. ■

The next construction shows that the lower bound of Lemma 358 is sharp for $osat(n,k)$ apart from a polynomial factor.

**Theorem 369 ( [422])** *For any large enough $k$ and $n \geq k^2+k$ there exists an oversaturated $k$-Sperner family $\mathcal{F} \subset 2^{[n]}$ with $|\mathcal{F}| \leq k^5 2^{k/2}$. In particular, we have $osat(k) = 2^{(1+o(1))k/2}$.*

**Proof.** The proof is based on the following lemma.

**Lemma 370** *If $k$ is large enough then for any integer $1 \leq t \leq k^2+k$, and set $X$ with $|X| = k^2+k$, there exist non-empty families $\mathcal{F}_t, \mathcal{G}_t \subseteq 2^X$ that have the following properties:*

*(a) For every $F \in \mathcal{F}_t$ and $G \in \mathcal{G}_t$, $|F|+|G| \geq k$,*

*(b) $|\mathcal{F}_t| + |\mathcal{G}_t| = O(k^2 2^{k/2})$,*

*(c) For every $S \subseteq X$ with $|S| = t$, there exists some $F \in \mathcal{F}_t$ and some $G \in \mathcal{G}_t$ such that $F \subsetneq S$ and $G \cap S = \emptyset$ hold.*

**Proof of Lemma.** Set $n = k^2+k$ and assume $X = [n]$. We need to define $\mathcal{F}_t$ and $\mathcal{G}_t$ for $1 \leq t \leq n$, but it is enough to do so for $t \leq n/2$ as for $t > n/2$ we can let $\mathcal{F}_t = \mathcal{G}_{n-t}$ and $\mathcal{G}_t = \mathcal{F}_{n-t}$. We distinguish two cases.

CASE I: $t \leq n/8$.

We let $\mathcal{F}_t = \{\emptyset\}$ and we pick $\mathcal{G}_t$ to be a subfamily of $\binom{[n]}{k}$ of size $2^{k/2}$, chosen uniformly at random. Therefore $|\mathcal{F}_t|+|\mathcal{G}_t| = 2^{k/2}+1$ and $|F|+|G| = k$ for all sets $F \in \mathcal{F}_t, G \in \mathcal{G}_t$; thus properties (a) and (b) are satisfied. For a set $S \in \binom{[n]}{t}$ let $B_S$ denote the event that no set in $\mathcal{G}_t$ is disjoint from $S$. Observe that

$$\mathbb{P}(B_S) = \left(1 - \prod_{i=0}^{k} \frac{n-t-i}{n-i}\right)^{2^{k/2}} \leq \left(1 - \left(\frac{k^2-t}{k^2}\right)^k\right)^{2^{k/2}}$$
$$\leq \left(1 - \left(\frac{7}{8} - \frac{1}{8k}\right)^k\right)^{2^{k/2}} < e^{-(\sqrt{2}(\frac{7}{8}-\frac{1}{8k}))^k} < e^{-\alpha^k},$$

for some constant $\alpha$ greater than 1. Therefore, the probability that there exists $S \in \binom{n}{t}$ for which $B_S$ holds is at most $2^n e^{-\alpha^k} \leq e^{-k^2\alpha^k}$ which tends to 0 as $k$ tends to infinity. Therefore if $k$ is large enough, then we can pick a $\mathcal{G}_t$ that satisfies property (c).

Case II: $n/8 < t \leq n/2$.

Let $p := \frac{t}{n}$ and let $a$ defined via $ak = \left\lfloor \frac{-k}{2\log p} + 1 \right\rfloor$. Note that $1/8 < p \leq 1/2$ implies $1/6 \leq a \leq 1/2 + 1/k < 4/7$.

We pick $\mathcal{F}_t$ and $\mathcal{G}_t$ randomly in the following way: $\mathcal{F}_t$ is a family of $8e^8k^22^{k/2}$ sets chosen with replacement from $\binom{[n]}{ak}$, while $\mathcal{G}_t$ is a family of $e^2k^22^{k/2}$ sets chosen with replacement from $\binom{[n]}{(1-a)k}$. Clearly, $|\mathcal{F}_t| + |\mathcal{G}_t| = O(k^22^{k/2})$ and $|F| + |G| = k$ hold for any $F \in \mathcal{F}_t, G \in \mathcal{G}_t$, so properties (a) and (b) are satisfied.

We will show that with positive probability property (c) holds; therefore it is possible to pick $\mathcal{F}_t$ and $\mathcal{G}_t$ as stated in the lemma. We again use the first moment method. The probability that a given $t$-subset $T$ of $[n]$ does not contain any set of $\mathcal{F}_t$ is

$$\begin{aligned}\left(1 - \prod_{i=0}^{ak-1} \frac{p(k^2+k)-i}{k^2+k-i}\right)^{|\mathcal{F}_t|} &\leq \left(1 - \left(\frac{p(k^2+k)-k}{k^2}\right)^{ak}\right)^{|\mathcal{F}_t|} \\ &= \left(1 - \left(1 - \frac{1-p}{pk}\right)^{ak} p^{ak}\right)^{|\mathcal{F}_t|}.\end{aligned} \tag{6.1}$$

Using $1 - x \geq e^{-2x}$ if $x$ is close enough to 0, we obtain that for large enough $k$ we have $1 - \frac{1-p}{pk} \geq e^{-\frac{2(1-p)}{pk}}$. Therefore $\left(1 - \frac{1-p}{pk}\right)^{ak} \geq e^{-\frac{2a(1-p)}{p}}$, which is at least $e^{-8}$ due to $a \leq 4/7$ and $p \geq 1/8$. Plugging this back into (6.1) we obtain the following upper bound on the probability of the event that a given $t$-subset $T$ of $[n]$ does not contain any set of $\mathcal{F}_t$:

$$(1 - e^{-8}p^{ak})^{|\mathcal{F}_t|} \leq e^{-e^{-8}p^{ak}8e^8k^22^{k/2}} = e^{-8p^{ak}k^22^{k/2}}.$$

Here, using the definition of $a$, we obtain that the exponent is at most

$$-8p^{ak}k^22^{k/2} \leq -8(p^{-\log_p\sqrt{2}+1/k})^k k^22^{k/2} \leq -8pk^2 \leq -k^2$$

as $p \geq 1/8$. Therefore the expected number of $t$-sets not containing any set of $\mathcal{F}_t$ is at most $\binom{k^2+k}{t}e^{-k^2}$. This tends to 0 as $k$ tends to infinity; therefore every $t$-subset of $[n]$ contains a set in $\mathcal{F}_t$ with probability tending to 1.

Similarly we prove that with probability tending to 1 as $k$ tends to infinity, every $t$-subset of $[n]$ is disjoint from a set in $\mathcal{G}_t$. For a given $t$-set the probability that it is not disjoint from any set in $\mathcal{G}_t$ is at most

$$\left(1-\prod_{i=0}^{(1-a)k-1}\frac{(1-p)(k^2+k)-i}{k^2+k-i}\right)^{|\mathcal{G}_t|} \leq \left(1-\left(1-\frac{p}{(1-p)k}\right)^{(1-a)k}p^{(1-a)k}\right)^{|\mathcal{G}_t|}. \tag{6.2}$$

Using $1-x \geq e^{-2x}$ for $x$ close enough to 0, we obtain that for large enough $k$ we have

$$\left(1-\frac{p}{(1-p)k}\right)^{(1-a)k} \geq e^{\frac{2p(1-a)}{1-p}} \geq e^{-2}$$

as $a \geq 1/6$ and $1/8 \leq p \leq 1/2$ hold. Plugging this back into (6.2) we obtain that the probability that a given $t$-set is not disjoint from any set in $\mathcal{G}_t$ is at most

$$(1-e^{-2}(1-p)^{(1-a)k})^{|\mathcal{G}_t|} \leq e^{-e^{-2}(1-p)^{(1-a)k}e^2k^22^{k/2}} = e^{-(1-p)^{(1-a)k}k^22^{k/2}}.$$

To bound the exponent from below, note that $a \geq -\frac{\log\sqrt{2}}{\log p} \geq 1+\frac{\log\sqrt{2}}{\log(1-p)}$ holds (the first inequality holds by the definition of $a$, the second by the fact that for $0<p<1$ the function $\log p \cdot \log(1-p)$ is maximized at $p=1/2$). Therefore we obtain

$$-(1-p)^{(1-a)k}k^22^{k/2} \leq -(1-p)^{-\frac{\log\sqrt{2}}{\log(1-p)}k}k^22^{k/2} \leq -k^2,$$

and thus the expected number of $t$-subsets of $[n]$ that are not disjoint from any set of $\mathcal{G}_t$ is at most

$$\binom{k^2+k}{t}e^{-k^2} \leq 2^{k^2+k}e^{-k^2},$$

which tends to 0 as $k$ tends to infinity. So $\mathcal{G}_t$ satisfies property (c) with probability tending to 1. □

To prove the theorem, set first $n := k^2+k$ and for $1 \leq t \leq n$ let $\mathcal{F}_t$ and $\mathcal{G}_t$ be families provided by Lemma 370. For each $F \in \mathcal{F}_t \cup \mathcal{G}_t$ let $i = i_F = \min\{k-1, |F|\}$ and let us pick an $(i+1)$-chain $F_1 \subsetneq F_2 \subsetneq \cdots \subsetneq F_i \subsetneq F$ which we denote by $\mathcal{C}_F$. Let us define

$$\mathcal{S} = \{C : \exists F \in \mathcal{F}_t, C \in \mathcal{C}_F\} \cup \{[n] \setminus C : \exists G \in \mathcal{G}_t, C \in \mathcal{C}_G\}.$$

By definition, $|\mathcal{C}_F| \leq k$ holds for all $t \leq k^2+k$ and $F \in \mathcal{F}_t \cup \mathcal{G}_t$; therefore, by property (b) of Lemma 370, we obtain

$$|\mathcal{S}| \leq \sum_{t=1}^{k^2+k} k(|\mathcal{F}_t|+|\mathcal{G}_t|) = O(k^52^{k/2}).$$

We have to show that for every $T \in 2^{[n]} \setminus \mathcal{S}$ there exists a $(k+1)$-chain $\mathcal{C}_T$ in $\mathcal{S} \cup \{T\}$ that contains $T$. Let $t$ denote the size of $T$. By property (c) of Lemma 370 there exists $F \in \mathcal{F}_t$ with $F \subsetneq T$ and $G \in \mathcal{G}_t$ with $G \cap T = \emptyset$. $T \notin \mathcal{S}$ implies that $T$ cannot be the complement of a set in $\mathcal{G}_t$; therefore $T \subsetneq [n] \setminus G$ holds. So the sets of $\mathcal{C}_F \cup \{T\} \cup \{[n] \setminus C : C \in \mathcal{C}_G\}$ form a chain of size $|\mathcal{C}_F| + |\mathcal{C}_G| + 1$. By property (a) of Lemma 370, this size is at least $k+1$.

For $m > n$ let us define

$$\mathcal{S}' = \mathcal{S} \cup \{S \cup ([m] \setminus [n]) : S \in \mathcal{S}\}.$$

Now for any $T' \in 2^{[m]} \setminus \mathcal{S}'$ let us define $T = T' \cap [n]$. If $T \notin \mathcal{S}$, then, using that we have already proven that $\mathcal{S}$ is oversaturated $k$-Sperner in $2^{[n]}$, we obtain a $(k+1)$-chain $\mathcal{C}$ in $\mathcal{S} \cup \{T\}$ containing $T$. Then replacing $T$ by $T'$ and the sets $C$ of $\mathcal{C}$ that contain $T$ by $C \cup ([m] \setminus [n])$ we obtain a $(k+1)$-chain $\mathcal{C}'$ in $\mathcal{S}' \cup \{T'\}$ containing $T'$.

If $T \in \mathcal{S}$, this implies $T'' = T' \cap ([m] \setminus [n]) \neq \emptyset$, and $T'' \neq [m] \setminus [n]$ by the definition of $\mathcal{S}'$. By property (c) of Lemma 370 we obtain $F \in \mathcal{F}_t$ and $G \in \mathcal{G}_t$ with $F \subsetneq T$ and $G \cap T = \emptyset$. Note that this time $G = [n] \setminus T$ is possible. Still, as $\emptyset \neq T'' \neq [m] \setminus [n]$, the sets of $\mathcal{C}_F \cup \{T'\} \cup \{([n] \setminus C) \cup ([m] \setminus [n]) : C \in \mathcal{C}_G\}$ are all distinct and, by property (a) of Lemma 370, form a chain of size at least $k+1$. ■

Let us finish this section with a couple of related problems. First, let us mention saturating flat antichains. There are several papers [276,287,328,329] in the literature dealing with these families.

**Theorem 371 (Grütmüller, Hartmann, Kalinowski, Leck, Roberts [287])** *If $\mathcal{F} \subseteq \binom{[n]}{2} \cup \binom{[n]}{3}$ is a saturating antichain, then the following holds*

$$|\mathcal{F}| \geq \binom{n}{2} - \left\lceil \frac{(n+1)^2}{8} \right\rceil .$$

The authors of [287] also determined all antichains for which equality holds and Gerbner et al. [257] proved a stability version of this result. They also considered flat saturating antichains in general. Denoting by $sat(n,l,l+1)$ the minimum size of a saturating antichain $\mathcal{F} \subseteq \binom{[n]}{l} \cup \binom{[n]}{l+1}$, they obtained the following result. (Remember that for a $k$-uniform hypergraph $\mathcal{H}$, $\pi_k(\mathcal{H})$ denotes the Turán-density of $\mathcal{H}$.)

**Theorem 372** *For any fixed integer $l$ we have*

$$\left(1 - \frac{l-1}{l}\pi_l(\mathcal{K}^l_{l+1}) - o(1)\right)\binom{n}{l} \leq sat(n,l,l+1) \leq \left(1 - \frac{(l-1)^{l-1}}{2l^{l-1}} + o(1)\right)\binom{n}{l}.$$

Finally, let us mention one more area of extremal finite set theory where saturation problems have been examined to some extent. In Chapter 7 we introduce forbidden subposet problems. Saturation problems for induced forbidden subposets have been studied in [181], where the authors established lower and upper bounds on the saturation number of small induced subposets, but plenty of problems remain open both in the induced and non-induced cases.

# 7

# *Forbidden subposet problems*

## CONTENTS



In Chapter 3, we considered variants of Sperner's theorem, and in Chapter 4 we saw random versions of Theorem 1. In this chapter we will consider an area of extremal set theory that can be described by the poset structure of $2^{[n]}$. Any family $\mathcal{F} \subseteq 2^{[n]}$ of sets is a poset under the inclusion relation, so any property described in the language of ordered sets can be translated to a property of set families. We say that a subposet $Q'$ of $Q$ is a *(weak) copy* of $P$, if there exists a bijection $f : P \to Q'$, such that for any $p, p' \in P$, the relation $p <_P p'$ implies $f(p) <_Q f(p')$. We say that $Q'$ is a *strong* or *induced* copy of $Q$, if, in addition to this, $f(p) <_Q f(p')$ also implies $p <_P p'$. If a poset $Q$ does not contain a weak copy of $P$, then it is $P$*-free*. Katona and Tarján [344] introduced the problem of determining

$$La(n, P) = \max\{|\mathcal{F}| : \mathcal{F} \subseteq 2^{[n]} \text{ is } P\text{-free}\}.$$

More generally, for a family $\mathcal{P}$ of posets, one can consider $La(n, \mathcal{P}) = \max\{|\mathcal{F}| : \mathcal{F} \subseteq 2^{[n]}$ is $P$-free for all $P \in \mathcal{P}\}$. In this language, Theorem 4 can be stated as

$$La(n, P_{k+1}) = \sum_{i=1}^{k} \binom{n}{\lfloor \frac{n-k}{2} \rfloor + i} = \Sigma(n, k),$$

where $P_k$ is the total order/path/chain on $k$ elements. As every poset $P$ is a subposet of $P_{|P|}$, applying the above result of Erdős yields $La(n, P) \leq (|P| - 1)\binom{n}{\lfloor n/2 \rfloor}$. On the other hand, for any poset $P$ containing a comparable pair, we have $La(n, P) \geq \binom{n}{\lfloor n/2 \rfloor}$ as shown by the family $\binom{[n]}{\lfloor n/2 \rfloor}$. Therefore, it is natural to introduce $\pi(P) = \lim_{n \to \infty} \frac{La(n,P)}{\binom{n}{\lfloor n/2 \rfloor}}$. It is still an open problem to show that $\pi(P)$ exists for every poset $P$.

The following conjecture was folklore for long time, before first being published in [281].

**Conjecture 373** *For any poset $P$, let $e(P)$ denote the largest integer $k$ such that for any $j$ and $n$, the family $\cup_{i=1}^{k}\binom{[n]}{j+i}$ is $P$-free. Then $\pi(P)$ exists and is equal to $e(P)$.*

We say that Conjecture 373 *strongly holds* for a poset $P$ if $La(n,P) = \Sigma(n, e(P))$ for every $n \geq e(P)$. This way we can reformulate Theorem 4 again: it states that Conjecture 373 strongly holds for the poset $P_{k+1}$.

Let us define and enumerate some classes of posets for which Conjecture 373 was shown to be true. The following general class of posets is called *complete multi-level posets*. Let $K_{r_1,r_2,\ldots,r_s}$ denote the poset on $\sum_{i=1}^{s} r_i$ elements $a_1^1, a_2^1, \ldots, a_{r_1}^1, a_1^2, a_2^2, \ldots, a_{r_2}^2, \ldots, a_1^s, a_2^s, \ldots, a_{r_s}^s$, with $a_h^i < a_l^j$ if and only if $i < j$. Special cases include $\bigvee = K_{1,2}$ and $\bigwedge = K_{2,1}$ (this special case of Conjecture 373 was first proven by Katona and Tarján [344]), or more generally $\bigwedge_r = K_{1,r}$ and $\bigvee = K_{1,r}$ (DeBonis and Katona [120]), the $r$-broom $K_{1,1,\ldots,1,r}$ (Thanh [528]), generalized diamonds $D_r = K_{1,r,1}$ (for most values of $r$, Griggs, Li, and Lu [280]). The general class of complete multi-level posets $K_{r_1,r_2,\ldots,r_s}$ was considered by Patkós [457]. Although he stated his results only in the three-level case $K_{r,s,t}$, his method generalizes for an arbitrary number of levels: he was able to prove Conjecture 373 for some values of $r_2, r_3, \ldots, r_{s-1}$. DeBonis, Katona, and Swanepoel [121] showed that Conjecture 373 strongly holds for the butterfly poset $B = K_{2,2}$.

Griggs and Katona [277] proved the conjecture for the $N$ poset ($K_{2,2}$ with one relation removed), and Bukh [95] obtained the nicest result of the area. To state it, we need to introduce the following notion. The *Hasse diagram* $H(P)$ of a poset $P$ is a graph with vertex set $P$ and two elements $p, q \in P$ are joined by an edge if $p < q$ and there is no element $r \in P$ with $p < r < q$. We say that $P$ is a *tree poset* if its Hasse diagram is a tree. Bukh proved Conjecture 373 for tree posets. Griggs and Lu [281] and Lu [393] studied crown posets (posets on two levels such that their Hasse diagram is a cycle).

The *height* $h(P)$ of a poset $P$ is the length of the longest chain in $P$. Obviously we have $e(P) \geq h(P) - 1$. In case of tree posets, it is easy to compute the parameter $e(T)$.

**Exercise 374** *If $T$ is a tree poset, then $e(T) = h(T) - 1$ holds.*

In general, the computation of $e(P)$ is much harder. It was shown by Pálvölgyi [452], that the problem of deciding whether $e(P)$ is at most $k$ is NP-complete.

The smallest poset for which Conjecture 373 has not yet been proven is the diamond poset $D_2$. The current best upper bound on $\pi(D_2)$ (improving earlier results by Griggs, Li, Lu [280], Axenovich, Manske, and Martin [24], Kramer, Martin and Young [384]) is due to Grósz, Methuku, and Tompkins [284]. They showed $\pi(D_2) \leq 2.20711$.

## 7.1 Chain partitioning and other methods

In this section we present the proofs of most special cases when Conjecture 373 is known to be true. We show two important methods to obtain results on forbidden subposet problems. Both of them rely on counting pairs $(F, \mathcal{C})$ where $\mathcal{C}$ is a maximal chain and $F$ is a set from $\mathcal{F} \cap \mathcal{C}$, with $\mathcal{F}$ being the family the size of which we try to bound.

The first method is due to Katona. For any family $\mathcal{P}$ of finite posets let $\mathcal{Q}(\mathcal{P})$ denote the family of all finite $\mathcal{P}$-free posets. If a family $\mathcal{F} \subseteq 2^{[n]}$ of sets is $\mathcal{P}$-free, consider its comparability graph $G$, and the connected components of $G$ will simply be called the *components* of $\mathcal{F}$. Then every component of $\mathcal{F}$ is a strong copy of some $Q \in \mathcal{Q}(\mathcal{P})$. For any family $\mathcal{G} \subseteq 2^{[n]}$ let $c(\mathcal{G})$ denote the number of maximal chains $\mathcal{C} \in \mathbf{C}_n$ with $\mathcal{C} \cap \mathcal{G} \neq \emptyset$. For any poset $Q$, let $c_n^*(Q) = \min\{c(\mathcal{G}) : \mathcal{G} \subseteq 2^{[n]}$ is a strong copy of $Q\}$. For example if $Q = P_2$, then a strong copy of $P_2$ is a pair $A \subset B$. If their sizes are $a$ and $b$, then $c(\{A, B\}) = a!(n-a) + b!(n-b) - a!(b-a)!(n-b)!$. It is an exercise to see that this is minimized when $a = \lfloor n/2 \rfloor$ and $b = \lfloor n/2 \rfloor + 1$, so $c_n^*(P_2) = \lfloor n/2 \rfloor! \lceil n/2 \rceil! + (\lfloor n/2 \rfloor + 1)!(\lceil n/2 \rceil - 1)! - \lfloor n/2 \rfloor!(\lceil n/2 \rceil - 1)!$.

**Theorem 375 (Katona [341])**

$$La(n, \mathcal{P}) \leq \frac{n!}{\inf_{Q \in \mathcal{Q}(\mathcal{P})} \frac{c_n^*(Q)}{|Q|}}.$$

**Proof.** Let $\mathcal{F} \subseteq 2^{[n]}$ be a $\mathcal{P}$-free family, and for any $Q \in \mathcal{Q}(\mathcal{P})$ let $\mathcal{F}_{Q,1}, \ldots, \mathcal{F}_{Q,i_Q}$ denote the components of $\mathcal{F}$ that are strong copies of $Q$. In particular, we have $\sum_{Q \in \mathcal{Q}(\mathcal{P})} i_Q |Q| = |\mathcal{F}|$. As every maximal chain in $[n]$ can hit at most one component of $\mathcal{F}$, we have

$$\sum_{Q \in \mathcal{Q}(\mathcal{P})} \sum_{j=1}^{i_Q} c(\mathcal{F}_{Q,j}) \leq n!.$$

By definition, we have $c_n^*(Q) \leq c(\mathcal{F}_{Q,j})$ for all $j = 1, \ldots, i_Q$; therefore we have

$$\sum_{Q \in \mathcal{Q}(\mathcal{P})} i_Q c_n^*(Q) \leq n!.$$

Using $\sum_{Q \in \mathcal{Q}(\mathcal{P})} i_Q |Q| = |\mathcal{F}|$ we obtain

$$|\mathcal{F}| \inf_{Q \in \mathcal{Q}(\mathcal{P})} \frac{c_n^*(Q)}{|Q|} = \inf_{Q \in \mathcal{Q}(\mathcal{P})} \frac{c_n^*(Q)}{|Q|} \sum_{Q \in \mathcal{Q}(\mathcal{P})} i_Q |Q| \leq \sum_{Q \in \mathcal{Q}(\mathcal{P})} i_Q c_n^*(Q) \leq n!.$$

■

**Exercise 376** *Use Theorem 375 to prove that* $La(n, \{\bigvee, \bigwedge\}) = 2\binom{n-1}{\lfloor\frac{n-1}{2}\rfloor}$.

**Exercise 377** *Use Theorem 375 to prove that* $La(n, \bigvee) = (1+\frac{2}{n}+o(\frac{1}{n}))\binom{n}{\lfloor\frac{n}{2}\rfloor}$.

The second method we present (introduced by Griggs, Li and Lu in [280]) is called the *partition method* and is motivated by the proof of the LYM-inequality (Lemma 3). For a family $\mathcal{F} \subseteq 2^{[n]}$ we counted the number of pairs $(F, \mathcal{C})$, where $\mathcal{C}$ is a maximal chain in $[n]$ and $F \in \mathcal{F} \cap \mathcal{C}$. Dividing by $n!$, we obtained the average number sets in the family contained by a maximal chain, or equivalently the value of $\mathbb{E}(\mathcal{F} \cap \mathcal{C})$, when $\mathcal{C}$ is a maximal chain taken uniformly at random among all maximal chains of $[n]$. On the left hand side of the LYM-inequality, we obtained the Lubell function

$$\lambda(\mathcal{F}) := \sum_{F \in \mathcal{F}} \frac{1}{\binom{n}{|F|}}.$$

**Lemma 378** *If for a family* $\mathcal{F} \subseteq 2^{[n]}$ *we have* $\lambda(\mathcal{F}) \leq m$, *then*

$$|\mathcal{F}| \leq m\binom{n}{\lfloor n/2 \rfloor}$$

*holds. Furthermore, if* $m$ *is an integer, then* $|\mathcal{F}| \leq \Sigma(n, m)$ *holds.*

**Proof.** The statement follows from the fact that every summand of the Lubell function is at least $\frac{1}{\binom{n}{\lfloor n/2 \rfloor}}$. The furthermore part follows similarly by considering how many summands can have the value $\frac{1}{\binom{n}{\lfloor n/2 \rfloor}}$, $\frac{1}{\binom{n}{\lfloor n/2 \rfloor - 1}}$, and so on. ■

To prove Conjecture 373 for a certain poset $P$, it is enough to show that for every $P$-free family $\mathcal{F} \subseteq 2^{[n]}$, we have $\lambda(\mathcal{F}) \leq e(P) + o(1)$. However, this is not always true. For the diamond poset $D_2$ we have $e(D_2) = 2$ and Griggs, Li, and Lu [280] found the following examples: if $S \in \binom{[n]}{\lfloor n/2 \rfloor}$ and $T = [n] \setminus S$, then

- $\mathcal{F}_1(S,T) = \{\emptyset\} \cup \binom{S}{1} \cup \binom{T}{2} \cup \{F \in \binom{[n]}{2} : |F \cap S| = |F \cap T| = 1\}$.
- $\mathcal{F}_2(S,T) = \{\emptyset\} \cup \binom{S}{2} \cup \binom{T}{2} \cup \{F \in \binom{[n]}{3} : |F \cap S| = 2, |F \cap T| = 1\} \cup \{F \in \binom{[n]}{3} : |F \cap S| = 1, |F \cap T| = 2\}$.
- $\mathcal{F}_3(S,T) = \binom{[n]}{1} \cup \binom{S}{2} \cup \binom{T}{2} \cup \{F \in \binom{[n]}{3} : |F \cap S| = 2, |F \cap T| = 1\} \cup \{F \in \binom{[n]}{3} : |F \cap S| = 1, |F \cap T| = 2\}$.

These families are all $D_2$-free and $\lambda(\mathcal{F}_i) = 2 + \frac{\lfloor n/2\rfloor \cdot \lceil n/2\rceil}{n(n-1)} = 2.25 + o(1)$. Kramer, Martin, and Young [384] showed that this example is best possible; i.e. if $\mathcal{F} \subseteq 2^{[n]}$ is $D_2$-free, then $\lambda(\mathcal{F}) \le 2.25 + o(1)$ holds.

In the proof of Theorem 4 one obtains, that every maximal chain $\mathcal{C}$ can contain at most $k$ sets of a $P_{k+1}$-free family $\mathcal{F}$ by definition, and thus $\lambda(\mathcal{F}) \le k$. Usually it is more complicated than that, but one may still try to bound the value of the Lubell function by partitioning the set $\mathbf{C}_n$ of all maximal chains into blocks $\mathbf{C}^1, \mathbf{C}^2, \ldots, \mathbf{C}^m$, in such a way that the average number of sets of $\mathcal{F}$ in a random $\mathcal{C} \in \mathbf{C}^j$ is bounded. The hint to the following exercise is to consider the *max-partition* of $\mathbf{C}_n$: given a family $\mathcal{F} \subseteq 2^{[n]}$ and a set $A \subseteq [n]$, let $\mathbf{C}_A$ consist of those maximal chains $\mathcal{C} \in \mathbf{C}_n$ that contain $A$, but do not contain any superset of $A$. Every $\mathcal{C} \in \mathbf{C}_n$ belongs to at most one block $\mathbf{C}_A$, and if we add the block of those chains that do not intersect $\mathcal{F}$, we obtain a partition of $\mathbf{C}_n$.

**Exercise 379 (Griggs, Li [279])** *If $\mathcal{F} \subseteq \binom{[n]}{\ge n/4} \cap \binom{[n]}{\le 3n/4}$ is a $\bigwedge_r$-free family, then $\lambda(\mathcal{F}) \le 1 + \frac{4(r-1)}{n}$ holds.*

The *min-max-partition* of $\mathbf{C}_n$ with respect to a family $\mathcal{F} \subseteq 2^{[n]}$ is $\{\mathbf{C}_{A,B} : A \subseteq B \subseteq [n]\} \cup \{\mathbf{C}_\emptyset\}$, where $\mathbf{C}_{A,B}$ consists of those maximal chains in $\mathbf{C}_n$ which have the following property: the smallest set that belongs to $\mathcal{F}$ is $A$ and the largest set that belongs to $\mathcal{F}$ is $B$. To obtain a partition of $\mathbf{C}_n$ one has to add $\mathbf{C}_\emptyset = \{\mathcal{C} \in \mathbf{C}_n : \mathcal{C} \cap \mathcal{F} = \emptyset\}$.

**Lemma 380 (Griggs, Li, Lu, [280])** *Let $\mathcal{F} \subseteq 2^{[n]}$ be a $D_s$-free family of sets with $s \ge 2$, and let $m_s := \lceil \log_2(s+2)\rceil$. If $s \in [2^{m_s-1}-1, 2^{m_s} - \binom{m_s}{\lceil \frac{m_s}{2}\rceil} - 1]$, then $\lambda(\mathcal{F}) \le m_s$.*

**Proof.** For any pair $A \subset B \subset [n]$, we will show that the expected size of $|\mathcal{F} \cap \mathcal{C}|$ for a random chain $\mathcal{C} \in \mathbf{C}_{A,B}$ is at most $m_s$. If $|B \setminus A| \le m_s - 1$, then this is obviously true, so we may assume $|B \setminus A| \ge m_s$. We calculate $\mathbb{E}(|\mathcal{F} \cap \mathcal{C}|)$ by adding the contributions of each subset $S \in \mathcal{F} \cap [A, B]$, which is $\frac{1}{\binom{|B\setminus A|}{|S\setminus A|}}$.

Since $\mathcal{F}$ is $D_s$-free and contains both $A$ and $B$, there are at most $s-1$ subsets $S \in \mathcal{F} \cap [A, B]$ besides $A$ and $B$. Then $\mathbb{E}(|\mathcal{F} \cap \mathcal{C}|)$ is maximized if we take the $s-1$ terms with largest contribution, i.e., with minimum $\binom{|B\setminus A|}{|S\setminus A|}$. This means taking the sets $S$ closest to the ends $A$ or $B$, so those with $|S \setminus A| = 1$ or those with $|S \setminus A| = |B \setminus A| - 1$, then those with $|S \setminus A| = 2$ or those with $|S \setminus A| = |B \setminus A| - 2$, and so on. The contribution from each full level we include is then one.

Let us consider first the case $|B \setminus A| = m_s$. Then we do not take any set $S$ with $|S \setminus A| = \lfloor m_s/2 \rfloor$, as there are $s-1$ other subsets in $[A, B]$ besides $A$ and $B$. This implies the Lubell function is at most $m_s$, including the terms for $A$ and $B$. For $|B \setminus A| > m_s$, since the levels working up from $A$ or down from $B$ are larger, there are more stes in the top and bottom $m_s$ levels, thus the Lubell function is strictly less than $m_s$. ■

**Corollary 381 (Griggs, Li, Lu, [280])** *With the notation and assumptions of Lemma 380, we have $La(n, D_s) = \Sigma(n, m_s)$. In particular Conjecture 373 strongly holds for the posets $D_s$, with $s$ being in the given range.*

One may generalize the above argument to complete multi-level posets. For the sake of simplicity we only consider the three-level case $K_{r,s,t}$ with $r, t \geq 2$. The next theorem can be derived from Exercise 379 and Lemma 380.

**Theorem 382 (Patkós, [457])** *Let $r, t \geq 2$ and $s \geq 4$. If $s - 2 \in [2^{m_s-1} - 1, 2^{m_s} - \binom{m_s}{\lceil \frac{m_s}{2} \rceil} - 1]$, then $\pi(K_{r,s,t}) = e(K_{r,s,t}) = m_s + 2$ holds.*

We give one more application of the chain partition method. A *harp poset* $H(l_1, \ldots, l_k)$ consists of chains $P_{l_1}, \ldots, P_{l_k}$ with their top elements identified and their bottom elements identified. For instance, in this notation we have $D_k = H(3, \ldots, 3)$ (with $k$ 3s).

**Theorem 383 (Griggs, Li, Lu, [280])** *If $l_1 > \cdots > l_k \geq 3$, then $La(n, H(l_1, \ldots, l_k)) = \Sigma(n, l_1 - 1)$. In particular, for such harps, Conjecture 373 strongly holds.*

**Proof.** We prove by induction on $k$ that for a $H(l_1, \ldots, l_k)$-free family $\mathcal{F}$, we have $\lambda(\mathcal{F}) \leq l_1 - 1$ and then Lemma 378 will imply the theorem. In the case $k = 1$, we know that $\mathcal{F}$ is $P_{l_1}$-free, so every maximal chain contains at most $l_1 - 1$ sets of $\mathcal{F}$, and thus $\lambda(\mathcal{F}) \leq l_1 - 1$, as claimed.

Suppose the statement is true for $k - 1$ and let us consider the min-max partition of $\mathbf{C}_n$ with respect to $\mathcal{F}$. We need to show that for any pair $A \subseteq B \subseteq [n]$ we have that the conditional expected value $\mathbb{E}(|\mathcal{F} \cap \mathcal{C}| \,|\mathcal{C} \in \mathbf{C}_{A,B})$ is at most $l_1 - 1$. By definition, chains in $\mathbf{C}_{A,B}$ contain sets of $\mathcal{F}$ only in $[A, B]$. If the largest chain in $\mathcal{F} \cap [A, B]$ has length at most $l_1 - 1$, then we are done. Otherwise let $S_1 \subsetneq S_2 \subsetneq \cdots \subsetneq S_j$ be a longest chain in $(\mathcal{F} \cap [A, B]) \setminus \{A, B\}$ with $j \geq l_1 - 2$. The probability, that a random chain in $\mathbf{C}_{A,B}$ contains $S_i$, is $\frac{1}{\binom{|B \setminus A|}{|S_i \setminus A|}}$; therefore the sum of these probabilities is still smaller than 1.

On the other hand $(\mathcal{F} \cap [A, B]) \setminus \{S_1, S_2, \ldots, S_j\}$ is $H(l_2, \ldots, l_k)$-free, as a copy together with the $S_i$'s would form a copy of $H(l_1, \ldots, l_k)$. Therefore, by the inductive hypothesis, we know that $\mathbb{E}(|(\mathcal{F} \setminus \{S_1, \ldots, S_{l_1-2}\}) \cap \mathcal{C}| \,|\mathcal{C} \in \mathbf{C}_{A,B}) \leq l_2 - 1$. By linearity of expectation we obtain $\mathbb{E}(|\mathcal{F} \cap \mathcal{C}| \,|\mathcal{C} \in \mathbf{C}_{A,B}) \leq l_1 - 1$, as needed. ■

**Exercise 384 (Patkós, [458])** *Let $\mathcal{F} \subset 2^{[n]} \setminus \{\emptyset, [n]\}$ be a butterfly-free family and let $\mathcal{M}$ be defined as $\{M \in \mathcal{F} : \exists F, F' \in \mathcal{F} \quad F \subset M \subset F'\}$. Then*

$$\sum_{F \in \mathcal{F}} \frac{1}{\binom{n}{|F|}} + \sum_{M \in \mathcal{M}} \left(1 - \frac{n}{|M|(n - |M|)}\right) \frac{1}{\binom{n}{|M|}} \leq 2.$$

We finish this section with the proof of Bukh's result that verifies Conjecture 373 for tree posets.

**Theorem 385 (Bukh [95])** *Let $T$ be a tree poset. Then $\Sigma(n, h(T)-1) \leq La(n,T) \leq (h(T)-1+O(\frac{1}{n}))\binom{n}{\lfloor \frac{n}{2} \rfloor}$ holds. In particular, $\pi(T) = e(T)$ holds for any tree poset $T$.*

**Proof.** The proof consists of several lemmas. To state them, we need the following definitions. A poset $P$ is called *saturated* if all its maximal chains have length $h(P)$. A *$k$-marked chain* is a $(k+1)$-tuple $(\mathcal{C}, F_1, F_2, \ldots, F_k)$, such that $\mathcal{C}$ is a maximal chain in $2^{[n]}$ and $F_i \in \mathcal{C}$ for all $i = 1, 2, \ldots, k$. The sets $F_i$ are called the *markers* of the $k$-marked chain.

**Lemma 386** *If $\mathcal{F} \subseteq 2^{[n]}$ is a family with $\lambda(\mathcal{F}) \geq (k-1+\varepsilon)$, then there are at least $(\varepsilon/k)n!$ $k$-marked chains whose markers belong to $\mathcal{F}$.*

**Proof of Lemma.** Let $C_i$ denote the number of maximal chains that contain exactly $i$ sets from $\mathcal{F}$. Then counting the number of pairs $(F, \mathcal{C})$ with $\mathcal{C}$ being a maximal chain and $F \in F \cap \mathcal{C}$, in two different ways, we obtain

$$\sum_{i=0}^{n} iC_i = \lambda(\mathcal{F})n! \geq (k-1+\varepsilon)n!.$$

This, and $\sum_{i=0}^{n} C_i = n!$ imply

$$\sum_{i=k}^{n} iC_i \geq \sum_{i=0}^{n} iC_i - (k-1)\sum_{i=0}^{k-1} C_i \geq \varepsilon n!.$$

Therefore the number of $k$-marked chains with markers in $\mathcal{F}$ is

$$\sum_{i=k}^{n} \binom{i}{k} C_i = \sum_{i=k}^{n} \binom{i-1}{k-1} \frac{i}{k} C_i \geq \frac{1}{k} \sum_{i=k}^{n} iC_i \geq (\varepsilon/k)n!.$$

□

**Lemma 387** *Every tree poset $T$ is an induced subposet of a saturated tree poset $T'$ with $h(T) = h(T')$.*

**Proof of Lemma.** Suppose the statement of the lemma is false and for some height $h$ there is a counterexample. Let $m$ be the number of maximal chains in it; then it has at most $mh$ elements. It implies there is only a finite number of posets; thus a finite number of counterexamples with height $h$ and $m$ maximal chains in $P$. Let $T$ be a counterexample with $|T|$ being maximal.

As $T$ is not saturated, it must contain a maximal chain $v_1 < v_2 \cdots < v_r$ with $r < h$. For $i = 0, 1, \ldots, r-1$, let $T_i$ denote the poset obtained from $T$ by adding a new element $v$ with $v_i < v < v_{i+1}$ (if $i = 0$ we only require $v < v_1$).

Clearly, $T_i$ is a tree poset and contains the same number of maximal chains as $T$ for all $i = 0, 1 \ldots, r-1$. Also $T$ is an induced subposet of $T_i$. Let us assume first $h(T_i) = h(T)$. If $T_i$ is an induced subposet of a saturated tree poset, then $T$ is also an induced subposet of that one, contradicting our assumption of $T$ being a counterexample. Hence $T_i$ is not an induced subposet of a saturated tree poset, but then it is a larger counterexample than $T$, a contradiction again.

Thus we can assume $h(T_i) > h(T)$ for every $i = 0, 1, \ldots, r-1$. We prove by induction on $i$ that $T$ contains a chain of length $h(T) - i$ starting at $v_{i+1}$ for every $i = 0, 1, \ldots, r-1$. If $i = 0$ then $T_0$ contains a chain of length $h(T)+1$ and this chain must start at $v$. As we only required the relation $v < v_1$, this chain must continue with $v_1$, so there exists a chain of length $h(T)$ starting at $v_1$. Suppose the statement is true for $i \leq j-1$; then there exists a chain of length $h(T) - j + 1$ starting at $v_j$. In particular, all chains *ending* at $v_j$ must have length at most $j$. So the chain of length $h(T)+1$ in $T_j$ must contain $v$, $v_j$, and $v_{j+1}$, and as the part till $v_j$ has length at most $j$, the part starting at $v_{j+1}$ must have length at least $h(T)-j$, as claimed. We obtained $T$ contains a chain of length $h(T) - r + 1 \geq 2$ starting at $v_r$, which contradicts the maximality of the chain $v_1 < v_2 < \cdots < v_r$ in $T$. □

**Lemma 388** *If $T$ is a saturated tree poset that is not a chain, then there exists $t \in T$ that is a leaf in $H(T)$ and there exists an interval $I \subset T$ containing $t$ such that $|I| < h(T)$ holds, and $T \setminus I$ is a saturated tree poset with $h(T) = h(T \setminus I)$.*

**Proof of Lemma.** In addition to the Hasse diagram $H(T)$, we will also use the comparability graph in this proof. Consider two leaves $t, t'$ in $T$ that maximize the graph distance in the comparability graph of $T$ and let $t = t_1, t_2, \ldots, t_r = t'$ be a shortest path between them. As $T$ is not a chain, we have $r > 2$. We may suppose that $t < t_2$. Note that by the assumption on $t$ and $t'$, all elements $t^*$ in $[t, t_2)$ have exactly one neighbor in $H(T)$ that is larger than $t^*$. Let $I \subseteq [t, t_2)$ be the largest interval that contains $t$, such that all other elements of $I$ have degree 2 in $H(T)$. So the smallest element $d$ of $[t, t_2]$ that does not belong to $I$ is either $t_2$. or have degree at least 3 in $H(T)$. In that case $d$ is larger than another leaf $t'' \neq t$. Note also that elements of $I$ are only comparable to elements of $\{f \in T : d \leq f\} \cup I$.

Let $M$ be a maximal chain in $T \setminus I$. We need to prove that $M$ is also maximal in $T$. If $M$ contains only elements from $\{f \in T : d \leq f\}$, then $M$ is not maximal in $T \setminus I$, as $t''$ could be added to $M$. If $M$ contains an element not in $\{f \in T : d \leq f\}$, then this element is incomparable to all elements of $I$, so indeed $M$ is maximal in $T$. □

**Lemma 389** *Let $T$ be a saturated tree poset of height $k$. Suppose $\mathcal{F} \subseteq 2^{[n]}$ is a set family, such that no chain contains more than $K$ sets from $\mathcal{F}$, and all the sets in $\mathcal{F}$ are of size between $n/4$ and $3n/4$. Moreover, suppose $\mathcal{L}$ is a* *family of k-marked chains with markers in $\mathcal{F}$, with $|\mathcal{L}| > \frac{\binom{|T|+1}{2}4^{K+1}}{n}n!$. Then there is a copy of T in $\mathcal{F}$ that contains only markers of some k-marked chains in $\mathcal{L}$.*

**Proof of Lemma.** We proceed by induction on $|T|$. If $T$ is a chain, then the $k$ markers of any $k$-marked chain contain a copy of $T$. In particular, it gives the base case of the induction. So suppose $T$ is not a chain. Then applying Lemma 388, there exists a leaf $t$ in $T$ and interval $I \subseteq T$ containing $t$ such that $h(T \setminus I) = k$ and $T \setminus I$ is a saturated tree poset. Our aim is to use induction to obtain a copy of $T \setminus I$ in $\mathcal{F}$ that can be extended to a copy of $T$. Finding a copy of $T \setminus I$ is immediate, but in order to be able to extend it, we need a copy satisfying some additional properties, described later.

By passing to the opposite poset of $T$ and considering $\overline{\mathcal{F}} = \{\overline{F} : F \in \mathcal{F}\}$, we may assume that $t$ is a minimal element of $T$. There exists a maximal chain $C$ in $T$ that contains $I$, and we have $|C| = k$ as $T$ is saturated. Then $s := |C \setminus I| = k - |I| \geq 1$.

We need several definitions. A chain $F_1 \supset F_2 \supset \cdots \supset F_s$ is a *bottleneck* if there exists a family $\mathcal{S} \subset \mathcal{F}$ with $|\mathcal{S}| < |T|$ such that for every $k$-marked chain $(\mathcal{C}, F_1, F_2, \ldots, F_s, F_{s+1}, \ldots, F_k)$ we have $\mathcal{S} \cap \{F_{s+1}, \ldots, F_k\} \neq \emptyset$. Such an $\mathcal{S}$ is a *witness* to the fact that $F_1, \ldots, F_s$ is a bottleneck (and we assume all sets of the witness are contained in $F_s$). We say that a $k$-marked chain $(\mathcal{C}, F_1, F_2, \ldots, F_k)$ is *bad* if $F_1, \ldots, F_s$ is a bottleneck. A $k$-marked chain is *good* if it is not bad. Observe that if there is a copy $\mathcal{F}_{T \setminus I}$ of $T \setminus I$ consisting of markers of good $k$-marked chains, then we can extend $\mathcal{F}_{T \setminus I}$ to a copy of $T$. Indeed, as the sets $F'_1, \ldots, F'_s$ representing $C \setminus I$ in $\mathcal{F}_{T \setminus I}$ do not form a bottleneck and $|\mathcal{F}_{T \setminus I}| < |T|$, there must be a good $k$-marked chain $(\mathcal{C}, F'_1, \ldots, F'_s, F'_{s+1}, \ldots, F'_k)$ such that $F'_{s+1}, \ldots, F'_k \notin \mathcal{F}_{T \setminus I}$; therefore $\mathcal{F}_{T \setminus I} \cup \{F'_{s+1}, \ldots, F'_k\}$ is a copy of $T$. Therefore all we need to prove is that there are enough good $k$-marked chains to obtain a copy of $T \setminus I$ by induction.

We count the number of bad $k$-marked chains in the following way. Consider $R \in \binom{[K]}{s}$ with elements $r_1 < r_2 < \cdots < r_s$. If $\mathcal{C} \in \mathbf{C}_n$ contains at least $r_s$ sets from $\mathcal{F}$, then let $F_1, \ldots, F_{r_s}$ denote the largest $r_s$ of these sets. The subchain $F_{r_1} \supset F_{r_2} \supset \cdots \supset F_{r_s}$ is denoted by $C_R(\mathcal{C})$. We say that $\mathcal{C}$ is *R-bad* if $C_R(\mathcal{C})$ is a bottleneck and there exists a $k$-marked chain in $\mathcal{L}$ with maximal chain $\mathcal{C}$ and top $s$ markers being $C_R(\mathcal{C})$.

Let us bound the number of $R$-bad chains. If $|\mathcal{C} \cap \mathcal{F}| < r_s$, then $\mathcal{C}$ cannot be $R$-bad. We partition chains in $\mathbf{C}_n$ according to their $r_s$th set $F_{r_s}$ from $\mathcal{F}$. As $C_R(\mathcal{C})$ must be a bottleneck, there is a witness $\mathcal{S}$ to this fact. This means that if $\mathcal{C}$ is $R$-bad, then $\mathcal{C}$ must meet $\mathcal{S}$ whose elements are all contained in $F_{r_s}$. But as $|\mathcal{S}| < |T|$ and all sets of $F_{r_s}$ have size between $n/4$ and $3n/4$, the proportion of those chains that do meet $\mathcal{S}$ is at most $4|T|/n$ (any proper non-empty subset of $F_{r_S}$ is contained in at most $1/|F_{r_s}|$ proportion of chains going through $F_{r_s}$). This holds independently of the choice of $F_{r_s}$; thus the number of $R$-bad chains is at most $\frac{4|T|}{n}n!$. There are $\binom{K}{s}$ many ways to choose $R$. Every chain that is $R$-bad for some $R$ can give rise to at most $\binom{K}{s}$ bad

$k$-marked chains; therefore the total number of bad $k$-marked chains is at most $\frac{4|T|\binom{K}{s}^2}{n}n! \leq \frac{|T|4^{K+1}}{n}n!$. So the number of good chains is at least

$$|\mathcal{L}| - \frac{|T|4^{K+1}}{n}n! \geq \frac{(\binom{|T|+1}{2} - |T|)4^{K+1}}{n}n! = \frac{\binom{|T|}{2}4^{K+1}}{n}n!.$$

As $|T \setminus I| < |T|$, the induction hypothesis implies the existence of a copy of $T \setminus I$ among the markers of good $k$-marked chains as required. □

Now we are ready to prove Theorem 385. By Lemma 387 we may suppose that $T$ is a saturated tree poset. Assume $\mathcal{F} \subseteq 2^{[n]}$ is a $T$-free family that contains more than $(h(T) - 1 + \frac{h(T)|T|^2 4^{|T|+2}}{n})\binom{n}{\lfloor n/2 \rfloor}$ sets. Let us define $\mathcal{F}' := \{F \in \mathcal{F} : |F - n/2| \leq n/4\}$. Note that $|\binom{[n]}{\leq n/4} \cup \binom{[n]}{\geq 3n/4}| \leq c^n$, where $c = -\frac{1}{4}\log\frac{1}{4} - \frac{3}{4}\log\frac{3}{4} < 2$ and therefore, assuming $n$ is large enough, we have $|\mathcal{F}'| \geq (h(T) - 1 + \frac{h(T)|T|^2 4^{|T|+1}}{n})\binom{n}{\lfloor n/2 \rfloor}$. As $\mathcal{F}'$ is $T$-free, it is also $P_{|T|}$-free.

Let $\varepsilon = h(T)|T|^2 4^{|T|+1}/n$. Then we can apply Lemma 386 to find $|T|^2 4^{T+1} n!/n$ $k$-marked chains. Then we can apply Lemma 389 with $K = |T|$ and $k = h(T)$, to obtain a copy of $T$ in $\mathcal{F}'$, contradicting the $T$-free property of $\mathcal{F}'$. ■

## 7.2 General bounds on $La(n, P)$ involving the height of $P$

In this section we bound $La(n, P)$ as a function of the size and the height of $P$. All proofs involve bounds on the Lubell function, obtained by an averaging argument over some subfamilies of $2^{[n]}$. Often, once the definition is given, the proofs are relatively standard; therefore we will post some of the details as exercises.

We start with a result of Burcsi and Nagy [97] that states that $\pi(P)$, if exists, is at most the arithmetic mean of the size and the height of $P$. The key ingredients for the proof are the following two definitions.

A *double chain* $\mathcal{D}$ in $2^{[n]}$ is an ordered pair $(\mathcal{C}, \mathcal{C}')$ of two maximal chains such that if $\mathcal{C} = \{C_0, C_1, \ldots, C_n\}$, then $\mathcal{C}' = \{\emptyset, C'_1, C'_2, \ldots, C'_{n-1}, [n]\}$ with $C'_i = C_{i-1} \cup (C_{i+1} \setminus C_i)$.

The *infinite double chain* is an infinite poset with elements $\{l_i, r_i : i \in \mathbb{Z}\}$ and $l_i < l_{i+1}, l_i < r_{i+1}, r_i < l_{i+1}$.

Note that any double chain in $2^{[n]}$ is a subposet of the infinite double chain. The following exercise contains two lemmas of the proof of the next theorem, due to Burcsi and Nagy [97].

**Exercise 390** ***(i)*** *If, for a family* $\mathcal{F} \subseteq 2^{[n]}$ *and positive real* $m$, *we have* $\sum_{\mathcal{D}} |\mathcal{F} \cap \mathcal{D}| \leq 2mn!$ *(where the summation goes over all the double chains in* $2^n$*), then* $\lambda(\mathcal{F}, n) \leq m$ *holds.*

***(ii)*** *Let* $m$ *be an integer or half of an integer and* $P$ *be a finite poset. Assume that any subposet of size* $2m+1$ *of the infinite double chain contains* $P$*. If* $\mathcal{F} \subseteq 2^{[n]}$ *is a* $P$*-free family, then* $\lambda(\mathcal{F}, n) \leq m$ *holds.*

**Theorem 391 (Burcsi, Nagy, [97])** *For any poset* $P$*, we have*

$$La(n, P) \leq \left(\frac{|P| + h(P)}{2} - 1\right)\binom{n}{\lfloor n/2 \rfloor}.$$

**Proof.** By Exercise 390 **(ii)**, it is enough to show that any finite poset $P$ is contained in any subposet $S$ of the infinite double chain, provided $|S| \geq |P| + h(P) - 1$. We prove this by induction on $h(P)$. If $h(P) = 1$, then $P$ is an antichain and is contained by any subposet of size at least $|P|$.

To prove the inductive step, let $P$ be a poset with $h(P) = h$ and partition $P = M \cup P'$ with $M$ being the minimal elements of $P$ and $P' = P \setminus M$. Suppose $S$ is a subposet of the infinite double chain with $|S| \geq |P| + h - 1$. Let $S_M$ be the first $m = |M|$ elements of $S$ in the following natural arrangement of the infinite double chain:

$$\ldots, l_{-2}, r_{-2}, l_{-1}, r_{-1}, l_0, r_0, l_1, r_1, l_2, r_2, \ldots$$

Observe that all elements of $S_M$ are smaller than all elements of $S \setminus S_M$ with one possible exception (the least $r_j$ after the last element of $S_M$). Therefore removing this possible exception from $S \setminus S_M$, we are left with a subposet $S'$ of the infinite double chain, with size at least $|P| + h - 1 - m - 1 = |P'| + (h-1) - 1$. As by definition $h(P') = h(P) - 1 = h - 1$, the inductive hypothesis yields a copy of $P'$ in $S'$. Clearly, this copy together with $S_M$ forms a copy of $P$. ■

Chen and Li [101] generalized the concept of double chain. They introduced the notion of *m-linkage*. Let $C_{0,0} \subsetneq C_{1,0} \subsetneq \cdots \subsetneq C_{n,0}$ be a maximal chain and for pairs $i, j$ with $m \leq i \leq n - m$ and $1 \leq j \leq m$ let $C_{i,j} = C_{i-1,0} \cup (C_{i+j,0} \setminus C_{i+j-1,0})$. If $m \leq n/2$ then these sets form an $m$-linkage $\mathcal{L}^{(m)} \subseteq 2^{[n]}$. As the $m$-linkage is determined by its *main chain*, the number of $m$-linkages in $2^{[n]}$ is $n!$.

**Exercise 392** *For a family* $\mathcal{F} \subseteq 2^{[n]}$*, the number of pairs* $(F, \mathcal{L}^{(m)})$ *with* $\mathcal{L}^{(m)}$ *being an* $m$*-linkage and* $F \in \mathcal{F} \cap \mathcal{L}^{(m)}$ *is*

$$\sum_{|F|<m \text{ or } |F|>n-m} |F|!(n - |F|)! + \sum_{m \leq |F| \leq n-m} (m+1)|F|!(n - |F|)!.$$

We define the lexicographic ordering $\prec$ on the sets of an $m$-linkage, i.e. $C_{i,j} \prec C_{i',j'}$ if either $i < i'$ or $i = i'$ and $j < j'$.

**Lemma 393** *Let $\mathcal{S}$ and $\mathcal{L}$ be two subfamilies of an $m$-linkage $\mathcal{L}^{(m)}$ such that for any $L \in \mathcal{L}$ an $S \in \mathcal{S}$ we have $L \prec S$, but for any $L \in \mathcal{L}$ there exists an $S \in \mathcal{S}$ with $L \not\subset S$. Then $|\mathcal{L}| \le \frac{1}{2}(m^2+m)+m$.*

**Proof.** Without loss of generality we can assume that $\mathcal{S} = \{S \in \mathcal{L}^{(m)} : C_{i,j} \preceq S\}$ for some $C_{i,j} \in \mathcal{L}^{(m)}$. Note that $C_{i-1,0}$ is contained in any set of $\mathcal{S}$ and $C_{i',j'} \subseteq C_{i-1,0}$ if and only if $i'+j' \le i-1$. The lemma now follows by observing that there are $j \le m$ sets $C_{i,h}$ with $h < j$ (these are not contained in $C_{i,j}$) and for any $i-m \le i' \le i-1$ there are $m-(i-1-i')$ sets $C_{i',j}$ with $C_{i',j} \not\subset C_{i-1,0}$. ■

**Lemma 394** *Let $\mathcal{F} \subseteq 2^{[n]}$ be a $P$-free family of sets. Then for any $m$-linkage $\mathcal{L}^{(m)}$ we have*

$$|\mathcal{F} \cap \mathcal{L}^{(m)}| \le |P| + \frac{1}{2}(m^2+3m-2)(h(P)-1)-1.$$

**Proof.** Let us suppose that $|\mathcal{F} \cap \mathcal{L}^{(m)}| \ge |P| + \frac{1}{2}(m^2+3m-2)(h(P)-1)$ holds. We will find a copy of $P$ in $\mathcal{F} \cap \mathcal{L}^{(m)}$. First, we partition the poset $P$ into antichains $A_1, A_2, \dots, A_{h(P)}$ such that for all $1 \le i \le h(P)$ the antichain $A_i$ consists of the maximal elements in $P \setminus (\cup_{j<i} A_j)$. We find a copy of $P$ in $h(P)$ rounds as follows: we set $\mathcal{M}_0 = \mathcal{F} \cap \mathcal{L}^{(m)}$ and let $\mathcal{F}_1 = \{F_{1,1} \prec F_{1,2} \prec \dots \prec F_{1,|A_1|}\}$ be the family of the largest sets of $\mathcal{F} \cap \mathcal{L}^{(m)}$ in the lexicographic order. The elements of $A_1$ are mapped to $\mathcal{F}_1$ to create this part of the copy of $P$. By Lemma 393, we obtain that for $\mathcal{R}_1 := \{F \in \mathcal{F} \cap \mathcal{L}^{(m)} : F \prec \mathcal{F}_1, \exists F' \in \mathcal{F}_1 \;\; F \not\subset F'\}$ we have $|\mathcal{R}_1| \le \frac{1}{2}(m^2+m)+m-1$. We set $\mathcal{M}_1 = (\mathcal{F} \cap \mathcal{L}^{(m)}) \setminus (\mathcal{F}_1 \cup \mathcal{R}_1)$.

Suppose the images $\mathcal{F}_1 \cup \dots \cup \mathcal{F}_j$ of sets in $A_1 \cup \dots \cup A_j$ have already been defined in a relation preserving way and $\mathcal{M}_j$ has been defined as $\{M \in \mathcal{F} \cap \mathcal{L}^{(m)} : \forall i \le j, F \in \mathcal{F}_i \;\; M \subset F\}$. Then we let $\mathcal{F}_{j+1} = \{F_{j+1,1} \prec F_{j+1,2} \prec \dots \prec F_{j+1,|A_{j+1}|}\}$ be the family of the largest sets of $\mathcal{M}_j$ in the lexicographic order. The elements of $A_{j+1}$ are mapped to $\mathcal{F}_{j+1}$ and we set $\mathcal{R}_{j+1} = \{F \in \mathcal{F} \cap \mathcal{L}^{(m)} : F \prec \mathcal{F}_{j+1}, \exists F' \in \mathcal{F}_{j+1} \;\; F \not\subset F'\}$. By Lemma 393, we have $|\mathcal{R}_{j+1}| \le \frac{1}{2}(m^2+m)+m-1$ and finally we let $\mathcal{M}_{j+1} = \mathcal{M}_j \setminus (\mathcal{F}_{j+1} \cup \mathcal{R}_{j+1})$.

As $\sum_{i=1}^{h(P)} |\mathcal{F}_i| = |P|$ and $\sum_{i=1}^{h(P)-1} |\mathcal{R}_i| \le (h(P)-1)\frac{1}{2}(m^2+3m-2)$ hold, we are able to define the $\mathcal{F}_i$'s for all $i = 1, 2, \dots, h(P)$. Therefore we can indeed find a copy of $P$ in $\mathcal{F} \cap \mathcal{L}^{(m)}$. ■

**Theorem 395 (Chen and Li, [101])** *For any poset $P$ and fixed integer $m \ge 1$, we have*

$$La(n,P) \le \frac{1}{m+1}\left(|P| + \frac{1}{2}(m^2+3m-2)(h(P)-1)-1\right)\binom{n}{\lfloor n/2 \rfloor}$$

*if $n$ is large enough. In particular, letting $m = \sqrt{\frac{|P|}{h(P)}}$, one obtains*

$$La(n,P) = O(\sqrt{|P|h(P)})\binom{n}{\lfloor n/2 \rfloor}.$$

**Proof.** We count the pairs $(F, \mathcal{L}^{(m)})$ in two ways, where $\mathcal{L}^{(m)}$ is an $m$-linkage in $2^{[n]}$ and $F \in \mathcal{F} \cap \mathcal{L}^{(m)}$. Using Exercise 392 and Lemma 394, and dividing by $n!$, we obtain

$$\sum_{|F|<m \text{ or } |F|>n-m} \frac{1}{\binom{n}{|F|}} + \sum_{m \leq |F| \leq n-m} \frac{(m+1)}{\binom{n}{|F|}} \leq |P| + \frac{1}{2}(m^2+3m-2)(h(P)-1)-1.$$

If $n \geq 2m^2$ holds, then $\frac{1}{\binom{n}{m}} \geq \frac{m+1}{\binom{n}{m+1}}$, so the left hand side of the above inequality is at least $\frac{(m+1)|\mathcal{F}|}{\binom{n}{\lfloor n/2 \rfloor}}$. This proves the theorem. ■

Further improvements were proved by Grósz, Methuku, and Tompkins. We do not include their proof as it is more involved and lengthier then the previous ones, yet it uses the same method of bounding the size of a $P$-free family in a simpler substructure of $2^{[n]}$ followed by an averaging argument. The structure they introduce and examine is the *k-interval chain*. Let us remind the reader that the interval $[A, B]$ for two sets $A \subset B$ is $\{C : A \subseteq C \subseteq B\}$. For a maximal chain $\mathcal{C} = \{C_0, C_1, \ldots, C_n\}$, let the $k$-interval chain $\mathcal{C}_k$ be $\bigcup_{i=0}^{n-k} [C_i, C_{i+k}]$.

**Theorem 396 (Grósz, Methuku, and Tompkins, [284])** *For any poset $P$ and fixed integer $k \geq 2$ we have*

$$La(n, P) \leq \frac{1}{2^{k-1}} \left(|P| + (3k-5)2^{k-2}(h(P)-1)-1\right) \binom{n}{\lfloor \frac{n}{2} \rfloor}.$$

*Choosing $k$ appropriately one obtains*

$$La(n, P) = O\left(h(P) \log\left(\frac{|P|}{h(P)} + 2\right)\right) \binom{n}{\lfloor \frac{n}{2} \rfloor}.$$

Let us note that $h(P)\binom{n}{\lfloor \frac{n}{2} \rfloor}$ is a trivial lower bound, while $|P|\binom{n}{\lfloor \frac{n}{2} \rfloor}$ is a trivial upper bound on $La(n, P)$. Theorem 391 shows their arithmetic mean is an upper bound. Theorem 395 improves it to geometric mean. Theorem 396 further improves it: $|P|$ is only in the logarithmic term. Using $|P|$ and $h(P)$ further improvement is not possible.

**Proposition 397 (Grósz, Methuku, and Tompkins, [284])** *For $P = K_{r,\ldots,r}$ we have*

$$La(n, P) \geq ((h(P)-2)\log r)\binom{n}{\lfloor \frac{n}{2} \rfloor} = \left((h(P)-2)\log \frac{|P|}{h(P)}\right)\binom{n}{\lfloor \frac{n}{2} \rfloor}.$$

## 7.3 Supersaturation

In this short section we gather known supersaturation results (without proof) for forbidden subposet problems. These minimize the number of copies of $P$ over all families $\mathcal{F} \subseteq 2^{[n]}$ with $|\mathcal{F}| = La(n,P)+X$ for some positive integer $X$. The prototype result of this sort is Theorem 172, the supersaturation version of Sperner's theorem by Kleitman. He proved that the number of 2-chains in $\mathcal{F}$ is minimized at *centered* families: if $F \in \mathcal{F}$, then for all $G \in 2^{[n]}$ with $||G|-n/2| < ||F|-n/2|$ we have $G \in \mathcal{F}$.

Kleitman [366] conjectured that the same holds for $k$-chains. This was first confirmed for $X \leq \Sigma(n,k) - \Sigma(n,k-1)$ by Das, Gan, Sudakov [115] and Dove, Griggs, Kang, Sereni [135]. Balogh and Wagner [48], improving on ideas of Das, Gan, and Sudakov, proved Kleitman's conjecture almost in full. Finally, Samotij [494] verified Kleitman's conjecture by proving a general statement about posets satisfying certain conditions.

**Theorem 398 (Samotij [494])** *For every $k \leq n$ and $M$, amongst families $\mathcal{F} \subseteq 2^{[n]}$ of size $M$, the number of $k$-chains in $\mathcal{F}$ is minimized by centered families.*

There is one more supersaturation result that we are aware of in the area of forbidden subposet problems. Patkós considered minimizing the number of copies of the butterfly poset $B$ in families of size $La(n,B)+X = \Sigma(n,2)+X$. He first used Lovász's version of the shadow theorem (Theorem 27), to obtain a stability result on 2-Sperner ($P_3$-free) families. Combining it with Exercise 384, he obtained the following stability theorem on butterfly-free families.

**Theorem 399 ( [458])** *Let $m$ be a non-negative integer with $m \leq \binom{\frac{2n}{3}-1}{\lceil n/2 \rceil}$ and let $\mathcal{F} \subseteq 2^{[n]}$ be a butterfly-free family such that $|\mathcal{F} \setminus \mathcal{F}^*| \geq m$ for every $\mathcal{F}^* \in \Sigma^*(n,2)$. Then the inequality $|\mathcal{F}| \leq \Sigma(n,2) - \frac{m}{4}$ holds if $n$ is large enough.*

Finally, he obtained the following supersaturation result. One can check that if one adds a new set to a family $\mathcal{F}$ in $\Sigma^*(n,2)$, then at least $f(n) = (\lceil n/2 \rceil + 1)\binom{\lceil n/2 \rceil}{2}$ many copies of butterflies are created. (Actually, these are special copies: either $a,b < c < d$ or $a < b < c,d$ hold for their elements.) The theorem states that this is best possible if $X$ is very small, and asymptotically best possible if $X$ is somewhat larger, but still much smaller than what was considered in Theorem 398.

**Theorem 400 ( [458])** *Let $f(n) = (\lceil n/2 \rceil + 1)\binom{\lceil n/2 \rceil}{2}$. Let $\mathcal{F} \subseteq 2^{[n]}$ be a family of sets with $|\mathcal{F}| = \Sigma(n,2)+X$.*

*(i) If $X = X(n)$ satisfies $\log X = o(n)$, then the number of butterflies contained by $\mathcal{F}$ is at least $(1-o(1))X \cdot f(n)$.*

***(ii)*** *Furthermore, if* $X \leq \frac{n}{100}$*, then the number of butterflies contained by* $\mathcal{F}$ *is at least* $X \cdot f(n)$.

## 7.4 Induced forbidden subposet problems

Recall that we say that a subposet $Q'$ of $Q$ is an *induced copy* of $P$ if there exists a bijection $f : P \to Q$, such that for any $p, p' \in P$ we have $p <_P p'$ if and only if $f(p) <_Q f(p')$. If a poset $Q$ does not contain an induced copy of $P$, then it is *induced* $P$*-free*. Similarly to $La(n, P)$ we define

$$La^*(n, P) = \max\{|\mathcal{F}| : \mathcal{F} \subseteq 2^{[n]} \text{ is induced } P\text{-free}\}.$$

and $\pi^*(P) = \lim_{n\to\infty} \frac{La^*(n,P)}{\binom{n}{\lfloor n/2 \rfloor}}$. Just as in the non-induced case, it is not known whether $\pi^*(P)$ exists for every poset $P$.

One can formulate the following analogue of Conjecture 373:

**Conjecture 401** *For any poset* $P$ *let* $e^*(P)$ *denote the largest integer* $k$ *such that for any* $j$ *and* $n$ *the family* $\cup_{i=1}^{k} \binom{[n]}{j+i}$ *is induced* $P$*-free. Then* $\pi^*(P)$ *exists and is equal to* $e^*(P)$.

Conjecture 401 was first proved for the poset $\bigvee_r$ in [99] and [341], while Patkós [457] verified Conjecture 401 for complete multi-level posets if the sizes of different levels satisfy some conditions. Here we present two more applications of the chain partition method. They are both based on the following lemma.

**Lemma 402 (Patkós, [457])** *Let* $s$ *and* $k$ *be natural numbers and* $\mathcal{G} \subseteq 2^{[k]}$ *be a family not containing* $s$ *sets of the same size.*

***(i)*** *If* $s$ *is fixed and* $k$ *tends to infinity, then* $\lambda(\mathcal{G}) \leq 2 + \frac{2(s-1)}{k} + O(1/k^2)$. *Furthermore, if* $\mathcal{G}$ *contains only sets of size from* $[\lceil k/2 \rceil, k-1]$*, then* $\lambda(\mathcal{G}) \leq \frac{(s-1)}{k} + O(1/k^2)$.

***(ii)*** *Let* $m_s^* = \min\{m : s \leq \binom{m}{\lfloor m/2 \rfloor}\}$*. There exists* $s_0$ *such that if* $s \geq s_0$*, then* $\lambda(\mathcal{G}) \leq m_s^* + 1$ *independently of* $k$.

***(iii)*** *For any constant* $c$ *with* $1/2 < c < 1$*, there exists an integer* $s_c$*, such that if* $s > s_c$ *and* $s \leq c\binom{m_s^*}{\lfloor m_s^*/2 \rfloor}$*, then* $\lambda(\mathcal{G}) \leq m_s^*$.

***(iv)*** *If* $s = 4$*, then* $\lambda(\mathcal{G}) \leq 4$.

**Proof.** As every $i$-set in $\mathcal{G}$ gives a summand $\frac{1}{\binom{k}{i}}$ to $\lambda(\mathcal{G})$, we have

$$\lambda(\mathcal{G}) \leq \sum_{i=0}^{k} \frac{\min\{s-1, \binom{k}{i}\}}{\binom{k}{i}} =: R_s(k).$$

If $k \geq s-1$, then $R_s(k) = 2+\frac{2(s-1)}{k}+\frac{2(s-1)}{\binom{k}{2}}+\sum_{i=3}^{k-3}\frac{s-1}{\binom{k}{i}} \leq 2+\frac{2(s-1)}{k}+\frac{2(s-1)}{\binom{k}{2}}+\frac{2k(s-1)}{\binom{k}{3}} = 2+\frac{2(s-1)}{k}+O(1/k^2)$, which proves the first part of **(i)**. If all sets of $\mathcal{G}$ have size from $[\lceil k/2\rceil, k-1]$, then $\lambda(\mathcal{G}) \leq \sum_{i=\lceil k/2\rceil}^{k-1}\frac{\min\{s-1,\binom{k}{i}\}}{\binom{k}{i}} = R_s(k)/2-1$.

As $R_s(k) \leq k+1$, the statement of **(ii)** is trivial if $k \leq m_s^*$. We will prove $R_s(k) \geq R_s(k+1)$ if $s$ is large enough and $k \geq m_s^*$, which will imply **(ii)**. $R_s(k)$ contains $k+1$ summands, while $R_s(k+1)$ contains $k+2$ summands. Observe that $\frac{1}{\binom{k}{i}} \geq \frac{1}{\binom{k+1}{i}}$ and $\frac{1}{\binom{k}{k-i}} \geq \frac{1}{\binom{k+1}{k+1-i}}$, and therefore the sum of the first $\lceil k/2\rceil-2$ summands of $R_s(k)$ is at least the sum of those in $R_s(k+1)$, and similarly in the case of the last $\lfloor k/2\rfloor - 1$ summands. Therefore, it is enough to compare the sum of the middle summands and show the inequality

$$\sum_{j=-1}^{1}\frac{\min\{s-1,\binom{k}{\lceil k/2\rceil+j}\}}{\binom{k}{\lceil k/2\rceil+j}} \geq \sum_{j=-1}^{2}\frac{\min\{s-1,\binom{k}{\lceil k/2\rceil+j}\}}{\binom{k+1}{\lceil k/2\rceil+j}}. \tag{7.1}$$

If $s$ is large enough, then so is $m_s^*$ and $k$. Therefore $s-1 < \binom{k}{\lceil k/2\rceil} = (1+o(1))\binom{k}{\lceil k/2\rceil+\alpha} = \frac{1}{2}\binom{k+1}{\lceil k/2\rceil+\alpha'}$ for any constants $\alpha,\alpha'$. This shows that the left hand side of (7.1) is $(3+o(1))\frac{s-1}{\binom{k}{\lceil k/2\rceil}}$ while the right hand side of (7.1) is $(2+o(1))\frac{s-1}{\binom{k}{\lceil k/2\rceil}}$.

In view of the above monotonicity, to prove **(iii)**, all we need to show is $R_s(m_s^*) \leq m_s^*$. We use that if $s$ and $m_s^*$ are large enough, then $\binom{m_s^*}{\lfloor m_s^*/2\rfloor} = (1+o(1))\binom{m_s^*}{\lfloor m_s^*/2\rfloor+j}$ if $|j| \leq \sqrt{m_s^*}/\log m_s^*$. By the assumption on $s$ for every such $j$ we have $\frac{s-1}{\binom{m_s^*}{\lfloor m_s^*/2\rfloor+j}} \leq \frac{1+c}{2} < 1$. But then

$$R_s(m_s^*) \leq m_s^*+1-\sum_{j=-\lfloor\sqrt{m_s^*}/\log m_s^*\rfloor}^{\lceil\sqrt{m_s^*}/\log m_s^*\rceil}\left(\frac{s-1}{\binom{m_s^*}{\lfloor m_s^*/2\rfloor+j}}-1\right) \leq$$

$$m_s^*+1-\sqrt{m_s^*}/\log m_s^*\frac{1+c}{2} \leq m_s^*$$

when $m_s^*$ is large enough.

Finally, if $s=4$, then for $k=2,3,4,5$ the sum $R_4(k)$ equals 3.5, 4, 4, 3.8, respectively. It is not hard to see that $R_4(k)$ is monotone decreasing if $k \geq 5$. ■

The first application we derive is the induced analogue of Exercise 379.

**Theorem 403** $La^*(n,\bigwedge_r) \leq \left(1+\frac{2(r-1)}{n}+o\left(\frac{1}{k}\right)\right)\binom{n}{\lfloor n/2\rfloor}$.

**Proof.** Let $\mathcal{F}\subseteq 2^{[n]}$ be an induced $\bigwedge_r$-free family. As $\left|\binom{[n]}{\leq n/2-n^{2/3}}\cup\binom{[n]}{\geq n/2+n^{2/3}}\right|$

$= o(\frac{1}{k})\binom{n}{\lfloor n/2 \rfloor}$, we can assume that all sets of $\mathcal{F}$ have size between $n/2 - n^{2/3}$ and $n/2 + n^{2/3}$. Let us consider the max-partition of $\mathbf{C}_n$ according to $\mathcal{F}$. By the $\bigwedge_r$-free property, we know that for any set $A \in \mathcal{F}$, the family $\mathcal{F}_A = \{F \in \mathcal{F} : F \subseteq A\}$ does not contain an antichain of size $r$, in particular for any integer $k$ we have $|\mathcal{F}_A \cap \binom{A}{k}| \leq r-1$. Therefore, by the furthermore part of Lemma 402 **(i)**, we obtain that average of $|\mathcal{C} \cap \mathcal{F}|$ over $\mathcal{C} \in \mathbf{C}_A$ is at most $1 + \frac{r-1}{|A|} + O_r(\frac{1}{|A|^2})$. By the assumption on the set sizes in $\mathcal{F}$, we obtain that $\lambda(\mathcal{F}) \leq 1 + \frac{2(r-1)}{n-2n^{-2/3}} + o(\frac{1}{k})$. The theorem follows from Lemma 378. ■

The other application is an analogue of Theorem 381.

**Theorem 404** *(i) If $\mathcal{F} \subseteq 2^{[n]}$ is induced $D_4$-free, then $\lambda(\mathcal{F}) \leq 4$. In particular, $La^*(n, D_4) = \Sigma(n, 4)$.*

***(ii)*** *If $s$ is as in Lemma 402* ***(iii)****, then for any induced $D_s$-free family $\mathcal{F} \subseteq 2^{[n]}$ we have $\lambda(\mathcal{F}) \leq m_s^*$. In particular, $La^*(n, D_s) = \Sigma(n, m_s^*)$.*

**Proof.** Let $\mathcal{F} \subseteq 2^{[n]}$ be induced $D_s$-free and consider the min-max partition of $\mathbf{C}_n$ with respect to $\mathcal{F}$. By definition, for any $\mathcal{C} \in \mathbf{C}_\emptyset$ or $\mathcal{C} \in \mathbf{C}_{A,A}$ we have $|\mathcal{C} \cap \mathcal{F}| \leq 1$. Let us consider $\mathbf{C}_{A,B}$ for some pair $A \subsetneq B$ of sets. $\mathcal{F}$ is induced $D_s$-free; therefore $\mathcal{F} \cap [A, B]$ does not contain $s$ sets of the same size. As no chain $\mathcal{C} \in \mathbf{C}_{A,B}$ contains a member $F$ of $\mathcal{F}$ with $F \subsetneq A$ or $B \subsetneq F$, Lemma 402 **(iii)** and **(iv)** imply that $\lambda(\mathcal{F}) \leq 4$ or $\lambda(\mathcal{F}) \leq m_s^*$, if $s = 4$ or if $s$ satisfies the condition of Lemma 402 **(iii)**. Lemma 378 implies both statements of the theorem. ■

The analogue of Bukh's result, Theorem 385 was proven with a somewhat weaker error term by Boehnlein and Jiang.

**Theorem 405 (Boehnlein, Jiang [56])** *Let $T$ be a poset whose Hasse diagram is a tree. Then $La^*(n, T) = (h(T) - 1 + o(1))\binom{n}{\lfloor n/2 \rfloor}$ holds.*

The function $La(n, P)$ is bounded by constant multiples of $\binom{n}{\lfloor n/2 \rfloor}$ for any poset $P$ containing a comparable pair. This is a simple consequence of Erdős's result, Theorem 4. The analoguous statement for the function $La^*(n, P)$ is much more complicated. Lu and Milans [394] proved this for posets with height 2 and later Methuku and Pálvölgyi obtained a proof of the general statement. In the remaining part of this section, we present their proof.

**Theorem 406 (Methuku, Pálvölgyi, [418])** *For every finite poset $P$ there exists a constant $c_P$ such that $La^*(n, P) \leq c_P\binom{n}{n/2}$ holds.*

The key of the proof is to establish connections between forbidden induced subposets and forbidden submatrix problems. To do that we need several definitions. We say that an ordering is *linear* if any two elements are comparable.

**Definition 407** *The dimension of a poset $P$ is the least integer $t$ such that there exist $t$ linear orderings $<_1, <_2, \ldots, <_t$ of the elements of $P$, such that for every $x, y \in P$, we have $x <_P y$ if and only if $x <_i y$ for all $i = 1, 2, \ldots, t$.*

Recall that a hypermatrix or $d$-matrix is an $n_1 \times n_2 \times \cdots \times n_d$ size ordered array. If all $n_i$'s equal $n$, then we say that $M$ is a $d$-matrix of size $n^d$. The entries of $M$ will be denoted by $M(\underline{i})$, where $\underline{i} = (i_1, i_2, \ldots, i_d)$, with $1 \leq i_j \leq n_j$ for all $1 \leq j \leq d$.

We say that a binary $d$-matrix $M$ (with entries 0 and 1) contains another binary $d$-matrix $A$ if it has a $d$-submatrix $M' \subseteq M$ that is of the same size as $A$ such that $A(\underline{i}) = 1$ implies $M'(\underline{i}) = 1$. If $M$ does not contain $A$ then we say that $M$ is $A$-free. The corresponding extremal function $ex_d(n_1 \times n_2 \times \cdots \times n_d, A)$ is defined to be the most number of 1 entries an $A$-free binary $d$-matrix $M$ of size $n_1 \times n_2 \times \cdots \times n_d$ can have. If all $n_i$'s equal $n$, then we write $ex_d(n, A)$.

**Lemma 408** *Suppose $m_i \leq n_i$ holds for all $1 \leq i \leq d$. Then we have*

$$ex_d(n_1 \times n_2 \times \cdots \times n_d, A) \leq ex_d(m_1 \times m_2 \times \cdots \times m_d, A) \prod_{i=1}^{d} \frac{n_i}{m_i}.$$

*In particular, if for all $n$ we have $ex_d(n, A) \leq Kn^{d-1}$, then for all $n_1, n_2, \ldots, n_d$ we have $ex_d(n_1 \times n_2 \times \cdots \times n_d, A) \leq K \frac{\prod_{i=1}^{d} n_i}{\min_j n_j}$.*

**Proof.** It is enough to prove the first statement as the second one follows by letting all $m_i$'s to be $\min_j n_j$. For the first statement, let $M$ be an $A$-free $d$-matrix of size $n_1 \times n_2 \times \cdots \times n_d$ with $ex_d(n_1 \times n_2 \times \cdots \times n_d, A)$ 1 entries and consider a randomly chosen submatrix $M'$ of size $m_1 \times m_2 \times \cdots \times m_d$. Clearly, $M'$ is $A$-free; therefore the number of its 1 entries is at most $ex_d(m_1 \times m_2 \times \cdots \times m_d, A)$. On the other hand the expected number of 1 entries of $M'$ is $\prod_{i=1}^{d} \frac{n_i}{m_i}$ times the number of 1 entries of $M$. ■

**Definition 409** *A $d$-matrix $M$ of size $k^d$ is a permutation $d$-matrix if the number of non-zero entries of $M$ is $k$, all of which is 1, and it contains exactly one 1 in each axis-parallel hyperplane.*

Generalizing the breakthrough result of Marcus and Tardos [406], Klazar and Marcus obtained the following theorem.

**Theorem 410 (Klazar, Marcus [362])** *If $A$ is a permutation $d$-matrix of size $k^d$, then $ex_d(n, A) = O(n^{d-1})$.*

**Lemma 411** *For every poset $P$ of size $k$ whose dimension is $d$, there is a permutation $d$-matrix $M_P$ of size $k^d$ and a bijection $f$ from the elements of the poset to the 1-entries of $M_P$ with the following property. Let $M_P(\underline{i}) = f(p)$ and $M_P(\underline{i'}) = f(p')$. Then $p < p'$ if and only if $i_j < i'_j$ for all $1 \leq j \leq d$.*

**Proof.** If $P$ is a poset of size $k$ and dimension $d$, then for any $p \in P$ let $\underline{i}_p = (i_{1,p}, i_{2,p}, \ldots, i_{d,p})$ where $i_{j,p}$ is the index of $p$ in the $j$th linear order $<_j$ assured by $dim(P) = d$. Observe that the binary $d$-matrix that have 1 entries $\{M(\underline{i}_p) : p \in P\}$ is the desired permutation $d$-matrix. ■

The next theorem and Theorem 410 together yield Theorem 406.

**Theorem 412** *Let $P$ be a poset of dimension $d$ and let $M_P$ be the corresponding permutation $d$-matrix. If $ex_d(n, M_P) \leq Kn^{d-1}$, then $La^*(n, P) \leq 2^{2d}K\binom{n}{\lfloor n/2 \rfloor}$ holds.*

The proof of Theorem 412 uses a double counting argument and we need further definitions. A *permutation $d$-partition* $\mathcal{Q} = Q_1|Q_2|\ldots|Q_d$ is an ordered partition of elements of $[n]$ into $d$ ordered parts. One can look at them as permutations partitioned into $d$ parts. The $i$th element of $Q_j$ is denoted by $Q_j(i)$. A *prefix* of the part $Q_j$ is a set of the form $\{Q_j(1), Q_j(2), \ldots, Q_j(i-1)\} =: Q_j[i)$. A *prefix union* of $\mathcal{Q}$ is a set of the form $\cup_j Q_j[i_j) =: Q[\underline{i})$.

Observe that the total number of permutation $d$-partitions is $\frac{(n+d-1)!}{(d-1)!}$ as they can be identified with permutations of elements of $[n]$ and $d-1$ separating elements, and these separating elements are indistinguishable.

**Proof of Theorem 412.**

**Claim 413** *For any $F \subseteq [n]$ we have*

$$|\{\mathcal{Q} : F \textit{ is a prefix union of } \mathcal{Q}\}| = \frac{(|F|+d-1)!}{(d-1)!}\frac{(n-|F|+d-1)!}{(d-1)!} = \frac{(n+2d-2)!}{((d-1)!)^2\binom{n+2d-2}{|F|+d-1}}.$$

**Proof of Claim.** If $F$ is a prefix union of $Q_1|Q_2|\ldots|Q_d = \mathcal{Q}$, then all ordered parts of $\mathcal{Q}$ start with a permutation of some elements of $F$ followed by a permutation of some elements of $\overline{F}$. Therefore $\mathcal{Q}$ is put together from a permutation $d$-partition $L_1|L_2|\ldots|L_d$ of $F$ and a permutation $d$-partition $R_1|R_2|\ldots|R_d$ of $\overline{F}$, i.e. $Q_i = L_iR_i$ for all $i = 1, 2, \ldots, d$. As mentioned earlier, $\frac{(|F|+d-1)!}{(d-1)!}$ and $\frac{(n-|F|+d-1)!}{(d-1)!}$ are the number of such permutation $d$-partitions. □

In the next claim we bound the number of sets in $F$ that can be prefix unions of the same permutation $d$-partition $\mathcal{Q} = Q_1|Q_2|\ldots|Q_d$. This is the step where the induced $P$-free property of $\mathcal{F}$ and the problem of forbidden matrices are connected. To this end we introduce the binary $d$-matrix $M_{\mathcal{Q}}(\mathcal{F})$ of size $(|Q_1|+1) \times (|Q_2|+1) \times \cdots \times (|Q_d|+1)$, where $|Q_i|$ is the number of elements in the ordered set $Q_i$. The entry $M_{\mathcal{Q}}(\mathcal{F})[\underline{i}]$ is 1 if and only if the prefix union $\mathcal{Q}[\underline{i}) \in \mathcal{F}$.

**Claim 414** *If $ex_d(n, M_P) \leq Kn^{d-1}$ and $\mathcal{F} \subseteq 2^{[n]}$ is an induced $P$-free family, then for any permutation $d$-partition $\mathcal{Q}$ we have*

$$|\{F \in \mathcal{F} | F \text{ is a prefix union of } \mathcal{Q}\}| \leq K \frac{(|Q_1|+1)(|Q_2+1)\dots(|Q_d|+1)}{\min_j(|Q_j|+1)} \leq K\left(\frac{n+d-1}{d-1}\right)^{d-1}.$$

**Proof of Claim.** Note that $\mathcal{Q}[\underline{i}] \subseteq \mathcal{Q}[\underline{i}')$ if and only if $i_j \leq i'_j$ holds for all $j = 1, 2, \dots, d$, with equality only if $\underline{i} = \underline{i}'$. Therefore $M_{\mathcal{Q}}(\mathcal{F})$ contains the $d$-matrix $M_P$ defined in Lemma 411 if and only if $\mathcal{F}$ contains an induced copy of $P$. As $\mathcal{F}$ is induced $P$-free, it follows that $M_{\mathcal{Q}}(\mathcal{F})$ is $M_P$-free; thus Lemma 408 and $ex_d(n, M_P) \leq Kn^{d-1}$ yield the statement of the claim. □

Using Claim 413 and Claim 414 and counting the pairs $(F, \mathcal{Q})$ with $F \in \mathcal{F}$ and $\mathcal{Q}$ being a permutation $d$-partition of $[n]$, we obtain

$$\sum_{F \in \mathcal{F}} \frac{(n+2d-2)!}{((d-1)!)^2\binom{n+2d-2}{|F|+d-1}} \leq \frac{(n+d-1)!}{(d-1)!} K \left(\frac{n+d-1}{d-1}\right)^{d-1}.$$

Rearranging gives

$$\frac{|\mathcal{F}|}{\binom{n+2d-2}{\lfloor \frac{n+2d-2}{2} \rfloor}} \leq \sum_{F \in \mathcal{F}} \frac{1}{\binom{n+2d-2}{|F|+d-1}} \leq K.$$

As $\binom{n+2d-2}{\lfloor \frac{n+2d-2}{2} \rfloor} \leq 2^{2d-2}\binom{n}{\lfloor \frac{n}{2} \rfloor}$, the theorem follows. ■

Theorem 406 was strengthened by Meroueh, as he proved that the value of the Lubell function of induced $P$-free families is bounded by a constant that depends only on $P$.

**Theorem 415 (Meroueh [414])** *For any finite poset $P$ there exists a constant $c(P)$ such that for any induced $P$-free family $\mathcal{F} \subseteq 2^{[n]}$ we have $\lambda(\mathcal{F}) \leq c(P)$.*

## 7.5 Other variants of the problem

The first variant we consider is maximizing the size of a family that is intersecting, in addition to being $P$-free. Let us denote by $La_I(n, P)$ the maximum cardinality of such a family. By definition we have $La_I(n, P) \leq La(n, P)$. On the other hand, a simple construction of an intersecting and $P$-free family is the union of $e(P)$ consecutive levels above $n/2$, i.e. $\binom{[n]}{\lfloor n/2 \rfloor+1} \cup \dots \binom{[n]}{\lfloor n/2 \rfloor+e(P)}$. Let

$\Sigma'(n, e(P))$ denote its size; then we have $\Sigma'(n, e(P)) = (1 + o(1))\Sigma(n, e(P))$; thus Conjecture 373 would imply that $\lim_{n\to\infty} \frac{La_I(n,P)}{\binom{n}{\lfloor n/2 \rfloor}}$ exists and equal to $e(P)$. Therefore, the interesting question is getting sharp results for some posets.

If $n$ is even, we can construct an even larger $P$-free intersecting family if we take a star on level $n/2$ and only the complement of a star on level $e(P)+n/2$ instead of the full level (more precisely, this has size larger than $\Sigma'(n, e(P))$ if $e(P) > 1$, and has exactly the same size if $e(P) = 1$).

$$\Sigma_I(n,k) := \begin{cases} \sum_{i=\frac{n+1}{2}}^{\frac{n+1}{2}+k-1} \binom{n}{i}, & \text{if } n \text{ is odd} \\ \binom{n-1}{\frac{n}{2}-1} + \sum_{i=\frac{n}{2}+1}^{\frac{n}{2}+k-1} \binom{n}{i} + \binom{n-1}{\frac{n}{2}+k}, & \text{if } n \text{ is even.} \end{cases} \tag{7.2}$$

Note that Section 1.3 was completely devoted to intersecting Sperner families, and Theorem 16 can be stated as $La_I(n, P_2) = \Sigma_I(n, 1)$. This result was generalized by Gerbner.

**Theorem 416 (Gerbner [256])** $La_I(n, P_{k+1}) = \Sigma_I(n, k)$.

Gerbner, Methuku and Tompkins [262] initiated the study of $La_I(n, P)$ for general posets. They characterized the cases of equality in the above theorem, and proved the following.

**Theorem 417 (Gerbner, Methuku, Tompkins [262])** *For the butterfly poset $B$, we have $La_I(n, B) = \Sigma_I(n, 2)$.*

In addition, they proved weighted inequalities, similar to Theorem 18 and Theorem 20.

**Exercise 418** *A double chain-complement pair is a double chain together with the complements of its members. Prove Theorem 416 and Theorem 417 for odd $n$ using double chain-complement pairs.*

Nagy [440] considered another variant of the forbidden subposet problem. He weakened the condition of the $P$-free property in the following way: a family $\mathcal{F}$ should not contain a copy of $P$ in which certain pairs of sets have the same size. Obviously, if $x <_P y$, then sets representing $x$ and $y$ cannot have the same size; therefore there is no point in imposing such a restriction; elements that are supposed to be represented by sets of the same size should form an antichain.

More formally the problem is the following: we are given a poset $P$ and a coloring $c$ of $P$ such that any color class is an antichain. Assume $\mathcal{F} \subseteq 2^{[n]}$ does not contain a copy of $P$ with the additional property that for any pair

$x, y \in P$ with $c(x) = c(y)$ the sets representing $x$ and $y$ have the same size. How large can $|\mathcal{F}|$ be? We denote the maximum size of such families by $La(n, (P, c))$. Obviously, one can consider multiple forbidden colored posets $(P_1, c_1), (P_2, c_2), \ldots, (P_k, c_k)$, where the "underlying" posets $P_1, P_2, \ldots, P_k$ can either be the same or different.

Let $P$ be a complete multi-level poset $K_{r_1, r_2, \ldots, r_s}$. There is only one coloring of $P$ with $s$ colors (i.e. elements of the same level receive the same color). We call this the canonical coloring $c_s$. A canonically colored copy of $P$ is an induced copy of $P$; therefore we obtain $La^*(n, K_{r_1, r_2, \ldots, r_s}) \leq La(n, (K_{r_1, r_2, \ldots, r_s}, c_s))$. Lemma 402 can be used to derive the next theorem a similar way as it was used when proving Theorem 403 and Theorem 404.

**Theorem 419** *(i)* $La(n, (\bigvee_s, c_2) \leq (1 + \frac{2(s-1)}{n} + o(\frac{1}{n}))\binom{n}{\lfloor n/2 \rfloor}$.

***(ii)*** $La(n, (D_s, c_3)) \leq \Sigma(n, m_s^*)$ *if* $s$ *and* $m_s^*$ *are as in Lemma 402.*

***(iii)*** $La(n, (K_{r,1,1\ldots,1,s}, c_h)) \leq (h - 2 + \frac{2(r+s-2)}{n} + o(\frac{1}{n}))\binom{n}{\lfloor n/2 \rfloor}$.

Nagy obtained the following strengthenings of the results of Katona and Tarján [344] on $La(n, \{\bigvee, \bigwedge\})$, and DeBonis, Katona and Swanepoel [121] on $La(n, B)$, where $B$ is the butterfly poset.

**Theorem 420 (Nagy [440])** $La(n, \{(\bigvee, c_2), (\bigwedge, c_2)\}) = 2\binom{n-1}{\lfloor \frac{n-1}{2} \rfloor}$.

**Theorem 421 (Nagy [440])** $La(n, \{(B, c'), (B, c'')\} = \Sigma(n, 2)$, *where* $c', c''$ *are the colorings of* $B$ *using 3 colors such that color classes are antichains.*

A general result states that for any poset $P$ and coloring $c$, the order of magnitude of the size of the extremal family is $\binom{n}{\lfloor n/2 \rfloor}$.

**Theorem 422 (Nagy [440])** *Let* $P$ *be a finite poset and* $c$ *be a coloring such that all color classes are antichains. Then there exists a constant* $C = C(P, c)$ *such* $La(n, (P, c)) \leq C\binom{n}{\lfloor n/2 \rfloor}$ *holds.*

**Proof.** The following lemma will imply the theorem easily.

**Lemma 423** *Let* $r_1, r_2, \ldots, r_k$ *be positive integers. Then there exists a constant* $C = C(r_1, r_2, \ldots, r_k)$ *such that if* $\mathcal{F} \subseteq 2^{[n]}$ *with* $|\mathcal{F}| > C\binom{n}{\lfloor n/2 \rfloor}$, *then one can find a canonically colored copy of* $K_{1, r_1, r_2, \ldots, r_k, 1}$.

**Proof of Lemma.** The proof is by induction on $k$. The base case $k = 1$ is covered by Theorem 419 **(ii)**, as $K_{1,r,1} = D_r$.

For the inductive step, let us define $\mathcal{G} = \{G \in \mathcal{F} : \exists F_1, \ldots, F_{r_k}, F \in \mathcal{F}$ with $|F_1| = \cdots = |F_{r_k}|, G \subset F_i \subset F$ for $i = 1, 2, \ldots, r_k\}$. Observe that $\mathcal{F} \setminus \mathcal{G}$ does not contain a canonically colored copy of $D_{r_k}$, as otherwise the bottom element of a copy would belong to $\mathcal{G}$. By Theorem 419 **(ii)**, there exists a $C'$ such that $|\mathcal{F} \setminus \mathcal{G}| \leq C'\binom{n}{\lfloor n/2 \rfloor}$.

If we define $C(r_1, \dots, r_k) = C' + C(r_1, \dots, r_{k-1})$, then $|\mathcal{F}| > C\binom{n}{\lfloor n/2 \rfloor}$ implies $|\mathcal{G}| > C(r_1, \dots, r_{k-1})\binom{n}{\lfloor n/2 \rfloor}$ By induction, we obtain a canonically colored copy $K$ of $K_{1,r_1,\dots,r_{k-1},1}$ in $\mathcal{G}$. If $T$ is the top set of this colored copy, then by definition of $\mathcal{G}$, there exist $F_1, \dots, F_{r_k}, F \in \mathcal{F}$ with $|F_1| = \cdots = |F_{r_k}|$ and $G \subset F_i \subset F$ for all $i = 1, 2, \dots, r_k$. Then removing $T$ and adding $F_1, \dots, F_{r_k}, F$ to $K$, we obtain a canonically colored copy of $K_{1,r_1,r_2,\dots,r_k,1}$ in $\mathcal{F}$. □

To prove the theorem, let $c : P \to [k]$ be a coloring of $P$ such that all color classes form antichains in $P$. By rearranging if necessary, we can assume that if $x <_P y$ with $c(x) = i, c(y) = j$, then $i < j$ holds. Let $r_i$ denote the size of the $i$th color class. Lemma 423 implies that if for a family $\mathcal{F} \subseteq 2^{[n]}$ we have $|\mathcal{F}| > C(r_1, \dots, r_k)\binom{n}{\lfloor n/2 \rfloor}$, then $\mathcal{F}$ contains a canonically colored copy of $K_{r_1,r_2,\dots,r_k}$, and therefore a $c$-colored copy of $P$. ■

## 7.6 Counting other subposets

In this section, we study another generalization of forbidden subposet problems: we are interested in the maximum number of copies of a given configuration $Q$ in families that do not contain a forbidden subposet $P$. More generally, we are given a poset $R$ and we are interested in the maximum number of copies of $Q$ in families that do not contain $P$. For a poset $P$ and a family $\mathcal{F}$, we denote by $c(P, \mathcal{F})$ the number of copies of $P$ in $\mathcal{F}$. The general parameter that one can define is

$$La(n, P, Q) = \max\{c(Q, \mathcal{F}) : \mathcal{F} \subseteq 2^{[n]}, c(P, \mathcal{F}) = 0\},$$

and for families of posets $\mathcal{P}, \mathcal{Q}$, we define

$$La(n, \mathcal{P}, \mathcal{Q}) = \max\left\{\sum_{Q \in \mathcal{Q}} c(Q, \mathcal{F}) : \mathcal{F} \subseteq 2^{[n]}, \forall P \in \mathcal{P} \;\; c(P, \mathcal{F}) = 0\right\}.$$

These quantities were introduced by Gerbner, Keszegh, and Patkós in [258], but there existed some earlier results that fit into this context. The first one is due to Katona. His approach will introduce the so-called *profile polytope method*.

**Theorem 424 (Katona [338])** *For any weight function $w : \{0, 1, \dots, n\} \to \mathbb{R}^+$ and antichain $\mathcal{F} \subseteq 2^{[n]}$ we have*

$$w(\mathcal{F}) = \sum_{F \in \mathcal{F}} w(|F|) \le \max_{0 \le i \le n} \left\{\binom{n}{i} w(i)\right\}.$$

**Proof.** Using Lemma 3 we obtain

$$\sum_{F\in\mathcal{F}} w(|F|) = \sum_{F\in\mathcal{F}} \frac{w(|F|)\binom{n}{|F|}}{\binom{n}{|F|}}$$
$$\leq \max_{0\leq i\leq n}\left\{\binom{n}{i}w(i)\right\}\sum_{F\in\mathcal{F}}\frac{1}{\binom{n}{|F|}} \leq \max_{0\leq i\leq n}\left\{\binom{n}{i}w(i)\right\}.$$

■

**Corollary 425 (Katona [338])** *For any $n$ we have $La(n,P_3,P_2) = \binom{n}{\lceil\frac{2n}{3}\rceil}\binom{\lceil\frac{2n}{3}\rceil}{\lceil\frac{n}{3}\rceil}$.*

Recall that for a family $\mathcal{F}$ of sets, its *canonical partition* is $\mathcal{F} = \mathcal{F}_1\cup\mathcal{F}_2\cup\cdots\cup\mathcal{F}_{h(\mathcal{F})}$, where $\mathcal{F}_i$ is the family of minimal sets in $\mathcal{F}\setminus\cup_{j=1}^{i-1}\mathcal{F}_j$, and $h(\mathcal{F})$ is the length of the largest chain $\mathcal{F}$, i.e. the height of $\mathcal{F}$ when considered as a poset ordered by inclusion.

**Proof.** Let $\mathcal{F}\subseteq 2^{[n]}$ be a $P_3$-free family and let us consider its canonical partition $\mathcal{F} = \mathcal{F}_1\cup\mathcal{F}_2$. Any copy of $P_2$ contains one set from $\mathcal{F}_1,\mathcal{F}_2$ each. For any $F\in\mathcal{F}_2$, the number of copies of $P_2$ that contain $F$ is $|\{F'\in\mathcal{F}_1 : F'\subseteq F\}|$. By Theorem 1, this is at most $\binom{|F|}{\lfloor |F|/2\rfloor}$. Summing up for all $F\in\mathcal{F}_2$ we obtain $c(P_2,\mathcal{F})\leq\sum_{F\in\mathcal{F}_2}\binom{|F|}{\lfloor |F|/2\rfloor}$. So we can apply Theorem 424 with $w(i)=\binom{i}{\lfloor i/2\rfloor}$ to obtain $c(P_3,\mathcal{F})\leq\max_{0\leq i\leq n}\{\binom{n}{i}\binom{i}{\lfloor i/2\rfloor}\}$.

Before maximizing $f(i)=\binom{n}{i}\binom{i}{\lfloor i/2\rfloor}$, let us observe that if $i^*$ is the value of $i$ that maximizes $f(i)$, then for the family $\mathcal{F}^* = \binom{[n]}{i^*}\cup\binom{[n]}{\lfloor i^*/2\rfloor}$ we have $c(P_2,\mathcal{F}^*) = f(i^*)$; therefore we obtained $La(n,P_3,P_2) = f(i^*)$.

To maximize $f(i)$ let us consider $f(i+1)/f(i)$. If $i=2k$, then we have $\frac{f(2k+1)}{f(2k)} = \frac{n-2k}{k+1}$, while if $i=2k-1$, then we have $\frac{f(2k)}{f(2k-1)} = \frac{n-2k+1}{k}$. It follows that this ratio is larger than 1 if $i<\lceil\frac{2n}{3}\rceil$, so $f(i)$ is maximized when $i=\lceil\frac{2n}{3}\rceil$. ■

The notion of the *profile vector* of a family $\mathcal{F}\subseteq 2^{[n]}$ was introduced by P.L. Erdős, Katona and Frankl [169]. It is a vector $f(\mathcal{F})$ of length $n+1$, indexed from 0 to $n$ with $f(\mathcal{F})_i = |\{F\in\mathcal{F} : |F| = i\}|$. Note that for any weight function $w:\{0,1,\dots,n\}\rightarrow\mathbb{R}^+$, if we use the weight vector $\mathbf{w}$ with $\mathbf{w}_i = w(i)$, we have $w(\mathcal{F}) = \sum_{F\in\mathcal{F}} w(|F|) = f(\mathcal{F})\cdot\mathbf{w}$. Suppose we want to maximize, just as in Theorem 424, the weight over a set $\mathbb{A}\subseteq 2^{2^{[n]}}$ of families. Then we can consider the convex hull $\langle\mu(\mathbb{A})\rangle$ of $\mu(\mathbb{A}) = \{f(\mathcal{F}) : \mathcal{F}\in\mathbb{A}\}$ which we call the *profile polytope* of $\mathbb{A}$. We know from linear programming that for any vector $v$, the scalar product $f(\mathcal{F})\cdot v$ is maximized at an *extreme point* of $\langle\mu(\mathbb{A})\rangle$ (a vector of $\mu(\mathbb{A})$ that cannot be expressed as a convex linear combination of other vectors), and for a vector $w$ with only non-negative coordinates the scalar product $f(\mathcal{F}) \cdot w$ is maximized at an *essential extreme point* of $\langle \mu(\mathbb{A}) \rangle$ (an extreme point that is maximal with respect to the coordinate-wise ordering). With these definitions, Theorem 424 states that if $\mathbb{S}_n$ denotes the set of all antichains in $2^{[n]}$, then the essential extreme points of $\langle \mu(\mathbb{S}_n) \rangle$ are the profiles of the full levels $f(\binom{[n]}{i})$, $i = 0, 1 \ldots, n$. Results on profile polytopes of different sets of families can be found in [143,144,169,172,174,209,256,493], and there is a complete chapter on the topic in Engel's book [142].

We can summarize the profile polytope method as follows: if we know the essential extreme points of $\langle \mu(\mathbb{A}) \rangle$ for a given set $\mathbb{A}$ of families, then we can maximize $w(\mathcal{F})$ over all families in $\mathbb{A}$ for any non-negative weight function. For many extremal problems on a set $\mathbb{B}$ of families we can assign a weight function and another set $\mathbb{A}$ of families such that the solution to the original extremal problem is bounded by or, if we are lucky enough, equal to the maximum weight over the families in $\mathbb{A}$. If we know $\mu(\mathbb{A})$, then we can determine the latter and hence solve or obtain bounds on the original problem. In the remainder of this section we will see further applications of this method, many of which use different types of profile vectors.

Corollary 425 determined the value of $La(n, P_3, P_2)$, so it is natural to ask for the value $La(n, P_k, P_l)$ for any pair $l < k$. This was done by Gerbner and Patkós in [266], where they introduced the notion of *$l$-chain profile vector* of a family $\mathcal{F} \subseteq 2^{[n]}$. This has $\binom{n+1}{l}$ coordinates, and the $\alpha$th coordinate $f_\alpha$, where $\alpha = (\alpha_1, ..., \alpha_l)$ with $0 \leq \alpha_1 < \alpha_2 < ... < \alpha_l \leq n$, denotes the number of $l$-chains in $\mathcal{F}$ in which the smallest set has size $\alpha_1$, the second smallest has size $\alpha_2$, and so on. Note that for $l = 1$ this is just the original notion of the profile vector.

Let $\mu_l(\mathbb{A})$ denote the set of all $l$-chain profile vectors of families in $\mathbb{A}$, $\langle \mu_l(\mathbb{A}) \rangle$ its convex hull, $\mathcal{E}_l(\mathbb{A})$ the extreme points of $\langle \mu_l(\mathbb{A}) \rangle$ and $E_l(\mathbb{A})$ the families from $\mathbb{A}$ with $l$-chain profile in $\mathcal{E}_l(\mathbb{A})$. Let furthermore $\mathcal{E}_l^*(\mathbb{A})$ denote the essential extreme points and $E_l^*(\mathbb{A})$ the corresponding families.

The next lemma will give us a tool to determine $l$-chain profile polytopes. This reduction method was first applied in [174] and is essentially the profile polytope version of the permutation method. We will need the following definitions. For any positive integer $l$ we consider $\mathbb{R}^{\binom{n+1}{l}}$, and index the coordinates by $l$-tuples of integers $(\alpha_1, \alpha_2, \ldots, \alpha_l)$, where $0 \leq \alpha_1 < \alpha_2 < \ldots < \alpha_l \leq n$. Let $T_\mathcal{C}^l$ denote the following linear operator acting on $\mathbb{R}^{\binom{n+1}{l}}$

$$T_\mathcal{C}^l : e \mapsto T_\mathcal{C}^l(e) \qquad \text{where} \qquad T_\mathcal{C}^l(e)_\alpha = \binom{n}{\alpha_l}\binom{\alpha_l}{\alpha_{l-1}} \cdots \binom{\alpha_2}{\alpha_1} e_\alpha.$$

For a family $\mathcal{F} \subseteq 2^{[n]}$ and a maximal chain $\mathcal{C}$ let $\mathcal{F}(\mathcal{C}) = \mathcal{F} \cap \mathcal{C}$ and for a set of families $\mathbb{A}$ let $\mathbb{A}(\mathcal{C}) = \{\mathcal{F}(\mathcal{C}) : \mathcal{F} \in \mathbb{A}\}$.

**Lemma 426 (Gerbner, Patkós [266])** *For any set of families* $\mathbb{A} \subseteq 2^{2^{[n]}}$, *if the extreme points* $e_1, e_2, ..., e_m$ *of* $\langle \mu_l(\mathbb{A}(\mathcal{C}))\rangle$ *do not depend on the choice of* $\mathcal{C}$*, then*

$$\langle \mu_l(\mathbb{A})\rangle \subseteq \langle \{T^l_{\mathcal{C}}(e_1), ..., T^l_{\mathcal{C}}(e_m)\}\rangle.$$

**Proof.** Let $\mathcal{F}$ be an element of $\mathbb{A}$ with $l$-chain profile $f = f(\mathcal{F}) = (..., f_\alpha, ...)$. For an $l$-chain $\mathbf{F} = \{F_1 \subset F_2 \subset ... \subset F_l\}$ with $|F_i| = \alpha_i$, $i = 1, ..., l$, let $\underline{w}(\mathbf{F})$ be the vector of length $\binom{n+1}{l}$ with $1/n!$ in the $\alpha$th coordinate and 0 everywhere else. Consider the sum $\sum \underline{w}(\mathbf{F})$ for all pairs $(\mathcal{C}, \mathbf{F})$, where $\mathcal{C}$ is a maximal chain and $\mathbf{F} \subset \mathcal{F} \cap \mathcal{C}$ is an $l$-chain. For a fixed $\mathcal{C}$ we have

$$\sum_{\mathbf{F} \in \mathcal{F}(\mathcal{C})} \underline{w}(\mathbf{F}) = \frac{1}{n!} f(\mathcal{F}(\mathcal{C})).$$

By definition, the profile $f(\mathcal{F}(\mathcal{C}))$ is a convex linear combination $\sum_{i=1}^m \lambda_i(\mathcal{C}) e_i$ of the $e_i$s. Therefore

$$\sum_{\mathcal{C},\mathbf{F}} \underline{w}(\mathbf{F}) = \sum_{\mathcal{C}} \sum_{\mathbf{F}} \underline{w}(\mathbf{F}) = \sum_{\mathcal{C}} 1/n! \sum_{i=1}^{m} \lambda_i(\mathcal{C}) e_i = \sum_{i=1}^{m} 1/n! \left( \sum_{\mathcal{C}} \lambda_i(\mathcal{C}) \right) e_i \tag{7.3}$$

holds, where $\sum_{\mathcal{C}} 1/n! \sum_{i=1}^m \lambda_i(\mathcal{C}) = 1$. Thus $\sum \underline{w}(\mathbf{F})$ is a convex linear combination of the $e_i$s.

Summing the other way around, we have

$$\sum_{\mathcal{C},\mathbf{F}} \underline{w}(\mathbf{F}) = \sum_{\mathbf{F}} \sum_{\mathcal{C}} \underline{w}(\mathcal{F}) =$$

$$\sum_{\mathbf{F}} (0, 0, ..., \frac{|F_1|!(|F_2| - |F_1|)!...(|F_l| - |F_{l-1}|)!(n - |F_l|)!}{n!}, ..., 0) = (..., \frac{f_\alpha}{\binom{n}{\alpha}}, ...), \tag{7.4}$$

since for a fixed $\mathbf{F} = \{F_1 \subset F_2 \subset ... \subset F_l\}$ there are exactly $|F_1|!(|F_2| - |F_1|)!...(|F_l| - |F_{l-1}|)!(n - |F_l|)!$ chains containing $\mathbf{F}$. So (7.3) and (7.4) give that this last vector is a convex linear combination of the $e_i$s, which implies that $f$ is the convex linear combination of $T^l_{\mathcal{C}}(e_1), ..., T^l_{\mathcal{C}}(e_m)$. ■

**Theorem 427 (Gerbner, Patkós [266])** *For any* $l \leq k$*, the extreme points of the* $l$*-chain profile polytope of* $k$*-Sperner families are the following:*

*the all zero vector*

$$\mathbf{0} = (0, \ldots, 0, \ldots, 0)$$

*and for all* $l \leq z \leq k$ *and* $\beta = \{\beta_1, ..., \beta_z\}$ *with* $0 \leq \beta_1 < ... < \beta_z \leq n$*, the vectors* $v_\beta$ *with*

$$(v_\beta)_\alpha = \begin{cases} \binom{n}{\alpha_l} \cdot \binom{\alpha_z}{\alpha_{l-1}} \cdot \cdots \cdot \binom{\alpha_2}{\alpha_1} & \text{if } \alpha \subseteq \beta \\ 0 & \text{otherwise.} \end{cases}$$

**Proof.** These vectors are $l$-chain profiles of the corresponding levels, and they are convex linearly independent.

A $k$-Sperner family on a maximal chain consists of at most $k$ sets; therefore its $l$-chain profile vector $f$ has ones in those coordinates $\alpha = (\alpha_1, ..., \alpha_l)$ for which there is an element in the family with size $\alpha_i$ for all $i = 1, ..., l$, and $f$ has zeros in the other coordinates. All these vectors are convex independent. Therefore they form the convex hull of the profile polytope on the chain, and Lemma 426 yields the statement of the theorem. ■

**Corollary 428** *For any $l > k$, the quantity $La(n, P_l, P_k)$ is attained for some family $\mathcal{F}$ that is the union of $l-1$ levels. Moreover, $La(n, P_{k+1}, P_k) = \binom{n}{i_k} \cdot \binom{i_k}{i_{k-1}} \cdot \dots \cdot \binom{i_2}{i_1}$, where $i_1 < i_2 < \dots < i_k < n$ are chosen arbitrarily such that the values $i_1, i_2 - i_1, i_3 - i_2, \dots, i_k - i_{k-1}, n - i_k$ differ by at most one.*

**Proof.** To obtain the general result we can apply Theorem 427 with the constant weight function.

To obtain the moreover part, let us consider a $P_{k+1}$-free family $\mathcal{F} \subseteq 2^{[n]}$. We count the number of pairs $(\mathbf{F}, \mathcal{C})$, where $\mathcal{C}$ is a maximal chain and $\mathbf{F} \subseteq \mathcal{F} \cap \mathcal{C}$ is a $k$-chain. As $\mathcal{F}$ is $P_{k+1}$-free, there is at most one such pair for fixed $\mathcal{C}$. On the other hand, for a fixed $k$-chain $\mathbf{F} = \{F_1 \subset F_2 \subset \dots \subset F_k\}$, the number of maximal chains $\mathcal{C}$ with $\mathbf{F} \subseteq \mathcal{C}$ is $|F_1|!(|F_2| - |F_1|)! \dots (|F_k| - |F_{k-1}|)!(n - |F_k|)!$. Dividing by $n!$ we obtain the LYM-type inequality

$$\sum_{\mathbf{F} \subset \mathcal{F}} \frac{1}{\binom{n}{|F_k|}\binom{|F_k|}{|F_{k-1}|} \cdots \binom{|F_2|}{|F_1|}} \leq 1.$$

The statement follows as $|F_1|!(|F_2| - |F_1|)! \dots (|F_k| - |F_{k-1}|)!(n - |F_k|)!$ is smallest if $|F_1|, (|F_2| - |F_1|), \dots, (|F_k| - |F_{k-1}|), (n - |F_k|)$ differ by at most one. ■

Corollary 428 shows that to make Conjecture 373 valid in this more general context, one has to remove at least the word consecutive. But there is another obstruction to extend Conjecture 373 in the general context. Let us consider $La(n, D_2, P_3)$. As *any* three levels of $2^{[n]}$ contain a copy $D_2$ and the union of two levels cannot contain a chain of length 3, we must have $c(P_3, \mathcal{F}) = 0$ for all $D_2$-free families $\mathcal{F}$ that are unions of full levels. Certainly, one can construct other $D_2$-free families with many copies of $P_3$.

**Exercise 429** *For any fixed pair $k > l$ of integers we have the following:*
*(i) $La(n-1, P_3, P_2) \leq La(n, D_2, P_3) \leq La(n, P_3, P_2)$,*
*(ii) $La(n, D_k, D_l) = \Theta(La(n-1, P_3, P_2))$.*

Having these counterexamples in mind, Gerbner, Keszegh and Patkós [258] proposed the following. (Let us remind the reader that the height $h(Q)$ of a poset $Q$ is the length of the longest chain in $Q$.)

**Conjecture 430 (Gerbner, Keszegh, Patkós, [258])** *For any pair $P, Q$ of posets with $h(P) > h(Q)$ there exists a $P$-free family $\mathcal{F} \subseteq 2^{[n]}$ that is the union of full levels, such that*

$$La(n, P, Q) = (1 + o(1))c(Q, \mathcal{F})$$

*holds.*

We say that for a pair $P, Q$ of posets Conjecture 430 *strongly holds* if for large enough $n$ we have $La(n, P, Q) = c(Q, \mathcal{F})$, and *almost holds* if $La(n, P, Q) \leq n^k c(Q, \mathcal{F})$ for some $k$ that depends only on $P$ and $Q$. In both cases the family $\mathcal{F}$ is $P$-free and the union of full levels. Note that the value of $La(n, P, Q)$ typically grows exponentially in $n$.

**Exercise 431 (Gerbner, Keszegh, Patkós [258])** *Use Exercise 376 and elementary methods to prove the following sharp results.*
***(i)*** $La(n, \bigvee, P_2) = La(n, \bigwedge, P_2) = \binom{n}{\lfloor n/2 \rfloor}$.
***(ii)*** $La(n, \{\bigvee, \bigwedge\}, P_2) = \binom{n-1}{\lfloor (n-1)/2 \rfloor}$.
***(iii)*** $La(n, B, D_r) = \binom{\binom{n}{\lfloor n/2 \rfloor}}{r}$.
***(iv)*** $La(n, \bigvee, \bigwedge_r) = La(n, \bigwedge, \bigvee_r) = \binom{\binom{n}{\lfloor n/2 \rfloor}}{r}$.

Parts **(i)**, **(iii)**, and **(iv)** of Exercise 431 show that Conjecture 430 strongly holds for those pairs of posets, while **(ii)** shows another difference between Conjecture 373 and Conjecture 430. Conjecture 373 would imply $La(n, \mathcal{P}, P_1) = (1 + o(1))\binom{n}{\lfloor n/2 \rfloor} \min\{e(P) : P \in \mathcal{P}\}$. As any family consisting of two full levels contains either $\bigvee$ or $\bigwedge$, while one full level does not contain a copy of $P_2$, we have $c(\mathcal{F}, P_2) = 0$ if $\mathcal{F}$ is $\{\bigvee, \bigwedge\}$-free and consists of full levels. We obtained that Conjecture 430 cannot hold in general for multiple forbidden subposets.

**Exercise 432 (Gerbner, Keszegh, Patkós [258])** *Use Theorem 427 to prove the following upper bounds. Recall that the poset $N$ contains four elements $a, b, c, d$ with $a \leq c$ and $b \leq c, d$. Hint: all lower bounds are based on families consisting of full levels of $2^{[n]}$.*
***(i)*** $La(n, P_3, \bigwedge_r) = La(n, P_3, \bigvee_r) = \binom{n}{i_r}\binom{\binom{i_r}{\lfloor i_r/2 \rfloor}}{r}$ *for some $i_r$ with* $i_r = (1 + o(1))\frac{2^r}{2^r+1}n$.
***(ii)*** $La(n, P_4, D_r) = \binom{n}{j_r}\binom{j_r}{i_r}\binom{\binom{j_r - i_r}{\lfloor (j_r - i_r)/2 \rfloor}}{r}$ *for some* $i_r = (1 + o(1))\frac{n}{2^r+2}$ *and either* $j_r = n - i_r$ *or* $j_r = n - i_r - 1$.
***(iii)*** $2^{(c+o(1))n} \leq La(n, P_3, N) = o(2^{3n})$, *where* $c = h(c_0) + 3c_0 h(c_0/(1 - c_0)) = 2.9502...$ *with $c_0$ being the real root of the equation* $0 = 7x^3 - 10x^2 + 5x - 1$ *and $h$ being the binary entropy function* $h(x) = -x\log_2 x - (1 - x)\log_2(1 - x)$.

Gerbner, Methuku, Nagy, Patkós and Vizer [260] considered forbidding any poset and counting a chain $P_k$, and proved the following theorem, showing Conjecture 430 almost holds if the forbidden poset has height more than $k$.

**Theorem 433 (Gerbner, Methuku, Nagy, Patkós, Vizer, [260])** *Let $l$ be the height of $P$. (i) If $l > k$, then*

$$La(n, P, P_k) = \Theta(La(n, P_{k+1}, P_k)).$$

*(ii) If $l \leq k$, then*

$$La(n, P, P_k) = O(n^{2k-1/2} La(n, P_l, P_{l-1})),$$

*and there exists a poset $P$ of height $l$ such that $La(n, P, P_k) = \Theta(La(n, P_l, P_{l-1}))$ holds.*
*(iii) For any connected poset $P$ of height 2 with at least 3 elements, we have*

$$\Omega\left(\binom{n}{\lfloor n/2 \rfloor}\right) = La(n, P, P_2) = O\left(n2^n\right).$$

Most theorems and exercises in this section so far were about $La(n, P_{h(Q)+1}, Q)$ for different posets $Q$. As all the considered posets are *ranked*, we knew the place of every element of every copy of $Q$ in the canonical partition of a $P_{h(Q)+1}$-free family. This helps tremendously to understand the structure of the copies of $Q$. In the rest of this section we describe results of Gerbner, Keszegh and Patkós [258] concerning these kinds of problems. Our aim is to prove that Conjecture 430 almost holds for any pair $P_{h(Q)+1}, Q$ with $Q$ being any complete multi-level poset.

We introduce the following binary operations of posets: for any pair $Q_1$, $Q_2$ of posets we define $Q_1 \otimes_r Q_2$ by adding an antichain of size $r$ between $Q_1$ and $Q_2$. More precisely, let us assume $Q_1$ consists of $q_1^1, \dots, q_a^1$ and $Q_2$ consists of $q_1^2, \dots, q_b^2$. Then $R = Q_1 \otimes_r Q_2$ consists of $q_1^1, \dots, q_a^1, m_1, m_2, \dots, m_r, q_1^2, \dots q_b^2$. We have $q_i^1 <_R q_j^1$ if and only if $q_i^1 <_{Q_1} q_j^1$ and similarly $q_i^2 <_R q_j^2$ if and only if $q_i^2 <_{Q_2} q_j^2$. Also we have $q_i^1 <_R m_k <_R q_j^2$ for every $i$, $k$, and $j$. Finally, the $m_k$'s form an antichain. Note that $h(Q_1 \otimes_r Q_2) = h(Q_1) + h(Q_2) + 1$. Let $Q \oplus r$ denote the poset $Q \otimes_r \mathbf{0}$, where $\mathbf{0}$ is the empty poset; i.e. $Q \oplus r$ is obtained from $Q$ by adding $r$ elements that form an antichain and that are all larger than all elements of $Q$. Similar operations of posets were considered first in the area of forbidden subposet problems by Burcsi and Nagy [97].

We will obtain bounds on $La(n, P_{h(Q_1 \otimes_r Q_2)+1}, Q_1 \otimes_r Q_2)$, involving bounds on $La(n, P_{h(Q_1)+1}, Q_1)$ and $La(n, P_{h(Q_2)+1}, Q_2)$. In order to do that, we will need two auxiliary statements that can be of independent interest. We will try to determine the maximum number of $r$-tuples $\{A_1, A_2, \dots, A_r\}$ that an antichain $\mathcal{A} \subseteq 2^{[n]}$ can contain, such that the intersection of the $A_i$'s is empty, and also those where the intersection of the $A_i$'s is empty and the union of the $A_i$'s is $[n]$. Actually, we will need a seemingly more general statement, but the general version reduces to the above mentioned two problems.

For a family $\mathcal{F} \subseteq 2^{[n]}$ of sets, let the *$r$-intersection profile vector* of $\mathcal{F}$ be $\beta^r(\mathcal{F}) = (\beta_0^r, \beta_1^r, \dots, \beta_{n-1}^r)$, where $\beta_i^r = \beta_i^r(\mathcal{F}) = |\{\{F_1, F_2, \dots, F_r\} : F_j \in \mathcal{F}$, these are $r$ different sets and $|\bigcap_{j=1}^r F_j| = i\}|$.

For a family $\mathcal{F} \subseteq 2^{[n]}$ of sets, let the *r-intersection-union profile vector* of $\mathcal{F}$ be $\gamma^r(\mathcal{F}) = (\gamma^r_{0,1}, \gamma^r_{0,2}, \dots, \gamma^r_{0,n}, \gamma^r_{1,2}, \dots, \gamma^r_{n-1,n})$, where $\gamma^r_{i,j} = \gamma^r_{i,j}(\mathcal{F}) = |\{\{F_1, \dots, F_r\} : F_1, \dots, F_r \in \mathcal{F}, \text{these are } r \text{ different sets}, |F_1 \cap \dots \cap F_r| = i, |F_1, \cup \dots \cup F_r| = j\}|$. Note that if $\mathcal{A} \subseteq 2^{[n]}$ is an antichain and $r \geq 2$, then $\gamma^r_{i,j}(\mathcal{A}) > 0$ implies $j - i \geq 2$; therefore the number of non-zero coordinates in $\gamma^r(\mathcal{A})$ is at most $\binom{n+1}{2} - n = \binom{n}{2} \leq n^2$.

We are not able to determine the convex hulls of $\gamma(\mathbb{S}_n) = \{\gamma(\mathcal{A}) : \mathcal{A} \subseteq 2^{[n]}$ is an antichain$\}$, $\beta(\mathbb{S}_n) = \{\beta(\mathcal{A}) : \mathcal{A} \subseteq 2^{[n]}$ is an antichain$\}$; we will only obtain upper bounds on the coordinates of these profile vectors.

**Theorem 434 (Gerbner, Keszegh, Patkós, [258])** ***(i)*** *If $\mathcal{F} \subseteq 2^{[n]}$ is an antichain and $j - i$ is even, then $\gamma^2_{i,j}(\mathcal{F}) \leq \gamma^2_{i,j}\left(\binom{[n]}{(i+j)/2}\right) = \frac{1}{2}\binom{n}{j}\binom{j}{i}\binom{j-i}{(j-i)/2}$. If $j - i$ is odd, then $\gamma^2_{i,j}(\mathcal{F}) \leq \binom{n}{j}\binom{j}{i}\binom{j-i-1}{\lfloor (j-i)/2 \rfloor - 1}$.*

***(ii)*** *If $\mathcal{F}$ is an antichain and $r \geq 3$, then $\gamma^r_{i,j}(\mathcal{F}) \leq n^{2r}\gamma^r_{i,j}\left(\binom{[n]}{\lfloor (i+j)/2 \rfloor}\right)$.*

During the proof we will use several times that the number of pairs $A \subset B \subset [n]$ with $|A| = a, |B| = b$ is $\binom{n}{b}\binom{b}{a} = \binom{n}{b-a}\binom{n-(b-a)}{a}$. The first calculation is obvious; for the second one we pick first $B \setminus A$ from $[n]$ and then $A$ from $[n] \setminus (B \setminus A)$.

**Proof.** To see **(i)**, we first consider the special case $i = 0, j = n$. Observe that $\gamma^2_{0,n}(\mathcal{F})$ is the number of complement pairs in $\mathcal{F}$. In an antichain, by Theorem 1, this is at most $|\mathcal{F}|/2 \leq \binom{n}{\lfloor n/2 \rfloor}/2$. If $n$ is even, then this is achieved when $\mathcal{F} = \binom{[n]}{n/2}$, while the case of odd $n$ was solved by Bollobás [64], who showed that the number of such pairs is at most $\binom{n-1}{\lfloor n/2 \rfloor - 1}$, and this is sharp, as shown by $\{F \in \binom{[n]}{\lfloor n/2 \rfloor} : 1 \in F\} \cup \{F \in \binom{[n]}{\lceil n/2 \rceil} : 1 \notin F\}$. Note that it is a corollary of Theorem 18.

To see the general statement, observe that for a pair $I \subset J$, writing $\mathcal{F}_{I,J} = \{F \in \mathcal{F} : I \subseteq F \subseteq J\}$, we have $\gamma^2_{i,j}(\mathcal{F}) = \sum_{I \in \binom{[n]}{i}, J \in \binom{[n]}{j}} \gamma^2_{0,j-i}(\mathcal{F}_{I,J})$. Therefore, if $j - i$ is even, we obtain $\gamma^2_{i,j}(\mathcal{F}) \leq \binom{n}{j}\binom{j}{i}\gamma^2_{0,j-i}\left(\binom{[j-i]}{(j-i)/2}\right) = \gamma^2_{i,j}\left(\binom{[n]}{(j+i)/2}\right)$, while if $j - i$ is odd, we obtain $\gamma^2_{i,j}(\mathcal{F}) \leq \binom{n}{j}\binom{j}{i}\binom{j-i-1}{\lfloor (j-i)/2 \rfloor - 1}$.

To show **(ii)**, it is enough to prove the statement for $i = 0$, $j = n$. Indeed, if that holds, then

$$\gamma^r_{i,j}(\mathcal{F}) = \sum_{I,J} \gamma^r_{0,j-i}(\mathcal{F}_{I,J}) \leq (j-i)^{2r} \sum_{I,J} \gamma^r_{0,j-i}\left(\binom{[j-i]}{\lfloor (j-i)/2 \rfloor}\right) \leq$$

$$n^{2r} \sum_{I,J} \gamma^r_{0,j-i}\left(\binom{[j-i]}{\lfloor (j-i)/2 \rfloor}\right) = n^{2r}\binom{n}{j}\binom{j}{i}\gamma^r_{0,j-i}\left(\binom{[j-i]}{\lfloor (j-i)/2 \rfloor}\right)$$

$$= n^{2r}\gamma^r_{i,j}\binom{[n]}{\lfloor (j+i)/2 \rfloor}.$$

We proceed by induction on $r$. We postpone the proof of the base case $r = 3$, and assume the statement holds for $r-1$ and any $i < j$. Let us consider $r-1$ sets $F_1, \dots, F_{r-1}$ of $\mathcal{F}$, and examine which sets in $\mathcal{F}$ can be added to them as $F_r$, to get empty intersection and $[n]$ as the union. Let $\mathcal{F}'$ be the subfamily of those sets. Let $A = \cap_{l=1}^{r-1} F_l$ and $B = \cup_{l=1}^{r-1} F_l$ with $a = |A|$ and $b = |B|$. Then members of $\mathcal{F}'$ contain the complement of $B$ and do not intersect $A$, and $\mathcal{F}'$ is an antichain. If we remove $\overline{B}$ from them, the resulting family is antichain on an underlying set of size $b-a$, thus have cardinality at most $\binom{b-a}{\lfloor (b-a)/2 \rfloor} = w(a,b)$. Note that we count every $r$-tuple $F_1, \dots, F_r$ exactly $r$ times. It implies

$$r\gamma_{0,n}^r(\mathcal{F}) \le \sum_{a<b} \gamma_{a,b}^{r-1}(\mathcal{F}) w(a,b) = \gamma^{r-1}(\mathcal{F}) \cdot \mathbf{w} \le n^2 \max_{a<b} \gamma_{a,b}^{r-1}(\mathcal{F}) w(a,b)$$

.

By induction this is at most $n^2 n^{2r-2} \max_{a<b} \gamma_{a,b}^{r-1}\left(\binom{[n]}{\lfloor (a+b)/2 \rfloor}\right) w(a,b)$. Let

$$\begin{aligned} f(a,b) =& w(a,b)\gamma_{a,b}^{r-1}\left(\binom{[n]}{\lfloor (a+b)/2 \rfloor}\right) \\ =& \binom{b-a}{\lfloor (b-a)/2 \rfloor}\binom{n}{b-a}\binom{n-(b-a)}{a}\gamma_{0,b-a}^{r-1}\left(\binom{[b-a]}{\lfloor (a+b)/2 \rfloor - a}\right). \end{aligned}$$

If we fix $b-a$ and consider $\frac{f(a,b)}{f(a+1,b+1)} = \frac{a+1}{n-(b-a)-a}$, we can see that the maximum is taken when $b+a = n$ or $b+a = n-1$, depending on the parity of $b-a$ and $n$.

Let $a^*, b^*$ be the values for which the above maximum is taken. Note that for any $a^* < p < b^*$ we have $r\gamma_{0,n}^r(\binom{[n]}{p}) \ge \binom{n}{b^*}\binom{b^*}{a^*}\gamma_{0,b^*-a^*}^{r-1}(\binom{[b^*-a^*]}{p-a^*})\binom{b^*-a^*}{p-n+b^*}$, by counting only those $r$-tuples where the first $r-1$ sets have intersection of size $a^*$ and union of size $b^*$. (This way we count those $r$-tuples at most $r$ times). This is exactly $f(a^*, b^*)$ if $p = \lfloor n/2 \rfloor = \lfloor (a^*+b^*)/2 \rfloor$, so we obtained $r\gamma_{0,n}^r(\mathcal{F}) \le n^{2r} r\gamma_{0,n}^r(\binom{[n]}{\lfloor n/2 \rfloor})$, as required.

For $r = 3$ we similarly consider two members of $\mathcal{F}$, and examine which sets can be added to them to get empty intersection and $[n]$ as the union. This leads to

$$3\gamma_{0,n}^3(\mathcal{F}) \le n^2 \max_{a<b} \gamma_{a,b}^2(\mathcal{F}) w(a,b).$$

Note that if the maximum is taken at $a'$ and $b'$ with $b'-a'$ even, then part **(i)** of the theorem gives $\gamma_{a',b'}^2(\mathcal{F}) \le \gamma_{a',b'}^2\left(\binom{[n]}{(b'+a')/2}\right)$ so $3\gamma_{0,n}^3(\mathcal{F}) \le n^2\gamma_{a',b'}^2\left(\binom{[n]}{(b'+a')/2}\right) w(a',b')$. This essentially lets us use $r = 2$ as the base case of induction, and finish the proof of this case similarly to the induction step above.

Let us choose $a^*, b^*$ that maximizes this upper bound with $b^* - a^* = b' - a'$. Similarly to the computation about $f(a,b)$ , we have $a^* + b^* = n$ or $n-1$ depending on the parity of $n$. Then we obtain $3\gamma_{0,n}^3(\mathcal{F}) \le$

$n^2\gamma^2_{a^*,b^*}\left(\binom{[n]}{(b^*+a^*)/2}\right)w(a^*,b^*) \leq n^2\binom{n}{b^*}\binom{b^*}{a^*}\gamma^2_{0,b^*-a^*}\left(\binom{[b^*-a^*]}{(b^*+a^*)/2}\right)w(a^*,b^*)$. The lower bound on $3\gamma^3_{0,n}\left(\binom{[n]}{\lfloor n/2\rfloor}\right)$ is $\binom{n}{b^*}\binom{b^*}{a^*}\gamma^2_{0,b^*-a^*}\left(\binom{[b^*-a^*]}{p-a^*}\right)\binom{b^*-a^*}{p-a^*}$ as in the inductive step.

However, if $b'-a'$ is odd, then $\gamma^2_{a',b'}\left(\binom{[n]}{\lfloor(a'+b')/2\rfloor}\right)=0$. But we know by part **(i)**

$$\gamma^2_{a',b'}(\mathcal{F})w(a',b') \leq \binom{b'-a'-1}{(b'-a'-1)/2-1}\binom{n}{b'}\binom{b'}{a'}w(a',b').$$

Similarly to the previous cases, if $b'-a'$ is fixed, then the maximum of the right hand side is taken for some $a^*,b^*$ with $b^*-a^*=b'-a'$ and $a^*+b^*=n$ if $n$ is odd, and $a^*+b^*=n-1$ or $a^*+b^*=n+1$ if $n$ is even. Thus we can assume $\lfloor n/2\rfloor=(a^*+b^*-1)/2$. On the other hand, since $b^*-a^*$ is odd, we have

$$3\gamma^3_{0,n}\left(\binom{[n]}{(a^*+b^*-1)/2}\right) \geq$$
$$\binom{n}{b^*-1}\binom{b^*-1}{a^*}\frac{1}{2}\binom{b^*-a^*-1}{(b^*-a^*-1)/2}\binom{b^*-1-a^*}{\frac{a^*+b^*-1}{2}-n+b^*-1},$$

by counting only those triples where two of the sets have intersection of size $a^*$ and union of size $b^*-1$. We can pick first the $(b^*-1)$-set $B$ and the $a^*$-set $A$ in $\binom{n}{b^*-1}\binom{b^*-1}{a^*}$ ways; then among $\{G\in\binom{[n]}{\lfloor(a^*+b^*)/2\rfloor}: A\subset G\subset B\}$ we can pick a pair $G_1,G_2$ with $G_1\cap G_2=A$, $G_1\cup G_2=B$ in $\binom{b^*-a^*-1}{(b^*-a^*-1)/2}/2$ ways, and then the third set contains the complement of $B$ and does not intersect $A$. Using that $\binom{b^*-a^*-1}{\frac{a^*+b^*-1}{2}-n+b^*-1}\geq\binom{b^*-a^*-1}{(b^*-a^*-1)/2-1}$, this implies

$$3\gamma^3_{0,n}(\mathcal{F}) \leq 3n^2\gamma^3_{0,n}\left(\binom{[n]}{(a^*+b^*-1)/2}\right)\frac{n-b^*+1}{(b^*-a^*-1)/2} \leq 3n^3\gamma^3_{0,n}\left(\binom{[n]}{\lfloor n/2\rfloor}\right),$$

as $b^*\geq n/2$. ■

**Theorem 435 (Gerbner, Keszegh, Patkós, [258])** *(i) For any antichain $\mathcal{F}\subseteq 2^{[n]}$, we have $\beta^2_i(\mathcal{F})\leq\beta^2_i\left(\binom{[n]}{j(i)}\right)$, where $j(i)=i+\lfloor(n-i)/3\rfloor$ if $n-i\equiv 0,1$ mod 3, and $j(i)=i+\lceil(n-i)/3\rceil$ if $n-i\equiv 2$ mod 3.*

***(ii)*** *For every $r\geq 3$ and $i\leq n$, there exists $j(r,i,n)$ such that $\beta^r_i(\mathcal{F})\leq n^{2r+1}\beta^r_i\left(\binom{[n]}{j(r,i,n)}\right)$ holds for any antichain $\mathcal{F}\subseteq 2^{[n]}$.*

**Proof.** First we prove **(i)** for the special case $i=0$. Let $\mathcal{F}\subseteq 2^{[n]}$ be an antichain, and let $\overline{\mathcal{F}}=\{\overline{F}:F\in\mathcal{F}\}$, where $\overline{F}=[n]\setminus F$. As $\mathcal{F}$ is an antichain, so is $\overline{\mathcal{F}}$, and thus $\mathcal{F}\cup\overline{\mathcal{F}}$ is $P_3$-free. Note that for every pair $F_1,F_2\in\mathcal{F}$ with $|F_1\cap F_2|=0$ and $F_1\cup F_2\neq[n]$, we have two 2-chains $F_1\subsetneq\overline{F_2}$ and $F_2\subsetneq\overline{F_1}$.

Also, every 2-chain in $\mathcal{F} \cup \overline{\mathcal{F}}$ comes from a pair $F_1, F_2 \in \mathcal{F}$ with $|F_1 \cap F_2| = 0$ and $F_1 \cup F_2 \neq [n]$.

Let us take the canonical partition of $\mathcal{F} \cup \overline{\mathcal{F}}$ into $\mathcal{F}_1 \cup \mathcal{F}_2$ and introduce the weight function $w(F) = \frac{1}{2}\binom{|F|}{\lceil |F|/2 \rceil}$ if $F \in \mathcal{F}_2, \overline{F} \notin \mathcal{F}_2$ and $w(F) = 1/2$ if $F \in \mathcal{F}_2, \overline{F} \in \mathcal{F}_2$. Then the number of disjoint pairs in $\mathcal{F}$ equals $\sum_{F \in F_2} w(F)$. This weight function does not depend only on the size of $F$, but $w'(f) = \frac{1}{2}\binom{|F|}{\lceil |F|/2 \rceil}$ does and obviously $w(F) \leq w'(F)$ holds for all $F$'s. As proved in Corollary 425, this weight function is maximized when $\mathcal{F}_2 = \binom{[n]}{\lceil 2n/3 \rceil}$. As $\mathcal{F}_2$ does not contain complement pairs, it also maximizes $w$.

To see the general statement of **(i)**, we can apply the special case to any $I \subseteq [n]$ and $\mathcal{F}_I = \{F \setminus I : I \subseteq F \in \mathcal{F}\}$. We obtain

$$\beta_i^2(\mathcal{F}) = \sum_{I \in \binom{[n]}{i}} \beta_0^2(\mathcal{F}_I) \leq \binom{n}{i} \beta_0^2\left(\binom{[n-i]}{j(i)-i}\right) = \beta_i^2\left(\binom{[n]}{j(i)}\right).$$

To see **(ii)**, let $\mathcal{F} \subseteq 2^{[n]}$ be an antichain. Observe

$$\begin{aligned}\beta_i^r(\mathcal{F}) = \sum_{j=i+1}^{n} \gamma_{i,j}^r(\mathcal{F}) &\leq n \max_{j: i+1 \leq j \leq n} \gamma_{i,j}^r(\mathcal{F}) \\ &\leq n^{2r+1} \max_{j: i+1 \leq j \leq n} \gamma_{i,j}^r\left(\binom{[n]}{\lfloor (i+j)/2 \rfloor}\right) \\ &\leq n^{2r+1} \beta_i^r\left(\binom{[n]}{j(r,i,n)}\right),\end{aligned}$$

where $j(r,i,n) = \lfloor (i + j^*)/2 \rfloor$ with $j^*$ being the value of $j$ that maximizes $\gamma_{i,j}^r\left(\binom{[n]}{\lfloor (i+j)/2 \rfloor}\right)$. The penultimate inequality follows from Theorem 434. ■

With Theorem 434 and Theorem 435 in hand, we can state our results on the operations $\otimes_r$ and $\oplus r$.

**Theorem 436 (Gerbner, Keszegh, Patkós, [258])** *Let $Q_1, Q_2$ be two non-empty posets.*

*(i) If $r \geq 2$, then we have*

$$La(n, P_{h(Q_1)+h(Q_2)+1}, Q_1 \otimes_r Q_2) \leq$$
$$n^{2r+2} \max_{0 \leq i < j \leq n} \left\{ \binom{n}{j}\binom{j}{i} \gamma_{0,j-i}^r\left(\binom{[j-i]}{\lfloor (j-i)/2 \rfloor}\right) La(i, P_{l(Q_1)+1}, Q_1) La(n-j, P_{l(Q_2)+1}, Q_2) \right\}.$$

*Furthermore, if $r \geq 3$ and Conjecture 430 almost holds for the pairs $P_{h(Q_1)+1}, Q_1$ and $P_{h(Q_2)+1}, Q_2$, then so it does for the pair $P_{h(Q_1 \otimes_r Q_2)+1}, Q_1 \otimes_r Q_2$.*

*(ii) If $r = 1$, then we have*

$$La(n, P_{h(Q_1)+h(Q_2)+1}, Q_1 \otimes_1 Q_2) \leq \max_{0 \leq j \leq n} \left\{ \binom{n}{j} La(j, P_{h(Q_1)+1}, Q_1) La(n-j, P_{h(Q_2)+1}, Q_2) \right\}.$$

*Furthermore, if Conjecture 430 strongly/almost holds for the pairs $P_{h(Q_1)+1}, Q_1$ and $P_{h(Q_2)+1}, Q_2$, then so it does for the pair $P_{h(Q_1 \otimes_1 Q_2)+1}, Q_1 \otimes_1 Q_2$.*

**Theorem 437 (Gerbner, Keszegh, Patkós, [258])** *Let $Q$ be a non-empty poset.*

*(i) If $r \geq 2$ and $n \in \mathbb{N}$, then there exists an $i = i(r,n)$ such that*

$$La(n, P_{h(Q)+2}, Q \oplus r) \leq n^{2r+2} \max_{0 \leq j \leq n} \left\{ \binom{n}{i} \beta_0^r \left( \binom{[n-i]}{j-i} \right) La(j, P_{h(Q)+1}, Q) \right\}.$$

*Furthermore, if Conjecture 430 almost holds for the pair $P_{h(Q)+1}, Q$, then so it does for the pair $P_{h(Q)+2}, Q_1 \oplus r$.*

*(ii) If $r = 1$, then we have*

$$La(n, P_{h(Q)+2}, Q \oplus 1) \leq \max_{0 \leq j \leq n} \left\{ \binom{n}{j} La(j, P_{h(Q)+1}, Q) \right\}.$$

*Furthermore, if Conjecture 430 strongly/almost holds for the pair $P_{h(Q)+1}, Q$, then so it does for the pair $P_{h(Q)+2}, Q \oplus 1$.*

**Proof of Theorem 436** Let $Q_1, Q_2$ be non-empty posets and let us consider the canonical partition of a $P_{h(Q_1 \otimes_r Q_2)+1}$-free family $\mathcal{F} \subseteq 2^{[n]}$. Then in any copy of $Q_1 \otimes_r Q_2$ in $\mathcal{F}$, if $F_1, \ldots, F_r$ correspond to the $r$ middle elements forming an antichain, we must have $F_1, \ldots, F_r \in \mathcal{F}_{l(Q_1)+1}$. Also, in that case the sets corresponding to the $Q_1$ part of $Q_1 \otimes_r Q_2$ must be contained in $\bigcap_{l=1}^r F_l$, while the sets corresponding to the $Q_2$ part of $Q_1 \otimes_r Q_2$ must contain $\bigcup_{l=1}^r F_l$. Therefore the number of copies of $Q_1 \otimes_r Q_2$ in $\mathcal{F}$ that contain $F_1, \ldots, F_r$ is at most $La(|\bigcap_{l=1}^r F_l|, P_{h(Q_1)+1}, Q_1) \cdot La(n - |\bigcup_{i=1}^r F_l|, P_{h(Q_2)+1}, Q_2)$. We obtained that the total number of copies of $Q_1 \otimes_r Q_2$ in $\mathcal{F}$ is at most

$$\sum_{F_1,\ldots,F_r \in \mathcal{F}_{h(Q_1)+1}} La\left( \left| \bigcap_{l=1}^r F_l \right|, P_{h(Q_1)+1}, Q_1 \right) \cdot La\left( n - \left| \bigcup_{l=1}^r F_l \right|, P_{h(Q_2)+1}, Q_2 \right). \tag{7.5}$$

If $r \geq 2$, then grouping the summands in (7.5) according to the pair $(\bigcap_{l=1}^r F_l, \bigcup_{l=1}^r F_l)$, we obtain

$$La(n, P_{h(Q_1 \otimes_r Q_2)+1}, Q_1 \otimes_r Q_2) \leq \gamma^r(\mathcal{F}_{h(Q_1)+1}) \cdot \mathbf{w},$$

where the $(i,j)$th coordinate of $\mathbf{w}$ is $La(i, P_{h(Q_1)}, Q_1) \cdot La(n-j, P_{h(Q_2)+1}, Q_2)$. Clearly, we have

$$\gamma^r(\mathcal{F}_{h(Q_1)+1}) \cdot \mathbf{w} \leq n^2 \max_{i,j} \gamma^r_{i,j}(\mathcal{F}_{h(Q_1)+1}) w(i,j) \leq$$
$$n^{2r+2} \max_{i,j} \gamma^r_{i,j} \left( \binom{[n]}{\lfloor (i+j)/2 \rfloor} \right) w(i,j),$$

where the last inequality follows from Theorem 434. We can calculate $\gamma^r_{i,j}\left(\binom{[n]}{\lfloor (i+j)/2 \rfloor}\right)$ by picking the union of size $j$ and the intersection of size $i$ first, which finishes the proof of the upper bound of part **(i)**.

To see the furthermore part, suppose that the above maximum is obtained when $i$ takes the value $i^*$ and $j$ takes the value $j^*$. We know that there exist two families $\mathcal{F}_{1,i^*} \subseteq 2^{[i^*]}$ and $\mathcal{F}_{2,n-j^*} \subseteq 2^{[n-j^*]}$, both unions of full levels, and there exist integers $k_1, k_2$ and constants $C_1, C_2$, such that $C_1(i^*)^{k_1} c(Q_1, \mathcal{F}_{1,i^*}) \geq La(i^*, P_{h(Q_1)+1}, Q_1)$ and $C_2(j^*)^{k_2} c(Q_2, \mathcal{F}_{2,n-j^*}) \geq La(n-j^*, P_{h(Q_2)+1}, Q_2)$ hold. Therefore, by the upper bound already proven, we know that $La(n, P_{h(Q_1 \otimes_r Q_2)+1}, Q_1 \otimes_r Q_2)$ is at most $n^{2r+2} C_1(i^*)^{k_1} C_2(j^*)^{k_2} \gamma^r_{i^*j^*}\left(\binom{n}{\lfloor (i^*+j^*)/2 \rfloor}\right) La(i^*, P_{h(Q_1)+1}, Q_1) La(n - i^*, P_{h(Q_2)+1}, Q_2)$.

If $\mathcal{F}_{1,i^*}$ consists of levels of set sizes $h_1, \ldots, h_{h(Q_1)}$ and $\mathcal{F}_{2,n-i^*}$ consists of levels of set sizes $h'_1, \ldots, h'_{h(Q_2)}$, then for the family

$$\mathcal{F} := \binom{[n]}{h_1} \cup \ldots \binom{[n]}{h_{h(Q_1)}} \cup \binom{[n]}{\lfloor (i^*+j^*)/2 \rfloor} \cup \binom{[n]}{j^* + h'_1} \cup \cdots \cup \binom{[n]}{j^* + h'_{h(Q_2)}},$$

we have $c(Q_1 \otimes Q_2, \mathcal{F}) \geq \gamma^r_{i^*j^*}\left(\binom{n}{\lfloor (i^*+j^*)/2 \rfloor}\right) La(i^*, P_{h(Q_1)+1}, Q_1) La(n - i^*, P_{h(Q_2)+1}, Q_2)$. Therefore with $k = 2r + 2 + k_1 + k_2$, the family $\mathcal{F}$ shows that Conjecture 430 almost holds for the pair $P_{h(Q_1 \otimes_r Q_2)+1}, Q_1 \otimes_r Q_2$.

If $r = 1$, then $\cup F_1 = \cap F_1 = F_1$, so (7.5) becomes

$$\sum_{F \in \mathcal{F}_{l(Q_1)+1}} La(|F|, P_{h(Q_1)+1}, Q_1) \cdot La(n - |F|, P_{h(Q_2)+1}, Q_2).$$

We can apply Theorem 424 and $w(i) = La(i, P_{h(Q_1)+1}, Q_1) \cdot La(i, P_{h(Q_2)+1}, Q_2)$ to obtain

$$La(n, P_{h(Q_1 \otimes Q_2)+1}, Q_1 \otimes Q_2)$$
$$\leq \max_{0 \leq i \leq n} \left\{ \binom{n}{i} La(i, P_{h(Q_1)+1}, Q_1) La(n-i, P_{h(Q_2)+1}, Q_2) \right\}$$

as required.

As the proofs are almost identical, we only show the "strongly holds" case of the furthermore part of **(ii)**. Suppose that the above maximum is obtained when $i$ takes the value $i^*$. We know that there exist two families $\mathcal{F}_{1,i^*} \subseteq 2^{[i^*]}$

and $\mathcal{F}_{2,n-i^*} \subseteq 2^{[n-i^*]}$, both unions of full levels, such that $c(Q_1, \mathcal{F}_{1,i^*}) = La(i^*, P_{h(Q_1)+1}, Q_1)$ and $c(Q_2, \mathcal{F}_{2,n-i^*}) = La(n-i^*, P_{h(Q_2)+1}, Q_2)$ hold. If $\mathcal{F}_{1,i^*}$ consists of levels of set sizes $j_1, \dots, j_{l(Q_1)}$ and $\mathcal{F}_{2,n-i^*}$ consists of levels of set sizes $j'_1, \dots, j'_{l(Q'_2)}$, then for the family

$$\mathcal{F} := \binom{[n]}{j_1} \cup \dots \binom{[n]}{j_{h(Q_1)}} \cup \binom{[n]}{i^*} \cup \binom{[n]}{i^* + j'_1} \cup \dots \cup \binom{[n]}{i^* + j'_{h(Q_2)}}$$

we have $c(Q_1 \otimes Q_2, \mathcal{F}) = \binom{n}{i^*} La(i^*, P_{h(Q_1)+1}, Q_1) La(n-i^*, P_{h(Q_2)+1}, Q_2)$. ■

**Proof of Theorem 437.** The proof goes very similarly to that of Theorem 436. Let us consider the canonical partition of a $P_{h(Q\otimes r)+1}$-free family $\mathcal{F} \subseteq 2^{[n]}$. Then in any copy of $Q \otimes r$ in $\mathcal{F}$, if $F_1, \dots, F_r$ correspond to the $r$ top elements forming an antichain, we must have $F_1, \dots, F_r \in \mathcal{F}_{h(Q)+1}$. Also, in that case the sets corresponding to the other elements of the poset must be contained in $\bigcap_{l=1}^r F_l$. Let $j = |\bigcap_{l=1}^r F_l|$. Then the number of copies of $Q \otimes r$ in $\mathcal{F}$ that contain $F_1, \dots, F_r$ is at most $La(j, P_{h(Q)+1}, Q)$. If $r \geq 2$, we obtain

$$c(Q \oplus r, \mathcal{F}) \leq \beta^r(\mathcal{F}_{h(Q)+1}) \cdot \mathbf{w},$$

where the $j$th coordinate of $\mathbf{w}$ is $La(j, P_{h(Q)+1}, Q)$. Clearly we have $\beta^r(\mathcal{F}_{h(Q)+1}) \cdot \mathbf{w} \leq n \max_i \beta_i^r(\mathcal{F}_{h(Q)+1}) w(i) \leq n^{2r+2} \max_i \beta_i^r \left(\binom{[n]}{j(r,i,n)}\right) w(i)$, with the last inequality following from Theorem 435, where $j(r,i,n)$ is defined. We have $\beta_i^r \left(\binom{[n]}{j(r,i,n)}\right) = \binom{n}{i} \beta_0^r \left(\binom{[n-i]}{j(r,i,n)-i}\right)$ by picking the intersection of size $i$ first.

To see the furthermore part of **(i)**, let $i^*$ be the value of $i$ for which the above maximum is attained. If $\mathcal{F}_{i^*} = \binom{[i^*]}{h_1} \cup \dots \cup \binom{[i^*]}{h_{h(Q)}}$ is a family with $Cn^k c(Q, \mathcal{F}_{i^*}) \geq La(i^*, P_{h(Q)+1}, Q)$, then for the family $\mathcal{F}^* = \binom{[n]}{h_1} \cup \dots \cup \binom{[n]}{h_{h(Q)}} \cup \binom{[n]}{j(r,i^*,n)}$, we have

$$c(Q \oplus r, \mathcal{F}^*) \geq \beta_{i^*}^r \left(\binom{[n]}{j(r,i^*,n)}\right) c(Q, \mathcal{F}_{i^*}) \geq$$
$$\binom{n}{i^*} \beta_0^r \left(\binom{[n-i^*]}{j(r,i^*,n)-i^*}\right) \frac{1}{Cn^k} La(n, i^*, Q);$$

therefore $\mathcal{F}^*$ with $C' = C$ and $k' = 2r+2+k$ shows that Conjecture 430 almost holds for the pair $P_{h(Q)+2}, Q \oplus r$.

If $r = 1$, then $|\cap F_1| = |F_1|$, so applying Theorem 424 we obtain

$$c(Q \oplus 1, \mathcal{F}) \leq \sum_{F \in \mathcal{F}_{l(Q)+1}} La(|F|, P_{h(Q)+1}, Q) \leq \max_{0 \leq i \leq n} \left\{ \binom{n}{i} La(i, P_{h(Q)+1}, Q) \right\}.$$

The proof of the furthermore part of **(ii)** is analogous to the previous ones and is left to the reader. ■

**Theorem 438 (Gerbner, Keszegh, Patkós, [258])** *Conjecture 430 almost holds for the pair $P_{s+1}, K_{r_1,r_2,\dots,r_s}$.*

**Proof.** We proceed by induction on the number of levels. The base case $s = 1$ is guaranteed by Sperner's Theorem 1. The inductive step follows by applying Theorem 437, as $K_{r_1,\dots,r_l} = K_{r_1,\dots,r_{l-1}} \oplus r_l$. ∎

**Theorem 439 (Gerbner, Keszegh, Patkós, [258])** *Conjecture 430 strongly holds for the pair $P_{s+1}, K_{r_1,r_2,\dots,r_s}$, if for every $i < s$, at least one of $r_i$ and $r_{i+1}$ is equal to 1.*

**Proof.** We proceed by induction on the number of levels. The base case $s = 1$ is guaranteed by Theorem 1 again. Suppose the statement has been proven for all complete multi-level posets satisfying the condition with height smaller than $l$ and consider $K_{r_1,r_2,\dots,r_l}$. We know that there exists an $i$ with $1 \le i \le l$ such that $r_i = 1$. If $1 < i < l$; then the inductive step will follow by applying the furthermore part of Theorem 436 to $Q_1 = K_{r_1,\dots,r_{i-1}}$ and $Q_2 = K_{r_{i+1},\dots,r_l}$. If $i = l$, then the inductive step will follow by applying the furthermore part of Theorem 437 to $Q = K_{r_1,\dots,r_{l-1}}$ and $r = 1$. The case $i = 1$ follows from $c(K_{r_1,\dots,r_l}, \mathcal{F}) = c(K_{r_l,\dots,r_1}, \overline{\mathcal{F}})$, where $\overline{\mathcal{F}} = \{[n] \setminus F : F \in \mathcal{F}\}$. ∎

Theorem 439 does not tell us anything about the set sizes in the family containing the most number of copies of $K_{r_1,r_2,\dots,r_s}$. The next theorem gives more insight for an even more special case.

**Theorem 440 (Gerbner, Keszegh, Patkós, [258])** *Let $K = K_{r,1,\dots,1,s}$ be the complete $(l+2)$-level poset where the middle $l$ levels contain only one element. The value of $La(n, P_{l+3}, K)$ is attained for a family $\mathcal{F} = \cup_{j=1}^{l+2} \binom{[n]}{i_j}$, where $i_1 = \lfloor i_2/2 \rfloor$, $i_{l+2} = \lfloor (n + i_{l+1})/2 \rfloor$ and $i_3 - i_2, i_4 - i_3, \dots, i_{l+1} - i_l$ differ by at most 1.*

**Proof.** Consider a $P_{l+3}$-free family $\mathcal{F} \subseteq 2^{[n]}$ and its canonical partition. If $l = 1$, then we count the number of copies of $K_{r,1,s}$ according to the set $F \in \mathcal{F}_2$ that plays the role of the middle element of $K_{r,1,s}$. The number of copies that contain $F$ is not more than $\binom{\binom{|F|}{\lfloor |F|/2 \rfloor}}{r}\binom{\binom{n-|F|}{\lfloor (n-|F|)/2 \rfloor}}{s}$. Applying Theorem 424 with $w(i) = \binom{\binom{i}{\lfloor i/2 \rfloor}}{r}\binom{\binom{n-i}{\lfloor (n-i)/2 \rfloor}}{s}$ yields $c(K_{r,1,s}, \mathcal{F}) \le \max_i \binom{n}{i}\binom{\binom{i}{\lfloor i/2 \rfloor}}{r}\binom{\binom{n-i}{\lfloor (n-i)/2 \rfloor}}{s}$. Let $i^*$ be the value of $i$ for which this maximum is attained. Then the family $\binom{[n]}{\lfloor i^*/2 \rfloor} \cup \binom{[n]}{i^*} \cup \binom{[n]}{\lfloor (n+i^*)/2 \rfloor}$ contains exactly that many copies of $K_{r,1,s}$.

If $l \ge 2$, then we count the number of copies of $K$ according to the sets $F_2 \in \mathcal{F}_2$ and $F_{l+1} \in \mathcal{F}_{l+1}$, playing the role of the elements on the second and $(l+1)$st level of $K$. For a fixed pair $F_2 \in \mathcal{F}_2$ and $F_{l+1} \in \mathcal{F}_{l+1}$ with $F_2 \subset F_{l+1}$, the number of copies of $K$ containing $F_2$ and $F_{l+1}$ is at most $\binom{\binom{|F_2|}{\lfloor |F_2|/2 \rfloor}}{r}\binom{\binom{n-|F_{l+1}|}{\lfloor (n-|F_{l+1}|)/2 \rfloor}}{s} La(|F_{l+1}| - |F_2|, P_{l-1}, P_{l-2})$. The value of $La(|F_{l+1}| -$

$|F_2|, P_{l-1}, P_{l-2})$ is given by Corollary 428. So we can apply Theorem 427 with $l = k = 2$ and $w(i,j) = \binom{\binom{i}{\lfloor i/2 \rfloor}}{r}\binom{\binom{n-j}{\lfloor (n-j)/2 \rfloor}}{s} La(j-i, P_{l-1}, P_{l-2})$ to obtain $c(K, \mathcal{F}) \leq \max_{i,j} \binom{n}{j}\binom{j}{i}\binom{\binom{i}{\lfloor i/2 \rfloor}}{r}\binom{\binom{n-j}{\lfloor (n-j)/2 \rfloor}}{s} La(j-i, P_{l-1}, P_{l-2})$. Let $i^*$ and $j^*$ be the values of $i$ and $j$ for which this maximum is attained. Then the family consisting of $\binom{[n]}{\lfloor i^*/2 \rfloor}, \binom{[n]}{i^*}, \binom{[n]}{j^*}, \binom{[n]}{\lfloor (n+j^*)/2 \rfloor}$ and the $l-2$ full levels determined by Theorem 427 contains exactly that many copies of $K$. ■

# 8

# Traces of sets

CONTENTS



The *trace* of a set $F$ on another set $X$ is $F \cap X$, and is denoted by $F|_X$. The trace of a family $\mathcal{F}$ of sets on $X$ is $\mathcal{F}|_X = \{F|_X : F \in \mathcal{F}\}$. No matter how many sets $F$ have the same trace, it is counted only once in $\mathcal{F}|_X$. We say that a family *traces* a set $X$ (the terminology *shatters* is also used very often), if $\mathcal{F}|_X = 2^X$. The collection of sets that are traced by $\mathcal{F}$ is denoted by $tr(\mathcal{F})$. The *Vapnik-Chervonenkis dimension* (or VC-dimension for short) of $\mathcal{F}$ is the size of the largest set in $tr(\mathcal{F})$, and is denoted by $\dim_{VC}(\mathcal{F})$.

The fundamental result concerning traces of families was proven in the early 1970s independently by Sauer [498], Shelah [504], and Vapnik and Chervonenkis [541]. It is very often referred to as the Sauer Lemma.

**Theorem 441 (Sauer, Shelah, Vapnik, Chervonenkis)** *If $\mathcal{F} \subseteq 2^{[n]}$ has VC-dimension at most $k$, then $|\mathcal{F}| \le \sum_{i=0}^{k} \binom{n}{i}$.*

The proof we present here was found by Alon [9] and Frankl [192] independently. It uses yet another version of shifting.

**Proof.** For a family $\mathcal{F} \subseteq 2^{[n]}$ and $i \in [n]$, let us define the *down-shifting* operation $D_i$ by

$$D_i(F) = \begin{cases} F \setminus \{i\} & \text{if } i \in F \text{ and } F \setminus \{i\} \notin \mathcal{F} \\ F & \text{otherwise.} \end{cases} \tag{8.1}$$

Let us write $D_i(\mathcal{F}) = \{D_i(F) : F \in \mathcal{F}\}$. By definition, we have $|\mathcal{F}| = |D_i(\mathcal{F})|$ for any $\mathcal{F}$ and $i$. Furthermore, writing $w(\mathcal{G}) = \sum_{G \in \mathcal{G}} |G|$, we have $w(D_i(\mathcal{F})) \le w(\mathcal{F})$, and if $D_i(\mathcal{F}) \neq \mathcal{F}$, then $w(D_i(\mathcal{F})) < w(\mathcal{F})$. Therefore, after a finite number of applications of some down-shifting operations, any family $\mathcal{F}$ can be transformed to a *down-shifted* family $\mathcal{F}^*$, i.e. one with the property that $\mathcal{F} = D_i(\mathcal{F})$ holds for any $i$.

Observe that a family is down-shifted if and only if it is a downset. Indeed, let us assume $X \subset Y$, $Y \in \mathcal{F}$ and $X \notin \mathcal{F}$. Then writing $Y \setminus X =$

$\{y_1, y_2, \ldots, y_m\}$ and $X_0 = X$, $X_i = X \cup \{y_1, \ldots, y_i\}$, we obtain a pair $X_i, X_{i+1}$ with $X_i \notin \mathcal{F}$, $X_{i+1} \in \mathcal{F}$. Then $X_i \in D_{y_{i+1}}(\mathcal{F})$ and $X_{i+1} \notin D_{y_{i+1}}(\mathcal{F})$, contradicting the downshifted property of $\mathcal{F}$.

The statement of the theorem is trivially true for downsets; therefore it is enough to show that $tr(D_i(\mathcal{F})) \subseteq tr(\mathcal{F})$ holds for any family $\mathcal{F}$ and $i \in [n]$. If $X \in tr(D_i(\mathcal{F}))$ and $i \notin X$, then for any subset $Y$ of $X$, there exists a $G \in D_i(\mathcal{F})$ with $Y = G \cap X = (G \cup \{i\}) \cap X$. As at least one of $G$ and $G \cup \{i\}$ belongs to $\mathcal{F}$, we have $X \in tr(\mathcal{F})$, as required. So let us suppose now $i \in X \in tr(D_i(\mathcal{F}))$, and consider subsets of $X$ in pairs $(Y, Y \cup \{i\})$ with $i \notin Y$. As $X \in tr(D_i(\mathcal{F}))$, there exists a $G \in D_i(\mathcal{F})$ with $G \cap X = Y \cup \{i\}$. As $D_i$ does not create sets containing $i$, we must have $G \in \mathcal{F}$, and the only reason for $G \in D_i(\mathcal{F})$ is $G \setminus \{i\} \in \mathcal{F}$. As obviously $(G \setminus \{i\}) \cap X = Y$, and the argument works for any pair $(Y, Y \cup \{i\})$, we obtain $X \in tr(\mathcal{F})$, as required. ■

**Theorem 442 (Pajor [450])** *For any finite family of sets we have* $|\mathcal{F}| \le |tr(\mathcal{F})|$.

Note that Theorem 442 implies Theorem 441, as $\mathcal{F} \subseteq 2^{[n]}$ can only trace subsets of $[n]$, and there are only $\sum_{i=0}^{k} \binom{n}{i}$ subsets of $[n]$ of size at most $k$.

**Proof.** We use induction on $|\mathcal{F}|$. The base case $|\mathcal{F}| = 1$ is true, as every family traces the empty set. To prove the inductive step, let $x$ be an element that is contained by at least one set in $\mathcal{F}$, but not by all sets in $\mathcal{F}$. We partition $\mathcal{F}$ into $\mathcal{F}_x \cup \mathcal{F}_{\overline{x}}$, where $\mathcal{F}_x = \{F \in \mathcal{F} : x \in F\}$ and $\mathcal{F}_{\overline{x}} = \{F \in \mathcal{F} : x \notin F\}$. As both $\mathcal{F}_x$ and $\mathcal{F}_{\overline{x}}$ are non-empty, by the inductive hypothesis we have $|tr(\mathcal{F}_x)| \ge |\mathcal{F}_x|$ and $|tr(\mathcal{F}_{\overline{x}})| \ge |\mathcal{F}_{\overline{x}}|$. We make two observations based on the definitions of $\mathcal{F}_x$ and $\mathcal{F}_{\overline{x}}$:

- if $x \in X$, then $X \notin tr(\mathcal{F}_x) \cup tr(\mathcal{F}_{\overline{x}})$,
- if $X \in tr(\mathcal{F}_x) \cap tr(\mathcal{F}_{\overline{x}})$, then $X \cup \{x\} \in tr(\mathcal{F})$.

These two observations imply $|tr(\mathcal{F})| \ge |tr(\mathcal{F}_x)| + |tr(\mathcal{F}_{\overline{x}})| \ge |\mathcal{F}_x| + |\mathcal{F}_{\overline{x}}| = |\mathcal{F}|$. ■

## 8.1 Characterizing the case of equality in the Sauer Lemma

In this short section we summarize results on extremal families in Theorem 441 and Theorem 442. The former direction seems much harder and attracted less researchers. Frankl [192] and Dudley [137] characterized families $\mathcal{F} \subseteq 2^{[n]}$ of size $n+1$ with $dim_{VC}(\mathcal{F}) = 1$. There are quite many known families of size $\sum_{i=1}^{d} \binom{n}{i}$ with VC-dimension $d$ (we will see some of them in later sections), but more results are available concerning families satisfying $|tr(\mathcal{F})| = |\mathcal{F}|$. A statement very similar to Theorem 442 was proved by Bollobás, Leader and Radcliffe [73]. We say that a family $\mathcal{F}$ *strongly traces* the set $X$, if there exists a set $S$ disjoint from $X$ such that $\{Y \cup S : Y \in 2^X\} \subseteq \mathcal{F}$ holds. The set $S$ is a *support* of $X$, and the family of supports is denoted by $\mathbf{S}(X)$. We write $str(\mathcal{F}) = \{X : \mathbf{S}(X) \neq \emptyset\}$.

**Theorem 443 (Bollobás, Leader, Radcliffe, [73])** *For any finite family of sets we have* $|\mathcal{F}| \geq |str(\mathcal{F})|$.

The next exercise immediately implies Theorem 443, as for any downset $\mathcal{D}$ we have $str(\mathcal{D}) = \mathcal{D}$.

**Exercise 444** *Prove that for any family* $\mathcal{F} \subseteq 2^{[n]}$ *and* $i \in [n]$ *we have* $str(D_i(\mathcal{F})) \supseteq str(\mathcal{F})$.

**Exercise 445** *For any family* $\mathcal{F} \subseteq 2^{[n]}$ *we have*

$$2^{[n]} \setminus tr(\mathcal{F}) = str(2^{[n]} \setminus \mathcal{F}).$$

When proving Theorem 441, we observed that a finite number of downshift operations transform a family into a downset. More is true: by definition, for any family $\mathcal{F}$ and $i$ we can see that $D_i(\mathcal{F})$ does not contain a set $D$ with $i \in D$, $D \setminus \{i\} \notin D_i(\mathcal{F})$. Therefore it is enough to apply the downshift $D_i$ once for every $i$. More precisely, if $\sigma$ is a permutation of $[n]$, then for any family $\mathcal{F} \subseteq 2^{[n]}$, the transformed family $D_\sigma(\mathcal{F}) := D_{\sigma(n)}(D_{\sigma(n-1)}(\dots(D_{\sigma(1)}(\mathcal{F})\dots)))$ is downshifted.

**Theorem 446 (Bollobás, Radcliffe [75])** *For any set system* $\mathcal{F} \subseteq 2^{[n]}$ *the following four properties are equivalent*

*(**i**)* $|\mathcal{F}| = |tr(\mathcal{F})|$,
*(**ii**)* $2^n - |\mathcal{F}| = |str(2^{[n]} \setminus \mathcal{F})|$,
*(**iii**) there is a unique downset* $\mathcal{D}$ *that can be achieved from* $\mathcal{F}$ *by shifting, i.e.* $D_\pi(\mathcal{F}) = D_\sigma(\mathcal{F})$ *for any permutations* $\sigma, \pi \in S_n$,
*(**iv**)* $|\mathcal{F}| = |str(\mathcal{F})|$.

**Proof.** First of all, **(i)** and **(ii)** are equivalent by Exercise 445.

In the proof of Theorem 441 we showed $tr(D_i(\mathcal{F})) \subseteq tr(\mathcal{F})$, and similarly in Exercise 444 one proves that $str(D_i(\mathcal{F})) \supseteq str(\mathcal{F})$. For any downset $\mathcal{D}$ we have $str(\mathcal{D}) = \mathcal{D} = tr(\mathcal{D})$; thus for any permutation $\sigma$ we have $tr(\mathcal{F}) \supseteq tr(D_\sigma(\mathcal{F})) = D_\sigma(\mathcal{F})$ and $str(\mathcal{F}) \subseteq str(D_\sigma(\mathcal{F})) = D_\sigma(\mathcal{F})$. Therefore, if $|\mathcal{F}| = |tr(\mathcal{F})|$, then we must have $tr(\mathcal{F}) = D_\sigma(\mathcal{F})$ and if $|\mathcal{F}| = |str(\mathcal{F})|$, then we must have $str(\mathcal{F}) = D_\sigma(\mathcal{F})$. Hence, for any pair of permutations $\sigma, \pi$ we must have $D_\sigma(\mathcal{F}) = D_\pi(\mathcal{F})$. So both **(i)** and **(iv)** implies **(iii)**.

To see that **(iii)** implies the other statements, let $I = \{i_1, i_2, \dots, i_k\}$, $\overline{I} = \{j_1, j_2, \dots, j_{n-k}\}$ and $\sigma$ be a permutation with $\sigma(1) = j_1, \dots, \sigma(n-k) =$

$j_{n-k}, \sigma(n-k+1) = i_1, \ldots, \sigma(n) = i_k$. We claim that $I \in tr(\mathcal{F})$ if and only if $I \in D_\sigma(\mathcal{F})$.

Indeed, if $I \in tr(\mathcal{F})$, then for any $X \subset I$ we have $F \in \mathcal{F}$ with $F \cap I = X$. First note that $X \in D_{j_1}(\ldots(D_{j_{n-k}}(\mathcal{F}))\ldots)$, as these downhsifts do not change $X$. This implies $2^I \subseteq D_{j_1}(\ldots(D_{j_{n-k}}(\mathcal{F}))\ldots)$ and then $2^I \subseteq D_\sigma(\mathcal{F})$. In particular $I \in D_\sigma(\mathcal{F})$. Conversely, as $D_\sigma(\mathcal{F})$ is a downset, $I \in D_\sigma(\mathcal{F})$ implies $2^I \subset D_\sigma(\mathcal{F})$. This means $2^I \subseteq D_{j_1}(\ldots(D_{j_{n-k}}(\mathcal{F}))\ldots)$ and therefore $I \in tr(\mathcal{F})$. Similarly, one can obtain $I \in str(\mathcal{F})$ if and only if $I \in D_\pi(\mathcal{F})$, where $\pi(1) = i_1, \ldots, \pi(k) = i_k, \pi(k+1) = j_1, \ldots, \pi(n) = j_{n-k}$.

Therefore, if **(iii)** holds, then $I \in tr(\mathcal{F})$ if and only if $I \in str(\mathcal{F})$. In particular, we have $|tr(\mathcal{F})| = |str(\mathcal{F})|$; thus **(i)** and **(iv)** hold. ■

Let us mention that different types of characterization of families with $|tr(\mathcal{F})| = |\mathcal{F}|$ were obtained (among others) in [49,136,383,417,420].

Let us finish this section with a recent theorem on the number of extremal families. Let $ExPajVC(n,k)$ denote the number of families $\mathcal{F} \subseteq 2^{[n]}$ of $VC$-dimension $k$ with $|\mathcal{F}| = |tr(\mathcal{F})|$ and let $ExSShVC(n,k)$ denote the number of families $\mathcal{F} \subseteq 2^{[n]}$ of $VC$-dimension $k$ with $|\mathcal{F}| = \sum_{i=0}^{k}\binom{n}{i}$. Theorem 441 implies that the latter families are among the former ones, and thus $ExSShVC(n,k) \le ExPajVC(n,k)$ holds. The function $ExSShVC(n,k)$ was first studied by Frankl [194], and better upper and lower bounds were obtained by Alon, Moran, and Yehudayoff [14]. Balogh, Mészáros, and Wagner determined the asymptotics of $\log ExSShVC(n,k)$ and $\log ExPajVC(n,k)$ if $k$ does not grow very quickly together with $n$.

**Theorem 447 (Balogh, Mészáros, Wagner, [42])** *If $k = k(n) = n^{o(1)}$, then*

$$n^{(1+o(1))\binom{n}{k}} \le ExSShVC(n,k) \le ExPajVC(n,k) \le n^{(1+o(1))\binom{n}{k}}$$

*holds.*

We give the proof of Theorem 447 in the following exercise. Let us remind the reader that $G_{n,k,k+1}$ denotes the bipartite graph having partite sets $\binom{[n]}{k}$ and $\binom{[n]}{k+1}$, with $\{F,G\} \in E(G_{n,k,k+1})$ if $F \in \binom{[n]}{k}, G \in \binom{[n]}{k+1}$ and $F \subset G$ hold. An induced matching $M$ in a graph $G$ is a set $\{e_1, e_2, \ldots, e_m\}$ of pairwise disjoint edges such that for any $i \neq j$ and $v_i \in e_i, v_j \in e_j$ the vertices $v_i$ and $v_j$ are not connected $G$.

**Exercise 448** *(i) Show that the following mapping $\phi$ is an injection from the induced matchings of $G_{n,k,k+1}$ to the set of families $\mathcal{F} \subseteq 2^{[n]}$ of $VC$-dimension*

*$k$ with $|\mathcal{F}| = \sum_{i=0}^{k} \binom{n}{i}$. Let $M$ be an induced matching; then*

$$\phi(M) = \binom{[n]}{\leq k-1} \cup \left\{ F \in \binom{[n]}{k} : \nexists e \in M : F \in e \right\} \cup \left\{ F \in \binom{[n]}{k+1} : \exists e \in M : F \in e \right\}. \quad (8.2)$$

***(ii)*** *Let $M = \{u_1v_1, u_2v_2, \ldots, u_mv_m\}$ be a matching with $u_i \in \binom{[n]}{k}$, $v_i \in \binom{[n]}{k+1}$. Assume that for some set $A \subseteq [n]$ and for every $j = 1, 2, \ldots, m$, we have $u_j \in \binom{[n]\setminus A}{k}$ and $v_j \in \binom{[n]}{k+1} \setminus \binom{[n]\setminus A}{k+1}$. Show that $M$ is an induced matching.*

***(iii)*** *Show that if $k = n^{o(1)}$ and $\varepsilon = o(1/k^2)$, then the number of possible matchings defined as in (ii) with $A = [\varepsilon n]$ is $n^{(1+o(1))\binom{n}{k}}$.*

***(iv)*** *Using the facts that (a) every family with $|\mathcal{F}| = |tr(\mathcal{F})|$ is a connected vertex subset of the hypercube $Q_n = \cup_{i=0}^{n-1} G_{n,i,i+1}$ (see [272]), (b) every graph with $N$ vertices and maximum degree $D$ contains at most $N(e(D-1))^{l-1}$ connected subgraphs of size $l$, show that $ExPajVC(n,k) \leq n^{(1+o(1))\binom{n}{k}}$ holds provided $k = n^{o(1)}$.*

## 8.2 The arrow relation

Sauer's lemma states that if a family $\mathcal{F}$ contains many sets, then there exists a $k$-subset $Y$ such that the trace $\mathcal{F}|_Y$ contains *all* subsets of $Y$. One way to generalize this is to weaken the *all* to *many.* More formally, Frankl [192] introduced the so-called *arrow relation*: $(m,n) \rightarrow (r,s)$ means that for any $\mathcal{F} \subseteq 2^{[n]}$ with $|\mathcal{F}| = m$, there exists an $s$-subset $Y \subseteq [n]$, such that $|\mathcal{F}|_Y| \geq r$ holds. Otherwise we write $(m,n) \not\rightarrow (r,s)$. With this notation Sauer's lemma states that $(m,n) \rightarrow (2^k, k)$ holds, whenever $m \geq 1 + \sum_{i=0}^{k-1} \binom{n}{i}$. Frankl observed that the downshifting argument works for the arrow relation as well.

**Theorem 449 (Frankl [192])** *The relation $(m,n) \rightarrow (r,s)$ holds if and only if for any downset $\mathcal{F} \subseteq 2^{[n]}$ with $|\mathcal{F}| = m$, there exists an $s$-subset $Y \subseteq [n]$, such that we have $|\mathcal{F}|_Y| \geq r$.*

**Proof.** Assume to the contrary that the statement holds for every downset, but does not hold for $\mathcal{F}$, i.e. there is an $s$-set $Y$ with $|\mathcal{F}|_Y| < r$. For any $i \in Y$, we partition the $2^s$ subsets of $Y$ to $2^{s-1}$ pairs $(Z, Z \cup \{i\})$. Then for any pair we have $|\mathcal{F} \cap \{Z, Z \cup \{i\}\}| \geq |D_i(\mathcal{F}) \cap \{Z, Z \cup \{i\}\}|$. This implies $D_i$ does not decrease the trace of $\mathcal{F}$ in $Y$ for $i \in Y$, and obviously $D_i$ does not change $\mathcal{F}|_Y$ for $i \notin Y$. Thus applying $D_i$ for every $i$, we obtain a downset $\mathcal{F}^*$ with $r \leq \mathcal{F}^*|_Y \leq \mathcal{F}|_Y < r$, a contradiction.

■

**Exercise 450 (Bondy [76])** *Prove that* $(m,n) \to (m,n-1)$ *holds if* $m \leq n$.

**Exercise 451 (Bollobás, see in [392])** *Prove that if* $m \leq \lfloor 3n/2 \rfloor$, *then* $(m,n) \to (m-1,n-1)$ *holds.*

The next exercise is a generalization of Mantel's theorem on the maximum number of edges in a triangle-free graph.

**Exercise 452 (Frankl [192])** *If* $m > \lfloor \frac{n^2}{4} \rfloor + n + 1$, *then* $(m,n) \to (7,3)$.

A similar result was obtained by Bollobás and Radcliffe.

**Theorem 453 ( [75])** *For any integer* $n \geq 4$ *and* $n \neq 6$, *if* $N > \binom{n}{2} + n + 1$, *then* $(N,n) \to (12,4)$.

Now we show a consequence of Theorem 23 that will be useful in further proofs. Recall that the colex order on $\binom{\mathbb{Z}^+}{k}$ is defined by $A < B$ if and only if $\max A \setminus B < \max B \setminus A$. This order can be extended to $\mathbb{Z}^+_{fin} = \cup_{k=0}^{\infty} \binom{\mathbb{Z}^+}{k}$ the following way: $A < B$ if and only if $\max A \setminus B < \max B \setminus A$ or $A \subsetneq B$. In this order $\emptyset$ is the smallest element. Furthermore, $2^{[k]}$ is an initial segment for any $k \geq 1$. The colex order's initial segment of size $N$ will be denoted by $\mathcal{S}_N$. By the above, any $\mathcal{S}_N$ is a downset and it can be obtained recursively as follows: if $m = m_N$ is the largest integer with $2^m \leq N$, then $\mathcal{S}_N = \mathcal{S}_m \cup \{S \cup \{m+1\} : S \in \mathcal{S}_{N-2^m}\}$.

**Theorem 454 (Ahlswede, Katona [2])** *Let* $w : \mathbb{N} \to \mathbb{R}$ *be a non-inccreasing weight function. If* $\mathcal{F}$ *is a downset with* $|\mathcal{F}| = N$, *then* $w(\mathcal{F}) = \sum_{F \in \mathcal{F}} w(|F|) \geq \sum_{G \in \mathcal{S}_N} w(|G|)$, *where* $\mathcal{S}_N$ *is the family of the first* $N$ *sets in colex ordering.*

**Proof.** Observe that writing $u_i(\mathcal{F}) = |\{F \in \mathcal{F} : |F| \geq i\}|$, we have $w(\mathcal{F}) = w(0) + \sum_{i=1}^{\infty} u_i(\mathcal{F})(w(i) - w(i-1))$. By the assumption on $w$ we have $w(i) - w(i-1) \leq 0$ for all $i$; therefore the following lemma proves the theorem.

**Lemma 455** *If* $\mathcal{F}$ *is a downset of size* $N$, *then* $u_i(\mathcal{F}) \leq u_i(\mathcal{S}_N)$ *for all* $i$.

**Proof of Lemma.** For $0 \leq l \leq k$, let $f(k,l,u)$ denote the bound given by Theorem 23 such that for any $k$-uniform family $\mathcal{F}$ of size $u$ we have $\Delta_l(\mathcal{F}) \geq f(k,l,u)$, and let us write $F(k,u) = \sum_{l=0}^{k} f(k,l,u)$ and $G(k,u) = \sum_{l=0}^{k-1} f(k,l,u)$. Observe that if $\mathcal{S}_N$ contains $s_i$ sets of size $i$, then the number of sets in $\mathcal{S}_N$ of size smaller than $i$ is between $G(i,s_i)$ and $G(i,s_i+1)$. Indeed, as $\mathcal{S}_N$ is an initial segment, its subfamily of $i$-element sets is $\mathcal{I}^i_{s_i}$, so by Theorem 23 we obtain the lower bound. If a family $\mathcal{F}_0$ contains more than $G(i,s_i+1)$ sets of size less than $i$, then it contains one not from the shadow of $\mathcal{I}^i_{s_i+1}$. But that set comes after $(s_i+1)$ $i$-sets in the lexicographic order, so $\mathcal{F}_0$ cannot be an initial segment, unless it contains at least $s_i+1$ sets of size $i$.

Let $\mathcal{F}$ be a downset of size $N$ such that $u_i(\mathcal{F})$ is maximal. If $2^i > N$, then a downset of size $N$ cannot contain a set of size $i$, so $u_i(\mathcal{F}) = u_i(\mathcal{S}_N) = 0$. Let $m$ be the largest integer with $2^m \leq N$. We prove the statement of the lemma for $i \leq m$ by backward induction on $i$. If $i = m$, then let $s$ be the maximum integer with $F(m, s) \leq N$. By the above, $u_i(\mathcal{F}) \leq s$ and $\mathcal{S}_N$ contains exactly $s$ sets of size $m$.

Suppose now that the lemma is proved for $i+1 \leq m$ and consider $u_i(\mathcal{F})$ for some downset $\mathcal{F}$ of size $N$. By induction, $u_{i+1}(\mathcal{F}) \leq u_{i+1}(\mathcal{S}_N)$. This implies that if $u_i(\mathcal{F}) > u_i(\mathcal{S}_N)$, then the number $f_i$ of $i$-sets in $\mathcal{F}$ is strictly larger than the number $s_i$ of $i$-sets in $\mathcal{S}_N$. But then $\mathcal{F}$ contains at least $G(i, f_i)$ sets of size less than $i$, and therefore $u_i(\mathcal{F}) \leq N - G(i, f_i) \leq N - G(i, s_i) \leq u_i(\mathcal{S}_N)$.
□■

**Theorem 456 (Frankl [192])** $(m, n) \to (m - 2^{t-1} + 1, n - 1)$ *whenever* $m \leq \lceil n(2^t - 1)/t \rceil$.

Note that if $t$ divides $n$, then the bound of the theorem is best possible, as shown by $\mathcal{F} = \cup_{i=0}^{n/t-1} 2^{[it+1,(i+1)t]}$, i.e. a family consisting of $n/t$ disjoint $t$-sets and their subsets.

**Proof.** By Theorem 449, it is enough to prove the statement for downsets. Assume towards a contradiction that $\mathcal{F}$ is a counterexample, so $|\mathcal{F}| \leq \lceil n(2^t - 1)/t \rceil$, and for every $i$ we have $|\mathcal{F}|_{[n]\setminus\{i\}}| \leq n - 2^{t-1}$, in particular $i$ belongs to at least $2^{t-1}$ members of $\mathcal{F}$. Let $L(i) = \{G \subseteq [n] \setminus \{i\} : G \cup \{i\} \in \mathcal{F}\}$ be the *link* of $i$. By the above, $L(i)$ is a downset of size at least $2^{t-1}$. Applying Theorem 454 with $w(x) = 1/(x + 1)$, and noticing that $\mathcal{S}_{2^{t-1}} = 2^{[t-1]}$, we obtain

$$\sum_{G \in L(i)} \frac{1}{|G| + 1} \geq \sum_{S \in \mathcal{S}_{|L(i)|}} \frac{1}{|S| + 1} \geq \sum_{S \in \mathcal{S}_{2^{t-1}}} \frac{1}{|S| + 1}$$
$$= \sum_{j=0}^{t-1} \frac{\binom{t-1}{j}}{j + 1} = \sum_{j=1}^{t} \frac{1}{t} \binom{t}{j} = (2^t - 1)/t.$$

Every downset contains $\emptyset$, so we obtain

$$|\mathcal{F}| = 1 + \sum_{F \in \mathcal{F}} \sum_{i \in F} \frac{1}{|F|} = 1 + \sum_{i \in F} \sum_{F \in \mathcal{F}} \frac{1}{|F|} = 1 + \sum_{i=1}^{n} \sum_{G \in L(i)} \frac{1}{|G| + 1} \geq 1 + n\frac{2^t - 1}{t}.$$

■

The next theorem summarizes results in the direction of Exercise 450, Exercise 451 and Theorem 456.

**Theorem 457** ***(i)*** *[545]* $(m, n) \to (m - 4, n - 1)$ *whenever* $m \leq 17n/6$.

***(ii)*** *[545]* $(m, n) \to (m - 5, n - 1)$ *whenever* $m \leq 13n/4$.

***(iii)*** *[545]* $(m, n) \to (m - 6, n - 1)$ *whenever* $m \leq 7n/2$.

***(iv)*** *[547]* $(m,n) \to (m-9, n-1)$ *whenever* $m \leq 5n$.
***(v)*** *[546]* $(m,n) \to (m-10, n-1)$ *whenever* $m \leq 5n$.
***(vi)*** *[546]* $(m,n) \to (m-13, n-1)$ *whenever* $m \leq \lceil 29n/5 \rceil$.

Now let us state a general result by Bollobás and Radcliffe.

**Theorem 458 (Bollobás, Radcliffe, [75])** *For any positive integers* $n$ *and* $k \in [n]$, $1 \leq m \leq 2^n$ *we have*

$$(m,n) \to (m^{k/n}, k).$$

**Proof.** For any family $\mathcal{F} \subseteq 2^{[n]}$ of size $m$, we consider the following subset of $\mathbb{R}^n$. For any $F \subset [m]$ let $B_F = \{\underline{x} \in [-1,1]^n : x_i \geq 0 \Leftrightarrow i \in F\}$ and let $B_{\mathcal{F}} = \cup_{F \in \mathcal{F}} B_F$. As the $B_F$'s are pairwise disjoint and have volume 1, we have $vol(B_{\mathcal{F}}) = m$. Let $e_1, e_2, \ldots, e_n$ be the vectors of the canonical basis of $\mathbb{R}^n$ and for a subset $I \subset [n]$ we denote by $P_I$ the orthogonal projection of $\mathbb{R}^n$ to the subspace spanned by $\{e_i : i \in I\}$. Then the theorem follows from the well-known inequality (see e.g. [96]) applied to $C = B_{\mathcal{F}}$: for any measurable set $C \subseteq \mathbb{R}^n$

$$vol_n(C)^{1/n} \leq \left( \prod_{I \in \binom{[n]}{k}} vol_k(P_I(C))^{1/k} \right)^{1/\binom{n}{k}}.$$

■

The next two theorems establish a positive and a negative result about traces on subsets of linear size.

**Theorem 459 (Bollobás, Radcliffe, [75])** *For any integer* $r \geq 2$ *and real* $0 < \alpha < 1$ *we have*

$$(n^r, n) \to ((1-o(1))n^{\lambda r}, \alpha n),$$

*where writing* $\lambda_0 = \log_2(1+\alpha)$, *we define* $\lambda$ *by*

$$\lambda = \begin{cases} \lambda_0 & \text{if } \sqrt{2}-1 \leq \alpha < 1 \\ -\frac{\lambda_0}{\lambda_0 \log_2 \lambda_0 + (1-\lambda_0)\log_2(1-\lambda_0)} & \text{otherwise.} \end{cases} \tag{8.3}$$

**Theorem 460 (Bollobás, Radcliffe, [75])** *For any integer* $r \geq 2$ *and real* $0 < \varepsilon < 1/2$, *there exists an* $n_0 = n_0(r, \varepsilon)$, *such that for every* $n \geq n_0$, *there exists a downset* $\mathcal{D} \subset 2^{[n]}$ *with* $|\mathcal{D}| \sum_{i=0}^{r} \binom{n}{i}$ *and* $|\mathcal{D}|_X| < \sum_{i=0}^{r} \binom{n/2}{i} - (1-\varepsilon)2^{-r}\binom{n}{r}$ *for all* $X \in \binom{[n]}{n/2}$, *i.e.*

$$\left( \sum_{i=0}^{r} \binom{n}{i}, n \right) \not\to \left( \sum_{i=0}^{r} \binom{n/2}{i} - (1-\varepsilon)2^{-r}\binom{n}{r}, n/2 \right).$$

Let us finish this section by a remark on multiplicities. Heuristically, it seems obvious that if $\mathcal{F}|_X$ contains many sets, then there should not exist a member of $\mathcal{F}|_X$ that occurs as a trace of many members of $\mathcal{F}$. For a formal definition let $m_{\mathcal{F}|_X} = \max_{Y \subseteq X} |\{F \in \mathcal{F} : F \cap X = Y\}|$. Wiener [549] introduced the following notation: $(m,n) \triangleright (r,s)$ holds if for every $\mathcal{F} \subseteq 2^{[n]}$ with $|\mathcal{F}| = m$, there exists an $s$-subset $X \subseteq [n]$ such that for any $R \subseteq \overline{X}$ we have $m_{\mathcal{F}|_{\overline{X}}}(R) \leq r$.

**Exercise 461** ( [549]) *Use the down shifting operation to prove that the following are equivalent*

*(i)* $(m,n) \triangleright (r,s)$,

*(ii) for every downset* $\mathcal{F} \subseteq 2^{[n]}$ *with* $|\mathcal{F}| = m$, *there exists an* $s$*-subset* $X \subseteq [n]$ *such that for any* $R \subseteq \overline{X}$, *we have* $m_{\mathcal{F}|_{\overline{X}}}(R) \leq r$,

*(iii) for any downset* $\mathcal{F} \subseteq 2^{[n]}$ *with* $|\mathcal{F}| = m$, *there exists an* $s$*-subset* $X \subseteq [n]$ *such that* $|\mathcal{F}|_X| \leq r$ *holds.*

**Theorem 462 (Wiener [549])** *If* $m \geq 2n$ *are positive integers and* $r = \lceil \frac{n^2}{2m-n-2} \rceil$, *then* $(m,n) \triangleright (r+1,r)$ *holds.*

## 8.3 Forbidden subconfigurations

In the previous section we considered the arrow notation and obtained results of the following form: if a family $\mathcal{F} \subseteq 2^{[n]}$ is large enough, then there exists a $k$-subset $X$ of $[n]$ such that $|\mathcal{F}|_X|$ is also large. In this section we will look for special configurations that appear as traces. Let $\mathcal{G}$ be a family with $|\cup_{G \in \mathcal{G}} G| = k$. We will write $\mathcal{F} \to \mathcal{G}$ ($\mathcal{F}$ *arrows/traces* $\mathcal{G}$), if there exists a $k$-set $X$, such that $\mathcal{F}|_X$ contains a subfamily isomorphic to $\mathcal{G}$. Otherwise we will write $\mathcal{F} \not\to \mathcal{G}$. For a collection $\mathbb{G}$ of families we write $\mathcal{F} \to \mathbb{G}$ if $\mathcal{F} \to \mathcal{G}$ for at least one $\mathcal{G} \in \mathbb{G}$. We introduce the parameters $Tr(n,\mathcal{G}) = \max\{|\mathcal{F}| : \mathcal{F} \subseteq 2^{[n]}, \mathcal{F} \not\to \mathcal{G}\}$ and $Tr(n,\mathbb{G}) = \max\{|\mathcal{F}| : \mathcal{F} \subseteq 2^{[n]}, \mathcal{F} \not\to G \text{ for all } G \in \mathbb{G}\}$. With this notation the Sauer Lemma can be stated as $Tr(n, 2^{[k]}) = \sum_{i=0}^{k-1} \binom{n}{i}$. Obviously, if $\mathcal{G} \subseteq \mathcal{G}'$, then $Tr(n,\mathcal{G}) \leq Tr(n,\mathcal{G}')$; thus for any family $\mathcal{G} \subseteq 2^{[k]}$ we have $Tr(n,\mathcal{G}) \leq \sum_{i=0}^{k-1} \binom{n}{i}$.

We start our investigations by showing some configurations $\mathcal{G} \subsetneq 2^{[k]}$ with $Tr(n, 2^{[k]}) = Tr(n,\mathcal{G})$. The interesting part of these theorems will be either a different construction with size $\sum_{i=0}^{k-1} \binom{n}{i}$, or that $\binom{[n]}{\leq k-1}$ and $\binom{[n]}{\geq n-k+1}$ will be the only optimal families.

**Theorem 463 (Füredi, Quinn [252])** *For any integers* $k$ *and* $l$ *with* $0 < l < k$, *we have* $Tr(n, \binom{[k]}{l}) = \sum_{i=0}^{k-1} \binom{n}{i}$.

**Proof.** As mentioned before, Theorem 441 shows that $Tr(n, \binom{[k]}{l}) \leq \sum_{i=0}^{k-1}\binom{n}{i}$; therefore we only need a construction. For any $X \in \binom{[n]}{\leq k-1}$ with $X = \{x_1, x_2, \ldots, x_i\}$, $x_1 < x_2 < \ldots x_i$, let us define

$$F_X = \begin{cases} X & \text{if } |X| < l \\ \{x_1, x_2, \ldots, x_l\} \cup ([x_l + 1, n] \setminus X) & \text{otherwise.} \end{cases} \tag{8.4}$$

Note that if $X, Y \in \binom{[n]}{\leq k-1}$, then $F_X \neq F_Y$. Indeed, as $|F_X| \geq |X|$ holds for all $X$ and $F_X = X$ for $X \leq l-1$, $F_X = F_Y$ would imply $|X|, |Y| \geq l$. If $\{y_1, \ldots, y_l\} \neq \{x_1, \ldots, x_l\}$, then $\min\{y_1, \ldots, y_l\} \triangle \{x_1, \ldots, x_l\}$ belongs to exactly one of $F_x$ and $F_y$. If the smallest $l$ elements of $X$ and $Y$ coincide, then by definition any element in $(Y \setminus \{y_1, \ldots, y_l\}) \triangle (X \setminus \{x_1, \ldots, x_l\})$ belongs to exactly one of $X$ and $Y$. Therefore $\mathcal{F} = \{F_X : X \in \binom{[n]}{\leq k-1}\}$ has size $\sum_{i=0}^{k-1}\binom{n}{i}$.

We are left to prove that $\mathcal{F} \not\rightarrow \binom{[k]}{l}$ holds. If $Z = \{z_1, z_2, \ldots, z_k\}$ is a $k$-subset of $[n]$, then we claim that there is no $F_X \in \mathcal{F}$ with $F_X \cap Z = \{z_1, z_2, \ldots, z_l\}$. Indeed, if $|X| < l$, then $F_X = X$ cannot have a trace larger than itself. If $x_l \leq z_l$, then as $|X| \leq k-1$, at least one of the $z_i$'s $(l+1 \leq i \leq k)$ is not in $X$; thus it must belong to $F_X$; hence $F_X \cap Z \neq \{z_1, z_2, \ldots, z_l\}$. Finally, if $x_l > z_l$, then $z_i \notin F_X$ for some $1 \leq i \leq l$. ■

Next we consider the case when the forbidden configuration is a chain. Note that forbidding the chain $\emptyset, \{1\}, [2], \ldots, [k]$ in the trace is equivalent to forbidding every chain of length $k+1$ in the trace. Here we use the notation $P_{k+1}$.

**Theorem 464 (Patkós [453])** *For any $k \leq n$ we have $Tr(n, P_{k+1}) = \sum_{i=0}^{k-1}\binom{n}{i}$. Furthermore, if for a family $\mathcal{F} \subseteq 2^{[n]}$ with $|\mathcal{F}| = \sum_{i=0}^{k-1}\binom{n}{i}$ we have $\mathcal{F} \not\rightarrow P_{k+1}$, then either $\mathcal{F} = \binom{[n]}{\leq k-1}$ or $\mathcal{F} = \binom{[n]}{\geq n-k+1}$.*

**Proof.** The upper bound on $Tr(n, P_{k+1})$ is straightforward from Theorem 441, and the lower bound is implied by the constructions given in the statement.

To prove the uniqueness of the extremal families, let us consider a family $\mathcal{F} \subseteq 2^{[n]}$ with $|\mathcal{F}| = \sum_{i=0}^{k-1}\binom{n}{i}$. By Theorem 442, we have $|tr(\mathcal{F})| \geq \sum_{i=0}^{k-1}\binom{n}{i}$. But if $\mathcal{F}$ traces a $k$-subset of $[n]$, then $\mathcal{F} \rightarrow P_{k+1}$, so $tr(\mathcal{F}) = \binom{[n]}{\leq k-1}$ and in particular, $|\mathcal{F}| = |tr(\mathcal{F})|$. From Theorem 446, it follows that $|\mathcal{F}| = |str(\mathcal{F})|$, and thus by $str(\mathcal{F}) \subseteq tr(\mathcal{F})$, we have that $str(\mathcal{F}) = tr(\mathcal{F}) = \binom{[n]}{\leq k-1}$.

Now let us consider a set $F \in \mathcal{F}$ of minimum size. If $|F| > n-k+1$, then $|\mathcal{F}| < Tr(n, P_{k+1})$ - a contradiction. Therefore $|F| \leq n-k+1$, so there exists $X \subseteq [n] \setminus F$ with $|X| = k-1$. By the paragraph above, we have $X \in str(\mathcal{F})$. Let us take an arbitrary $S(X) \in \mathbf{S}(X)$. We claim that there is no element $s \in S(X) \setminus F$. Indeed, if there is, then let us consider $\mathcal{F}|_{X \cup \{s\}}$. Since $F \in \mathcal{F}$ and $s \notin F$, we have $\emptyset \in \mathcal{F}|_{X \cup \{s\}}$. Since $X \in str(\mathcal{F})$ and $s \in S(X)$, there is a chain in $\mathcal{F}|_{X \cup \{s\}}$ of length $k$ with set sizes $1, 2, ..., k$, which together with the empty set form a chain of length $k+1$ - a contradiction. Thus $S(X) \subseteq F$, which implies $S(X) = F$. Indeed, by the definition of support $S(X) \cup \emptyset = S(X) \in \mathcal{F}$ and $F$ is of minimum size in $\mathcal{F}$. This implies $F + 2^X \subseteq \mathcal{F}$. As $X$ was chosen arbitrarily, we obtain that for any $Y$ with $Y \cap F = \emptyset$ and $|Y| = k-1$, we have $F + 2^Y \subseteq \mathcal{F}$.

We claim that for any such $Y$, the set $F \cup Y$ is maximal in $\mathcal{F}$. Indeed, if not, then $F \cup Y \cup A \in \mathcal{F}$ for some non-empty $A$. Therefore, for some $a \in A$, the trace $\mathcal{F}|_{Y \cup \{a\}}$ contains a chain of length $k+1$ (the trace of $F \cup Y \cup A$ is $Y \cup \{a\}$ and from the trace of $F + 2^Y$ we can pick the other $k$ sets) - a contradiction.

We claim that for any $Y' \subseteq F \cup Y$ with $|Y'| = k-1$ we have $F \cup Y \setminus Y' + 2^{Y'} \subseteq \mathcal{F}$ (and $F \cup Y \setminus Y'$ is minimal in $\mathcal{F}$). To see this, observe that the only support of $Y'$ is $S(Y') = F \cup Y \setminus Y'$. Indeed, if there was an element $s \in (F \cup Y \setminus Y' \setminus S(Y')$, then we would have a chain of length $k+1$ in $\mathcal{F}|_{Y' \cup s}$. Thus $S(Y') \supseteq F \cup Y \setminus Y'$, and $S(Y') \supset F \cup Y \setminus Y'$ would contradict the maximality of $F \cup Y \cup Y'$ as $S(Y') \cup Y' \in \mathcal{F}$ by definition. The minimality of $F \cup Y \setminus Y'$ follows just as the maximality of $F \cup Y$.

We obtained that for any $Y, Y'$ with $Y' \subseteq F \cup Y$ and $|Y| = |Y'| = k-1$ we have $F \cup Y \setminus Y' + 2^{Y'} \subseteq \mathcal{F}$ and $F \cup Y \setminus Y'$ is minimal in $\mathcal{F}$, so we could have started with $F \cup Y \setminus Y'$ in place of $F$. Thus we get, that for any $Y_1, Y_1', Y_2, Y_2', ..., Y_m, Y_m'$ the set $F' = ((((F \cup Y_1 \setminus Y_1') \cup Y_2 \setminus Y_2')...) \cup Y_m \setminus Y_m')$ is minimal in $\mathcal{F}$ and $F' + 2^{Y_m'} \subseteq \mathcal{F}$. That is for any $G \subseteq [n]$ with $|F| \leq |G| \leq |F| + k - 1$, we have $G \in \mathcal{F}$. But as $Tr(n, P_{k+1}) = \sum_{i=0}^{k-1} \binom{n}{i}$, it is possible if and only if $\mathcal{F} = \binom{[n]}{\leq k-1}$ or $\mathcal{F} = \binom{[n]}{\geq n-k+1}$. ■

Gerbner, Patkós and Vizer considered forbidding other posets in traces. They determined $Tr(n, P)$ for every poset containing a unique maximum element, and proved the following.

**Theorem 465 (Gerbner, Patkós, Vizer, [267])** *For the butterfly poset $B$, we have $Tr(n, B) = \lceil 3n/2 \rceil - 1$.*

We will omit the proof of the uniqueness part of the next theorem. Note that the upper bound follows from Exercise 452.

**Theorem 466 (Balogh, Keevash, Sudakov, [41])** $Tr(n, \{\binom{[3]}{1}, \binom{[3]}{2}\}) = \lfloor n^2/4 \rfloor + n + 1$ *and the unique optimal family consists of the empty set and all intervals $I \subset [n]$ that contain at least one of $\lfloor n/2 \rfloor$ and $\lfloor n/2 \rfloor + 1$.*

The next theorem gives an example of a collection of three configurations, with the property that the size of a family $\mathcal{F}$ not tracing any of them is bounded by a constant, independent of $| \cup_{F \in \mathcal{F}} F|$.

**Theorem 467 (Balogh, Bollobás [36], Balogh, Keevash, Sudakov [41])** *There exists a function $f(k, l, m)$ such that $Tr(n, \{\binom{[k]}{1}, \binom{[l]}{l-1}, P_m\}) < f(k, l, m)$.*

We present the proof of Balogh and Bollobás [36].

**Proof.** Observe first that it is enough to prove that there exists a function $f'(k,l,m)$, such that for any *antichain* $\mathcal{F}$ with $|\mathcal{F}| > f'(k,l,m)$, we have either $\mathcal{F} \to \binom{[k]}{1}$ or $\mathcal{F} \to \binom{[l]}{l-1}$ or $\mathcal{F} \to P_m$. Indeed, by Lemma 5, if $|\mathcal{F}| > (m-1)f'(k,l,m)$, then $\mathcal{F}$ either contains a chain of length $m$ and thus $\mathcal{F} \to P_m$ or $\mathcal{F}$ contains an antichain larger than $f'(k,l,m)$.

To prove the existence of $f'(k,l,m)$, we proceed by induction. It is easy to verify that $f'(2,l,m) = f'(k,2,l) = f'(k,l,2) = 2$: if two sets $A,B$ are different, then for any $x \in A\triangle B$ we have $\{A,B\}|_{\{x\}} = P_2$ (so even $f(k,l,2) = 2$), while if $A,B$ are not in containment, then for $x \in A\setminus B, y \in B\setminus A$ we have $\{A,B\}|_{\{x,y\}} = \binom{[2]}{1} = \binom{[2]}{2-1}$.

To finish the induction, we will prove

$$f'(k,l,m) \leq 2(m-2)(f'(k-1,l,m)-1)(f'(k,l-1,m)-1)+2.$$

Suppose to the contrary that there exists an antichain $\mathcal{F}$ with $|\mathcal{F}| = 2(m-2)(f'(k-1,l,m)-1)(f'(k,l-1,m)-1)+2$ and $\mathcal{F} \not\to \binom{[k]}{1}$, $\mathcal{F} \not\to \binom{[l]}{l-1}$, $\mathcal{F} \not\to P_m$. Let $\mathcal{F}$ be such that $|\cup_{F\in\mathcal{F}}F|$ is minimal among the counterexamples. Note that if $\mathcal{F}$ is a counterexample, then so is $\overline{\mathcal{F}} = \{V\setminus F : F \in \mathcal{F}\}$ where $V = \cup_{F\in\mathcal{F}}F$. Therefore, we may assume that the average set size in $\mathcal{F}$ is at least $|V|/2$.

Observe that there is no $x \in \cap_{F\in\mathcal{F}}F$ as then $\{F\setminus\{x\} : F \in \mathcal{F}\}$ would also be a counterexample, contradicting the minimality of $\mathcal{F}$. So by the above, there exists an $x \in V$ such that

$$|\mathcal{F}| > |\mathcal{F}_x| > |\mathcal{F}|/2 = (m-2)(f'(k-1,l,m)-1)(f'(k,l-1,m)-1)+1,$$

where $\mathcal{F}_x = \{F \in \mathcal{F} : x \in F\}$. Let us fix $A \in \mathcal{F}$ with $x \notin A$ and for all $B \in \mathcal{F}_x|_A$ define $\mathcal{F}_B = \{F \in \mathcal{F}_x : F\cap A = B\}$.

CASE I: There exists a $B \in \mathcal{F}_x|_A$ with $|\mathcal{F}_B| \geq f'(k-1,l,m)$.

Then either $\mathcal{F}_B \to \binom{[k-1]}{1}$ or $\mathcal{F}_B \to \binom{[l]}{l-1}$ or $\mathcal{F}_B \to P_m$. In the latter two cases we are done, as $\mathcal{F}_B \subset \mathcal{F}$. If $\mathcal{F}_B \to \binom{[k-1]}{1}$, then there exists $F_1,F_2,\dots,F_{k-1} \in \mathcal{F}_B$, $X = \{x_1,x_2,\dots,x_{k-1}\}$ with $x_i \in F_j$ if and only if $i=j$. As $k-1 \geq 2$ and all the $F_i$'s are from $\mathcal{F}_B$, we must have $X\cap A = \emptyset$. As $\mathcal{F}$ is an antichain, there must exist $y \in A\setminus F_i = A\setminus B$. Then we have $\{F_1,F_2,\dots,F_{k-1},A\}|_{X\cup\{y\}} = \binom{[k]}{1}$.

CASE II: $|\mathcal{F}_B| < f'(k-1,l,m)$ for all $B \in \mathcal{F}_x|_A$.

Then $|\mathcal{F}_x|_A| \geq (m-2)(f'(k,l-1,m)-1)+1$. If all antichains in $\mathcal{F}_x|_A$ have size less than $f'(k,l-1,m)$, then by Lemma 5, it contains a chain of size $m-1$. Note that $A \notin \mathcal{F}_x|_A$, as $\mathcal{F}$ is an antichain. Therefore, if $F_1|_A \subset F_2|_A \subset \dots F_{m-1}|_A$ is a chain of length $m-1$ with $F_i \in F_x$, then $\{F_1,F_2,\dots,F_{m-1},A\}|_A = P_m$.

Therefore, we can assume that $\mathcal{F}_x|_A$ contains an antichain $\mathcal{F}'$ of size $f'(k,l-1,m)$. If $\mathcal{F}' \to \binom{[k]}{1}$ or $\mathcal{F}' \to P_m$, then we are done, otherwise $\mathcal{F}' \to$

$\binom{[l-1]}{l-2}$. This means there exist $a_1, a_2, \ldots, a_{l-1} \in A$ and $F_1, F_2, \ldots, F_{l-1} \in \mathcal{F}_x$, such that $a_i \notin F_j$ if and only if $i = j$ holds. Let $B = \{a_1, a_2, \ldots, a_{l-1}, x\}$; then $\{F_1, F_2, \ldots, F_{l-1}, A\}|_B = \binom{[l]}{l-1}$. ■

We finish this section by a brief introduction to the area of forbidden subconfigurations, which is a slight generalization of the $\mathcal{F} \to \mathcal{G}$ relation. We do not go into detail; the interested reader is referred to the recent dynamic survey by Anstee [20], and all citations within.

If $F$ is a $k \times l$ binary matrix, then we say a binary matrix $A$ *has* $F$ *as a configuration* if there is a submatrix of $A$ which is a row and column permutation of $F$. If $A = A_{\mathcal{F}}$ is the incidence matrix of a family $\mathcal{F} \subseteq 2^{[n]}$, i.e. columns of $A$ are the characteristic vectors of sets in $\mathcal{F}$, then $F$ is a configuration of $A$ if and only if there is a $k$-set $X \subseteq [n]$ such that $\mathcal{F}|_X$ (this time counted with multiplicities!) contains sets with incidence matrix $F$. A matrix is *simple* if it does not contain repeated columns; thus a binary matrix is simple if and only if it is the incidence matrix of a family of sets. Let $forb(n, F)$ denote the largest integer $m$, such that there exists a simple $m \times n$ binary matrix $A$ that does not have $F$ as a configuration. The slight generalization compared to $\mathcal{F} \not\to \mathcal{G}$ consists of the facts that $F$ might contain an all zero row (i.e. the corresponding vertex does not belong to any traces) and $F$ does not need to be simple, so it might allow some trace to exist but not as many times as it occurs in $F$.

Note that the Sauer lemma can be formulated as $forb(n, K_k) = \sum_{i=0}^{k-1} \binom{n}{k-1}$, where $K_k$ is the $k \times 2^k$ binary matrix containing all possible columns once. This implies that for every fixed $k \times l$ binary matrix $F$, we have $forb(n, F) = O(n^k)$. Indeed, if for a $k \times l$ matrix we write $t \cdot A$ to denote the $k \times tl$ matrix obtained by repeating the columns of $A$ $t$ times, then $F$ is a subconfiguration of $l \cdot K_k$. Thus it is enough to show $forb(n, l \cdot K_k) = 1 + (l-1)\binom{n}{k}2^k + \sum_{i=0}^{k-1} \binom{n}{i} = O(n^k)$. If $A$ has more columns than $1 + (l-1)\binom{n}{k}2^k + \sum_{i=0}^{k-1} \binom{n}{i}$, then using Theorem 441 we can pick $2^k$ columns, such a way that there exist $k$ rows that form a $K_k$ with those columns. Then we can similarly pick $2^k$ columns from the remaining ones, and so on, we can repeatedly pick $(l-1)\binom{n}{k} + 1$ pairwise disjoint $2^k$-tuples of columns with the following property. For each of these $2^k$-tuples there exists a set of $k$ rows such that the submatrix formed by these columns and rows is $K_k$. Then there exist $k$ rows that belong to at least $l$ such $2^k$-tuples of columns. The corresponding submatrix is $l \cdot K_k$.

Let us formulate the main conjecture of the area due to Anstee and Sali [22] concerning the order of magnitude of $forb(n, F)$. We need to introduce a new operation.

Let $A$ be an $m_1 \times n_1$ simple matrix and let $B$ be an $m_2 \times n_2$ simple matrix. Then $A \times B$ denotes the $(m_1 + m_2) \times (n_1 n_2)$ simple matrix where each column consists of a column of $A$ placed on a column of $B$, and this is done in all possible ways. More generally, the $p$-fold product $A = A_1 \times A_2 \times \cdots \times A_p$

is an $(\sum_{i=1}^{p} m_i) \times (\prod_{i=1}^{p} n_i)$ simple matrix, where each column consists of $p$ columns placed on top of each other, in such a way that the $i$th of them from the top is a column of $A_i$. In the language of set families, if $A_i$ is the incidence matrix of $\mathcal{F}_i$, such that the underlying sets of the $\mathcal{F}_i$'s are pairwise disjoint, then $A$ is the incidence matrix of the family $\mathcal{F} = \{F_1 \cup F_2 \cup \cdots \cup F_t : F_i \in \mathcal{F}_i\}$.

For a binary matrix $A$, its *binary complement* is obtained by replacing each 1-entry by 0 and each 0-entry by 1. Note that if $A$ is the incidence matrix of a family $\mathcal{F}$, then its binary complement is the incidence matrix of the complement family $\overline{\mathcal{F}}$. Let $I_k$ denote the $k \times k$ identity matrix, $I_k^c$ denote the binary complement of $I_k$ and $T_k$ denote the $k \times k$ upper triangular binary matrix. As subconfigurations, these correspond to arrowing $\binom{[k]}{1}$, $\binom{[k]}{k-1}$ and $P_k^-$, respectively, where $P_k^-$ is a maximal chain in $2^{[k]}$ with the empty set removed. For any $k \times l$ binary matrix $F$, let $X(F)$ be the smallest $p$, so that for any large enough $t$, the matrix $F$ is a configuration in $A_1 \times A_2 \times \cdots \times A_p$ for every choice of $A_i$ as either $I_t$, $I_t^c$ or $T_t$. By definition, for large $t$, there exists a $(X(F)-1)t \times t^{X(F)-1}$ simple matrix $A$ not having $F$ as a subconfiguration; therefore the important part of the following conjecture is to prove (or disprove) the matching upper bound.

**Conjecture 468** ( [22]) *For any binary matrix $F$, we have $forb(n, F) = \Theta(n^{X(F)-1})$.*

The decision problem on the parameter $X(F)$ was shown to be NP-complete by Raggi [472].

Note that any simple binary matrix is a subconfiguration of $K_k$; therefore $forb(n, F) \le \sum_{i=1}^{k-1} \binom{n}{i} = O(n^{k-1})$. We finish this section with a theorem of Anstee and Fleming, that characterizes those matrices $F$ for which $forb(n, F)$ has maximum order of magnitude. The incidence matrices of $\binom{n}{i}$, $\binom{n}{\ge i}$, $\binom{n}{\le i}$ are denoted by $K_k^i$, $K_k^{\ge i}$, and $K_k^{\le i}$.

**Theorem 469 (Anstee, Fleming, [23])** ***(i)*** *If $F$ is a simple $k \times l$ matrix with the property that there is a pair of rows of $F$ that do not contain $K_2^0$, a pair of rows of $F$ that do not contain $K_2^2$, and a pair of rows of $F$ that do not contain the configuration $K_2^1 = I_2$, then $forb(n, F) = O(n^{k-2})$.*

***(ii)*** *If $F$ is a simple $k \times l$ matrix with the property that either every pair of rows has $K_2^0$, or every pair of rows has $K_2^2$, or every pair of rows has $K_2^1$ as a subconfiguration, then $forb(n, F) = \Theta(n^{k-1})$.*

**Proof.** The proof is based on the following lemma.

**Lemma 470** *Let $k$ be given with $k \ge 2$. Let $F_1$, $F_2$, $F_3$ be $k$-rowed simple not necessarily distinct matrices with the following property. There exist two rows $i_1$, $j_1$ such that $F_1|_{\{i_1,j_1\}}$ contains no configuration $K_2^2$, and there exist two rows $i_2$, $j_2$ such that $F_2|_{\{i_2,j_2\}}$ contains no configuration $K_2^0$, and there* *exist two rows $i_3, j_3$ such that $F_3|_{\{i_3,j_3\}}$ contains no configuration $K_2^1$. Then $forb(n, \{F_1, F_2, F_3\}) \leq 2\sum_{i=0}^{k-2}\binom{n}{i}$.*

**Proof of Lemma.** For the case $k = 2$ we proceed by induction on $n$. The base case $n = 1$ is obvious, as a simple 1-rowed matrix can have at most 2 columns, while $2\sum_{i=0}^{k-2}\binom{n}{i} = 2$ for $k = 2$. For larger $n$, we can assume that $F_1$ is $K_2^{\leq 1}$, as this is the only maximal configurations without $K_2^2$. By similar reasoning, $F_2$ is $K_2^{\geq 1}$ and $F_3$ is the incidence matrix of $P_3$.

Note that if a simple matrix $A$ contains an all 0 row, or an all 1 row, then deleting these rows will not affect the simplicity of $A$. Similarly, if $A$ contains a pair of rows, one of which is the binary complement of the other (i.e. we can get one from the other by replacing every 0 by 1 and every 1 by 0), then deleting one of them will leave the remainder of $A$ simple. If $A$ does not have $F_1, F_2, F_3$ as subconfiguration, then any pair of rows of $A$ can contain only two types of the four possible columns $\begin{bmatrix}0\\0\end{bmatrix}$, $\begin{bmatrix}0\\1\end{bmatrix}$, $\begin{bmatrix}1\\0\end{bmatrix}$, and $\begin{bmatrix}1\\1\end{bmatrix}$. It is easy to check that in each of the $6 = \binom{4}{2}$ cases, either one of the rows is all 0, or one of the rows is all 1, or the two rows are binary complements of each other; thus we can remove one of them, to obtain a simple binary matrix $A'$ with $n-1$ rows. The induction hypothesis proves that $A'$ and therefore $A$ cannot have more than 2 columns.

If $k > 2$ and $n \leq k$, then we are done, since $2\sum_{i=0}^{k-2}\binom{n}{i} \geq 2^n$ holds, and as $A$ is simple, it cannot have more than $2^n$ columns. Then we proceed by induction on $n$ and $k$. If $A$ is an $n$-rowed simple matrix not containing $F_1, F_2, F_3$ as subconfigurations, then we partition the columns of $A$ as we did in the proof of Theorem 442 (there we used sets instead of columns), by deleting the first row and grouping all repeated columns in the matrix $B_2$:

$$A = \begin{bmatrix} 00000 & 000000 & 111111 & 1111 \\ B_1 & B_2 & B_2 & B_3 \end{bmatrix}$$

Then $B_1B_2B_3$ is a simple $(n-1)$-rowed matrix not containing $F_1, F_2, F_3$ as subconfiguration, so by induction it cannot have more than $2\sum_{i=0}^{k-2}\binom{n-1}{i}$ columns. To bound the number of columns in $B_2$, observe that as $k > 2$, we can delete a row from each $F_j$ $j = 1, 2, 3$ to obtain $F_j'$ that still possess the properties required in the conditions of the lemma. By deleting repeated columns, we can make sure that the $F_j'$s are simple and $(k-1)$-rowed. By induction, we obtain that the number of columns in $B_2$ is at most $\sum_{i=0}^{k-3}\binom{n-1}{i}$. So the number of columns in $A$ is at most $2\sum_{i=0}^{k-2}\binom{n-1}{i} + \sum_{i=0}^{k-3}\binom{n-1}{i} = 2\binom{n-1}{0} + 2\sum_{i=1}^{k-2}(\binom{n-1}{i} + \binom{n-1}{i-1}) = 2\sum_{i=0}^{k-2}\binom{n}{i}$. □

Now **(i)** follows by applying Lemma 470 with $F_1 = F_2 = F_3 = F$.

To obtain **(ii)**, observe that $K_2^2$ is not a subconfiguration of $I_t$, $K_2^0$ is not a subconfiguration of $I_t^c$ and $K_2^1$ is not a subconfiguration of $T_t$. If $F$ is a simple $k$-rowed matrix such that every pair of rows has $K_2^2$, then consider the $(k-1)$-fold product $P = I_{\lfloor n/(k-1)\rfloor} \times I_{\lfloor n/(k-1)\rfloor} \times \ldots I_{\lfloor n/(k-1)\rfloor}$. We claim that $P$ does not contain $F$ as a subconfiguration. Indeed, otherwise at least two rows of the subconfiguration $F$ would belong to the same factor of the product, but that contradicts the choice of $H$ and the property of $F$. As $P$ has $\Omega(n^{k-1})$ rows, we obtain $forb(n, F) = \Omega(n^{k-1})$, and we have already mentioned $forb(n, F) = O(n^{k-1})$ holds for all simple $k$-rowed matrix. If every pair of rows has $K_2^0$ or $K_2^1$ instead, then replacing $I_{\lfloor n/(k-1) \rfloor}$ by $I^c_{\lfloor n/(k-1) \rfloor}$ or $T_{\lfloor n/(k-1) \rfloor}$ in the $(k-1)$-fold product gives the same result. ■

## 8.4 Uniform versions

In this section we consider uniform versions of problems in the previous sections. We will write $Tr^{(k)}(n, \mathcal{G}) = \max\{|\mathcal{F}| : \mathcal{F} \subseteq \binom{[n]}{k}, \mathcal{F} \not\to \mathcal{G}\}$ and $Tr^{(k)}(n, \mathbb{G}) = \max\{|\mathcal{F}| : \mathcal{F} \subseteq \binom{[n]}{k}, \forall \mathcal{G} \in \mathbb{G}\, \mathcal{F} \not\to \mathcal{G}\}$. Let us start with trying to find a uniform Sauer lemma.

**Theorem 471 (Frankl, Pach [218])** *If a family $\mathcal{F} \subseteq \binom{[n]}{k}$ has VC-dimension smaller than d, then $|\mathcal{F}| \le \binom{n}{d-1}$.*

**Proof.** Let $\mathcal{F} = \{F_1, F_2, \dots, F_{|\mathcal{F}|}\}$, and let $Y_1, Y_2, \dots, Y_{\binom{n}{d-1}}$ be an enumeration of $\binom{[n]}{d-1}$. Let us define the $|\mathcal{F}| \times \binom{n}{d-1}$ *inclusion matrix* $A$ with entries $a_{ij}$ by

$$a_{ij} = \begin{cases} 1 & \text{if } F_i \supset Y_j \\ 0 & \text{otherwise.} \end{cases}$$

If $|\mathcal{F}| > \binom{n}{d-1}$ holds, then the rows cannot be linearly independent; therefore there exist real numbers $\alpha_1, \alpha_2, \dots, \alpha_{|\mathcal{F}|}$, not all of which are zero, such that for any $1 \le j \le \binom{n}{d-1}$, we have

$$\sum_{Y_j \subset F_i} \alpha_i = 0.$$

Let $X$ be a subset of $[n]$ that is minimal with respect to the property $\sum_{X \subseteq F_i} \alpha_i \neq 0$. By the choice of the $\alpha_i$'s, we have $|X| \ge d$. We claim that $X \in tr(\mathcal{F})$ holds, which would give the desired contradiction. The claim $X \in tr(\mathcal{F})$ will follow from the statement that for every $Y \subseteq X$, we have

$$\sum_{F_i \cap X = Y} \alpha_i = (-1)^{|X|-|Y|} \sum_{F_i \supseteq X} \alpha_i \neq 0, \tag{8.5}$$

as this implies that there exists an $F_i$ with $F_i \cap X = Y$.

The equation (8.5) is trivial for $Y = X$ and we proceed by induction on $|X| - |Y|$.

$$\sum_{F_i \cap X = Y} \alpha_i = \sum_{F_i \supseteq Y} \alpha_i - \sum_{Z: Y \subset Z \subseteq X} \sum_{F_i \cap X = Z} \alpha_i$$
$$= 0 - \sum_{Z: Y \subset Z \subseteq X} (-1)^{|X|-|Z|} \sum_{F_i \supseteq X} \alpha_i = (-1)^{|X|-|Y|} \sum_{F_i \supseteq X} \alpha_i.$$

■

In the special case of $d = k$, the star $\mathcal{F}_1 = \{F \in \binom{[n]}{k} : 1 \in F\}$ shows that there is a $k$-uniform family of size $\binom{n-1}{k-1}$ with $VC$-dimension smaller than $k$. Indeed, if a $k$-set $X$ does not contain 1, then there is no set $F \in \mathcal{F}_1$ with $X \cap F = X$, while if $1 \in X$, then there is no set $F \in \mathcal{F}_1$ with $X \cap F = \emptyset$. Frankl and Pach conjectured that this is the optimal construction; i.e every $\mathcal{F} \subseteq \binom{[n]}{k}$ with $|\mathcal{F}| > \binom{n-1}{k-1}$ has $VC$-dimension $k$. A counterexample of size $\binom{n-1}{k-1} + \binom{n-4}{k-3}$ was given by Ahlswede and Khatchatrian [6]. Mubayi and Zhao found many more families of the same size with $VC$-dimension smaller than $k$.

**Proposition 472 (Mubayi and Zhao [437])** *Let $P(n, r)$ denote the number of non-isomorphic $r$-graphs on $[n]$. Then for $k \geq 3$, there are at least $P(n-4, k-1)/2$ non-isomorphic families $\mathcal{F} \subseteq \binom{[n]}{k}$, such that $|\mathcal{F}| = \binom{n-1}{k-1} + \binom{n-4}{k-3}$ and $\dim_{VC} < k$ hold.*

**Proof.** The constructed families $\mathcal{F}$ consist of three parts $\mathcal{F}_0 \cup \mathcal{F}_1 \cup \mathcal{F}_2$, such that $\mathcal{F}_0 = \{F \in \binom{[n]}{k} : 1, 2 \in F\}$, $\mathcal{F}_1 \subseteq \{F \in \binom{[n]}{k} : 1 \in F, 2 \notin F\}$, $\mathcal{F}_2 \subseteq \{F \in \binom{[n]}{k} : 2 \in F, 1 \notin F\}$. To specify $\mathcal{F}_1$ and $\mathcal{F}_2$, we will describe some properties that the *link graphs* $\mathcal{G}_i = \{F \setminus \{i\} : F \in \mathcal{F}_i\}$ for $i = 1, 2$ should possess. Note that $\mathcal{G}_i \subseteq \binom{[3,n]}{k-1}$.

1. $\mathcal{G}_1 \cup \mathcal{G}_2 = \binom{[3,n]}{k-1}$.
2. $\mathcal{G}_1 \cap \mathcal{G}_2 = \{G \in \binom{[3,n]}{k-1} : 3, 4 \in G\}$.
3. $\mathcal{G}_1 \supseteq \{G \in \binom{[3,n]}{k-1} : 3 \in G, 4 \notin G\}$ and $\mathcal{G}_2 \supseteq \{G \in \binom{[3,n]}{k-1} : 4 \in G, 3 \notin G\}$.

If $\mathcal{F}$ is defined in such a way that the above three properties of $G_1$ and $G_2$ hold, then

$$|\mathcal{F}| = |\mathcal{F}_0| + |\mathcal{F}_1| + |\mathcal{F}_2| = |\mathcal{F}_0| + |\mathcal{G}_1| + |\mathcal{G}_2| = |\mathcal{F}_0| + |\mathcal{G}_1 \cup \mathcal{G}_2| + |\mathcal{G}_1 \cap \mathcal{G}_2|$$
$$= \binom{n-2}{k-2} + \binom{n-2}{k-1} + \binom{n-4}{k-3} = \binom{n-1}{k-1} + \binom{n-4}{k-3}.$$

We claim that for these families we have $\dim_{CV}(\mathcal{F}) < k$. Suppose not and $E \in tr(\mathcal{F}) \cap \binom{[n]}{k}$. Clearly $E \in \mathcal{F}$, as otherwise $E \notin \mathcal{F}|_E$. All sets in $\mathcal{F}$ contain at least one of 1 and 2; therefore so does $E$. It also implies that $\{1,2\} \not\subset E$ as otherwise $\emptyset \notin \mathcal{F}|_E$. So without loss of generality we may assume that $1 \in E, 2 \notin E$. As $E \in tr(\mathcal{F})$, we must have $E \setminus \{1\} \in \mathcal{F}|_E$. By definition of $\mathcal{F}$, this can only happen if $(E \setminus \{1\}) \cup \{2\} \in \mathcal{F}$. This implies $E \setminus \{1\} \in \mathcal{G}_1 \cap \mathcal{G}_2$, so $\{3,4\} \subseteq E$ holds. As $\{1,3,4\} \subseteq E \in tr(\mathcal{F})$, we have $E \setminus \{1,4\} \in \mathcal{F}|_E$. The set $F \in \mathcal{F}$ with $E \setminus \{1,4\} = F|_E$ must belong to $\mathcal{F}_2$, as otherwise it would contain 1. But this is impossible by the third condition on $\mathcal{G}_2$. This contradiction proves that $\dim_{CV}(\mathcal{F}) < k$.

Finally, let us see how many non-isomorphic families we obtained by the above construction. We have $\mathcal{G}_1 \triangle \mathcal{G}_2 = \binom{[5,n]}{k-1}$ and these sets can be put into any of $\mathcal{G}_1$ or $\mathcal{G}_2$. Note that $deg_{\mathcal{F}}(1), \deg_{\mathcal{F}}(2) > deg_{\mathcal{F}}(3) = deg_{\mathcal{F}}(4) > deg_{\mathcal{F}}(x)$ for any $x \geq 5$. Indeed, $deg_{\mathcal{F}}(1), \deg_{\mathcal{F}}(2)$ is at least $\binom{n-2}{k-2} + \binom{n-3}{k-3}$, $deg_{\mathcal{F}}(3) = deg_{\mathcal{F}}(4) = \binom{n-3}{k-3} + \binom{n-3}{k-2} + \binom{n-4}{k-3}$, while if $x > 5$, then $deg_{\mathcal{F}}(x) = \binom{n-3}{k-3} + \binom{n-3}{k-2}$. Therefore, if two constructions $\mathcal{F}$ and $\mathcal{F}'$ are isomorphic, then $\mathcal{G}_1|_{[5,n]}$ is isomorphic to either $\mathcal{G}'_1|_{[5,n]}$ or to $\mathcal{G}'_2|_{[5,n]}$. Thus we can pick the pair $\mathcal{G}_1|_{[5,n]}$, $\mathcal{G}_2|_{[5,n]}$ in $P(n-4, k-1)/2$ many non-isomorphic ways. ■

Mubayi and Zhao also managed to improve on the upper bound of Theorem 471 for some special values of $n$ and $k$.

**Theorem 473 ( [437])** *Let $p$ be a prime, $t$ be a positive integer, $k = pt+1$, and $n \geq n_0(k)$. If $\mathcal{F} \subseteq \binom{[n]}{k}$ with $|\mathcal{F}| > \binom{n}{k-1} - \log_p n + k!k^k$, then $\dim_{VC}(\mathcal{F}) = k$.*

Patkós [454] observed that for the star $\mathcal{F}_1$ we have $\mathcal{F}_1 \not\rightarrow P_{k+1}$, where $P_{k+1}$ is a maximal chain in $[k]$. He showed that if $n$ is large enough, then $Tr^{(k)}(n, P_{k+1}) = \binom{n-1}{k-1}$ holds, and the only extremal families are stars. He conjectured that $Tr^{(k)}(n, P_{r+1}) = \binom{n-k+r-1}{r-1}$ holds for any $2 \leq r \leq k$. This was verified for large enough values of $n$ by Tan.

**Theorem 474 (Tan [524])** *For any $1 \leq r, k$ with $r-1 \leq k$, there exists $n_0(k,r)$ such that if $n \geq n_0(k,r)$, then $Tr^{(k)}(n, P_{r+1}) = \binom{n-k+r-1}{r-1}$ holds, and the only families $\mathcal{F} \subseteq \binom{[n]}{k}$ with $\mathcal{F} \not\rightarrow P_{r+1}$ and $|\mathcal{F}| = Tr^{(k)}(n, P_{r+1})$ are of the form $\{F \in \binom{[n]}{k} : D \subseteq F\}$ for some fixed $(k-r+1)$-subset $D$.*

**Proof.** Let the *almost maximal chain* $P_r^- \subseteq 2^{[r]}$ be a chain $X_1 \subseteq X_2 \subseteq \cdots \subseteq X_r$ of length $r$ with $|X_i| = i$ for all $i = 1, 2, \ldots, r$; i.e. an almost maximal chain is a maximal chain with the empty set removed. We start with the following lemma.

**Lemma 475** *For $k, r \geq 2$ we have $Tr^{(k)}(n, P_r^-) \leq \frac{n}{k} Tr^{(k)}(n, P_{r-1})$.*

**Proof of Lemma.** Let $\mathcal{F} \subseteq \binom{[n]}{k}$ be a family with $\mathcal{F} \not\rightarrow P_r^-$ and for any $x = 1, 2, \ldots, n$, let us define $\mathcal{F}_x = \{F \in \mathcal{F} : x \in F\}$. It is enough to prove $|\mathcal{F}_x| \leq Tr^{(k-1)}(n-1, P_r)$, as then $k|\mathcal{F}| = \sum_{x \in [n]} |\mathcal{F}_x| \leq nTr^{(k-1)}(n-1, P_r)$.

Suppose not, and for some $x$, the $(k-1)$-uniform family $\mathcal{F}'_x = \{F \setminus \{x\} : F \in \mathcal{F}_x\}$ has size more than $Tr^{(k-1)}(n-1, P_r)$. Then there exists an $(r-1)$-set $Y$ with $\mathcal{F}'_x|_Y$ containing a maximal chain. But then the traces of the corresponding sets show that $\mathcal{F}|_{Y \cup \{x\}} \supseteq \mathcal{F}_x|_{Y \cup \{x\}} \to P_r^-$ - a contradiction. □

To prove the theorem, we proceed by induction on $r+k$. Clearly, for $k < r$ (and in particular for $k = r-1$), we have $Tr^{(k)}(n, P_{r+1}) = \binom{n}{k}$. Also, $Tr^{(k)}(n, P_2) = 1$, as if $x \in A \setminus B$, then $\{A, B\}|_{\{x\}}$ is a maximal chain. To prove the inductive step, let $r \leq k$ and $\mathcal{F} \subseteq \binom{[n]}{k}$ with $\mathcal{F} \not\to P_{r+1}$. We distinguish two cases.

Case I: $\mathcal{F}$ is intersecting.

If there exists an $x$ with $x \in \cap_{F \in \mathcal{F}} F$ and $n \geq n_0(k-1, r)$, then $|\mathcal{F}| = |\mathcal{F}_x| = |\mathcal{F}'_x| \leq Tr^{(k)}(n-1, k-1, P_{r+1}) = \binom{n-k+r-1}{r-1}$, as required. So we may assume $\cap_{F \in \mathcal{F}} F = \emptyset$. We define $l = \min\{|F \cap G| : F, G \in \mathcal{F}\}$. As $\mathcal{F}$ is intersecting, we have $l \geq 1$. Let us fix two sets $F, G \in \mathcal{F}$ with $|F \cap G| = l$, and partition $\mathcal{F}$ into $\mathcal{F}_1 \cup \mathcal{F}_2$ with $\mathcal{F}_1 = \{F' \in \mathcal{F} : F \cap G \subseteq F'\}$ and $\mathcal{F}_2 = \mathcal{F} \setminus \mathcal{F}_1$. Then $\cap_{F \in \mathcal{F}} F = \emptyset$ implies $\mathcal{F}_2 \neq \emptyset$. By definiton, for any $H \in \mathcal{F}_2$ we have $(F \cap G) \setminus H \neq \emptyset$.

We claim that $|\mathcal{F}_1| \leq 8^k \binom{n-k}{r-2}$ holds. For a fixed $H \in \mathcal{F}_2$ and $S \subseteq F \cup G \cup H$, we define $\mathcal{F}_S = \{F' \in \mathcal{F}_1 : F' \cap (F \cup G \cup H) = S\}$ and $\mathcal{G}_S = \{F' \setminus S : F' \in \mathcal{F}_S\}$. By the definition of $\mathcal{F}_1$, we know that $\mathcal{F}_S$ is non-empty only if $S \supset F \cap G$. Suppose $|\mathcal{F}_S| > Tr^{(k-|S|)}(n - |F \cup G \cup H|, P_r)$. Then $\mathcal{G}_S \to P_r$, so there exists an $(r-1)$-subset $X \subseteq [n] \setminus (F \cup G \cup H)$, such that the trace $\mathcal{G}_S|_X$ contains a maximal chain of $X$. For any element $y \in (F \cap G) \setminus H$, we obtain that $\mathcal{F}_S|_{X \cup \{y\}}$ contains an almost maximal chain of $X \cup \{y\}$, which together with $H|_{X \cup \{y\}}$ form a maximal chain on $X \cup \{y\}$, contradicting $\mathcal{F} \not\to P_{r+1}$. Hence, $|\mathcal{F}_S| \leq Tr^{(k-|S|)}(n-k, P_r) \leq \binom{n-k}{r-2}$. As $|\mathcal{F}_1| = \sum_{S \subseteq F \cup G \cup H} |\mathcal{F}_S|$ and $|F \cup G \cup H| \leq 3k$, we have $|\mathcal{F}_1| \leq 8^k \binom{n-k}{r-2}$.

Next we prove that $|\mathcal{F}_2| \leq 4^k \binom{n-k}{r-2}$. For the same $F$ and $G$ and $S \subseteq F \cup G$, we define $\mathcal{F}'_S = \{F' \in \mathcal{F}_2 : F' \cap (F \cup G) = S\}$ and $\mathcal{G}'_S = \{F' \setminus S : F' \in \mathcal{F}'_S\}$. Suppose $|\mathcal{F}'_S| = |\mathcal{G}'_S| > Tr^{(k)}(n - |F \cup G|, k - |S|, P_r)$. Then $\mathcal{G}'_S \to Pr$, so there exists an $(r-1)$-subset $X \subseteq [n] \setminus (F \cup G)$, such that $\mathcal{F}'_S|_X$ contains a maximal chain of $X$. The minimality of $l$ implies that $\mathcal{F}'_S$ is non-empty if $S \cap (F \setminus G) \neq \emptyset$, so fix $y \in S \cap (F \setminus G)$. Then $\mathcal{F}'_S|_{X \cup \{y\}}$ contains $P_r^-$, which together with $G|_{X \cup \{y\}}$ forms a $P_{r+1}$, contradicting $\mathcal{F} \not\to P_{r+1}$. By the inductive hypothesis, we obtain $|\mathcal{F}'_S| \leq Tr^{(k-|S|)}(n - |F \cup G|, P_r) \leq \binom{n-k}{r-2}$, and therefore $|\mathcal{F}_2| = \sum_{S \subseteq F \cup G} |\mathcal{F}'_S| \leq 4^k \binom{n-k}{r-2}$.

The above statements yield $|\mathcal{F}| = |\mathcal{F}_1| + |\mathcal{F}_2| \leq (8^k + 4^k)\binom{n-k}{r-2} \leq \binom{n-k+r-1}{r-1}$, if $n$ is large enough.

Case II: $\mathcal{F}$ is not intersecting.

Let us fix $F, G \in \mathcal{F}$ with $F \cap G = \emptyset$, and partition $\mathcal{F}$ into $\mathcal{F}_1 \cup \mathcal{F}_2$ with $\mathcal{F}_1 = \{F' \in \mathcal{F} : F' \cap F \neq \emptyset\}$ and $\mathcal{F}_2 = \mathcal{F} \setminus \mathcal{F}_1$. Observe that $\mathcal{F}_2 \not\to P_r^-$ as otherwise $\mathcal{F} \supseteq \{F\} \cup \mathcal{F}_2 \to P_{r+1}$. Applying Lemma 475, we obtain

$$\begin{aligned}|\mathcal{F}_2| \leq & Tr^{(k)}(n-k, P_r^-) \leq \frac{n-k}{k} Tr^{(k-1)}(n-k-1, P_r)\\ \leq & \frac{n-k}{k} Tr^{(k-1)}(n, P_r) = \frac{n-k}{k}\binom{n-k+r-1}{r-2}\\ = & \frac{r-1}{k}\binom{n-k+r-1}{r-1}.\end{aligned}$$

We claim that $|\mathcal{F}_1| \leq 4^k\binom{n-k}{r-2}$. As in the second claim of CASE I, for any $S \subseteq F \cup G$ we define $\mathcal{F}_S = \{F' \in \mathcal{F}_1 : F' \cap (F \cup G) = S\}$ and $\mathcal{G}_S = \{F' \setminus S : F' \in \mathcal{F}_S\}$. Suppose $|\mathcal{G}_S| = |\mathcal{F}_S| > Tr^{(k-|S|)}(n - |F \cup G|, P_r)$, and thus $\mathcal{G}_S \to P_r$. Therefore, there exists an $(r-1)$-subset $X \subseteq [n] \setminus (F \cup G)$ with $\mathcal{F}_S|_X$ containing a $P_r$. As $\mathcal{G}_S$ is non-empty only if $S \cap F \neq \emptyset$, we can pick $y \in S \cap F$. Then as $G \cap (X \cup \{y\} = \emptyset$, we have $\mathcal{F}_S \cup \{G\} \subseteq \mathcal{F} \to P_{r+1}$ - a contradiction. By the inductive hypothesis, we obtain $|\mathcal{F}_S| \leq Tr^{(k-|S|)}(n - |F \cup G|, P_r) \leq \binom{n-k}{r-2}$ and as there are at most $2^{|F \cup G|}$ possibilities for $S$, we have $|\mathcal{F}_1| \leq 4^k\binom{n-k}{r-2}$ as claimed.

The above bounds yield $|\mathcal{F}| = |\mathcal{F}_1| + |\mathcal{F}_2| \leq \frac{r-1}{k}\binom{n-k+r-1}{r-1} + 4^k\binom{n-k}{r-2} < \binom{n-k+r-1}{r-1}$ if $n$ is large enough, as $r - 1 < k$. This proves CASE II.

We obtained that either $\mathcal{F} < Tr^{(k)}(n, P_{r+1})$ or there exists $x \in \cap_{F \in \mathcal{F}} F$ and $\mathcal{F}'_x = \{F \setminus \{x\} : F \in \mathcal{F}\} \not\to P_{r+1}$ with $|\mathcal{F}'_x| = |\mathcal{F}| = \binom{n-k+r-1}{r-1} = Tr^{(k-1)}(n - 1, P_{r+1})$. So inductively we obtain that either $|\mathcal{F}| < Tr^{(k)}(n, P_{r+1})$, or there exists a $(k-r+1)$set $D$, such that $D \subseteq \cap_{F \in \mathcal{F}} F$, as required. ■

If $X_1, X_2, \ldots, X_k$ are pairwise disjoint non-empty sets, then let $\mathcal{K}(X_1, X_2, \ldots, X_k) = \{K : |X_1 \cap K| = 1 \text{ for all } i = 1, 2, \ldots, k\}$ be the complete $k$-partite family with parts $X_1, X_2, \ldots, X_k$. Clearly, $\mathcal{K}(X_1, X_2, \ldots, X_k)$ is $k$-uniform. Let us fix $x_i \in X_i$ and let $X = \{x_1, x_2, \ldots, x_k\}$. We claim that $\mathcal{B}(X_1, X_2, \ldots, X_k) = \{B \in \mathcal{K}(X_1, X_2, \ldots, X_k) : B \cap X \neq \emptyset\}$ has $VC$-dimension less than $k$. Indeed, if $K \in \mathcal{K}(X_1, X_2, \ldots, X_k)$, then $\mathcal{B}(X_1, X_2, \ldots, X_k)|_K$ does not contain $K \setminus X$, while if $K \notin \mathcal{K}(X_1, X_2, \ldots, X_k)$, then $K \notin \mathcal{B}(X_1, X_2, \ldots, X_k)$ and thus $\mathcal{B}(X_1, X_2, \ldots, X_k)|_K$ does not contain $K$. Frankl and Watanabe showed that this construction is best possible; i.e. $\mathcal{B}(X_1, X_2, \ldots, X_k)$ is the largest $k$-partite family with $VC$-dimension less than $k$.

**Theorem 476 (Watanabe, Frankl [223])** *If $\mathcal{F} \subseteq \mathcal{K}(X_1, X_2, \ldots, X_k)$ with $dim_{VC}(\mathcal{F}) < k$, then $|\mathcal{F}| \leq |\mathcal{B}(X_1, X_2, , \ldots, X_k)|$.*

**Proof.** The proof will be derived from the following two claims.

**Claim 477** *For any $\mathcal{F} \subseteq \mathcal{K}(X_1, X_2, \ldots, X_k)$ with $\dim_{VC}(\mathcal{F}) < k$, there exists an enumeration of $\mathcal{F} = \{F_1, F_2, \ldots, F_m\}$ and a family $\mathcal{G} = \{G_1, G_2, \ldots, G_m\}$ of $(k-1)$-sets, such that $G_i \subseteq F_i$ for all $i = 1, 2, \ldots, m$ and $G_i \not\subseteq F_j$ for all pairs $1 \leq i < j \leq m$.*

**Proof of Claim.** We define a sequence of families. We start by letting $\mathcal{F}_0 = \mathcal{F}$. If $\mathcal{F}_0, \mathcal{F}_1, \ldots, \mathcal{F}_t$ are defined and there exists an $F_{t+1} \in \mathcal{F}_t$ and a set $G_{t+1} \in \binom{F_{t+1}}{k-1}$, such that no $F' \in \mathcal{F}_t$ with $F' \neq F_{t+1}$ contains $G_{t+1}$, then we let $\mathcal{F}_{t+1} = \mathcal{F}_t \setminus \{F_{t+1}\}$. Clearly, if we can define all $\mathcal{F}_0, \mathcal{F}_1, \ldots, \mathcal{F}_m$, then the enumeration $\mathcal{F} = \{F_1, F_2, \ldots, F_m\}$ and the corresponding $G_i$'s satisfy the claim.

So suppose that for some $t < m$ for every pair $G \subset F \in \mathcal{F}_t$ with $|G| = k-1$, there exists an $F' \neq F$ in $\mathcal{F}_t$ with $G \subset F'$. We claim that for any $F \in \mathcal{F}_t$ we have $F \in tr(\mathcal{F}_t)$. Suppose not, and let $H \subsetneq F$ be a maximal set such that there is no $F' \in \mathcal{F}_t$ with $F \cap F' = H$. If we write $F = \{x_1, x_2, \ldots, x_k\}$ with $x_i \in X_i$, then we can assume without loss of generality that $H = \{x_1, x_2, \ldots, x_j\}$ for some $j < k$ (as $F = F \cap F$.) By choice of $H$, there exists a set $F_J \in \mathcal{F}_t$ such that $F \cap F_J = \{x_1, x_2, \ldots, x_{j+1}\}$. Also, by the property of $\mathcal{F}_t$ we know that there exists $F^* \in \mathcal{F}_t$ such that $F^* \neq F_J$ and $F_J \setminus \{x_{j+1}\} \subset F^*$. But then $x_{j+1} \notin F^*$, so $F^* = F_J \setminus \{x_{j+1}\} \cup \{x'_{j+1}\}$ for some $x'_{j+1} \in X_{j+1}$ with $x'_{j+1} \neq x_{j+1}$, and therefore $F^* \cap F = H$. This contradiction proves the claim. □

Now let us consider the $\binom{\sum_i |X_i|}{k-1} \times |\mathcal{F}|$ inclusion matrix $M_{\mathcal{F}}$, that we used already in Theorem 471. The rows of $M_{\mathcal{F}}$ are indexed with $(k-1)$-subsets of $\cup_i X_i$ and the columns of $M_{\mathcal{F}}$ are indexed with sets of $\mathcal{F}$, such that $M_{\mathcal{F}}(G, F) = 1$ if $G \subset F$, and 0 otherwise. Claim 477 states that the submatrix corresponding to the rows $G_i$ is a lower triangular matrix with diagonal entries 1; therefore the columns are linearly independent. Obviously, $M_{\mathcal{F}}$ is a submatrix of $M_{\mathcal{K}(X_1, X_2, \ldots, X_k)}$, so the next claim will finish the proof of the theorem.

**Claim 478** *The rank of $M_{\mathcal{K}(X_1, X_2, \ldots, X_k)}$ over any field is $|\mathcal{B}(X_1, X_2, \ldots, X_k)|$.*

**Proof of Claim.** The lower bound follows from Claim 477 and that fact that $\dim_{VC}(\mathcal{B}(X_1, X_2, \ldots, X_k)) < k$.

To show the upper bound it is enough to prove that if we add any column $C$ to the columns indexed by $\mathcal{B} = \mathcal{B}(X_1, X_2, \ldots, X_k)$, then they are not linearly independent anymore; therefore the column space of $M_{\mathcal{K}(X_1, X_2, \ldots, X_k)}$ is spanned by the columns indexed with sets of $\mathcal{B}$. As $C \notin \mathcal{B}$, we have $C \cap X = \emptyset$, and $C$ is the only $k$-subset of $C \cup X$ with this property. So for the family $\mathcal{G} = \{G \in \mathcal{K}(X_1, X_2, \ldots, X_k) : G \subseteq C \cup X\}$ we have $\mathcal{G} \subseteq \mathcal{B} \cup \{C\}$ and $|\mathcal{G}| = 2^k$, since for any $i$ and $G \in \mathcal{G}$ we have that $G \cap X_i$ is either $\{x_i\}$ or $C \cap X_i$. We claim that if $v_G$ denotes the column of $M_{\mathcal{K}(X_1, X_2, \ldots, X_k)}$ indexed by $G$, then $\sum_{G \in \mathcal{G}} (-1)^{|G \cap X|} v_G$ is the 0 vector. Indeed, if a $(k-1)$-subset $H$ is not contained in $X \cup C$ or contains at least 2 elements of some $X_i$, then all $v_G$ have 0 in their $H$-coordinate. If $H \subset X \cup C$ with $|H \cap X_i| \leq 1$ for all $1 \leq i \leq k$, then there are exactly two sets $G_1, G_2 \in \mathcal{G}$ containing $H$: if $i$ is the only index with $H \cap X_i = \emptyset$, then $G_1 = H \cup \{x_i\}$ and $G_2 = H \cup (C \cap X_i)$. As $|G_1 \cap X| = |G_2 \cap X| + 1$, the signs of the $H$-coordinates of $v_{G_1}$ and $v_{G_2}$ are different. □■

**Theorem 479 (Frankl, Pach [218])** *For any $r \geq 1$ $k \geq 2$ and $n \geq r+k-1$ we have $ex_{k-1}(r+k-1, \mathcal{K}^k_{k-1}) \leq Tr^{(r)}(n, \binom{[k]}{1}) \leq \binom{r+k-1}{k-1}$.*

**Proof.** To see the lower bound, let $\mathcal{F} \subseteq \binom{[r+k-1]}{k-1}$ be a family not containing a $\mathcal{K}^k_{k-1}$. Then clearly for the family $\overline{\mathcal{F}} = \{[r+k-1] \setminus F : F \in \mathcal{F}\}$, we have $\overline{\mathcal{F}} \not\to \binom{[k]}{1}$.

For the upper bound, let $\mathcal{F} = \{F_1, F_2, \dots, F_m\}$ be an $r$-uniform family with $\mathcal{F} \not\to \binom{[k]}{1}$. For every $1 \leq i \leq m$, let $M_i$ be a minimal subset of $\cup_{j=1}^m F_j \setminus F_i$ with $M_i \cap F_j \neq \emptyset$ for all $j \neq i$. Observe that $(\mathcal{F} \setminus \{F_j\})|_{M_j}$ contains $\binom{|M_j|}{1}$ by the minimality of $M_j$ , so $|M_j| \leq k-1$ holds. Also, by definition, the pairs $\{(F_j, M_j)\}_{j=1}^m$ form an ISPS. By Theorem 6, we have $\sum_{i=1}^m \frac{1}{\binom{|F_i|+|M_i|}{|M_i|}} \leq 1$, so the upper bound follows. ■

Note that Theorem 242 implies that for any fixed $k$, the ratio of the upper and lower bounds is bounded by a constant as $r$ tends to infinity. Frankl and Pach also determined the exact values of $Tr^{(r)}(n, \binom{[3]}{1})$ and $Tr^{(2)}(n, \binom{[k]}{1})$ (both independent of $n$), and all extremal families.

**Theorem 480 (Balogh, Keevash, Sudakov [41])** *For any $k \leq r$ and $n \geq r+k-2$, we have the following:*

***(i)*** $\binom{r+k-2}{k-2} \leq Tr^{(r)}(n, \{\binom{[k]}{1}, \binom{[k]}{2}, \dots, \binom{[k]}{k-1}\}) \leq Tr^{(r)}(n, \{\binom{[k]}{1}, \binom{[k]}{k-1}\}) < kr^{k-2}$.

***(ii)*** $Tr^{(r)}(n, \{\binom{[k]}{1}, \binom{[k]}{2}, \dots, \binom{[k]}{k-2}\}) = \Omega(r^{k-1})$.

**Proof.** The lower bound of **(i)** is given by the following construction: let $\mathcal{F} = \{\cup_{i=0}^{k-2}[ir, ir + a_i - 1] : a_i \geq 0, \sum_{i=0}^{k-2} a_i = r\}$, so $\mathcal{F}$ is a family of those $r$-sets that are unions of at most $k-1$ intervals whose left endpoints are multiples of $r$. We have $|\mathcal{F}| = \binom{r+k-2}{k-2}$ as every $F \in \mathcal{F}$ corresponds to a $(k-1)$-tuple $(a_0, a_1, \dots, a_{k-2}$ with $\sum_{i=0}^{k-2} = r$ and $a_i \in \mathbb{N}$ for all $i = 0, 1, \dots, k-2$. This is the definition of combination with repetition.

Furthermore, $\cup_{F \in \mathcal{F}} = [0, (k-1)r - 1]$, so if a $k$-set $K$ is not a subset of $[0, (k-1)r - 1]$, then $\mathcal{F}|_K$ cannot contain $\binom{K}{i}$ for any $i > 0$. Let $K$ be a $k$-subset of $[0, (k-1)r - 1]$. Then there exists $1 \leq i \leq k-1$ such that the $[ir, i(r+1) - 1]$ contains two points $a, b$ of $K$. If $a < b$, then all sets in $\mathcal{F}$ that contain $b$ also contain $a$. Thus $\mathcal{F}|_K$ cannot contain a full level of $2^K$ (apart from $\emptyset$ and $K$) as those separate every pair of points.

To see the upper bound of **i**, let $\mathcal{F} = \{F_1, F_2, \dots, F_m\}$ be $r$-uniform with $\mathcal{F} \not\to \binom{[k]}{1}$ and $|\mathcal{F}| \geq kr^{k-2}$. We will show that $\mathcal{F} \to \binom{[k]}{k-1}$. As in the proof of

Theorem 479, let $M_j$ be a minimal subset of $\cup_{i=1}^m F_i \setminus F_j$ such that $M_j \cap F_i \neq \emptyset$ for all $i \neq j$. As observed there, $|M_j| \leq k-1$ for all $j$. We have $F_1 \cap M_j \neq \emptyset$ for all $1 < j \leq n$. Consider $x_1 \in F_1$ that is contained in the most sets $M_j$, and let $I_j$ be the set of indices of those sets. Thus $I_1 \subset [m]$ with $|I_1| \geq kr^{k-3}$ and $x_1 \in M_j$ for all $j \in I_1$. Repeating this, we can obtain the sequences $x_1, x_2, \ldots, x_{k-2}$ and $I_1 \supset I_2 \supset \cdots \supset I_{k-2}$ such that for any $1 \leq t \leq k-2$ we have $\{x_1, x_2, \ldots, x_t\} \subset M_j$ for $j \in I_t$ and $|I_t| \geq kr^{k-2-t}$. Observe that the sets $M_j$ form an antichain, as $M_j$ is the only one that is disjoint from $F_j$. We have $|I_{k-2}| \geq k$, and if $j \in I_{k-2}$, then $x_1, x_2, \ldots, x_{k-2} \in M_j$. As $|M_j| \leq k-1$, we have $M_j = \{x_1, x_2, \ldots, x_{k-2}, y_j\}$. As $F_j$ is disjoint only from $M_j$, it must contain $y_i$ with $i \neq j$. This implies that letting $Y = \{y_j : j \in I_{k-2}\}$ we obtain $F_j|Y = Y \setminus \{y_j\}$ and thus for any $k$-subset $Y'$ of $Y$ we have $\mathcal{F}|_{Y'} \supset \binom{Y'}{k-1}$.

To prove **(ii)**, consider the following construction: let the $(r+k-1)$-set $X$ be the union of the pairwise disjoint sets $X_1, X_2, \ldots, X_{k-1}$, such that $|X_i| = \lfloor \frac{r+k-2+i}{k-1} \rfloor$. This is a partition with the following properties: $||X_i| - |X_j|| \leq 1$ for all $i, j$, and $\prod_{i=1}^{k-1} |X_i|$ is maximized over all partitions. Let $\mathcal{F} = \{X \setminus T : |T \cap X_i| = 1 \text{ for all } i = 1, 2, \ldots, k-1\}$; i.e $\mathcal{F}$ consists of the complements of all covers of the $X_i$'s. Then $|\mathcal{F}| = \prod_{i=1}^{k-1} \lfloor \frac{r+k-2+i}{k-1} \rfloor = \Omega(r^{k-1})$ if $k$ is fixed and $r$ tends to infinity. If a $k$-set $K$ is not contained in $X$, then $\mathcal{F}|_K \neq \binom{K}{j}$ for $j > 0$ as sets not contained in $X$ do not occur as trace. If $K \subset X$, then there exist $i$ such that $|K \cap X_i| \geq 2$. So let $a, b \in K \cap X_i$. Then any set in $\mathcal{F}$ contains at least one of $a$ and $b$, but for any $l \leq k-2$ there exists a set in $L \in \binom{K}{l}$ with $a, b \notin L$, so $\mathcal{F}|_K \neq \binom{K}{l}$. ■

Let us mention that for $k = 3$ Balogh, Keevash and Sudakov obtained the exact value of $Tr^{(r)}(n, \{\binom{[3]}{1}, \binom{[3]}{2}\})$ and determined all extremal families.

**Theorem 481 (Balogh, Keevash, Sudakov, [41])** *(i) If $k \geq 3$, then $Tr^{(r)}(n, \{\binom{[k]}{1}, P_3\}) = \max\{k-1, r+1\}$.*

*(ii) If $r \geq k-2$, then $\binom{r+k-2}{k-2} \leq Tr^{(r)}(n, \{\binom{[k]}{1}, P_k\}) \leq \binom{(k-1)r}{k-2}$.*

**Proof.** We start with three lemmas.

**Lemma 482** *Suppose $\mathcal{F}$ is $r$-uniform with $\mathcal{F} \not\to P_3$, and let $F_1, F_2 \in \mathcal{F}$ maximize $|F_1 \cup F_2|$ over all pairs in $\mathcal{F}$. We say that $F \in \mathcal{F}$ is of type 1 if $F \cap (F_1 \cup F_2) = F_1 \cap F_2$, and of type 2 if $F_1 \triangle F_2 \subset F \subset F_1 \cup F_2$. Then any $F \neq F_1, F_2$ in $\mathcal{F}$ is of type 1 or type 2, and all sets in $\mathcal{F}$ are of the same type.*

**Proof of Lemma.** Let $F_1, F_2$ as in the lemma. Observe first that there cannot exist $F'$ of type 1 and $F''$ of type 2, because then for $x \in F_1 \setminus F_2, y \in F_2 \setminus F_1$ we have $F'|_{\{x.y\}} = \emptyset, F_1|_{\{x.y\}} = \{x\}, F''|_{\{x.y\}} = \{x, y\}$, thus $\mathcal{F} \to P_3$.

Suppose $F \in \mathcal{F} \setminus \{F_1, F_2\}$ is disjoint from $F_1 \triangle F_2$. Then $F_1 \cap F \subseteq F_1 \cap F_2$. Observe that the maximality of $|F_1 \cup F_2| = 2r - |F_1 \cap F_2|$ implies the minimality of $|F_1 \cap F_2|$; hence we must have $F \cap F_1 = F_1 \cap F_2$. With a similar argument we have $F \cap F_2 = F_1 \cap F_2$; therefore $F \cap (F_1 \cup F_2) = F_1 \cap F_2$, i.e. $F$ is of type 1.

Suppose now $F \cap (F_1 \Delta F_2) \neq \emptyset$, say, $x \in F \cap (F_1 \setminus F_2)$. There cannot exist $y \in F_1 \setminus (F \cup F_2)$ as then $\{F_2, F, F_1\}|_{\{x,y\}} = P_3$ would hold. This means $F_1 \setminus F_2 \subset F$, so by maximality of $|F_1 \cup F_2|$ we must have $F \subsetneq F_1 \cup F_2$. But as $F \neq F_1$, we must have $F \cap (F_2 \setminus F_1) \neq \emptyset$. Repeating the above argument we obtain $F_2 \setminus F_1 \subset F$, and therefore $F_1 \Delta F_2 \subset F \subset F_1 \cup F_2$; i.e. $F$ is of type 2. □

**Lemma 483** *Suppose $\mathcal{F}$ is $r$-uniform with $\mathcal{F} \not\rightarrow P_3$ and $\mathcal{F}$ does not contain a sunflower of size 3. Then $|\mathcal{F}| \leq r+1$, with equality only if $\mathcal{F} = \binom{[r+1]}{r}$.*

**Proof of Lemma.** We proceed by induction on $r$, with the base case $r = 1$ being trivial. Let $F_1, F_2 \in \mathcal{F}$ maximize $|F_1 \cup F_2|$ over all pairs in $\mathcal{F}$. There is no $F \in \mathcal{F}$ with $F \cap (F_1 \cup F_2) = F_1 \cap F_2$, as then $F_1$, $F_2$ and $F$ form a sunflower. Therefore, by Lemma 482 for any $F \in \mathcal{F} \setminus \{F_1, F_2\}$ we have $F_1 \Delta F_2 \subset F \subset F_1 \cup F_2$. If $|F_1 \cup F_2| = r+1$, then clearly $|\mathcal{F}| \leq r+1$ with equality only if $\mathcal{F} = \binom{F_1 \cup F_2}{r}$. If $|F_1 \cup F_2| > r+1$, then $|F_1 \Delta F_2| \geq 4$ and we can apply the induction hypothesis to $\mathcal{F}' = \{F \setminus (F_1 \Delta F_2) : F \in \mathcal{F} \setminus \{F_1, F_2\}\}$. By the above, $\mathcal{F}'$ is $s$-uniform with $s \leq r-4$, so $|\mathcal{F}| = |\mathcal{F}'|+2 \leq ((r-4)+1)+2 < r+1$. □

**Lemma 484** *If $\mathcal{F}$ is $r$-uniform, $\mathcal{F} \not\rightarrow \binom{[k]}{1}$, and $F_1, F_2, \ldots, F_{k-1}$ maximizes the size of the union over all $(k-1)$-tuples, then every $F \in \mathcal{F}$ is contained in $F_1 \cup F_2 \cup \cdots \cup F_{k-1}$.*

**Proof of Lemma.** Assume there exists an $F \in \mathcal{F}$ with $x \in F \setminus \cup_{i=1}^{k-1} F_i$. Observe that for every $1 \leq i \leq k-1$ there exists $x_i \in F_i \setminus (F \cup \cup_{i=1}^{k-1} F_i)$, as otherwise $|\cup_{i=1}^{k-1} F_i| < |F \cup \cup_{j \neq i} F_j|$ would contradict the maximality of $|\cup_{i=1}^{k-1} F_i|$. But then $\{F, F_1, F_2, \ldots, F_{k-1}\}|_{\{x, x_1, \ldots, x_{k-1}\}}$ is isomorphic to $\binom{[k]}{1}$. □

Now we can prove the theorem. The lower bound of **(i)** is given by the larger of the constructions $\binom{[r+1]}{r}$ and the family of $k-1$ pairwise disjoint $r$-sets.

For the upper bound we proceed by induction on $r$, with the base case $r = 1$ being trivial. Let $\mathcal{F} \subseteq \binom{[n]}{r}$ with $\mathcal{F} \not\rightarrow \binom{[k]}{1}$ and $\mathcal{F} \not\rightarrow P_3$, and let us fix $F_1, F_2$ that maximizes the size of their union over all pairs in $\mathcal{F}$. Then by Lemma 482, all other sets in $\mathcal{F}$ are either of type 1 or type 2.

In the type 1 case, $\mathcal{F}$ is a sunflower with center $F_1 \cap F_2$, and as $\mathcal{F} \not\rightarrow \binom{[k]}{1}$ we must have $|\mathcal{F}| \leq k-1$. In the type 2 case, we claim that $\mathcal{F}$ cannot contain a sunflower with three petals. $F_1$ and $F_2$ cannot be both in the sunflower, as other sets in $\mathcal{F}$ are contained in $F_1 \cup F_2$, so they must contain $F_1 \Delta F_2$. Let $F', F'', F'''$ be the sunflower with center $C$; then $C \supset F_1 \cap F_2$. Consider $x \in F' \setminus C, y \in F'' \setminus C$; then by the sunflower property we have $x, y \notin F'''$, while the type 2 property assures that $x, y \in (F' \cup F'') \setminus F''' \subset F_1 \cap F_2$. Thus

$\{F''', F', F_1\}|_{\{x,y\}} = P_3$ - a contradiction. As $\mathcal{F}$ contains no sunflower of size 3, Lemma 483 implies $|\mathcal{F}| \leq r+1$.

The lower bound of **(ii)** is given by the construction $\binom{[r+k-2]}{r}$. If a $k$-set $X$ is not contained in $[r+k-2]$, then $x \in X \setminus [r+k-2]$ cannot occur as a singleton trace, while if $X \subset [r+k-2]$, then all traces are of size at least 2, so no singleton trace occurs. If $Y$ is a $(k-1)$-set and $Y \not\subset [r+k-2]$, then $Y$ cannot occur as trace, while if $Y \subset [r+k-2]$, then all traces have size at least one; therefore $\binom{[r+k-2]}{r} \not\to P_k$. To see the upper bound, observe that by Lemma 484, if $\mathcal{F} \not\to \binom{[k]}{1}$ is $r$-uniform, then $|\cup_{i=1}^{k-1} F_i| \leq r(k-1)$. As $\mathcal{F} \not\to P_k$ implies $\mathcal{F} \not\to 2^{[k-1]}$, we can apply Theorem 471 with $d = k-1$ to obtain that $|\mathcal{F}| \leq \binom{r(k-1)}{k-2}$.

■

# 9

# *Combinatorial search theory*

## CONTENTS



## 9.1 Basics

How many yes-no questions do we need to ask to find an unknown element of a set $X$ of size $n$? This is the starting point of combinatorial search theory. We assume that the set of elements is $[n]$, and the questions are of the form whether the unknown element belongs to a subset $F$ of $[n]$. The answer is YES or NO. By halving the possibilities with each question, one may get an algorithm that finds the unknown element with $\lceil \log n \rceil$ questions, but the questions of this algorithm depend on the answers we obtained to previous questions. These type of algorithms are called *adaptive*. One can achieve the same goal with the same number of questions without relying on previous answers (a *non-adaptive algorithm*), by asking the sets $F_i = \{x \in [n] : \text{the } i\text{th bit of } x \text{ in its binary form is } 1\}$. One cannot do better neither adaptively, nor non-adaptively, as if $k$ questions are asked, then the number of possible sequences of answers is $2^k$. To be able to distinguish between the cases when $x$ or $y$ is the unknown element, the corresponding sequences of answers should differ; therefore we must have $2^k \geq n$.

The above simple problem can be made more complicated if we cannot ask all possible subsets of $[n]$, or there are more unknown elements and we have to find all of them (or a percentage of them), and there exist lots of variants. The unknown elements are usually called *defective elements* and questions are called *queries*. The above argument to obtain a lower bound works in a more general setting. If the number of possible answers to a query is $a$, and we have $N$ objects out of which one is defective, then the minimum number $m$ of queries needed to find the defective object satisfies the inequality $a^m \geq N$. The bound obtained in this way is sometimes called the *information theoretical* lower bound. For example, if we have $d$ defective elements, and we have to find all of them in such a way that the answer to a query set $Q$ is the number of defective elements in $Q$, then we have $a = d+1$ and $N = \binom{n}{d}$, therefore the information theoretical lower bound is $\lceil \log_{d+1} \binom{n}{d} \rceil$.

Many problems can be described in this setting. In the case of comparison based sorting, we can take $d = 1$ and $X$ to be the set of all $n!$ total orders on $n$ labeled elements. Comparing $x$ and $y$ is equivalent to asking the query consisting of all total orders where $x < y$. The information theoretical lower bound gives that the minimum number of queries needed is $\log n! = \Omega(n \log n)$ and there are well-known adaptive sorting algorithms that use $O(n \log n)$ comparisons.

Let us consider the original problem and try to describe all families $\mathcal{F}$ of query sets that do find the one defective element. If there exist 2 elements $x, y$ of $X$ with the property that for every $F \in \mathcal{F}$, the element $x$ belongs to $F$ if and only if so does $y$, then $\mathcal{F}$ cannot distinguish between $x$ and $y$. This reasoning also works in the other direction, showing that a query set $\mathcal{F} \subset 2^X$ finds the only defective element if and only if $\mathcal{F}$ is *separating*, i.e. for every pair $x, y \in X$ there exists an $F \in \mathcal{F}$ that contains exactly one of $x$ and $y$. Equivalently, $|F \cap \{x, y\}| = 1$ and we also say $F$ *separates* $x$ and $y$. The reasoning from the first paragraph shows that the minimum size of a separating family is $\lceil \log_2 X \rceil$.

Although it does not have any immediate connection to combinatorial search, it is quite natural to change the requirement of being separable to *totally* (or completely) *separable*: for every ordered pair $(x, y) \in \mathcal{F} \times \mathcal{F}$, $x \neq y$ there exists a set $F \in \mathcal{F}$ with $x \in F, y \notin F$.

What do totally separable families look like? To describe them, it is useful to define the *dual family* $\mathcal{F}'$ of $\mathcal{F}$. We obtain $\mathcal{F}'$ from $\mathcal{F}$ by switching the roles of elements and sets. So the number of sets in $\mathcal{F}'$ is $|X|$, and the underlying set of $\mathcal{F}'$ contains $|\mathcal{F}|$ elements. Formally, if $\mathcal{F}$ has incidence matrix $M$, then $\mathcal{F}'$ is the family with incidence matrix $M^T$.

**Exercise 485 (Spencer [519])** *Show that a family $\mathcal{F} \subseteq 2^{[n]}$ is totally separating if and only if its dual family is Sperner. As a consequence, show that the minimum size of a totally separating family $\mathcal{F} \subseteq 2^{[n]}$ is $\min\{m : \binom{m}{\lfloor m/2 \rfloor} \geq n\}$.*

Let us consider now the non-adaptive case of more defective elements. Here the answer to a query $F$ is YES if and only if it contains at least 1 defective element. What should a family $\mathcal{F}$ of non-adaptive query sets look like if it finds all defective elements, if there is an upper bound $d$ on the number of defectives (known before we ask the queries)? As in the case $d = 1$, we have to be able to distinguish every possible pair $D, D' \subset X$ with $|D|, |D'| \leq d$. That is, for every such pair there should exist $F \in \mathcal{F}$ such that $F$ is disjoint from exactly one of $D$ and $D'$ and intersects the other. We call such families *d-separable.* What property of the dual family $\mathcal{F}'$ is equivalent to $\mathcal{F}$ being $d$-separable? If a query set $F \in \mathcal{F}$ separates $D$ and $D'$, say $F \cap D = \emptyset$, $F \cap D' \neq \emptyset$, then in the dual family $\mathcal{F}'$ the union $\cup_{d\in D}F_d$ will not contain the element corresponding to $F$, while $\cup_{d'\in D'}F_{d'}$ will. We obtain the following observation.

**Proposition 486** *A family $\mathcal{F} \subseteq 2^X$ is d-separable if and only if for any subfamilies $\mathcal{D}, \mathcal{D}' \subseteq \mathcal{F}'$ with $|\mathcal{D}|, |\mathcal{D}'| \le d$, $\mathcal{D} \neq \mathcal{D}'$ we have $\cup_{D\in\mathcal{D}}D \neq \cup_{D'\in\mathcal{D}'}D'$.*

Families with the property of $\mathcal{F}'$ in Proposition 486 are called *d-superimposed* or *d-union free.* A related concept is the notion of cover-free families. We say that $\mathcal{F}$ is *d-cover-free* if there do not exist $d+1$ distinct sets $F_1, F_2, \ldots, F_{d+1} \in \mathcal{F}$ such that $F_{d+1} \subseteq \cup_{i=1}^d F_i$ holds.

**Exercise 487** *Show that if a family $\mathcal{F}$ is d-superimposed, then it is $(d-1)$-cover-free and if $\mathcal{F}$ is d-cover-free, then it is d-superimposed.*

Let $T(n,r)$ denote the size of a largest $r$-cover-free family $\mathcal{F} \subseteq 2^{[n]}$. Then Proposition 486 and Exercise 487 imply that the minimum size of a non-adaptive query set that finds all of at most $d$ defectives is not larger than $\min\{m : F(m, d-1) \ge n\}$.

Dyachkov and Rykov proved [139]] $\frac{\log T(n,r)}{n} = O(\frac{\log r}{r^2})$. Then Ruszinkó [490] gave the first purely combinatorial proof of this. Here we present a proof due to Füredi.

**Theorem 488 (Füredi [232])** *If $\mathcal{F} \subseteq 2^{[n]}$ is r-cover-free, then*

$$|\mathcal{F}| \le r + \binom{n}{\lceil \frac{n-r}{\binom{r+1}{2}} \rceil}$$

*holds.*

Observe that the inequality $\binom{n}{k} \le (\frac{en}{k})^k$ and the above theorem imply $\frac{\log T(n,r)}{n} = O(\frac{\log r}{r^2})$.

**Proof.** For any $t$ with $0 \le t \le n/2$, let us define $\mathcal{F}_t \subseteq \mathcal{F}$ by

$$\mathcal{F}_t = \left\{F \in \mathcal{F} : \exists T \subseteq \binom{F}{t} \text{ such that } T \not\subseteq F' \text{ for all } F \neq F' \in \mathcal{F}\right\}.$$

Furthermore, set $\mathcal{F}_0 = \{F \in \mathcal{F} : |F| < t\}$. Clearly, $\mathcal{F}$ is an antichain; therefore $\mathcal{F}_t$ and $\nabla_t(\mathcal{F}_0)$ are disjoint. The first proof of Theorem 1 and $t \le n/2$ imply that $|\mathcal{F}_0| \le |\nabla(\mathcal{F}_0)|$. This yields $|\mathcal{F}_0| + |\mathcal{F}_t| \le |\nabla(\mathcal{F}_0)| + |\mathcal{F}_t| \le \binom{n}{t}$.

Writing $\mathcal{F}' = \mathcal{F} \setminus (\mathcal{F}_0 \cup \mathcal{F}_t)$ we claim that for any $F' \in \mathcal{F}'$ and $F_1, F_2, \ldots, F_i \in \mathcal{F}$, we have $|F' \setminus \cup_{j\le i}F_j| > t(r-i)$. Indeed, assume $F' \setminus \cup_{j\le i}F_j$ can be covered by at most $r-i$ many $t$-subsets $A_{i+1}, A_{i+2}, \ldots, A_r$ of $F'$. Then, as $F' \in \mathcal{F}'$, every $A_j$ is also a subset of some $F_j \in \mathcal{F}$ $j = i+1, i+2, \ldots, r$; hence $F' \subseteq \cup_{j=1}^r F_j$, which contradicts the $r$-cover-free property of $\mathcal{F}$.

Using the above inequality we obtain that for any $F_0, F_1, \dots, F_r \in \mathcal{F}'$ we have

$$|\cup_{j=0}^r F_j| = |F_0| + \sum_{j=1}^r |F_j \setminus \sum_{l=0}^{j-1} F_l| > \sum_{j=0}^r (1 + t(r-j)) = r + 1 + t\binom{r+1}{2}.$$

If we substitute $t = \lceil \frac{n-r}{\binom{r+1}{2}} \rceil$, then the right hand side is larger than $n$, which is impossible. This implies $|\mathcal{F}'| \le r$ and thus $|\mathcal{F}| \le |\mathcal{F}_0| + |\mathcal{F}_t| + r \le r + \binom{n}{\lceil \frac{n-r}{\binom{r+1}{2}} \rceil}$, as required. ■

Constructions of $r$-cover-free families $\mathcal{F} \subseteq 2^{[n]}$ with $\log|\mathcal{F}| = \Theta(\frac{n}{r^2})$ appear in the literatute. The following greedy algorithm that outputs such a family is due to Hwang and Sós.

**Theorem 489 (Hwang, Sós, [315])** *For any $n$ and $r$, there exists an $r$-cover-free family $\mathcal{F} \subseteq 2^{[n]}$ with $|\mathcal{F}| > \frac{1}{2}\left(1 + \frac{1}{4r^2}\right)^n$.*

**Proof.** We will pick $k$-subsets $F_1, F_2, \dots, F_s$ of $[n]$ one by one with $k = 4r\lfloor \frac{n}{4r^2} \rfloor$, such that $|F_i \cap F_j| < \frac{k}{r}$ holds. The latter condition clearly implies the $r$-cover-free property. We pick $F_1 \in \binom{[n]}{k}$ arbitrarily, and then define $\mathcal{B}_1 = \{B \in \binom{n}{k} : |F_1 \cap B| \ge \frac{k}{r}\}$. Then pick $F_2$ from $\binom{[n]}{k} \setminus \mathcal{B}_1$. Similarly, for any $F_i$ we define $\mathcal{B}_i = \{B \in \binom{n}{k} : |F_i \cap B| \ge \frac{k}{r}\}$, and once $F_1, F_2, \dots, F_i$ is already defined, we pick $F_{i+1}$ from $\binom{[n]}{k} \setminus \cup_{j=1}^i \mathcal{B}_j$ arbitrarily. As $|\mathcal{B}_i| = \sum_{j=\frac{k}{r}}^k \binom{k}{j}\binom{n-k}{k-j}$, the number $s$ of sets at the end of the process is at least

$$\frac{\binom{n}{k}}{\sum_{j=\frac{k}{r}}^k \binom{k}{j}\binom{n-k}{k-j}}.$$

Observe that

$$\frac{\binom{k}{j}\binom{n-k}{k-j}}{\binom{k}{j-1}\binom{n-k}{k-j+1}} = \frac{(k-j+1)^2}{j(n-2k+j)}$$

is monotone decreasing in $j$. Therefore, if $j \ge 3\lfloor \frac{n}{4r^2} \rfloor =: 3t$, then its value is at most

$$\frac{t^2(4r-3)^2}{3t(4r^2t - 8rt + 3t)} = \frac{(4r-3)^2}{3(4r^2 - 8r + 3)} < \frac{1}{3}.$$

As $\frac{k}{r} = 4t$, it follows that

$$\sum_{j=\frac{k}{r}}^k \binom{k}{j}\binom{n-k}{k-j} = \sum_{j=4t}^k \binom{k}{j}\binom{n-k}{k-j} \le 2\binom{k}{4t}\binom{n-k}{k-4t}$$
$$< 2\binom{k}{3t}\binom{n-k}{k-3t} \cdot 3^{-t}.$$

Finally, observe that

$$3^t \geq 3^{\frac{n}{(4r^2)}-1} > e^{\frac{n}{(4r^2)}} = \left(e^{1/(4r^2)}\right)^n \geq \left(1+\frac{1}{(4r)^2}\right)^n,$$

and thus

$$s \geq \frac{\binom{n}{k}}{\sum_{j=\frac{k}{r}}^{k}\binom{k}{j}\binom{n-k}{k-j}} \geq \frac{\binom{k}{3t}\binom{n-k}{k-3t}}{\sum_{j=\frac{k}{r}}^{k}\binom{k}{j}\binom{n-k}{k-j}}$$
$$\geq \frac{\binom{k}{3t}\binom{n-k}{k-3t}}{2\binom{k}{3t}\binom{n-k}{k-3t}\cdot 3^{-t}} \geq \frac{1}{2}\left(1+\frac{1}{(4r)^2}\right)^n.$$

■

Proposition 486 stated that finding the minimum number of queries needed to find all the at most $d$ defectives non-adaptively, is equivalent to finding the maximum size $SUP(m,d)$ of a $d$-superimposed family $\mathcal{F} \subseteq 2^{[m]}$. Theorem 488 and Theorem 489 imply that $2^{\Omega(\frac{m}{d^2})} \leq SUP(m,d) \leq 2^{\Omega(\frac{m\log d}{d^2})}$ holds.

Does the situation change significantly if we only want to find one of the defectives? Csűrös and Ruszinkó [112] defined *r-single user tracing superimposed* families. These have the property that if we are given the union of $r$ sets of the family, then we can recover at least one of them. Formally, $\mathcal{F}$ is $r$-single user tracing superimposed ($r$-SUT for short) if for any $k$ and any collection $\mathcal{F}_1, \mathcal{F}_2, \ldots, \mathcal{F}_k$ of subfamilies of $\mathcal{F}$ with $\cup_{F\in\mathcal{F}_1}F = \cup_{F\in\mathcal{F}_2}F = \cdots = \cup_{F\in\mathcal{F}_k}F$ and $1 \leq |\mathcal{F}_i| \leq r$ for all $i = 1,2,\ldots,r$, we have $\cap_{i=1}^{k}\mathcal{F}_i \neq \emptyset$. By taking the dual family, this corresponds to the property that if the number of defectives is known to be at most $r$, then using these queries one can identify at least one of the defectives. In the remainder of this section we will obtain bounds on the maximum possible size of an $r$-SUT family $\mathcal{F} \subseteq 2^{[n]}$.

For the upper bound we will need the following definition. A family $\mathcal{F}$ is *locally r-thin* if for any sets $F_1, F_2, \ldots, F_r \in \mathcal{F}$ there exists an element $x$ that belongs to exactly one of the $F_i$'s. A family is *locally $\leq r$-thin* if it is locally $t$-thin for all $t \leq r$.

**Theorem 490 (Alon, Fachini, Körner [13])** *If $\mathcal{F} \subseteq 2^{[n]}$ is locally r-thin, then*

$$|\mathcal{F}| \leq \begin{cases} 2^{O(\frac{n}{r})} & \text{if } r \text{ is even} \\ 2^{O(\frac{n\log r}{r})} & \text{if } r \text{ is odd} \end{cases} \tag{9.1}$$

**Theorem 491 (Csűrös, Ruszinkó, [112])** *If $\mathcal{F} \subseteq 2^{[n]}$ is an r-SUT family, then $|\mathcal{F}| \leq 2^{O(\frac{n}{r})}$ holds.*

**Proof.** First we claim that any $r$-SUT family is locally $\leq (r+1)$-thin. Indeed, if not, then there exist $F_1, F_2, \ldots, F_k$ with $k \leq r+1$, such that every element

$x \in \cup_{i=1}^k F_i$ belongs to at least two of the $F_i$'s. But then the families $\mathcal{F}_i = \{F_1, F_2, \ldots, F_{i-1}, F_{i+1}, \ldots, F_k\}$ are of size $k-1 \leq r$; their unions are the same and their intersection is empty. This would contradict the assumption that $\mathcal{F}$ is an $r$-SUT.

Now the theorem follows as if $r$ is odd, then $r+1$ is even and one can use the bound of Theorem 490 for $r+1$, while if $r$ is even, then one can use the bound of Theorem 490 for $r$. ■

**Theorem 492 (Alon, Asodi [11])** *(i) For any $r \geq 2$ and $n \geq 20r$, there exists an $r$-SUT family of subsets of $[n]$ of size $\lfloor 2^{n/20r} \rfloor$.*

*(ii) For all $n$, there exists a 2-SUT family of subsets of $[n]$ of size $\lfloor \frac{3}{8} 2^{n/3} \rfloor$.*

**Proof.** To prove **(i)** and **(ii)** we need to show two constructions. Both of them are random and can be derandomized using methods that are standard in this area, but which are beyond the scope of this book.

To prove **(i)**, fix $r$ and $n \geq 20r$. Let us define $m = \lfloor 2^{\frac{n}{20r}} \rfloor$ and $p = \frac{1}{r}$. We consider the random family $\mathcal{F} = \{F_1, F_2, \ldots, F_m\} \subseteq 2^{[n]}$, that is formed randomly such that for any $x \in [n]$ and $1 \leq i \leq m$, we include $x$ in $F_i$ with probability $p$, independently of any other inclusion $y \in F_j$. We are going to show that with positive probability $\mathcal{F}$ is an $r$-SUT. (Note that this includes the fact that all $F_i$'s are different.) We need to prove that with positive probability, for any set $\mathcal{F}_1, \mathcal{F}_2, \ldots, \mathcal{F}_k$ of subfamilies of $\mathcal{F}$ with $|\mathcal{F}_i| \leq r$ for all $1 \leq i \leq k$ and $\emptyset = \cap_{i=1}^k \mathcal{F}_i$, the unions $\cup_{F \in \mathcal{F}_i} F$ are not equal.

**Claim 493** *The following statement holds with probability greater than $1/2$. For any $s < 2r$ and any $F_1, F_2, \ldots, F_s \in \mathcal{F}$, there exists $x \in [n]$ that belongs to exactly one of the $F_i$'s.*

**Proof of Claim.** Let us fix $s < 2r$ and sets $F_1, F_2, \ldots, F_s$. The probability that these sets do not satisfy the statement, i.e. no $x \in [n]$ belongs to exactly one of them is

$$(1 - sp(1-p)^{s-1})^n \leq \left[1 - \frac{s}{r}\left(1 - \frac{1}{r}\right)^{2r-2}\right]^n \leq \left(1 - \frac{s}{r}e^{-2}\right)^n$$
$$\leq e^{-e^{-2}\frac{sn}{r}} < 2^{-0.15\frac{sn}{r}},$$

where we used $(1-\frac{1}{n})^n < e^{-1}$, $1 - x \leq e^{-x}$ and $e^{-e^{-2}} < 2^{-0.15}$. This implies, that the expected value of the total number of $s$-tuples for all $s < 2r$ with the above property is at most

$$\sum_{s=1}^{2r-1} \binom{m}{s} 2^{-0.15\frac{sn}{r}} \leq \sum_{s=1}^{2r-1} m^s 2^{-0.15\frac{sn}{r}} \leq \sum_{s=1}^{2r-1} 2^{\frac{ns}{20r}} 2^{-0.15\frac{sn}{r}}$$
$$= \sum_{s=1}^{2r-1} 2^{-0.1\frac{sn}{r}} \leq \sum_{s=1}^{2r-1} 2^{-2s} < \frac{1}{2}.$$

As the total number of bad $s$-tuples is a non-negative integer valued random variable, the Markov inequality yields the claim. □

**Claim 494** *The following holds with probability greater than* $\frac{1}{2}$. *For all* $t \leq r$ *and for all distinct* $F_1, F_2, \ldots, F_r, G_1, G_2, \ldots, G_t \in \mathcal{F}$, *we have* $\cup_{i=1}^r F_i \not\subseteq \cup_{j=1}^t G_j$.

**Proof of Claim.** For one particular choice $t \leq r$ and $F_1, F_2, \ldots, F_r, G_1, G_2, \ldots, G_t$ and $x \in [n]$ we have

$$\mathbb{P}(x \in \cup_{i=1}^r F_i \setminus \cup_{j=1}^t G_j) = (1-(1-p)^r)(1-p)^t$$
$$\geq \left[1 - \left(1 - \frac{1}{r}\right)^r\right]\left(1 - \frac{1}{r}\right)^r \geq \frac{1}{2}(1-e^{-1})e^{-1} > 0.1.$$

As a consequence we obtain

$$\mathbb{P}(\cup_{i=1}^r F_i \subseteq \cup_{j=1}^t G_j) < 0.9^n.$$

Therefore, the expected number of possible bad choices is at most

$$m^{2r} \cdot 0.9^n \leq 2^{\frac{n}{10}} 0.9^n < \frac{1}{2}.$$

The Markov inequality yields the claim. □

There is a positive probability that $\mathcal{F}$ satisfies the properties of both Claim 493 and Claim 494. Therefore, to finish the proof of **(i)**, we need to show that any such family $\mathcal{F}$ is an $r$-SUT. Let $\mathcal{F}_1, \mathcal{F}_2, \ldots, \mathcal{F}_k$ be subfamilies of $\mathcal{F}$ with $|\mathcal{F}_i| \leq r$ for all $i = 1, 2, \ldots, k$ and $\cap_{i=1}^k \mathcal{F}_i = \emptyset$. We need to show that the unions $\cup_{F \in \mathcal{F}_i} F$ are not all equal. We distinguish two cases according to $|\cup_{i=1}^k \mathcal{F}_i|$.

Case I: $|\cup_{i=1}^k \mathcal{F}_i| < 2r$.

As $\mathcal{F}$ satisfies the property ensured by Claim 493, there exists an $x \in [n]$ that belongs to exactly one set $F$ from $\cup_{i=1}^k \mathcal{F}_i$. As $\cap_{i=1}^k \mathcal{F}_i = \emptyset$, there exists a $j$ such that $F \notin \mathcal{F}_j$ holds. This implies that $x \notin \cup_{G \in \mathcal{F}_j} G$, while for any $j'$ with $F \in \mathcal{F}_{j'}$ we have $x \in \cup_{H \in \mathcal{F}_{j'}} H$.

Case II: $|\cup_{i=1}^k \mathcal{F}_i| \geq 2r$.

Let $\mathcal{F}_1 = \{G_1, G_2, \ldots, G_t\}$. As $t \leq r$ and $|\cup_{i=1}^k \mathcal{F}_i| \geq 2r$, we have $|(\cup_{i=2}^k \mathcal{F}_i) \setminus \mathcal{F}_1| \geq r$. Pick $r$ distinct sets $F_1, F_2 \ldots, F_r \in (\cup_{i=2}^k \mathcal{F}_i) \setminus \mathcal{F}_1$. If all unions $\cup_{F \in \mathcal{F}_i} F$ were equal, then we should have $\cup_{i=1}^r F_i \subseteq \cup_{j=1}^t G_j$, which contradicts the assumption that $\mathcal{F}$ satisfies the property ensured by Claim 494.

The two cases finish the proof of **(i)**.

To prove **(ii)**, we consider a random family $\mathcal{F} = \{F_1, F_2, \ldots, F_m\}$ of sets with $m = \lfloor \frac{1}{2} 2^{n/3} \rfloor$, such that for every $x \in [n]$ and $1 \leq i \leq m$, we include $x$ in $F_i$ with probability $p = 1 - \frac{1}{\sqrt{2}}$, independently of any other inclusion $y \in F_j$. Observe that a family is 2-SUT if and only if

1. for any distinct four sets $A, B, C, D$ of the family we have $A \cup B \neq C \cup D$ and
2. for any distinct three sets of the family we have $A \cup B$, $A \cup C$, and $B \cup C$ are not all equal.

Let us calculate the expected number of four-tuples and triples of sets in $\mathcal{F}$ that fail the above conditions. (Observe again, that if $\mathcal{F}$ satisfies the second property, then all sets in $\mathcal{F}$ are distinct.) For every distinct $A, B, C, D \in \mathcal{F}$ we have

$$\mathbb{P}(A \cup B = C \cup D) = ((1-p)^4 + (1-(1-p)^2)^2)^n,$$

as for every $x \in [n]$, we have either $x \notin A \cup B \cup C \cup D$ or $x \in (A \cup B) \cap (C \cup D)$. Substituting $p = 1 - \frac{1}{\sqrt{2}}$, we obtain that this probability is $2^{-n}$, and thus the expected number of four-tuples failing the first condition is at most

$$m^4 2^{-n} \leq \frac{1}{16} 2^{4n/3} 2^{-n} = \frac{1}{16} 2^{n/3} \leq \lceil \frac{m}{8} \rceil.$$

Similarly, the probability that a fixed triple $A, B, C \in \mathcal{F}$ fails the second condition is $(1 - 3p(1-p)^2)^n$ as an $x \in [n]$ would show this failure if it belonged to exactly one of $A, B$, and $C$. Substituting $p = 1 - \frac{1}{\sqrt{2}}$, we obtain

$$\mathbb{P}(A \cup B = A \cup C = B \cup C) < 2^{-4n/5},$$

and thus the expected number of triples failing the second condition is at most

$$m^3 2^{-4n/5} \leq \frac{1}{8} 2^{n/5} < \frac{m}{8}.$$

Therefore, there exists a choice of $\mathcal{F}$ such that the number of bad triples and four-tuples combined is at most $\lceil \frac{m}{8} \rceil + \frac{m}{8} = \lceil \frac{m}{4} \rceil$. Removing one set from each such triple and four-tuple will leave us a 2-SUT family of size $\lfloor \frac{3m}{4} \rfloor = \lfloor \frac{3}{8} 2^{n/3} \rfloor$. ■

**Exercise 495** *A family $\mathcal{F} \subseteq 2^{[n]}$ is minimal separating if it is separating but no $\mathcal{G} \subsetneq \mathcal{F}$ is separating. Show that for any minimal separating family $\mathcal{F} \subseteq 2^{[n]}$ we have $|\mathcal{F}| \leq n-1$. Show that this statement is equivalent to Exercise 450.*

## 9.2 Searching with small query sets

In the previous section we considered only the basic case when to find one or all of the defective elements, we are allowed to use any subset as query set. In most real life applications not all subsets are available. To model this, one is given a subfamily $\mathcal{H}$ of subsets of $X$, and queries can be picked only from $\mathcal{H}$. In order to be able to find the defective element, $\mathcal{H}$ needs to be separating (or in case of multiple defectives $(\leq d)$-separable). If this is satisfied, one can ask to determine the minimum number of queries required to determine the/an/all/some defective element(s) in an adaptive or non-adaptive search. In this section we present the very natural and important case, when $\mathcal{H} = \binom{X}{\leq k}$ for some integer $k$.

**Theorem 496 (Katona [334])** *If $k \leq n/2$ and a family $\mathcal{F} \subseteq \binom{[n]}{\leq k}$ is separating, then*

$$\frac{\ln n}{\ln(en/k)}\frac{n}{k} \leq \frac{\log n}{\frac{k}{n}\log\frac{k}{n} + \frac{n-k}{n}\log\frac{n}{n-k}} \leq |\mathcal{F}|$$

*holds.*

**Proof.** Let $\mathcal{F} = \{F_1, F_2, \ldots, F_m\} \subseteq \binom{[n]}{\leq k}$ be a separating family and let $x$ be an element of $[n]$, chosen uniformly at random. Let $X_i$ denote the indicator variable of the random event $x \in F_i$. For the entropy of each these variables we have

$$H(X_i) = \frac{|F_i|}{n}\log\frac{|F_i|}{n} + \frac{n-|F_i|}{n}\log\frac{n-|F_i|}{n} \leq \frac{k}{n} + \log\frac{k}{n} + \frac{n-k}{n}\log\frac{n-k}{n}.$$

Let us consider the joint distribution $(X_1, X_2, \ldots, X_m)$. As $\mathcal{F}$ is separating, for every possible value of $x$, the variable $(X_1, X_2, \ldots, X_m)$ takes a different value. So for the joint entropy we have $H(X_1, X_2, \ldots, X_m) = \log n$. The subadditivity of the entropy implies

$$\log n = H(X_1, X_2, \ldots, X_m) \leq \sum_{i=1}^{m} H(X_i) \leq |\mathcal{F}|(\frac{k}{n} + \log\frac{k}{n} + \frac{n-k}{n}\log\frac{n-k}{n}),$$

which gives the second inequality. Simple calculation gives the first inequality. ■

**Exercise 497** *Let us assume that $X$ is the edge set $E(G)$ of a graph $G$. Show that if the family $\mathcal{H} \subseteq 2^X$ of possible queries consists of the edge sets of paths in $G$, then any separating subfamily $\mathcal{F} \subseteq \mathcal{H}$ has size at least $\frac{|E(G)|\ln|E(G)|}{|V(G)|\ln(e|V(G)|/2)}$.*

Upper bounds on the smallest separating subfamily for different graphs $G$ were obtained in [39] and [177].

**Exercise 498** *Let us assume that $X$ is the set of all 1-subspaces of an $n$-dimensional vector space $V$ over the finite field of size $q$. Show that if the family $\mathcal{H} \subseteq 2^X$ of possible queries is $\{\{x : x \in U\} : U$ is a subspace of $V\}$, then any separating subfamily $\mathcal{F} \subseteq \mathcal{H}$ has size at least $(1 - o(1))nq$ if $q$ tends to infinity.*

Upper bounds on the smallest separating subfamily for different values of $n$ and $q$ were obtained in [306].

## 9.3 Parity search

In this section we consider the search problem of finding all of at most $d$ defective elements, in such a way that the answer to the query set $Q$ is the number of defectives in $Q$ modulo 2. We think of $d$ as a fixed integer, while the number $n$ of all elements tends to infinity. The information theoretical lower bound implies that the minimum number $m$ of queries needed satisfies $2^m \geq \sum_{i=0}^{d} \binom{n}{i}$; therefore we have $m \geq d \log n - O_d(1)$. We will present a construction due to Reiher that shows that this bound can be achieved asymptotically, even in the non-adaptive case. First, let us observe that the sets of a family $\mathcal{F} \subseteq 2^{[n]}$ can serve as non-adaptive queries that find all defective elements, if and only if for every pair $X, Y \in \binom{[n]}{\leq d}$, there exists a set $F \in \mathcal{F}$ such that $|F \cap X|$ and $|F \cap Y|$ have different parity. Let us denote by $f(n, d)$ the minimum size of a family $\mathcal{F}$ that satisfies this property.

**Theorem 499 (Reiher [476])** *If $d, m$, and $n$ denote three positive integers with $dm \leq n < 2^m$, then $f(n, d) \leq dm$. In particular, $f(n, d) \leq d \log n$.*

**Proof.** The proof is done in the following three lemmas.

**Lemma 500** *Let $A$ be a finite subset of a field of characteristic 2 such that there is a nonzero element $a \in A$. Then for some odd positive integer $k \leq |A|$ we have $\sum_{x \in A} x^k \neq 0$.*

**Proof of Lemma.** Consider the expression

$$\sum_{x \in A} x \prod_{b \in A \setminus \{a\}} (x - b).$$

On the one hand it is equal to $= a \prod_{b \in a \setminus \{a\}} (a - b) \neq 0$. On the other hand if we expand the products, we obtain

$$\sum_{i=1}^{|A|} \alpha_i \sum_{x \in A} x^i,$$

where the $\alpha_i$'s are from our base field. It implies that there exists a positive integer $k' \leq |A|$ with $\sum_{x \in A} x^{k'} \neq 0$. Finally, observe that the least such integer $k$ must be odd. Indeed, the field is of characteristic 2; hence if $k$ is even we have

$$\left( \sum_{x \in A} x^{k/2} \right)^2 = \sum_{x \in A} x^k \neq 0;$$

thus $k/2$ would be an even smaller positive integer with the above property. □

**Lemma 501** *For any two positive integers $d$ and $m$, the vector space $\mathbb{F}_2^{dm}$ has a generating subset $B$ of size at least $2^m - 1$, such that each vector admits at most one representation as the sum of at most $d$ distinct members of $B$.*

**Proof of Lemma.** Note first that because of the one-to-one correspondence between vectors of $\mathbb{F}_2^{dm}$ and $\mathbb{F}_{2^m}^d$, we can find our subset $B$ in $\mathbb{F}_{2^m}^d$. For any nonzero element $\alpha$ of $\mathbb{F}_{2^m}$ let us define the vector $v_\alpha = (\alpha, \alpha^3, \ldots, \alpha^{2d-1}) \in \mathbb{F}_{2^m}^d$. We claim that the set

$$B^* = \{v_\alpha : \alpha \in \mathbb{F}_{2^m}^\times\}$$

satisfies the representation property required by the lemma. Indeed, if $D \neq D'$ are two subsets of $B^*$ with $|D|, |D'| \leq d$ and $\sum_{v \in D} v = \sum_{v' \in D'} v'$, then for the symmetric difference $D \triangle D'$, we have $0 < |D \triangle D'| \leq 2d$ and $\sum_{v \in D \triangle D'} v = 0$. That is for the set $A = \{\alpha \in \mathbb{F}_{2^m}^\times : v_\alpha \in D \triangle D'\}$ we have $0 = \sum_{\alpha \in A} v_\alpha = \sum_{\alpha \in A} (\alpha, \alpha^3, \ldots, \alpha^{2d-1})$, contradicting Lemma 500.

Finally, observe that to enlarge $B^*$ to a set $B$ that possesses the generating property and still has the representation property, we can add the basis of a subspace $U$ that is disjoint from $\langle B^* \rangle$ and satisfies $\langle B^* \rangle \oplus U = \mathbb{F}_{2^m}^d$. □

The *weight* of a binary vector is the number of its 1-entries.

**Lemma 502** *Let $d, m$, and $n$ denote three positive integers such that $dm \leq n < 2^m$. Then there is some vector subspace of $\mathbb{F}_2^n$ of dimension $n - dm$, such that every nonzero vector in it has weight more than $2d$.*

**Proof of Lemma.** Let $\mathbb{F}_2^n = \mathbb{F}_2^{dm} \oplus \mathbb{F}_2^{n-dm}$ and consider its subspaces $W_1 = \{0\} \oplus \mathbb{F}_2^{n-dm}$ and $W_2 = \mathbb{F}^{dm} \oplus \{0\}$. Applying Lemma 501 to $W_2$, we can obtain a sequence $b_1, b_2, \ldots, b_n$ of distinct vectors in $W_2$, such that $b_1, b_2, \ldots, b_{dm}$ form a basis of $W_2$, and every vector of $W_2$ can be represented as a sum of at most $d$ of the $b_i$'s in at most one way. Let $v_1 = v_2 = \cdots = v_{dm} = 0$ and $v_{dm+1}, v_{dm+2}, \ldots, v_n$ be a basis of $W_1$. By definition, the vectors $u_1, u_2, \ldots, u_n$ with $u_i = b_i + v_i$ form a basis of $\mathbb{F}_2^n$, and no vector of $W_1$ can be obtained as the sum of at most $2d$ distinct $u_i$'s. Indeed, assume $v = \sum_{j=1}^{d'} u_{i_j}$, where $v \in W_1$ and $d' \leq 2d$. Then we have $\sum_{j=1}^{d'} b_{i_j} = 0$ by $v \in W_1$ , which implies

$$\sum_{j=1}^{\lfloor d'/2 \rfloor} b_{i_j} = \sum_{j=\lfloor d'/2 \rfloor + 1}^{d'} b_{i_j};$$

thus we obtained two different representations of the same vector in $W_2$.

Now if $\phi$ is an automorphism of $\mathbb{F}_2^n$ that maps the $u_i$'s to the standard basis, then $W' = \phi(W_1)$ has the desired property. □

With the above lemmas at hand, we can prove Theorem 499. We need to find a family $\mathcal{F} \subseteq 2^{[n]}$ with $|\mathcal{F}| \leq dm$, such that for any pair $X, Y \in \binom{[n]}{\leq d}$, there exists a set $F \in \mathcal{F}$ with the property that $|F \cap X|$ and $|F \cap Y|$

have different parity. Using characteristic vectors this translates to finding $dm$ vectors $v_1, v_2, \ldots, v_{dm} \in \mathbb{F}_2^n$, such that for any pair of vectors $x, y$ of weight at most $d$, there exists an $i$ with the property that $v_i \cdot x \neq v_i \cdot y$. Let $W$ be the subspace of $\mathbb{F}_2^n$ obtained by Lemma 502, and let $v_1, v_2, \ldots, v_{dm}$ be a basis of $W$'s orthogonal complement. We claim that these vectors possess the required property. Indeed, if $x$ and $y$ both have weight at most $d$, then their difference $x - y = x + y$ has weight at most $2d$, and therefore $x - y \notin W$, so there exists an $i$ such that $v_i \cdot (x - y) \neq 0$ as claimed. ∎

## 9.4 Searching with lies

In this section we consider versions of some of the models mentioned so far, with one modification: some of the answers to queries might be erroneous. The error might come as the effect of noise when the answers are sent back through some not perfect channel, or due to the imperfection of testing samples, etc. In these cases, the errors are assumed to be random. Therefore models that address this problem fix a probability $p$ and every query is answered incorrectly with probability $p$ independently of all other queries. The task then is to provide an adaptive/ non-adaptive algorithm finding all defectives with probability at least, say, $2/3$, using as few queries as possible. We will not describe any such problem or algorithm; we refer the reader to [460].

Another possibility of how erroneous answers might appear is, if we think that the person providing the answers is a malicious adversary. It is common and sometimes even useful to think of combinatorial search problems as two-player games, where one player wants to finish the game (find all defectives) as soon as possible, while the the other player, the adversary wants to prevent this by giving answers "as confusing as possible". It is a major help to the adversary if we allow him to lie, that is, to give faulty answers at most $k$ times, where $k$ is a fixed number known to both players.

Let us examine when the game of searching for the one and only defective element is over in this case. Assume we have asked the queries $Q_1, Q_2, \ldots Q_m$ and received the answers to them. Now all the candidates $x \in X$ for being defective have the property, that $x \in Q_i$ holds if and only if the answer to the $Q_i$ query is YES, with the exception of at most $k$ queries. Thus we can find the defective element if there is only one $x \in X$ with the above property.

For non-adaptive algorithms this observation leads us to the very much studied notion of error correcting codes: a family $\mathcal{F} \subseteq 2^{[n]}$ is $e$-error correcting if for any pair $F, F' \in \mathcal{F}$ the balls $B_e(F)$ and $B_e(F')$ in the Hamming graph are disjoint. We do not elaborate on this huge topic (interested readers are referred to [399] and the references therein); we only pose the following easy exercise that establishes the connection between the two areas.

**Exercise 503** *Prove that a family $\mathcal{H} \subseteq 2^{[n]}$ finds the defective element against at most $k$ lies if and only if its dual $\mathcal{H} \subseteq 2^{[m]}$ is $k$ error-correcting.*

The adaptive version of the problem to find the one defective element against a fixed number of lies is called the Rényi-Ulam game, as they started investigating this and some similar other problems involving randomness in [478] and [540]. The problem was independently introduced by Berlekamp in his PhD thesis [53]. In the remainder of this section we try to summarize the main approach to the problem. For a more detailed introduction to the Rényi-Ulam game and many related problems, we recommend the surveys by Pelc [461] and Deppe [131].

An element $x \in X$ can be ruled out to be the defective element if and only if there have been at least $k+1$ queries asked for which the answer, if true, would imply $x$ is not the defective. So a state during the game can be described by a vector $\underline{v} = (v_1, \ldots, v_n)$, where $v_i$ is $k$ minus the number of false answers in case $x_i$ is the defective element. Thus the $i$th entry (if non-negative) tells how many more lies are allowed if $x_i$ is the defective element. Note that the start state is described by the vector $(k, k, \ldots, k)$ and the game is over if and only if the vector corresponding to the actual state contains exactly one non-negative entry. As two elements with the same entry in this vector are interchangeable, one only needs to store the number $s_j$ of elements with entry $j$, for every $0 \leq j \leq k$. Let $S_j$ denote the set of these elements. The numbers $s_j$ define another vector $\underline{s} = (s_k, s_{k-1}, \ldots, s_1, s_0)$, that describes all important information of the current state of the game. The vector of the start state is $(n, 0, \ldots, 0)$, and the game is over if and only if $\sum_{j=0}^{k} s_j = 1$.

The next important observation is that the information provided by the answer to a query $Q$ in a state with vector $\underline{s} = (s_k, s_{k-1}, \ldots, s_1, s_0)$ depends only on the sizes $q_j := |S_j \cap Q|$ $j = 0, 1, \ldots, k$. Conversely, for any vector $\underline{q} = (q_k, q_{k-1}, \ldots, q_0)$ with $q_j \leq s_j$ for all $j = 0, 1, \ldots, k$, we can find a query $Q$ with $q_j = |S_j \cap Q_j|$. Let $y_j = q_j + s_{j+1} - q_{j+1}$, $z_j = s_j - q_j + q_{j+1}$, $\underline{y} = (y_k, y_{k-1}, \ldots, y_0)$ and $\underline{z} = (z_k, z_{k-1}, \ldots, z_0)$. If the answer to $Q$ is YES, then the vector of the new state becomes $\underline{y}$, while if the answer to $Q$ is NO, then the vector of the new state becomes $\underline{z}$. We say that the vector $\underline{s}$ *reduces* to $\underline{y}$ and $\underline{z}$.

The following quantity, the $m$th volume of a state was introduced by Berlekamp as

$$V_m(\underline{s}) := \sum_{j=0}^{k} s_j \sum_{i=0}^{j} \binom{m}{i}.$$

**Theorem 504 (Berlekamp [51])** *If a vector $\underline{s}$ reduces to $\underline{y}$ and $\underline{z}$, then we have*

$$V_m(\underline{s}) = V_{m-1}(\underline{y}) + V_{m-1}(\underline{z}).$$

**Proof.** Observe that $y_j + z_j = s_j + s_{j+1}$ holds for any $j = 0, 1, \ldots, k$. So we need to prove

$$\sum_{j=0}^{k} s_j \sum_{i=0}^{j} \binom{m}{i} = \sum_{j=0}^{k} (s_j + s_{j+1}) \sum_{i=0}^{j} \binom{m-1}{i}.$$

As the $s_j$'s are arbitrary, this is equivalent to

$$\sum_{i=0}^{j} \binom{m}{i} = \sum_{i=0}^{j} \binom{m-1}{i} + \sum_{i=0}^{j-1} \binom{m-1}{i}.$$

This latter holds by Pascal's identity $\binom{m}{i} = \binom{m-1}{i} + \binom{m-1}{i-1}$. ■

**Corollary 505** *If starting from a state, the game can be over within $m$ queries, then the vector $\underline{s}$ of the state satisfies $V_m(\underline{s}) \leq 2^m$. In particular, if $m$ is the minimum number of queries used in an adaptive strategy for the Rényi-Ulam game on $X$ with $k$ lies, then $|X| \leq \frac{2^m}{\sum_{j=0}^{k} \binom{m}{j}}$ holds.*

**Proof.** For $m = 0$, the statement of the corollary follows from the fact that the game is over (i.e. no more queries are needed) if and only if $\sum_{j=0}^{k} s_j = 1$, so $V(\underline{s})$ is either 0 or 1. For larger values of $m$, the statement follows by induction using Theorem 504. The second statement is obtained applying the Corollary to the start state $(|X|, 0, 0, \ldots, 0)$. ■

Let us introduce the quantity $ch(\underline{s}) = \min\{m : V_m(\underline{s}) \leq 2^m\}$, the *character* of $\underline{s}$. Then Corollary 505 states that $ch(\underline{s})$ is a lower bound on the minimum number of queries needed in the Rényi-Ulam game started from state $\underline{s}$. Let $A_k(n)$ denote the minimum number of queries needed to find the only defective element from $X$ with $|X| = n$ in the Rényi-Ulam game with at most $k$ lies. $A_k(n)$ was determined for all values of $n$ if $k = 1, 2, 3$ by Pelc [459], Guzicki [292], and Deppe [129], respectively. Then Deppe obtained an almost sharp upper bound for large enough values of $n$ and fixed but arbitrary $k$.

**Theorem 506 (Deppe [130])** *Let us write $ch_k(n) = ch(\underline{s})$ for the vector $\underline{s}$ with $s_k = n$, $s_j = 0$ for $j \neq k$. For any positive integer $k$, there exists $n_0 = n_0(k)$, such that if $n \geq n_0$, then $ch_k(n) \leq A_k(n) \leq ch_k(n) + 1$.*

## 9.5 Between adaptive and non-adaptive algorithms

An adaptive algorithm can use all the answers to previous queries to decide what query set to ask next. At the other extremity we have non-adaptive algorithms that ask all queries at once. In many real life applications testing takes too long, so adaptive algorithms cannot be used efficiently time-wise, but non-adaptive algorithms might use many more queries, so testing could turn out to be too costly. Bridging between these two extremities, one might introduce *searching in $k$ rounds*. Formally, if $X$ is the set of all tested elements and $\mathcal{H} \subseteq 2^X$ is the family of allowed query sets, then in a $k$-round search algorithm one asks families $\mathcal{F}_1, \mathcal{F}_2, \ldots, \mathcal{F}_k \subseteq \mathcal{H}$ of queries such that $\mathcal{F}_j$ might depend on the answers to queries in $\cup_{i<j}\mathcal{F}_i$. The aim is to minimize $\sum_{i=1}^{k} |\mathcal{F}_i|$, the total number of queries used in the worst case. Note that rounds are also called *stages* in the literature. All previously considered problems can be of interest in this model, too.

First, we consider the basic problem of finding the one and only defective element. For any positive integer $k$, let $c_k(\mathcal{H})$ denote the minimum number of queries needed in a $k$ round algorithm, if all queries must belong to $\mathcal{H}$. Observe that the case $k = 1$ corresponds to non-adaptive algorithms, the case $k \geq |\mathcal{H}|$ corresponds to adaptive search, and, by definition, we have $c_1(\mathcal{H}) \geq c_2(\mathcal{H}) \geq \cdots \geq c_{|\mathcal{H}|}(\mathcal{H})$. Note that in all the models we consider, when there is only one defective, asking a query set or its complement is equivalent. A permutation on the underlying set does not change anything either. Thus we say that a family $\mathcal{N}$ is a *representation* of another family $\mathcal{H}$, if we can obtain it from $\mathcal{H}$ using these two operations.

Our first aim is to characterize *minimal* separating families. That is, separating families $\mathcal{H} \subseteq 2^{[n]}$, such that for any $H \in \mathcal{H}$ the subfamily $\mathcal{H} \setminus \{H\}$ is not separating, or equivalently, families satisfying $c_1(\mathcal{H}) = |\mathcal{H}|$. Recall that Exercise 495 states that for *any* separating family $\mathcal{H} \subseteq 2^{[n]}$, we have $c_1(\mathcal{H}) \leq n-1$. The next lemma characterizes the case of equality.

**Lemma 507 (Wiener [549])** *For any $k \geq 2$ and separating family $\mathcal{H} \subseteq 2^{[n]}$ we have $c_k(\mathcal{H}) = n - 1$ if and only if $\mathcal{N} = \{\{1\}, \{2\}, \ldots, \{n-1\}\}$ is a representation of $\mathcal{H}$.*

**Proof.** Clearly, $c_k(\mathcal{N}) = n - 1$ for any $k$, so we only need to show that if $c_k(\mathcal{H}) = n - 1$ holds for a separating family $\mathcal{H}$ and integer $k \geq 2$, then $\mathcal{N}$ is a representation of $\mathcal{H}$.

We prove this by contradiction. So suppose $\mathcal{N}$ is not a representation of $\mathcal{H}$, and thus there exists a set $H \in \mathcal{H}$ with $2 \leq |H| \leq n-2$. We use this set to present a two-round algorithm that finds the defective element and uses at most $n-2$ sets, showing $c_2(\mathcal{H}) \leq n-2$. The algorithm first asks if the defective element belongs to $H$, so formally, we have $\mathcal{F}_1 = \{H\}$. As $H$ does not separate any pair $x, y \in H$ or $x, y \notin H$, we must have that the trace $\mathcal{H} \setminus \{H\}|_H$ is a separating subfamily of $2^H$ and the trace $\mathcal{H} \setminus \{H\}|_{[n]\setminus H}$ is a separating subfamily of $2^{[n]\setminus H}$. We have seen that any minimal separating subfamily of $2^X$ has size at most $|X|-1$, so $\mathcal{H} \setminus \{H\}$ must contain a family $\mathcal{F}_{2,H}$ of size at most $|H| - 1 \leq n-3$, such that $\mathcal{F}_{2,H}|_H$ is a separating subfamily of $2^H$ and a family $\mathcal{F}_{2,[n]\setminus H}$ of size at most $n - |H| - 1 \leq n-3$ such that $\mathcal{F}_{2,[n]\setminus H}|_{[n]\setminus H}$ is a separating subfamily of $2^{[n]\setminus H}$. So our algorithm asks $\mathcal{F}_{2,H}$ in the second round if the answer to the only query of the first round was YES, and asks $\mathcal{F}_{2,[n]\setminus H}$ in the second round if the answer to the only query of the first round was NO. In both cases the total number of queries asked was at most $n-2$, as claimed. ■

Next we prove a necessary condition for $c_k(\mathcal{H}) = |\mathcal{H}|$ to hold if $k \geq 3$. The proof is based on the following variant of Theorem 462.

**Theorem 508 (Wiener, [549])** *Suppose $\mathcal{A} \subseteq 2^{[m]}$ is a family with $|\mathcal{A}| = n$ and the property that for any $x \in [m]$ we have $|\mathcal{A}|_{[m]\setminus\{x\}}| < |\mathcal{A}|$. Then there exists a subset $X \subset [m]$ of size $\lceil \frac{m^2}{2n-m-2} \rceil$ such that the multiplicity of every subset $S \subseteq [m] \setminus X$ in $\mathcal{H}|_{[m]\setminus X}$ is at most $\lceil \frac{m^2}{2n-m-2} \rceil + 1$.*

**Theorem 509 (Wiener, [549])** *If $\mathcal{H} \subseteq 2^{[n]}$ is a separating family with $c_k(\mathcal{H}) = |\mathcal{H}| = m$ for some $k \geq 3$, then there exists a representation $\mathcal{H}^*$ such that the dual of $\mathcal{H}^*$ contains a set $A$ and $\lceil \frac{m^2}{2n-m-2} \rceil$ other sets each containing $A$ and exactly one other element.*

**Proof.** As $|\mathcal{H}| = m$, we will assume that the dual family $\mathcal{H}^*$ is a subfamily of $2^{[m]}$. By the monotonicity of $c_k(\mathcal{H})$ in $k$, we know that $c_1(\mathcal{H}) = c_3(\mathcal{H}) = |\mathcal{H}|$ holds. In particular, $\mathcal{H}$ is a minimal separating family. This is equivalent to the condition that the dual family $\mathcal{H}^*$ possesses the property of the condition of Theorem 508. So applying that theorem, we obtain a subset $X \subseteq [m]$ of size $r := \lceil \frac{m^2}{2n-m-2} \rceil$, such that the multiplicity of every set $S \subseteq [m] \setminus X$ in $\mathcal{H}^*|_{[m]\setminus X}$ is at most $r+1$. The subset $X$ of $[m]$ corresponds to a subfamily $\mathcal{X} \subseteq \mathcal{H}$.

Let us consider 3-round search algorithms that in the first round ask all sets in $\mathcal{H} \setminus \mathcal{X}$, thus $m - r$ queries. The "multiplicity property" ensured by Theorem 508 means that no matter what answers we obtain to these queries, for any $a \in [n]$ there are at most $r+1$ elements in $[n]$, that are contained in exactly those sets of $\mathcal{H} \setminus \mathcal{X}$ as $a$. Let us assume $a$ is the defective element and denote the set of elements contained in exactly those sets of $\mathcal{H} \setminus \mathcal{X}$ as $a$ by $Y$. Then all we know after the first round is that one of the elements of $Y$ is the defective element.

As $c_3(\mathcal{H}) = |\mathcal{H}|$, no matter how we continue our algorithm, in the remaining two rounds we must ask all sets in $\mathcal{X}$ to determine the defective element. That is, we have $c_2(\mathcal{X}|_Y) = |\mathcal{X}|$. By Lemma 507, this is possible if and only if $|Y| = r+1$ and $\mathcal{X}|_Y$ has a representation consisting of $r$ singletons of $Y$. Translating this to $\mathcal{H}^*$, we obtain that the sets corresponding to elements of $Y$ possess the property stated in the theorem. Indeed, if $H_y$ denotes the set in $\mathcal{H}^*$ corresponding to $y \in Y$, then the trace $H_y|_{[m]\setminus X}$ is the same for all $y \in Y$, and $\{H_y|_X : y \in Y\} = \{\emptyset\} \cup \{\{x\} : x \in X\}$. ■

**Exercise 510** *Consider the family*

$$\mathcal{H}_0 = \{\{1,2,3,4\},\{5,6,7\},\{1,8,9\},\{2,5,10\},\{3,6,8\}\} \subseteq 2^{[11]}.$$

*Show that* $c_3(\mathcal{H}_0) = 5$*, but* $c_4(\mathcal{H}_0) = 4$.

In the remainder of this section, we show for any $r$ an asymptotially optimal $r$-round solution to a variant of the basic model, first considered by Katona [342]. Any subset of the underlying set can be asked as a query, and the goal is to find *a* defective element, but we do not know if there is any, and if there is, then how many. Thus the output of the algorithm is either a defective element, or the proof that there is none. Let $P(n,r)$ denote the total number of queries needed in case of $r$ rounds. Katona determined $P(n,2)$ exactly, as well as the number of queries needed in the adaptive and the non-adaptive version. He also gave an algorithm for the $r$-round version and conjectured it is close to optimal. This conjecture was verified by Gerbner and Vizer [268].

**Theorem 511 (Gerbner, Vizer [268])** *For any* $r, n \geq 1$*, we have*

$$rn^{1/r} \geq |P(n,r)| \geq rn^{1/r} - 2r + 1.$$

**Proof.** For the upper bound we describe the algorithm by Katona [342]. Let $X_1 = X$, and in the $i$th round, for $i < r$, consider a partition of $X_1$ into $\lceil n^{1/r} \rceil$ parts such that their sizes differ by at most one. Let $\mathcal{F}_i$ consist of these parts except for $F_i$, which is one of the smaller parts. After the answers we pick one of the sets in $\mathcal{F}_i$ that received a YES answer, thus one that contains a defective element. If there is no such set, we pick $F_i$. Let $X_{i+1}$ be the picked set. In the $r$th round the above step would mean asking all but one of the singletons; instead we ask all the singletons.

For the lower bound we describe a strategy for an adversary. First we introduce some notation. Let $\mathcal{F}_i^Y \subset \mathcal{F}_i$ be the family of queries that are answered YES by the adversary, and let $\mathcal{F}_i^N \subset \mathcal{F}_i$ be the family of those queries that are answered NO (and so $\mathcal{F}_i^N = \mathcal{F}_i \setminus \mathcal{F}_i^Y$). Let

$$G_i := \bigcup (\cup_{j=1}^{i} \mathcal{F}_j^N)$$

be the set of those elements that are known to be not defective after round $i$. Informally we can forget about them, and restrict the underlying set to $[n] \setminus G_i$ after round $i$. More precisely let $\mathcal{G}_i := \cup_{j=1}^{i} (\mathcal{F}_j^Y \setminus G_i)$, the set of the queries answered YES during the first $i$ rounds, restricted to $[n] \setminus G_i$.

Let $|\mathcal{F}_i| = k_i$ and $m_i := \min\{|G| : G \in \mathcal{G}_i\}$, the cardinality of the smallest set in $\mathcal{G}_i$. Let $n_0 = n$ and $n_i := \lfloor n_{i-1}/(k_i + 1) \rfloor \geq n/\Pi_{j=1}^{i}(k_j+1) - i$ (the latter inequality is an easy consequence of the fact that $k_i \geq 0$). We remark that when we describe how the adversary answers the queries in round $i$, we use only information that the adversary has at that point. For example, $k_1, \dots, k_i$ are known, but $k_{i+1}$ is not known at that point.

**Lemma 512** *The adversary can answer $\mathcal{F}_1, \ldots, \mathcal{F}_{r-1}$ such a way that, for all $1 \leq t \leq r-1$, we have $n_t \leq m_t - 1$, or all the answers are NO in the first $t$ rounds and $|G_t| \leq n - n_t$.*

**Proof of Lemma.** We use induction on $t$. If $t = 1$, then the adversary orders the elements of $\mathcal{F}_1$ in the following way. Let $H_1 := F_1$ be one of the smallest sets in $\mathcal{F}_1$, and for $2 \leq i \leq |\mathcal{F}_1|$ let $F_i \in \mathcal{F}_1 \setminus \{F_1, F_2, ..., F_{i-1}\}$ be such that the cardinality of $H_i := F_i \setminus \cup_{j=i}^{i-1} F_j$ is as small as possible. (Note that the sets $H_i$ are disjoint from each other.)

After this, if there is no $i$ with $|H_i| \geq n_1 + 1$, then the adversary answers NO for all queries in $\mathcal{F}_1$, and we clearly have $|G_1| \leq n - n/(k_1+1) \leq n - n_1$. If there is an $i$ with $|H_i| \geq n_1 + 1$, then the adversary chooses the smallest such $i$, and answers NO to $F_j$ if $j < i$ and YES if $j \geq i$. So each query in $\mathcal{F}_1^Y$ contains a least $|H_i| \geq n_1 + 1$ elements not in $\cup_{j=1}^{i-1} H_j (= G_1)$, and we are done with the case $t = 1$.

So assume that $t \geq 2$, and first consider the case when the adversary answered in the previous rounds only NO answers. Then - by induction - there are at least $n_{t-1}$ elements we do not know anything about. The adversary restricts the queries to those elements, and does the same as in the first round. That results in either that $m_t - 1 \geq n_{t-1}/(k_t+1) \geq n_t$, or the adversary answers only NO and at least $n_{t-1}/(k_t+1) \geq n_t$ many elements still do not appear in any queries.

Now we assume that the adversary answered YES at least once in the first $t-1$ rounds, and then every element of $\mathcal{G}_{t-1}$ has size at least $n_{t-1}+1$. In this case the adversary essentially does the same as in the first round, so orders the elements of $\mathcal{F}_t$ the following way (note that every element of $\mathcal{F}_t$ is disjoint from $G_{t-1}$). Let $H_1 := F_1$ be one of the smallest sets in $\mathcal{F}_t$, and for $2 \leq i \leq |\mathcal{F}_t|$ let $F_i \in \mathcal{F}_t \setminus \{F_1, F_2, ..., F_{i-1}\}$ be such, that the cardinality of $H_i := F_i \setminus \cup_{j=i}^{i-1} F_j$ is as small as possible. (Note again that the sets $H_i$ are disjoint from each other.)

Let us assume first that there is an $i$ with $|H_i| \geq n_t + 1$, and consider the smallest such $i$. Then the adversary answers NO to $F_j$ if $j < i$ and YES if $j \geq i$. Then each query in $\mathcal{F}_t^Y$ contains a least $|H_i| \geq n_t + 1$ elements not in $\cup_{j=1}^{i-1} H_j$. This means those members of $\mathcal{G}_t$ that correspond to queries in round $t$ have indeed size at least $n_t + 1$. The other members - by induction - had size at least $n_{t-1}+1$ before the round, and at most $| \cup_{j=1}^{i} H_j| \leq k_t n_t$ elements were moved to $G_t$, thus deleted from them in the current ($t^{th}$) round. Then at least $n_t + 1$ remains in each.

If there is no such $i$, then the adversary answers NO to every query. As earlier there was a YES answer, we still have to show that $n_t \leq m_t - 1$, but this time we do not have to deal with the new queries. For the earlier queries the same argument works: at most $| \cup_{j=1}^{i} H_j| \leq k_t n_t$ elements were deleted from each set in $\mathcal{G}_{t-1}$ and we are done with the proof of Lemma 512.

□

**Lemma 513** *For $r \geq 2$, to be able to find an defective element, in the $r^{th}$ round we need to ask at least $m_{r-1} - 1$ queries if there is at least one yes answer in the first $r-1$ rounds, and at least $n - |G_{r-1}|$ queries if all the answers were no in the first $r-1$ rounds.*

**Proof of Lemma.** We prove by induction on $n - |G_{r-1}| + m_{r-1}$. If $m_{r-1} = 0$ (so there were no YES answers during the first $r-1$ rounds), then we are done by the result of Katona [342] on the non-adaptive version of this problem. If $m_{r-1} = 1$, then we are also done, since there is a one-element query containing exactly one defective element. Using these observations and $n - |G_{r-1}| \geq m_{r-1}$, we are done with the cases $n - |G_{r-1}| + m_{r-1} = 1, 2, 3$; thus we can suppose $n - |G_{r-1}| + m_{r-1} \geq 4$ and $m_{r-1} \geq 2$.

First we show that at least one 1-element set has to be asked in the last round. Assume all the queries have size at least two and the adversary answers YES to all of them. Then, for every $x \in X \setminus G_{r-1}$, it is possible that $X \setminus G_{r-1} \setminus \{x\}$ is the set of defective elements. Indeed, it does not contradict the previous answers as $m_{r-1} \geq 2$, and does not contradict the answers in the last round, as each query contains a defective element and was answered YES. Thus no $x$ can be shown as a defective element, a contradiction.

Let us assume $\{x\}$ is asked in the $r^{th}$ round. But then the adversary can say NO to $\{x\}$ first (this is compatible with the answers in the first $r-1$ rounds, since $m_{r-1} \geq 2$), and consider it as if it were asked during the first $r-1$ rounds, and delete $x$ from the remaining queries asked in the $r^{th}$ round. Note that in this new scenario $m_{r-1}$ can decrease by at most 1. As $|[n] \setminus (G_{r-1} \cup x)| < n - |G_{r-1}|$ since $x \notin G_{r-1}$, by induction we know that at least $m_{r-1} - 1$ further queries are needed, and we are done with the proof of Lemma 513.

□

Lemma 513 and Lemma 512 show that

$$k_1 + ... + k_{r-1} + \frac{n}{(k_1+1)...(k_{r-1}+1)} - r$$

is a lower bound on $|P(n,r)|$. Using some reorganization and the inequality of arithmetic and geometric means we have:

$$k_1 + ... + k_{r-1} + \frac{n}{(k_1+1)...(k_{r-1}+1)} - r \geq r(n^{1/r} - \frac{r-1}{r} - 1),$$

which finishes the proof. ■

Gerbner and Vizer also showed that if the goal is to find $d$ defectives (or show that there are less than $d$ defectives altogether), then for the number $P(n,d,r)$ of queries we have $r\lceil (d^{r-1}n)^{1/r} \rceil + (r-1)d \geq |P(n,d,r)| \geq r(dn)^{1/r} - 2d - r(d+1) + 2$.

# *Bibliography*


[1] R. Ahlswede. Simple hypergraphs with maximal number of adjacent pairs of edges. *Journal of Combinatorial Theory, Series B*, 28(2):164–167, 1980.

[2] R. Ahlswede and G. O. H. Katona. Contributions to the geometry of Hamming spaces. *Discrete Mathematics*, 17(1):1–22, 1977.

[3] R. Ahlswede and G. O. H. Katona. Graphs with maximal number of adjacent pairs of edges. *Acta Mathematica Hungarica*, 32(1-2):97–120, 1978.

[4] R. Ahlswede and L. H. Khachatrian. The complete nontrivial-intersection theorem for systems of finite sets. *Journal of Combinatorial Theory, Series A*, 76(1):121–138, 1996.

[5] R. Ahlswede and L. H. Khachatrian. The complete intersection theorem for systems of finite sets. *European Journal of Combinatorics*, 18(2):125–136, 1997.

[6] R. Ahlswede and L. H. Khachatrian. Counterexample to the Frankl-Pach conjecture for uniform, dense families. *Combinatorica*, 17(2):299–301, 1997.

[7] V. Alekseev. On the number of intersection semilattices (in Russian). *Diskretnaya Matematika*, 1:129–136, 1989.

[8] P. Allen, J. Böttcher, O. Cooley, and R. Mycroft. Tight cycles in hypergraphs. *Electronic Notes in Discrete Mathematics*, 49:675–682, 2015.

[9] N. Alon. On the density of sets of vectors. *Discrete Mathematics*, 46(2):199–202, 1983.

[10] N. Alon. An extremal problem for sets with applications to graph theory. *Journal of Combinatorial Theory, Series A*, 40(1):82–89, 1985.

[11] N. Alon and V. Asodi. Tracing a single user. *European Journal of Combinatorics*, 27(8):1227–1234, 2006.

[12] N. Alon, L. Babai, and H. Suzuki. Multilinear polynomials and Frankl-Ray-Chaudhuri-Wilson type intersection theorems. *Journal of Combinatorial Theory, Series A*, 58(2):165–180, 1991.

[13] N. Alon, E. Fachini, and J. Körner. Locally thin set families. *Combinatorics, Probability and Computing*, 9(6):481–488, 2000.

[14] N. Alon, S. Moran, and A. Yehudayoff. Sign rank, VC dimension and spectral gaps. In *Electronic Colloquium on Computational Complexity (ECCC)*, volume 21, page 10, 2014.

[15] N. Alon, L. Rónyai, and T. Szabó. Norm-graphs: variations and applications. *Journal of Combinatorial Theory, Series B*, 76(2):280–290, 1999.

[16] N. Alon and A. Shapira. On an extremal hypergraph problem of Brown, Erdős and Sós. *Combinatorica*, 26(6):627–645, 2006.

[17] N. Alon and C. Shikhelman. Many $T$ copies in $H$-free graphs. *Journal of Combinatorial Theory, Series B*, 121:146–172, 2016.

[18] N. Alon, A. Shpilka, and C. Umans. On sunflowers and matrix multiplication. *Computational Complexity*, 22(2):219–243, 2013.

[19] N. Alon and J. H. Spencer. *The probabilistic method.* John Wiley & Sons, 4 edition, 2016.

[20] R. Anstee. A survey of forbidden configuration results. *Electronic Journal of Combinatorics*, 1000:DS20–Jan, 2013.

[21] R. Anstee and S. Salazar. Forbidden Berge hypergraphs. *Electronic Journal of Combinatorics*, 24(1), 2017. P1.59.

[22] R. Anstee and A. Sali. Small forbidden configurations IV: The 3 rowed case. *Combinatorica*, 25(5):503–518, 2005.

[23] R. P. Anstee and B. Fleming. Two refinements of the bound of Sauer, Perles and Shelah, and of Vapnik and Chervonenkis. *Discrete Mathematics*, 310(23):3318–3323, 2010.

[24] M. Axenovich, J. Manske, and R. Martin. $Q_2$-free families in the Boolean lattice. *Order*, 29(1):177–191, 2012.

[25] H. Aydinian, É. Czabarka, P. L. Erdős, and L. A. Székely. A tour of $M$-part $L$-Sperner families. *Journal of Combinatorial Theory, Series A*, 118(2):702–725, 2011.

[26] H. Aydinian and P. L. Erdős. All maximum size two-part Sperner systems: In short. *Combinatorics, Probability and Computing*, 16(04):553–555, 2007.

[27] L. Babai. A short proof of the nonuniform Ray-Chaudhuri-Wilson inequality. *Combinatorica*, 8(1):133–135, 1988.

[28] L. Babai and P. Frankl. *Linear Algebra Methods in Combinatorics: With Applications to Geometry and Computer Science.* Department of Computer Science, University of Chicago, preliminary version 2 edition, 1992.

[29] R. Baber. *Some results in extremal combinatorics.* PhD thesis, UCL (University College London), 2011.

[30] R. Baber and J. Talbot. Hypergraphs do jump. *Combinatorics, Probability and Computing*, 20(2):161–171, 2011.

[31] R. Baber and J. Talbot. New Turán densities for 3-graphs. *Electronic Journal of Combinatorics*, 19(2):P22, 2012.

[32] N. Balachandran and S. Bhattacharya. On an extremal hypergraph problem related to combinatorial batch codes. *Discrete Applied Mathematics*, 162:373–380, 2014.

[33] I. Balla, B. Bollobás, and T. Eccles. Union-closed families of sets. *Journal of Combinatorial Theory, Series A*, 120(3):531–544, 2013.

[34] J. Balogh, T. Bohman, B. Bollobás, and Y. Zhao. Turán densities of some hypergraphs related to $K_{k+1}^k$. *SIAM Journal on Discrete Mathematics*, 26(4):1609–1617, 2012.

[35] J. Balogh, T. Bohman, and D. Mubayi. Erdős–Ko–Rado in random hypergraphs. *Combinatorics, Probability and Computing*, 18(05):629–646, 2009.

[36] J. Balogh and B. Bollobás. Unavoidable traces of set systems. *Combinatorica*, 25(6):633–643, 2005.

[37] J. Balogh, B. Bollobás, R. Morris, and O. Riordan. Linear algebra and bootstrap percolation. *Journal of Combinatorial Theory, Series A*, 119(6):1328–1335, 2012.

[38] J. Balogh, B. Bollobás, and B. P. Narayanan. Transference for the Erdős–Ko–Rado theorem. In *Forum of Mathematics, Sigma*, volume 3, page e23. Cambridge Univ Press, 2015.

[39] J. Balogh, B. Csaba, R. R. Martin, and A. Pluhár. On the path separation number of graphs. *Discrete Applied Mathematics*, 213:26–33, 2016.

[40] J. Balogh, S. Das, M. Delcourt, H. Liu, and M. Sharifzadeh. Intersecting families of discrete structures are typically trivial. *Journal of Combinatorial Theory, Series A*, 132:224–245, 2015.

[41] J. Balogh, P. Keevash, and B. Sudakov. Disjoint representability of sets and their complements. *Journal of Combinatorial Theory, Series B*, 95(1):12–28, 2005.

[42] J. Balogh, T. Mészáros, and A. Z. Wagner. Two results about the hypercube. *arXiv preprint arXiv:1710.08509*, 2017.

[43] J. Balogh, R. Morris, and W. Samotij. Independent sets in hypergraphs. *Journal of the American Mathematical Society*, 28(3):669–709, 2015.

[44] J. Balogh, R. Morris, and W. Samotij. The method of hypergraph containers. *arXiv preprint arXiv:1801.04584*, 2018.

[45] J. Balogh and D. Mubayi. A new short proof of a theorem of Ahlswede and Khachatrian. *Journal of Combinatorial Theory, Series A*, 115(2):326–330, 2008.

[46] J. Balogh and D. Mubayi. Almost all triangle-free triple systems are tripartite. *Combinatorica*, 32(2):143–169, 2012.

[47] J. Balogh, R. Mycroft, and A. Treglown. A random version of Sperner's theorem. *Journal of Combinatorial Theory, Series A*, 128:104–110, 2014.

[48] J. Balogh and A. Z. Wagner. Kleitman's conjecture about families of given size minimizing the number of $k$-chains. *arXiv preprint arXiv:1609.02262*, 2016.

[49] H.-J. Bandelt, V. Chepoi, A. Dress, and J. Koolen. Combinatorics of lopsided sets. *European Journal of Combinatorics*, 27(5):669–689, 2006.

[50] C. Berge. *Hypergraphs: combinatorics of finite sets, Vol. 45*. Amsterdam: North-Holland Holl, 1989.

[51] E. Berlekamp. Block coding for the binary symmetric channel with noiseless, delayless feedback. In H. Mann, editor, *Error Correcting Codes*, pages 61–85. Wiley, 1968.

[52] E. Berlekamp. On subsets with intersections of even cardinality. *Canad. Math. Bull*, 12(4):471–477, 1969.

[53] E. R. Berlekamp. *Block coding with noiseless feedback.* PhD thesis, Massachusetts Institute of Technology, 1964.

[54] J. Bermond and P. Frankl. On a conjecture of Chvátal on $m$-intersecting hypergraphs. *Bulletin of the London Mathematical Society*, 9(3):310–312, 1977.

[55] A. Blokhuis. A new upper bound for the cardinality of 2-distance sets in Euclidean space. *North-Holland Mathematics Studies*, 87:65–66, 1984.

[56] E. Boehnlein and T. Jiang. Set families with a forbidden induced subposet. *Combinatorics, Probability and Computing*, 21(04):496–511, 2012.

[57] T. Bohman, C. Cooper, A. Frieze, R. Martin, and M. Ruszinkó. On randomly generated intersecting hypergraphs. *Journal of Combinatorics*, 10(3):R29, 2003.

[58] T. Bohman, A. Frieze, R. Martin, M. Ruszinkó, and C. Smyth. Randomly generated intersecting hypergraphs II. *Random Structures & Algorithms*, 30(1-2):17–34, 2007.

[59] T. Bohman, A. Frieze, D. Mubayi, and O. Pikhurko. Hypergraphs with independent neighborhoods. *Combinatorica*, 30(3):277–293, 2010.

[60] T. Bohman, A. Frieze, M. Ruszinkó, and L. Thoma. $G$-intersecting families. *Combinatorics, Probability and Computing*, 10(05):367–384, 2001.

[61] T. Bohman and R. R. Martin. A note on $G$-intersecting families. *Discrete Mathematics*, 260(1-3):183–188, 2003.

[62] B. Bollobás. On generalized graphs. *Acta Mathematica Hungarica*, 16(3-4):447–452, 1965.

[63] B. Bollobás. Weakly $k$-saturated graphs. In *Beiträge zur Graphentheorie (Kolloquium, Manebach, 1967)*, pages 25–31, 1968.

[64] B. Bollobás. Sperner systems consisting of pairs of complementary subsets. *Journal of Combinatorial Theory, Series A*, 15(3):363–366, 1973.

[65] B. Bollobás. Three-graphs without two triples whose symmetric difference is contained in a third. *Discrete Mathematics*, 8(1):21–24, 1974.

[66] B. Bollobás. *Random Graphs.* Number 73 in Cambridge studies in advanced mathematics. Cambridge University Press, 2001.

[67] B. Bollobás. *Extremal Graph Theory.* Courier Corporation, 2004.

[68] B. Bollobás, D. Daykin, and P. Erdős. Sets of independent edges of a hypergraph. *Quarterly Journal of Mathematics: Oxford Series(2)*, 27(105):25–32, 1976.

[69] B. Bollobás and E. Győri. Pentagons vs. triangles. *Discrete Math.*, 308(19):4332–4336, 2008.

[70] B. Bollobás and I. Leader. Compressions and isoperimetric inequalities. *Journal of Combinatorial Theory, Series A*, 56(1):47–62, 1991.

[71] B. Bollobás and I. Leader. Set systems with few disjoint pairs. *Combinatorica*, 23(4):559–570, 2003.

[72] B. Bollobás, I. Leader, and C. Malvenuto. Daisies and other Turán problems. *Combinatorics, Probability and Computing*, 20(5):743–747, 2011.

[73] B. Bollobás, I. Leader, and A. Radcliffe. Reverse Kleitman inequalities. *Proceedings of the London Mathematical Society*, 3(1):153–168, 1989.

[74] B. Bollobás, B. P. Narayanan, and A. M. Raigorodskii. On the stability of the Erdős–Ko–Rado theorem. *Journal of Combinatorial Theory, Series A*, 137:64–78, 2016.

[75] B. Bollobás and A. Radcliffe. Defect Sauer results. *Journal of Combinatorial Theory, Series A*, 72(2):189–208, 1995.

[76] J. A. Bondy. Induced subsets. *Journal of Combinatorial Theory, Series B*, 12(2):201–202, 1972.

[77] P. Borg. A short proof of a cross-intersection theorem of Hilton. *Discrete Mathematics*, 309(14):4750–4753, 2009.

[78] P. Borg. On cross-intersecting uniform sub-families of hereditary families. *Electronic Journal of Combinatorics*, 17(1):R60, 2010.

[79] P. Borg. Cross-intersecting sub-families of hereditary families. *Journal of Combinatorial Theory, Series A*, 119(4):871–881, 2012.

[80] P. Borg. The maximum product of sizes of cross-$t$-intersecting uniform families. *Australasian Journal of Combinatorics*, 60(1):69–78, 2014.

[81] P. Borg. The maximum sum and the maximum product of sizes of cross-intersecting families. *European Journal of Combinatorics*, 35:117–130, 2014.

[82] P. Borg. A cross-intersection theorem for subsets of a set. *Bulletin of the London Mathematical Society*, 47(2):248–256, 2015.

[83] P. Borg. The maximum product of weights of cross-intersecting families. *Journal of the London Mathematical Society*, 94(3):993–1018, 2016.

[84] R. C. Bose. A note on Fisher's inequality for balanced incomplete block designs. *The Annals of Mathematical Statistics*, 20(4):619–620, 1949.

[85] I. Bošnjak and P. Markovic. The 11-element case of Frankl's conjecture. *Electronic Journal of Combinatorics*, 15(1):R88, 2008.

[86] E. Boyer, D. Kreher, S. Radziszowski, J. Zhou, and A. Sidorenko. On $(n,5,3)$-Turán systems. *Ars Combinatoria*, 37:1–19, 1994.

[87] A. Brace and D. E. Daykin. A finite set covering theorem. *Bulletin of the Australian Mathematical Society*, 5(02):197–202, 1971.

[88] A. Brandt, D. Irwin, and T. Jiang. Stability and Turán numbers of a class of hypergraphs via Lagrangians. *Combinatorics, Probability and Computing*, 26(3):367–405, 2017.

[89] L. Brankovic, P. Lieby, and M. Miller. Flattening antichains with respect to the volume. *Electronic Journal of Combinatorics*, 6(R1):2, 1999.

[90] A. Brouwer, C. Mills, W. Mills, and A. Verbeek. Counting families of mutually intersecting sets. *Electronic Journal of Combinatorics*, 20(2):P8, 2013.

[91] W. Brown, P. Erdős, and V. Sós. Some extremal problems on $r$-graphs. In *New directions in the theory of graphs (Proc. Third Ann Arbor Conf., Univ. Michigan, Ann Arbor, Mich, 1971)*, pages 53–63, 1973.

[92] H. Bruhn and O. Schaudt. The journey of the union-closed sets conjecture. *Graphs and Combinatorics*, 31(6):2043–2074, 2015.

[93] M. Bucić, S. Letzter, B. Sudakov, and T. Tran. Minimum saturated families of sets. *arXiv preprint arXiv:1801.05471*, 2018.

[94] C. Bujtás and Z. Tuza. Turán numbers and batch codes. *Discrete Applied Mathematics*, 186:45–55, 2015.

[95] B. Bukh. Set families with a forbidden subposet. *Electronic Journal of Combinatorics*, 16(1):R142, 2009.

[96] Y. D. Burago and V. A. Zalgaller. *Geometric Inequalities.* Springer Verlag, Berlin, New York, 1986.

[97] P. Burcsi and D. T. Nagy. The method of double chains for largest families with excluded subposets. *Electronic Journal of Graph Theory and Applications*, 1(1), 2013.

[98] N. Bushaw and N. Kettle. Turán numbers for forests of paths in hypergraphs. *SIAM Journal on Discrete Mathematics*, 28(2):711–721, 2014.

[99] T. Carroll and G. O. H. Katona. Bounds on maximal families of sets not containing three sets with $A \cap B \subset C$, $A \not\subset B$. *Order*, 25(3):229–236, 2008.

[100] Y. Chang, Y. Peng, and Y. Yao. Connection between a class of polynomial optimization problems and maximum cliques of non-uniform hypergraphs. *Journal of Combinatorial Optimization*, 31(2):881–892, 2016.

[101] H.-B. Chen and W.-T. Li. A note on the largest size of families of sets with a forbidden poset. *Order*, 31(1):137–142, 2014.

[102] F. Chung and L. Lu. An upper bound for the Turán number $t_3(n,4)$. *Journal of Combinatorial Theory, Series A*, 87(2):381–389, 1999.

[103] F. R. Chung and P. Frankl. The maximum number of edges in a 3-graph not containing a given star. *Graphs and Combinatorics*, 3(1):111–126, 1987.

[104] F. R. K. Chung. Unavoidable stars in 3-graphs. *Journal of Combinatorial Theory, Series A*, 35(3):252–262, 1983.

[105] R. Church. Numerical analysis of certain free distributive structures. *Duke Mathematical Journal*, 6(3):732–734, 1940.

[106] V. Chvátal. An extremal set-intersection theorem. *Journal of the London Mathematical Society*, 2(2):355–359, 1974.

[107] S. M. Cioaba. Bounds on the Turán density of $PG(3,2)$. *Electronic Journal of Combinatorics*, 11(3):1, 2004.

[108] C. Collier-Cartaino, N. Graber, and T. Jiang. Linear Turán numbers of $r$-uniform linear cycles and related Ramsey numbers. *arXiv preprint arXiv:1404.5015*, 2014.

[109] D. Coppersmith and J. B. Shearer. New bounds for union-free families of sets. *Electronic Journal of Combinatorics*, 5(1):R39, 1998.

[110] E. Croot, V. F. Lev, and P. P. Pach. Progression-free sets in $\mathbb{Z}_4^n$ are exponentially small. *Annals of Mathematics*, 185(1):331–337, 2017.

[111] R. Csákány and J. Kahn. A homological approach to two problems on finite sets. *Journal of Algebraic Combinatorics*, 9(2):141–149, 1999.

[112] M. Csűrös and M. Ruszinkó. Single-user tracing and disjointly superimposed codes. *IEEE Transactions on Information Theory*, 51(4):1606–1611, 2005.

[113] G. Czédli. On averaging Frankl's conjecture for large union-closed-sets. *Journal of Combinatorial Theory, Series A*, 116(3):724–729, 2009.

[114] G. Czédli, M. Maróti, and E. T. Schmidt. On the scope of averaging for Frankl's conjecture. *Order*, 26(1):31–48, 2009.

[115] S. Das, W. Gan, and B. Sudakov. Sperner's theorem and a problem of Erdős, Katona and Kleitman. *Combinatorics, Probability and Computing*, 24(04):585–608, 2015.

[116] S. Das, W. Gan, and B. Sudakov. The minimum number of disjoint pairs in set systems and related problems. *Combinatorica*, pages 1–38, 2016.

[117] S. Das and B. Sudakov. Most probably intersecting hypergraphs. *Electronic Journal of Combinatorics*, 22(1):P1–80, 2015.

[118] S. Das and T. Tran. Removal and stability for Erdős–Ko–Rado. *SIAM Journal on Discrete Mathematics*, 30(2):1102–1114, 2016.

[119] A. Davoodi, E. Győri, A. Methuku, and C. Tompkins. An Erdős-Gallai type theorem for hypergraphs. *arXiv preprint arXiv:1608.03241*, 2016.

[120] A. De Bonis and G. O. H. Katona. Largest families without an $r$-fork. *Order*, 24(3):181–191, 2007.

[121] A. De Bonis, G. O. H. Katona, and K. J. Swanepoel. Largest family without $A \cup B \subseteq C \cap D$. *Journal of Combinatorial Theory, Series A*, 111(2):331–336, 2005.

[122] N. de Bruijn and P. Erdős. On a combinatorial problem. *Proc. Akad. Wet. Amsterdam*, 51:1277–1279, 1948.

[123] D. de Caen. Extension of a theorem of Moon and Moser on complete subgraphs. *Ars Combinatoria*, 16:5–10, 1983.

[124] D. de Caen. The current status of Turán's problem on hypergraphs. *Extremal Problems for Finite Sets*, 3:187–197, 1991.

[125] D. de Caen and Z. Füredi. The maximum size of 3-uniform hypergraphs not containing a Fano plane. *Journal of Combinatorial Theory, Series B*, 78(2):274–276, 2000.

[126] D. de Caen, D. Kreher, and J. Wiseman. On the constructive upper bounds for the Turán numbers $t(n, 2r+1, 2r)$. *Congressus Numerantum*, 65:277–280, 1988.

[127] M. K. de Carli Silva, F. Mário de Oliveira Filho, and C. M. Sato. Flag Algebras: A First Glance. *ArXiv preprint arXiv:1607.04741*, July 2016.

[128] R. Dedekind. *Über Zerlegungen von Zahlen Durch Ihre Grössten Gemeinsamen Theiler*, chapter 1, pages 1–40. Vieweg+Teubner Verlag, Wiesbaden, 1897.

[129] C. Deppe. Solution of Ulam's searching game with three lies or an optimal adaptive strategy for binary three-error-correcting codes. *Discrete Mathematics*, 224(1-3):79–98, 2000.

[130] C. Deppe. Strategies for the rényi–ulam game with fixed number of lies. *Theoretical Computer Science*, 314(1-2):45–55, 2004.

[131] C. Deppe. Coding with feedback and searching with lies. *Entropy, Search, Complexity*, pages 27–70, 2007.

[132] P. Devlin and J. Kahn. On "stability" in the Erdős–Ko–Rado theorem. *SIAM Journal on Discrete Mathematics*, 30(2):1283–1289, 2016.

[133] M. Deza, P. Frankl, and N. Singhi. On functions of strength $t$. *Combinatorica*, 3(3):331–339, 1983.

[134] I. Dinur and E. Friedgut. Intersecting families are essentially contained in juntas. *Combinatorics, Probability and Computing*, 18(1-2):107–122, 2009.

[135] A. P. Dove, J. R. Griggs, R. J. Kang, and J.-S. Sereni. Supersaturation in the Boolean lattice. *Integers*, 14:2, 2014.

[136] A. Dress. Towards a theory of holistic clustering. *DIMACS Series in Discrete Mathematics and Theoretical Computer Science*, 37:271–289, 1997.

[137] R. Dudley. The structure of some Vapnik-Chervonenkis classes. In *Proceedings of the Berkeley Conference in Honor of Jerzy Neyman*, volume 2, pages 495–507, 1985.

[138] R. A. Duke and P. Erdős. Systems of finite sets having a common intersection. In *Proceedings, 8th SE Conf. Combinatorics, Graph Theory and Computing*, pages 247–252, 1977.

[139] A. G. D'yachkov and V. V. Rykov. Bounds on the length of disjunctive codes. *Problemy Peredachi Informatsii*, 18(3):7–13, 1982.

[140] T. Eccles. A stability result for the union-closed size problem. *Combinatorics, Probability and Computing*, 25(3):399–418, 2016.

[141] J. S. Ellenberg and D. Gijswijt. On large subsets of $\mathbb{F}_q^n$ with no three-term arithmetic progression. *Annals of Mathematics*, 185(1):339–343, 2017.

[142] K. Engel. *Sperner Theory*, volume 65. Cambridge University Press, 1997.

[143] K. Engel and P. L. Erdős. Sperner families satisfying additional conditions and their convex hulls. *Graphs and Combinatorics*, 5(1):47–56, 1989.

[144] K. Engel and P. L. Erdős. Polytopes determined by complementfree Sperner families. *Discrete Mathematics*, 81(2):165–169, 1990.

[145] S. English, D. Gerbner, A. Methuku, and M. Tait. Linearity of Saturation for Berge Hypergraphs. Manuscript, 2018.

[146] S. English, N. Graber, P. Kirkpatrick, A. Methuku, and E. C. Sullivan. Saturation of Berge hypergraphs. *arXiv preprint arXiv:1710.03735*, 2017.

[147] P. Erdős. Some theorems on graphs. *Riveon lematematika*, 9:13–17, 1955.

[148] P. Erdős. Problems and results in combinatorial analysis. Combinatorics, Proc. Sympos. Pure Math. 19, 77-89 (1971)., 1971.

[149] P. Erdős. On a lemma of Littlewood and Offord. *Bulletin of the American Mathematical Society*, 51(12):898–902, 1945.

[150] P. Erdős. Extremal problems in graph theory. In *Theory of Graphs and its Applications (Proceedings of the Symposium in Smolenice)*. Publishing House of the Czechoslovak Academy of Sciences, Prague, 1964.

[151] P. Erdős. On extremal problems of graphs and generalized graphs. *Israel Journal of Mathematics*, 2(3):183–190, 1964.

[152] P. Erdős. A problem on independent $r$-tuples. *Annales Universitatis Scientarium Budapestinensis de Rolando Eötvös Nominatae Sectio Mathematica*, 8:83–95, 1965.

[153] P. Erdős. On some new inequalities concerning extremal properties of graphs. In *Theory of Graphs (Proceedings of Colloquium, Tihany, 1966)*, pages 77–81, 1968.

[154] P. Erdős. On some extremal problems on $r$-graphs. *Discrete Mathematics*, 1(1):1–6, 1971.

[155] P. Erdős. Topics in combinatorial analysis. In *Proc. Second Louisiana Conf. on Comb., Graph Theory and Computing (RC Mullin et al., eds.)*, pages 2–20, 1971.

[156] P. Erdős. Problems and results in graph theory and combinatorial analysis. *Proc. British Combinatorial Conj., 5th*, pages 169–192, 1975.

[157] P. Erdős, P. Frankl, and V. Rödl. The asymptotic number of graphs not containing a fixed subgraph and a problem for hypergraphs having no exponent. *Graphs and Combinatorics*, 2(1):113–121, 1986.

[158] P. Erdős, Z. Füredi, and Z. Tuza. Saturated r-uniform hypergraphs. *Discrete Mathematics*, 98(2):95–104, 1991.

[159] P. Erdős and T. Gallai. On maximal paths and circuits of graphs. *Acta Mathematica Hungarica*, 10(3-4):337–356, 1959.

[160] P. Erdős and N. Hindman. Enumeration of intersecting families. *Discrete Mathematics*, 48(1):61–65, 1984.

[161] P. Erdős and D. J. Kleitman. On coloring graphs to maximize the proportion of multicolored k-edges. *Journal of Combinatorial Theory*, 5(2):164–169, 1968.

[162] P. Erdős and D. J. Kleitman. Extremal problems among subsets of a set. *Discrete Mathematics*, 8(3):281–294, 1974.

[163] P. Erdős, C. Ko, and R. Rado. Intersection theorems for systems of finite sets. *The Quarterly Journal of Mathematics*, 12(1):313–320, 1961.

[164] P. Erdős and L. Lovász. Problems and results on 3-chromatic hypergraphs and some related questions. *Infinite and Finite Sets*, 10(2):609–627, 1975.

[165] P. Erdős and L. Moser. Problem 35. In *Proceedings on the Conference of Combinatorial Structures and their Applications, Calgary*, page 506, 1969.

[166] P. Erdős and R. Rado. Intersection theorems for systems of sets. *Journal of the London Mathematical Society*, 1(1):85–90, 1960.

[167] P. Erdős and M. Simonovits. Supersaturated graphs and hypergraphs. *Combinatorica*, 3(2):181–192, 1983.

[168] P. Erdős and E. Szemerédi. Combinatorial properties of systems of sets. *Journal of Combinatorial Theory, Series A*, 24(3):308–313, 1978.

[169] P. L. Erdős, P. Frankl, and G. O. H. Katona. Intersecting Sperner families and their convex hulls. *Combinatorica*, 4(1):21–34, 1984.

[170] P. L. Erdős, Z. Füredi, and G. O. H. Katona. Two-part and $k$-Sperner families: new proofs using permutations. *SIAM Journal on Discrete Mathematics*, 19(2):489–500, 2005.

[171] P. L. Erdős and G. O. H. Katona. All maximum 2-part Sperner families. *Journal of Combinatorial Theory, Series A*, 43(1):58–69, 1986.

[172] P. L. Erdős and G. O. H. Katona. Convex hulls of more-part Sperner families. *Graphs and Combinatorics*, 2(1):123–134, 1986.

[173] P. L. Erdős and G. O. H. Katona. A 3-part Sperner theorem. *Studia Scientiarum Mathematicarum Hungarica*, 22:383–393, 1987.

[174] P. L. Erdős, G. O. H. Katona, and P. Frankl. Extremal hypergraph problems and convex hulls. *Combinatorica*, 5(1):11–26, 1985.

[175] B. Ergemlidze, E. Győri, and A. Methuku. Asymptotics for Turán numbers of cycles in 3-uniform linear hypergraphs. *arXiv preprint arXiv:1705.03561*, 2017.

[176] V. Falgas-Ravry. *Thresholds in probabilistic and extremal combinatorics.* PhD thesis, Queen Mary University of London, 2012.

[177] V. Falgas-Ravry, T. Kittipassorn, D. Korándi, S. Letzter, and B. P. Narayanan. Separating path systems. *Journal of Combinatorics*, 5(3):335–354, 2014.

[178] V. Falgas-Ravry and E. R. Vaughan. Turán $H$-densities for 3-graphs. *Electronic Journal of Combinatorics*, 19(3):P40, 2012.

[179] V. Falgas-Ravry and E. R. Vaughan. Applications of the semi-definite method to the Turán density problem for 3-graphs. *Combinatorics, Probability and Computing*, 22(1):21–54, 2013.

[180] J. R. Faudree, R. J. Faudree, and J. R. Schmitt. A survey of minimum saturated graphs. *Electronic Journal of Combinatorics*, 1000:DS19–Jul, 2011.

[181] M. Ferrara, B. Kay, L. Kramer, R. R. Martin, B. Reiniger, H. C. Smith, and E. Sullivan. The saturation number of induced subposets of the Boolean lattice. *Discrete Mathematics*, 340(10):2479–2487, 2017.

[182] R. A. Fisher. An examination of the different possible solutions of a problem in incomplete blocks. *Annals of Human Genetics*, 10(1):52–75, 1940.

[183] P. Frankl. On Sperner families satisfying an additional condition. *Journal of Combinatorial Theory, Series A*, 20(1):1–11, 1976.

[184] P. Frankl. On families of finite sets no two of which intersect in a singleton. *Bulletin of the Australian Mathematical Society*, 17(1):125–134, 1977.

[185] P. Frankl. On the minimum number of disjoint pairs in a family of finite sets. *Journal of Combinatorial Theory, Series A*, 22(2):249–251, 1977.

[186] P. Frankl. The Erdős-Ko-Rado theorem is true for $n = ckt$. In *Combinatorics (Proceedings of the Fifth Hungarian Colloquium, Keszthely, 1976)*, volume 1, pages 365–375, 1978.

[187] P. Frankl. An extremal problem for 3-graphs. *Acta Mathematica Hungarica*, 32(1-2):157–160, 1978.

[188] P. Frankl. On intersecting families of finite sets. *Journal of Combinatorial Theory, Series A*, 24(2):146–161, 1978.

[189] P. Frankl. On intersecting families of finite sets. *Bulletin of the Australian Mathematical Society*, 21(03):363–372, 1980.

[190] P. Frankl. On a problem of Chvátal and Erdős on hypergraphs containing no generalized simplex. *Journal of Combinatorial Theory, Series A*, 30(2):169–182, 1981.

[191] P. Frankl. An extremal problem for two families of sets. *European Journal of Combinatorics*, 3(2):125–127, 1982.

[192] P. Frankl. On the trace of finite sets. *Journal of Combinatorial Theory, Series A*, 34(1):41–45, 1983.

[193] P. Frankl. The shifting technique in extremal set theory. *Surveys in Combinatorics*, 123:81–110, 1987.

[194] P. Frankl. Traces of antichains. *Graphs and Combinatorics*, 5(1):295–299, 1989.

[195] P. Frankl. Asymptotic solution of a Turán-type problem. *Graphs and Combinatorics*, 6(3):223–227, 1990.

[196] P. Frankl. Multiply-intersecting families. *Journal of Combinatorial Theory, Series B*, 53(2):195–234, 1991.

[197] P. Frankl. Improved bounds for Erdős' matching conjecture. *Journal of Combinatorial Theory, Series A*, 120(5):1068–1072, 2013.

[198] P. Frankl. On the maximum number of edges in a hypergraph with given matching number. *Discrete Applied Mathematics*, 216:562–581, 2017.

[199] P. Frankl. Some exact results for multiply intersecting families. *Manuscript*, 2017.

[200] P. Frankl and Z. Füredi. A short proof for a theorem of Harper about Hamming-spheres. *Discrete Mathematics*, 34(3):311–313, 1981.

[201] P. Frankl and Z. Füredi. An exact result for 3-graphs. *Discrete Mathematics*, 50:323–328, 1984.

[202] P. Frankl and Z. Füredi. Union-free hypergraphs and probability theory. *European Journal of Combinatorics*, 5(2):127–131, 1984.

[203] P. Frankl and Z. Füredi. Forbidding just one intersection. *Journal of Combinatorial Theory, Series A*, 39(2):160–176, 1985.

[204] P. Frankl and Z. Füredi. Non-trivial intersecting families. *Journal of Combinatorial Theory, Series A*, 41(1):150–153, 1986.

[205] P. Frankl and Z. Füredi. Union-free families of sets and equations over fields. *Journal of Number Theory*, 23(2):210–218, 1986.

[206] P. Frankl and Z. Füredi. Exact solution of some Turán-type problems. *Journal of Combinatorial Theory, Series A*, 45(2):226–262, 1987.

[207] P. Frankl and Z. Füredi. Extremal problems and the Lagrange function of hypergraphs. *Bulletin of the Institute of Mathematics, Academia Sinica*, 16:305–313, 1988.

[208] P. Frankl and Z. Füredi. Extremal problems whose solutions are the blowups of the small Witt-designs. *Journal of Combinatorial Theory, Series A*, 52(1):129–147, 1989.

[209] P. Frankl and G. O. H. Katona. Polytopes determined by hypergraph classes. *European Journal of Combinatorics*, 6(3):233–243, 1985.

[210] P. Frankl and A. Kupavskii. A size-sensitive inequality for cross-intersecting families. *European Journal of Combinatorics*, 62:263–271, 2017.

[211] P. Frankl and A. Kupavskii. Uniform $s$-cross-intersecting families. *Combinatorics, Probability and Computing*, 26(4):517–524, 2017.

[212] P. Frankl and A. Kupavskii. Counting Intersecting and Pairs of Cross-Intersecting Families. *Combinatorics, Probability and Computing*, 27(1):60–68, 2018.

[213] P. Frankl and A. Kupavskii. Proof of the Erdős Matching Conjecture in a New Range. *arXiv preprint arXiv:1806.08855*, 2018.

[214] P. Frankl, S. J. Lee, M. Siggers, and N. Tokushige. An Erdős–Ko–Rado theorem for cross $t$-intersecting families. *Journal of Combinatorial Theory, Series A*, 128:207–249, 2014.

[215] P. Frankl, T. Łuczak, and K. Mieczkowska. On matchings in hypergraphs. *Electronic Journal of Combinatorics*, 19(2):P42, 2012.

[216] P. Frankl, K. Ota, and N. Tokushige. Uniform intersecting families with covering number four. *Journal of Combinatorial Theory, Series A*, 71(1):127–145, 1995.

[217] P. Frankl, K. Ota, and N. Tokushige. Covers in uniform intersecting families and a counterexample to a conjecture of Lovász. *Journal of Combinatorial Theory, Series A*, 74(1):33–42, 1996.

[218] P. Frankl and J. Pach. On disjointly representable sets. *Combinatorica*, 4(1):39–45, 1984.

[219] P. Frankl and V. Rödl. Extremal problems on set systems. *Random Structures & Algorithms*, 20(2):131–164, 2002.

[220] P. Frankl, V. Rödl, and A. Ruciński. On the maximum number of edges in a triple system not containing a disjoint family of a given size. *Combinatorics, Probability and Computing*, 21:141–148, 3 2012.

[221] P. Frankl and N. Tokushige. Some best possible inequalities concerning cross-intersecting families. *Journal of Combinatorial Theory, Series A*, 61(1):87–97, 1992.

[222] P. Frankl and N. Tokushige. Some inequalities concerning cross-intersecting families. *Combinatorics, Probability and Computing*, 7(3):247–260, 1998.

[223] P. Frankl and M. Watanabe. Density results for uniform families. *Combinatorica*, 14(1):115–119, 1994.

[224] P. Frankl and R. M. Wilson. Intersection theorems with geometric consequences. *Combinatorica*, 1(4):357–368, 1981.

[225] E. Friedgut. On the measure of intersecting families, uniqueness and stability. *Combinatorica*, 28(5):503–528, 2008.

[226] Z. Füredi. Geometrical solution of an intersection problem for two hypergraphs. *European Journal of Combinatorics*, 5(2):133–136, 1984.

[227] Z. Füredi. Hypergraphs in which all disjoint pairs have distinct unions. *Combinatorica*, 4(2):161–168, 1984.

[228] Z. Füredi. Problem session. Kombinatorik geordneter Mengen, Oberwolfach, B.R.D., 1985.

[229] Z. Füredi. A Ramsey-Sperner theorem. *Graphs and Combinatorics*, 1(1):51–56, 1985.

[230] Z. Füredi. Turán type problems. In *Surveys in Combinatorics*, volume 166. Cambridge Univ. Press Cambridge, 1991.

[231] Z. Füredi. New asymptotics for bipartite Turán numbers. *Journal of Combinatorial Theory, Series A*, 75(1):141–144, 1996.

[232] Z. Füredi. On $r$-cover-free families. *Journal of Combinatorial Theory, Series A*, 73(1):172–173, 1996.

[233] Z. Füredi. An upper bound on Zarankiewicz'problem. *Combinatorics, Probability and Computing*, 5(1):29–33, 1996.

[234] Z. Füredi. Linear paths and trees in uniform hypergraphs. *Electronic Notes in Discrete Mathematics*, 38:377–382, 2011.

[235] Z. Füredi. 2-cancellative hypergraphs and codes. *Combinatorics, Probability and Computing*, 21(1-2):159–177, 2012.

[236] Z. Füredi. Linear trees in uniform hypergraphs. *European Journal of Combinatorics*, 35:264–272, 2014.

[237] Z. Füredi, J. R. Griggs, A. M. Odlyzko, and J. B. Shearer. Ramsey–Sperner theory. *Discrete Mathematics*, 63(2-3):143–152, 1987.

[238] Z. Füredi, K.-W. Hwang, and P. M. Weichsel. A proof and generalizations of the Erdős-Ko-Rado theorem using the method of linearly independent polynomials. In *Topics in Discrete Mathematics*, pages 215–224. Springer, 2006.

[239] Z. Füredi and T. Jiang. Hypergraph Turán numbers of linear cycles. *Journal of Combinatorial Theory, Series A*, 123(1):252–270, 2014.

[240] Z. Füredi and T. Jiang. Turán numbers of hypergraph trees. *arXiv preprint arXiv:1505.03210*, 2015.

[241] Z. Füredi, T. Jiang, A. Kostochka, D. Mubayi, and J. Verstraëte. Hypergraphs not containing a tight tree with a bounded trunk. *arXiv preprint arXiv:1712.04081*, 2017.

[242] Z. Füredi, T. Jiang, A. Kostochka, D. Mubayi, and J. Verstraëte. Tight paths and matchings in convex geometric hypergraphs. *arXiv preprint arXiv:1709.01173*, 2017.

[243] Z. Füredi, T. Jiang, A. Kostochka, D. Mubayi, and J. Verstraëte. The extremal number for (a, b)-paths and other hypergraph trees. Manuscript, 2018.

[244] Z. Füredi, T. Jiang, and R. Seiver. Exact solution of the hypergraph Turán problem for $k$-uniform linear paths. *Combinatorica*, 34(3):229–322, 2014.

[245] Z. Füredi, R. Luo, and A. Kostochka. Avoiding long Berge cycles. *arXiv preprint arXiv:1805.04195*, 2018.

[246] Z. Füredi, D. Mubayi, and O. Pikhurko. Quadruple systems with independent neighborhoods. *Journal of Combinatorial Theory, Series A*, 115(8):1552–1560, 2008.

[247] Z. Füredi and L. Özkahya. Unavoidable subhypergraphs: a-clusters. *Journal of Combinatorial Theory, Series A*, 118(8):2246–2256, 2011.

[248] Z. Füredi and L. Özkahya. On 3-uniform hypergraphs without a cycle of a given length. *Discrete Applied Mathematics*, 216:582–588, 2017.

[249] Z. Füredi, O. Pikhurko, and M. Simonovits. The Turán density of the hypergraph $\{abc, ade, bde, cde\}$. *Electronic Journal of Combinatorics*, 10(1):R18, 2003.

[250] Z. Füredi, O. Pikhurko, and M. Simonovits. On triple systems with independent neighbourhoods. *Combinatorics, Probability and Computing*, 14(5-6):795–813, 2005.

[251] Z. Füredi, O. Pikhurko, and M. Simonovits. 4-books of three pages. *Journal of Combinatorial Theory, Series A*, 113(5):882–891, 2006.

[252] Z. Füredi and F. Quinn. Traces of finite sets. *Ars Combin*, 18:195–200, 1984.

[253] Z. Füredi and M. Simonovits. Triple systems not containing a Fano configuration. *Combinatorics, Probability and Computing*, 14(4):467–484, 2005.

[254] Z. Füredi and B. Sudakov. Extremal set systems with restricted $k$-wise intersections. *Journal of Combinatorial Theory, Series A*, 105(1):143–159, 2004.

[255] M. M. Gauy, H. Hàn, and I. C. Oliveira. Erdős–ko–rado for random hypergraphs: asymptotics and stability. *Combinatorics, Probability and Computing*, 26(3):406–422, 2017.

[256] D. Gerbner. Profile polytopes of some classes of families. *Combinatorica*, 33(2):199–216, 2013.

[257] D. Gerbner, B. Keszegh, N. Lemons, C. Palmer, D. Pálvölgyi, and B. Patkós. Saturating Sperner families. *Graphs and Combinatorics*, 29(5):1355–1364, 2013.

[258] D. Gerbner, B. Keszegh, and B. Patkós. Generalized forbidden subposet problems. *arXiv preprint arXiv:1701.05030*, 2017.

[259] D. Gerbner, N. Lemons, C. Palmer, B. Patkós, and V. Szécsi. Almost intersecting families of sets. *SIAM Journal on Discrete Mathematics*, 26(4):1657–1669, 2012.

[260] D. Gerbner, A. Methuku, D. T. Nagy, B. Patkós, and M. Vizer. On the number of containments in $p$-free families. *arXiv preprint arXiv:1804.01606*, 2018.

[261] D. Gerbner, A. Methuku, and C. Palmer. A general lemma for Berge-Turán problems. Manuscript, 2018.

[262] D. Gerbner, A. Methuku, and C. Tompkins. Intersecting $P$-free families. *Journal of Combinatorial Theory, Series A*, 151:61–83, 2017.

[263] D. Gerbner, A. Methuku, and M. Vizer. Asymptotics for the Turán number of Berge-$K_{2,t}$. *arXiv preprint arXiv:1705.04134*, 2017.

[264] D. Gerbner and C. Palmer. Extremal results for berge hypergraphs. *SIAM Journal on Discrete Mathematics*, 31(4):2314–2327, 2017.

[265] D. Gerbner and C. Palmer. Counting copies of a fixed subgraph in $f$-free graphs. *arXiv preprint arXiv:1805.07520*, 2018.

[266] D. Gerbner and B. Patkós. $l$-chain profile vectors. *SIAM Journal on Discrete Mathematics*, 22(1):185–193, 2008.

[267] D. Gerbner, B. Patkós, and M. Vizer. Forbidden subposet problems for traces of set families. *arXiv preprint arXiv:1706.01212*, 2017.

[268] D. Gerbner and M. Vizer. Rounds in a combinatorial search problem. *arXiv preprint arXiv:1611.10133*, 2016.

[269] G. R. Giraud. Remarques sur deux problemes extrémaux. *Discrete Mathematics*, 84(3):319–321, 1990.

[270] J. Goldwasser and R. Hansen. The exact Turán number of $F(3,3)$ and all extremal configurations. *SIAM Journal on Discrete Mathematics*, 27(2):910–917, 2013.

[271] W. T. Gowers. Hypergraph regularity and the multidimensional Szemerédi theorem. *Annals of Mathematics*, pages 897–946, 2007.

[272] G. Greco. Embeddings and the trace of finite sets. *Information Processing Letters*, 67(4):199–203, 1998.

[273] C. Greene, G. O. H. Katona, and D. J. Kleitman. Extensions of the Erdős-Ko-Rado theorem. *Studies in Applied Mathematics*, 55(1):1–8, 1976.

[274] J. R. Griggs. Another Three Part Sperner Theorem. *Studies in Applied Mathematics*, 57(2):181–184, 1977.

[275] J. R. Griggs. The Littlewood-Offord problem: tightest packing and an M-part Sperner theorem. *European Journal of Combinatorics*, 1(3):225–234, 1980.

[276] J. R. Griggs, S. Hartmann, T. Kalinowski, U. Leck, and I. T. Roberts. Full and maximal squashed flat antichains of minimum weight. *arXiv preprint arXiv:1704.00067*, 2017.

[277] J. R. Griggs and G. O. H. Katona. No four subsets forming an N. *Journal of Combinatorial Theory, Series A*, 115(4):677–685, 2008.

[278] J. R. Griggs and D. J. Kleitman. A three part Sperner theorem. *Discrete Mathematics*, 17(3):281–289, 1977.

[279] J. R. Griggs and W.-T. Li. The partition method for poset-free families. *Journal of Combinatorial Optimization*, 25(4):587–596, 2013.

[280] J. R. Griggs, W.-T. Li, and L. Lu. Diamond-free families. *Journal of Combinatorial Theory, Series A*, 119(2):310–322, 2012.

[281] J. R. Griggs and L. Lu. On families of subsets with a forbidden subposet. *Combinatorics, Probability and Computing*, 18(05):731–748, 2009.

[282] J. R. Griggs, A. M. Odlyzko, and J. B. Shearer. $k$-color Sperner theorems. *Journal of Combinatorial Theory, Series A*, 42(1):31–54, 1986.

[283] V. Grolmusz and B. Sudakov. On $k$-wise set-intersections and $k$-wise Hamming-distances. *Journal of Combinatorial Theory, Series A*, 99(1):180–190, 2002.

[284] D. Grósz, A. Methuku, and C. Tompkins. An improvement of the general bound on the largest family of subsets avoiding a subposet. *Order*, pages 1–13, 2016.

[285] D. Grósz, A. Methuku, and C. Tompkins. Uniformity thresholds for the asymptotic size of extremal Berge-$F$-free hypergraphs. *Electronic Notes in Discrete Mathematics*, 61:527–533, 2017.

[286] V. Gruslys, I. Leader, and I. Tomon. Partitioning the Boolean lattice into copies of a poset. *arXiv preprint arXiv:1609.02520*, 2016.

[287] M. Grüttmüller, S. Hartmann, T. Kalinowski, U. Leck, and I. T. Roberts. Maximal flat antichains of minimum weight. *Electronic Journal of Combinatorics*, 16(1):R69, 2009.

[288] A. Grzesik. *Flag Algebras in Extremal Graph Theory*. PhD thesis, Jagiellonian University, Krakow, 2014.

[289] R. Gu, X. Li, Y. Peng, and Y. Shi. Some Motzkin–Straus type results for non-uniform hypergraphs. *Journal of Combinatorial Optimization*, 31(1):223–238, 2016.

[290] R. Gu, X. Li, and Y. Shi. Hypergraph Turán numbers of vertex disjoint cycles. *arXiv preprint arXiv:1305.5372*, 2013.

[291] K. Gunderson and J. Semeraro. Tournaments, 4-uniform hypergraphs, and an exact extremal result. *Journal of Combinatorial Theory, Series B*, 2017.

[292] W. Guzicki. Ulam's searching game with two lies. *Journal of Combinatorial Theory, Series A*, 54(1):1–19, 1990.

[293] E. Győri. Triangle-free hypergraphs. *Combinatorics, Probability and Computing*, 15(1-2):185–191, 2006.

[294] E. Győri, G. Y. Katona, and N. Lemons. Hypergraph extensions of the Erdős-Gallai theorem. *European Journal of Combinatorics*, 58:238–246, 2016.

[295] E. Győri and N. Lemons. Hypergraphs with no odd cycle of given length. In *European Conference on Combinatorics, Graph Theory and Applications (EuroComb 2009)*, volume 34 of *Electronic Notes in Discrete Mathematics*, pages 359–362. Elsevier Sci. B. V., Amsterdam, 2009.

[296] E. Győri and N. Lemons. 3-uniform hypergraphs avoiding a given odd cycle. *Combinatorica*, 32(2):187–203, 2012.

[297] E. Győri and N. Lemons. Hypergraphs with no cycle of a given length. *Combinatorics, Probability and Computing*, 21(1-2):193–201, 2012.

[298] E. Győri and N. Lemons. Hypergraphs with no cycle of length 4. *Discrete Mathematics*, 312(9):1518–1520, 2012.

[299] P. Hall. On representatives of subsets. *Journal of the London Mathematical Society*, 1(1):26–30, 1935.

[300] A. Hamm and J. Kahn. On Erdős-Ko-Rado for random hypergraphs I. *arXiv preprint arXiv:1412.5085*, 2014.

[301] A. Hamm and J. Kahn. On Erdős-Ko-Rado for random hypergraphs II. *arXiv preprint arXiv:1406.5793*, 2014.

[302] G. Hansel. Sur le nombre des fonctions booléennes monotones de $n$ variables. *Comptes rendus hebdomadeires des seances de l'Academia des sciences, Seria A*, 262(20):1088, 1966.

[303] L. H. Harper. Optimal numberings and isoperimetric problems on graphs. *Journal of Combinatorial Theory*, 1(3):385–393, 1966.

[304] L. H. Harper. On a problem of Kleitman and West. *Discrete Mathematics*, 93(2):169–182, 1991.

[305] D. Hefetz and P. Keevash. A hypergraph Turán theorem via Lagrangians of intersecting families. *Journal of Combinatorial Theory, Series A*, 120(8):2020–2038, 2013.

[306] T. Héger, B. Patkós, and M. Takáts. Search problems in vector spaces. *Designs, Codes and Cryptography*, 76(2):207–216, 2015.

[307] A. J. Hilton. The Erdős–Ko–Rado theorem with valency conditions. *Unpublished Manuscript*, 1976.

[308] A. J. Hilton. An intersection theorem for a collection of families of subsets of a finite set. *Journal of the London Mathematical Society*, 2(3):369–376, 1977.

[309] A. J. Hilton and E. C. Milner. Some intersection theorems for systems of finite sets. *The Quarterly Journal of Mathematics*, 18(1):369–384, 1967.

[310] K. A. Hogenson. *Random and deterministic versions of extremal poset problems*. PhD thesis, Iowa State University, 2016.

[311] Y. Hu. On the union-closed sets conjecture. *arXiv preprint arXiv:1706.06167*, 2017.

[312] H. Huang, P.-S. Loh, and B. Sudakov. The size of a hypergraph and its matching number. *Combinatorics, Probability and Computing*, 21(03):442–450, 2012.

[313] H. Huang and J. Ma. On tight cycles in hypergraphs. *arXiv preprint arXiv:1711.07442*, 2017.

[314] M. N. Huxley and H. Iwaniec. Bombieri's theorem in short intervals. *Mathematika*, 22(2):188194, 1975.

[315] F. Hwang and V. T. Sós. Non-adaptive hypergeometric group testing. *Studia Scientiarum Mathematicarum Hungarica*, 22:257–263, 1987.

[316] K.-W. Hwang and Y. Kim. A proof of Alon–Babai–Suzuki's conjecture and multilinear polynomials. *European Journal of Combinatorics*, 43:289–294, 2015.

[317] E. Jackowska, J. Polcyn, and A. Ruciński. Turán numbers for 3-uniform linear paths of length 3. *Electronic Journal of Combinatorics*, 23(2):P2–30, 2016.

[318] S. Janson, T. Łuczak, and A. Ruciński. *Random Graphs*, volume 45. John Wiley & Sons, 2011.

[319] T. Jiang and J. Ma. Cycles of given lengths in hypergraphs. *arXiv preprint arXiv:1609.08212*, 2016.

[320] T. Jiang, Y. Peng, and B. Wu. Turán numbers of extensions of some sparse hypergraphs via Lagrangians. *arXiv preprint arXiv:1609.08983*, 2016.

[321] J. R. Johnson and J. Talbot. $G$-intersection theorems for matchings and other graphs. *Combinatorics, Probability and Computing*, 17(04):559–575, 2008.

[322] T. Johnston and L. Lu. Strong jumps and Lagrangians of non-uniform hypergraphs. *arXiv preprint arXiv:1403.1220*, 2014.

[323] T. Johnston and L. Lu. Turán problems on non-uniform hypergraphs. *Electronic Journal of Combinatorics*, 21(4):P4–22, 2014.

[324] J. Kahn. On a problem of Erdős and Lovász: random lines in a projective plane. *Combinatorica*, 12(4):417–423, 1992.

[325] J. Kahn. On a problem of Erdős and Lovász. II. $n(r) = o(r)$. *Journal of the American Mathematical Society*, 7(1):125–143, 1994.

[326] J. Kahn. Entropy, independent sets and antichains: a new approach to Dedekind's problem. *Proceedings of the American Mathematical Society*, 130(2):371–378, 2002.

[327] G. Kalai. Intersection patterns of convex sets. *Israel Journal of Mathematics*, 48(2-3):161–174, 1984.

[328] T. Kalinowski, U. Leck, C. Reiher, and I. T. Roberts. Minimizing the regularity of maximal regular antichains of 2-sets and 3-sets. *Australasian Journal of Combinatorics*, 64(2):277–288, 2016.

[329] T. Kalinowski, U. Leck, and I. T. Roberts. Maximal antichains of minimum size. *Electronic Journal of Combinatorics*, 20(2):P3, 2013.

[330] I. Karpas. Two results on union-closed families. *arXiv preprint arXiv:1708.01434*, 2017.

[331] G. Katona, T. Nemetz, and M. Simonovits. Újabb bizonyítás a Turán-féle gráftételre és megjegyzések bizonyos általánosításaira (in Hungarian). *Matematikai Lapok*, 15:228–238, 1964.

[332] G. O. H. Katona. Intersection theorems for systems of finite sets. *Acta Mathematica Hungarica*, 15(3-4):329–337, 1964.

[333] G. O. H. Katona. On a conjecture of Erdős and a stronger form of Sperner's theorem. *Studia Scientarum Mathematicarum Hungarica*, 1:59–63, 1966.

[334] G. O. H. Katona. On separating systems of a finite set. *Journal of Combinatorial Theory*, 1(2):174–194, 1966.

[335] G. O. H. Katona. A theorem of finite sets. In *Theory of Graphs, Proceedings of Colloquium held at Tihany, Hungary*, pages 187–207, Budapest, 1968. Akadémia Kiadó.

[336] G. O. H. Katona. A generalization of some generalizations of Sperner's theorem. *Journal of Combinatorial Theory, Series B*, 12(1):72–81, 1972.

[337] G. O. H. Katona. A simple proof of the Erdős-Chao Ko-Rado theorem. *Journal of Combinatorial Theory, Series B*, 13(2):183–184, 1972.

[338] G. O. H. Katona. A three part Sperner theorem. *Studia Scientiarum Mathematicarum Hungarica*, 8:379–390, 1973.

[339] G. O. H. Katona. Extremal problems for hypergraphs. In *Combinatorics*, pages 215–244. Springer, 1975.

[340] G. O. H. Katona. A simple proof of a theorem of Milner. *Journal of Combinatorial Theory, Series A*, 83(1):138–140, 1998.

[341] G. O. H. Katona. Forbidden intersection patterns in the families of subsets (introducing a method). In *Horizons of Combinatorics*, pages 119–140. Springer, 2008.

[342] G. O. H. Katona. Finding at least one excellent element in two rounds. *Journal of Statistical Planning and Inference*, 141(8):2946–2952, 2011.

[343] G. O. H. Katona, G. Y. Katona, and Z. Katona. Most probably intersecting families of subsets. *Combinatorics, Probability and Computing*, 21(1-2):219–227, 2012.

[344] G. O. H. Katona and T. G. Tarján. Extremal problems with excluded subgraphs in the $n$-cube. In *Graph Theory*, pages 84–93. Springer, 1983.

[345] P. Keevash. The Turán problem for hypergraphs of fixed size. *Electronic Journal of Combinatorics*, 12(1):N11, 2005.

[346] P. Keevash. The Turán problem for projective geometries. *Journal of Combinatorial Theory, Series A*, 111(2):289–309, 2005.

[347] P. Keevash. Shadows and intersections: stability and new proofs. *Advances in Mathematics*, 218(5):1685–1703, 2008.

[348] P. Keevash. Hypergraph Turán problems. *Surveys in Combinatorics*, 392:83–140, 2011.

[349] P. Keevash. The existence of designs. *arXiv preprint arXiv:1401.3665*, 2014.

[350] P. Keevash, J. Lenz, and D. Mubayi. Spectral extremal problems for hypergraphs. *SIAM Journal on Discrete Mathematics*, 28(4):1838–1854, 2014.

[351] P. Keevash and D. Mubayi. Stability theorems for cancellative hypergraphs. *Journal of Combinatorial Theory, Series B*, 92(1):163–175, 2004.

[352] P. Keevash and D. Mubayi. Set systems without a simplex or a cluster. *Combinatorica*, 30(2):175–200, 2010.

[353] P. Keevash and D. Mubayi. The Turán number of $F_{3,3}$. *Combinatorics, Probability and Computing*, 21(3):451–456, 2012.

[354] P. Keevash, D. Mubayi, and R. M. Wilson. Set systems with no singleton intersection. *SIAM Journal on Discrete Mathematics*, 20(4):1031–1041, 2006.

[355] P. Keevash and B. Sudakov. The exact Turán number of the Fano plane. *Combinatorica*, 25(5):561–574, 2005.

[356] P. Keevash and B. Sudakov. On a hypergraph Turán problem of Frankl. *Combinatorica*, 25(6):673–706, 2005.

[357] N. Keller and N. Lifshitz. The junta method for hypergraphs and Chvátal's simplex conjecture. *arXiv preprint arXiv:1707.02643*, 2017.

[358] T. Kővári, V. Sós, and P. Turán. On a problem of K. Zarankiewicz. In *Colloquium Mathematicae*, volume 3, pages 50–57, 1954.

[359] K. H. Kim and F. W. Roush. On a problem of Turán. In *Studies in Pure Mathematics*, pages 423–425. Springer, 1983.

[360] Z. Király, Z. L. Nagy, D. Pálvölgyi, and M. Visontai. On families of weakly cross-intersecting set-pairs. *Fundamenta Informaticae*, 117(1-4):189–198, 2012.

[361] Á. Kisvölcsey. Flattening antichains. *Combinatorica*, 26(1):65–82, 2006.

[362] M. Klazar and A. Marcus. Extensions of the linear bound in the Füredi–Hajnal conjecture. *Advances in Applied Mathematics*, 38(2):258–266, 2007.

[363] D. Kleitman. On Dedekind's problem: the number of monotone Boolean functions. *Proceedings of the American Mathematical Society*, 21(3):677–682, 1969.

[364] D. Kleitman and G. Markowsky. On Dedekind's problem: the number of isotone Boolean functions. II. *Transactions of the American Mathematical Society*, 213:373–390, 1975.

[365] D. J. Kleitman. On a lemma of Littlewood and Offord on the distribution of certain sums. *Mathematische Zeitschrift*, 90(4):251–259, 1965.

[366] D. J. Kleitman. A conjecture of Erdős-Katona on commensurable pairs among subsets of an $n$-set. In *Theory of Graphs, Proceedings of Colloquium held at Tihany, Hungary*, pages 187–207, 1966.

[367] D. J. Kleitman. Extremal properties of collections of subsets containing no two sets and their union. *Journal of Combinatorial Theory, Series A*, 20(3):390–392, 1976.

[368] D. J. Kleitman and K. J. Winston. On the number of graphs without 4-cycles. *Discrete Mathematics*, 41(2):167–172, 1982.

[369] E. Knill. Graph generated union-closed families of sets. *arXiv preprint math/9409215*, 1994.

[370] Y. Kohayakawa and B. Kreuter. The width of random subsets of Boolean lattices. *Journal of Combinatorial Theory, Series A*, 100(2):376–386, 2002.

[371] Y. Kohayakawa, B. Nagle, V. Rödl, and M. Schacht. Weak hypergraph regularity and linear hypergraphs. *Journal of Combinatorial Theory, Series B*, 100(2):151–160, 2010.

[372] J. Komlós, A. Shokoufandeh, M. Simonovits, and E. Szemerédi. The regularity lemma and its applications in graph theory. In *Theoretical Aspects of Computer Science*, pages 84–112. Springer, 2002.

[373] J. Komlós and M. Simonovits. Szemerédi's regularity lemma and its applications in graph theory. Technical report, DIMACS, 1996.

[374] D. König. Gráfok és alkalmazásuk a determinánsok és a halmazok elméletére. *Mathematikai és Természettudományi Értesítő*, 34:104–119, 1916.

[375] J. Körner and B. Sinaimeri. On cancellative set families. *Combinatorics, Probability and Computing*, 16(5):767–773, 2007.

[376] V. Korobkov. On monotone functions of the algebra of logic. *Problemy Kibernetiki*, 13:5–28, 1965.

[377] A. D. Korshunov. The number of monotone Boolean functions (in Russian). *Problemy Kibernetiki*, 38:5–108, 1981.

[378] A. Kostochka, D. Mubayi, and J. Verstraëte. Turán problems and shadows I: Paths and cycles. *Journal of Combinatorial Theory, Series A*, 129:57–79, 2015.

[379] A. Kostochka, D. Mubayi, and J. Verstraëte. Turán problems and shadows III: expansions of graphs. *SIAM Journal on Discrete Mathematics*, 29(2):868–876, 2015.

[380] A. Kostochka, D. Mubayi, and J. Verstraëte. Turán problems and shadows II: trees. *Journal of Combinatorial Theory, Series B*, 122:457–478, 2017.

[381] A. V. Kostochka. A class of constructions for Turán's $(3,4)$-problem. *Combinatorica*, 2(2):187–192, 1982.

[382] A. V. Kostochka. Extremal problems on $\Delta$-systems. In *Numbers, Information and Complexity*, pages 143–150. Springer, 2000.

[383] L. Kozma and S. Moran. Shattering, graph orientations and connectivity. *Electronic Journal of Combinatorics*, 20(3), 2013.

[384] L. Kramer, R. R. Martin, and M. Young. On diamond-free subposets of the Boolean lattice. *Journal of Combinatorial Theory, Series A*, 120(3):545–560, 2013.

[385] J. Kruskal. The optimal number of simplices in a complex. *Mathematical Optimization Techniques*, pages 251–268, 1963.

[386] F. Lazebnik and J. Verstraëte. On hypergraphs of girth five. *Electron. J. Combin.*, 10:Research Paper 25, 15 pp. (electronic), 2003.

[387] P. Lieby. The Separation Problem. Master's thesis, (Honours thesis), Northern Territory University, 1994.

[388] P. Lieby. *Extremal Problems in Finite Sets.* PhD thesis, Northern Territory University, 1999.

[389] P. Lieby. Antichains on three levels. *Electronic Journal of Combinatorics*, 11(1):32, 2004.

[390] J. Liu and X. Liu. Set systems with positive intersection sizes. *Discrete Mathematics*, 340(10):2333–2340, 2017.

[391] L. Lovász. On minimax theorems of combinatorics (in Hungarian). *Matematikai Lapok*, 26:209–264, 1975.

[392] L. Lovász. *Combinatorial Problems and Exercises.* Akadémiai Kiadó - North Holland, 1979.

[393] L. Lu. On crown-free families of subsets. *Journal of Combinatorial Theory, Series A*, 126:216–231, 2014.

[394] L. Lu and K. G. Milans. Set families with forbidden subposets. *Journal of Combinatorial Theory, Series A*, 136:126–142, 2015.

[395] L. Lu and L. Székely. Using Lovász Local Lemma in the space of random injections. *Electronic Journal of Combinatorics*, 14(1):R63, 2007.

[396] L. Lu and Y. Zhao. An exact result for hypergraphs and upper bounds for the Turán density of $K_{r+1}^r$. *SIAM Journal on Discrete Mathematics*, 23(3):1324–1334, 2009.

[397] D. Lubell. A short proof of Sperner's lemma. *Journal of Combinatorial Theory*, 1(2):299, 1966.

[398] T. Łuczak and K. Mieczkowska. On Erdős' extremal problem on matchings in hypergraphs. *Journal of Combinatorial Theory, Series A*, 124:178–194, 2014.

[399] F. J. MacWilliams and N. J. A. Sloane. *The Theory of Error-Correcting codes*. Elsevier, 1977.

[400] L. Maherani and M. Shahsiah. Turán numbers of complete 3-uniform Berge-hypergraphs. *Graphs and Combinatorics*, pages 1–14, 2016.

[401] F. Maire. On the flat antichain conjecture. *Australasian Journal of Combinatorics*, 15:241–246, 1997.

[402] K. N. Majumdar. On some theorems in combinatorics relating to incomplete block designs. *The Annals of Mathematical Statistics*, 24(3):377–389, 1953.

[403] K. Majumder. A new construction of non-extendable intersecting families of sets. *Electronic Journal of Combinatorics*, 23(3):P3–28, 2016.

[404] K. Majumder. On the maximum number of points in a maximal intersecting family of finite sets. *Combinatorica*, 37(1):87–97, 2017.

[405] W. Mantel. Problem 28. *Wiskundige Opgaven*, 10(60-61):320, 1907.

[406] A. Marcus and G. Tardos. Excluded permutation matrices and the Stanley–Wilf conjecture. *Journal of Combinatorial Theory, Series A*, 107(1):153–160, 2004.

[407] P. Marković. An attempt at Frankl's conjecture. *Publications de l'Institut Mathematique*, 81(95):29–43, 2007.

[408] K. Markström. Extremal hypergraphs and bounds for the Turán density of the 4-uniform $K_5$. *Discrete Mathematics*, 309(16):5231–5234, 2009.

[409] K. Markström and J. Talbot. On the density of 2-colorable 3-graphs in which any four points span at most two edges. *Journal of Combinatorial Designs*, 18(2):105–114, 2010.

[410] J. Maßberg. The union-closed sets conjecture for small families. *Graphs and Combinatorics*, 32(5):2047–2051, 2016.

[411] M. Matsumoto and N. Tokushige. The exact bound in the Erdős-Ko-Rado theorem for cross-intersecting families. *Journal of Combinatorial Theory, Series A*, 52(1):90–97, 1989.

[412] M. Matsumoto and N. Tokushige. A generalization of the Katona theorem for cross $t$-intersecting families. *Graphs and Combinatorics*, 5(1):159–171, 1989.

[413] U. Matthias. *Hypergraphen ohne vollständige $r$-partite Teilgraphen.* PhD thesis, Heidelberg, 1994.

[414] A. Méroueh. A LYM inequality for induced posets. *Journal of Combinatorial Theory, Series A*, 155:398–417, 2018.

[415] L. D. Meshalkin. Generalization of Sperner's theorem on the number of subsets of a finite set. *Theory of Probability & Its Applications*, 8(2):203–204, 1963.

[416] A. Mészáros. New bounds for 3-part Sperner families. *Moscow Journal of Combinatorics and Number Theory*, 5(4):255–273, 2015.

[417] T. Mészáros and L. Rónyai. Shattering-extremal set systems of VC dimension at most 2. *Electronic Journal of Combinatorics*, 21(4), 2014.

[418] A. Methuku and D. Pálvölgyi. Forbidden hypermatrices imply general bounds on induced forbidden subposet problems. *Combinatorics, Probability and Computing*, pages 1–10, 2017.

[419] E. Milner. A combinatorial theorem on systems of sets. *Journal of the London Mathematical Society*, 1(1):204–206, 1968.

[420] S. Moran and M. K. Warmuth. *Labeled Compression Schemes for Extremal Classes*, pages 34–49. Springer International Publishing, 2016.

[421] R. Morris. FC-families and improved bounds for Frankl's conjecture. *European Journal of Combinatorics*, 27(2):269–282, 2006.

[422] N. Morrison, J. A. Noel, and A. Scott. On Saturated $k$-Sperner Systems. *Electronic Journal of Combinatorics*, 21(3), 2014.

[423] G. Moshkovitz and A. Shapira. Exact bounds for some hypergraph saturation problems. *Journal of Combinatorial Theory, Series B*, 111:242–248, 2015.

[424] T. S. Motzkin and E. G. Straus. Maxima for graphs and a new proof of a theorem of Turán. *Canadian Journal of Mathematics*, 17(4):533–540, 1965.

[425] D. Mubayi. Some exact results and new asymptotics for hypergraph Turán numbers. *Combinatorics, Probability and Computing*, 11(3):299–309, 2002.

[426] D. Mubayi. On hypergraphs with every four points spanning at most two triples. *Electronic Journal of Combinatorics*, 10(1):N10, 2003.

[427] D. Mubayi. A hypergraph extension of Turán's theorem. *Journal of Combinatorial Theory, Series B*, 96(1):122–134, 2006.

[428] D. Mubayi. Counting substructures II: hypergraphs. *Combinatorica*, 33(5):591–612, 2013.

[429] D. Mubayi and O. Pikhurko. A new generalization of Mantel's theorem to $k$-graphs. *Journal of Combinatorial Theory, Series B*, 97(4):669–678, 2007.

[430] D. Mubayi and V. Rödl. On the Turán number of triple systems. *Journal of Combinatorial Theory, Series A*, 100(1):136–152, 2002.

[431] D. Mubayi and J. Talbot. Extremal problems for $t$-partite and $t$-colorable hypergraphs. *Electronic Journal of Combinatorics*, 15(1):R26, 2008.

[432] D. Mubayi and J. Verstraëte. A hypergraph extension of the bipartite Turán problem. *Journal of Combinatorial Theory, Series A*, 106(2):237–253, 2004.

[433] D. Mubayi and J. Verstraëte. Proof of a conjecture of Erdős on triangles in set-systems. *Combinatorica*, 25(5):599–614, 2005.

[434] D. Mubayi and J. Verstraëte. Minimal paths and cycles in set systems. *European Journal of Combinatorics*, 28(6):1681–1693, 2007.

[435] D. Mubayi and J. Verstraëte. A survey of Turán problems for expansions. In *Recent Trends in Combinatorics*, pages 117–143. Springer, 2016.

[436] D. Mubayi and Y. Zhao. Non-uniform Turán-type problems. *Journal of Combinatorial Theory, Series A*, 111(1):106–110, 2005.

[437] D. Mubayi and Y. Zhao. On the VC-dimension of uniform hypergraphs. *Journal of Algebraic Combinatorics*, 25(1):101–110, 2007.

[438] B. Nagle and V. Rödl. Regularity properties for triple systems. *Random Structures & Algorithms*, 23(3):264–332, 2003.

[439] B. Nagle, V. Rödl, and M. Schacht. The counting lemma for regular $k$-uniform hypergraphs. *Random Structures & Algorithms*, 28(2):113–179, 2006.

[440] D. T. Nagy. Forbidden subposet problems with size restrictions. *Journal of Combinatorial Theory, Series A*, 155:42 – 66, 2018.

[441] Z. L. Nagy and B. Patkós. On the number of maximal intersecting $k$-uniform families and further applications of Tuza's set pair method. *Electronic Journal of Combinatorics*, 22(1):P1–83, 2015.

[442] E. Naslund and W. Sawin. Upper bounds for sunflower-free sets. In *Forum of Mathematics, Sigma*, volume 5. Cambridge University Press, 2017.

[443] M. C. Neto and R. Morris. Maximum-size antichains in random set-systems. *Random Struct. Algorithms*, 2016.

[444] V. Nikiforov. An analytic theory of extremal hypergraph problems. *arXiv preprint arXiv:1305.1073*, 2013.

[445] V. Nikiforov. Analytic methods for uniform hypergraphs. *Linear Algebra and its Applications*, 457:455–535, 2014.

[446] J. A. Noel, A. Scott, and B. Sudakov. Supersaturation in posets and applications involving the container method. *Journal of Combinatorial Theory, Series A*, 154:247–284, 2018.

[447] S. Norin and L. Yepremyan. Turán number of generalized triangles. *Journal of Combinatorial Theory, Series A*, 146:312–343, 2017.

[448] S. Norin and L. Yepremyan. Turán numbers of extensions. *Journal of Combinatorial Theory, Series A*, 155:476–492, 2018.

[449] D. Osthus. Maximum antichains in random subsets of a finite set. *Journal of Combinatorial Theory, Series A*, 90(2):336–346, 2000.

[450] A. Pajor. *Sous-espaces $l_1^n$ des espaces de Banach.* Hermann, 1985.

[451] C. Palmer, M. Tait, C. Timmons, and A. Z. Wagner. Turán numbers for Berge-hypergraphs and related extremal problems. *arXiv preprint arXiv:1706.04249*, 2017.

[452] D. Pálvölgyi. Weak embeddings of posets to the boolean lattice. *Discrete Mathematics & Theoretical Computer Science*, 20(1), 2018.

[453] B. Patkós. $l$-trace $k$-Sperner families of sets. *Journal of Combinatorial Theory, Series A*, 116(5):1047–1055, 2009.

[454] B. Patkós. Traces of uniform families of sets. *Electronic Journal of Combinatorics*, 16:N8, 2009.

[455] B. Patkós. On randomly generated non-trivially intersecting hypergraphs. *Electronic Journal of Combinatorics*, 17(1):R26, 2010.

[456] B. Patkós. A note on traces of set families. *Moscow Journal of Combinatorics and Number Theory*, 2(1):47–55, 2012.

[457] B. Patkós. Induced and non-induced forbidden subposet problems. *Electronic Journal of Combinatorics*, 22(1):P1–30, 2015.

[458] B. Patkós. Supersaturation and stability for forbidden subposet problems. *Journal of Combinatorial Theory, Series A*, 136:220–237, 2015.

[459] A. Pelc. Solution of Ulam's problem on searching with a lie. *Journal of Combinatorial Theory, Series A*, 44(1):129–140, 1987.

[460] A. Pelc. Searching with known error probability. *Theoretical Computer Science*, 63(2):185–202, 1989.

[461] A. Pelc. Searching games with errors - fifty years of coping with liars. *Theoretical Computer Science*, 270(1):71–109, 2002.

[462] Y. Peng, H. Peng, Q. Tang, and C. Zhao. An extension of the Motzkin–Straus theorem to non-uniform hypergraphs and its applications. *Discrete Applied Mathematics*, 200:170–175, 2016.

[463] O. Pikhurko. The minimum size of saturated hypergraphs. *Combinatorics, Probability and Computing*, 8(5):483–492, 1999.

[464] O. Pikhurko. Weakly saturated hypergraphs and exterior algebra. *Combinatorics, Probability and Computing*, 10(5):435–451, 2001.

[465] O. Pikhurko. An exact Turán result for the generalized triangle. *Combinatorica*, 28(2):187–208, 2008.

[466] O. Pikhurko. The minimum size of 3-graphs without a 4-set spanning no or exactly three edges. *European Journal of Combinatorics*, 32(7):1142–1155, 2011.

[467] O. Pikhurko. Exact computation of the hypergraph Turán function for expanded complete 2-graphs. *Journal of Combinatorial Theory, Series B*, 103(2):220–225, 2013.

[468] O. Pikhurko and J. Verstraëte. The maximum size of hypergraphs without generalized 4-cycles. *Journal of Combinatorial Theory, Series A*, 116(3):637–649, 2009.

[469] J. Polcyn. One more Turán number and Ramsey number for the loose 3-uniform path of length three. *Discussiones Mathematicae Graph Theory*, 37(2):443–464, 2017.

[470] J. Polcyn and A. Ruciński. Refined Turán numbers and Ramsey numbers for the loose 3-uniform path of length three. *Discrete Mathematics*, 340(2):107–118, 2017.

[471] L. Pyber. A new generalization of the Erdős-Ko-Rado theorem. *Journal of Combinatorial Theory, Series A*, 43(1):85–90, 1986.

[472] M. Raggi. Forbidden configurations: Finding the number predicted by the Anstee-Sali conjecture is $NP$-hard. *Ars Mathematica Contemporanea*, 10(1):1–8, 2014.

[473] D. K. Ray-Chaudhuri and R. M. Wilson. On $t$-designs. *Osaka Journal of Mathematics*, 12(3):737–744, 1975.

[474] A. A. Razborov. Flag algebras. *The Journal of Symbolic Logic*, 72(4):1239–1282, 2007.

[475] A. A. Razborov. On 3-hypergraphs with forbidden 4-vertex configurations. *SIAM Journal on Discrete Mathematics*, 24(3):946–963, 2010.

[476] C. Reiher. The parity search problem. *Moscow Journal of Combinatorics and Number Theory*, 3(1):78–83, 2013.

[477] D. Reimer. An average set size theorem. *Combinatorics, Probability and Computing*, 12(01):89–93, 2003.

[478] A. Rényi. On a problem of information theory (in hungarian). *MTA Matematikai Kutató Intézetének Közleményei B*, 6:505–516, 1961.

[479] I. Roberts. The flat antichain conjecture for small average set size. Technical report, Northern Territory University, 1999.

[480] I. T. Roberts and J. Simpson. A note on the union-closed sets conjecture. *Australasian Journal of Combinatorics*, 47:265–267, 2010.

[481] V. Rödl. On a packing and covering problem. *European Journal of Combinatorics*, 6(1):69–78, 1985.

[482] V. Rödl, B. Nagle, J. Skokan, M. Schacht, and Y. Kohayakawa. The hypergraph regularity method and its applications. *Proceedings of the National Academy of Sciences of the United States of America*, 102(23):8109–8113, 2005.

[483] V. Rödl and M. Schacht. Regular partitions of hypergraphs: counting lemmas. *Combinatorics, Probability and Computing*, 16(6):887–901, 2007.

[484] V. Rödl and M. Schacht. Regular partitions of hypergraphs: regularity lemmas. *Combinatorics, Probability and Computing*, 16(6):833–885, 2007.

[485] V. Rödl and J. Skokan. Regularity lemma for $k$-uniform hypergraphs. *Random Structures & Algorithms*, 25(1):1–42, 2004.

[486] V. Rödl and J. Skokan. Applications of the regularity lemma for uniform hypergraphs. *Random Structures & Algorithms*, 28(2):180–194, 2006.

[487] A. Ruciński, E. Jackowska, and J. Polcyn. Multicolor Ramsey numbers and restricted Turán numbers for the loose 3-uniform path of length three. *Electronic Journal of Combinatorics*, 24(3):P3–5, 2017.

[488] P. A. Russell. Compressions and probably intersecting families. *Combinatorics, Probability and Computing*, 21(1-2):301–313, 2012.

[489] P. A. Russell and M. Walters. Probably intersecting families are not nested. *Combinatorics, Probability and Computing*, 22(01):146–160, 2013.

[490] M. Ruszinkó. On the upper bound of the size of the $r$-cover-free families. *Journal of Combinatorial Theory, Series A*, 66(2):302–310, 1994.

[491] I. Z. Ruzsa and E. Szemerédi. Triple systems with no six points carrying three triangles. *Combinatorics (Keszthely, 1976), Coll. Math. Soc. J. Bolyai*, 18:939–945, 1978.

[492] A. Sali. Stronger form of an $M$-part Sperner theorem. *European Journal of Combinatorics*, 4(2):179–183, 1983.

[493] A. Sali. A note on convex hulls of more-part Sperner families. *Journal of Combinatorial Theory, Series A*, 49(1):188–190, 1988.

[494] W. Samotij. Subsets of posets minimising the number of chains. *arXiv preprint arXiv:1708.02436*, 2017.

[495] A. A. Sapozhenko. The number of antichains in multilayered ranked sets. *Diskretnaya Matematika*, 1(2):110–128, 1989.

[496] G. N. Sárközy and S. Selkow. An extension of the Ruzsa-Szemerédi theorem. *Combinatorica*, 25(1):77–84, 2004.

[497] G. N. Sárközy and S. Selkow. On a Turán-type hypergraph problem of Brown, Erdős and T. Sós. *Discrete Mathematics*, 297(1):190–195, 2005.

[498] N. Sauer. On the density of families of sets. *Journal of Combinatorial Theory, Series A*, 13(1):145–147, 1972.

[499] D. Saxton and A. Thomason. Hypergraph containers. *Inventiones mathematicae*, 201(3):925–992, 2015.

[500] A. Scott and E. Wilmer. Hypergraphs of bounded disjointness. *SIAM Journal on Discrete Mathematics*, 28(1):372–384, 2014.

[501] A. D. Scott. Another simple proof of a theorem of Milner. *Journal of Combinatorial Theory, Series A*, 87(2):379–380, 1999.

[502] S. Shahriari. On the structure of maximum 2-part Sperner families. *Discrete Mathematics*, 162(1):229–238, 1996.

[503] J. B. Shearer. A new construction for cancellative families of sets. *Electronic Journal of Combinatorics*, 3(1):R15, 1996.

[504] S. Shelah. A combinatorial problem; stability and order for models and theories in infinitary languages. *Pacific Journal of Mathematics*, 41(1):247–261, 1972.

[505] L. Shi. On Turán densities of small triple graphs. *European Journal of Combinatorics*, 52:95–102, 2016.

[506] A. F. Sidorenko. Systems of sets that have the $T$-property. *Moscow University Mathematics Bulletin*, 36:19–22, 1981.

[507] A. F. Sidorenko. The maximal number of edges in a homogeneous hypergraph containing no prohibited subgraphs. *Mathematical Notes*, 41(3):247–259, 1987.

[508] A. F. Sidorenko. Asymptotic solution for a new class of forbidden $r$-graphs. *Combinatorica*, 9(2):207–215, 1989.

[509] A. F. Sidorenko. Extremal combinatorial problems in spaces with continuous measure (in Russian). *Issledovanie Operatsiy i ASU*, 34:34–40, 1989.

[510] A. F. Sidorenko. An analytic approach to extremal problems for graphs and hypergraphs. In *Proceedings of the Conference on Extremal Problems for Finite Sets*, 1991.

[511] A. F. Sidorenko. What we know and what we do not know about Turán numbers. *Graphs and Combinatorics*, 11(2):179–199, 1995.

[512] A. F. Sidorenko. Upper bounds for Turán numbers. *Journal of Combinatorial Theory, Series A*, 77(1):134–147, 1997.

[513] M. Simonovits. A method for solving extremal problems in graph theory, stability problems. In *Theory of Graphs (Proceedings of Colloquium, Tihany, 1966)*, pages 279–319, 1968.

[514] M. Simonovits. On colour-critical graphs. *Studia Scientiarum Mathematicarum Hungarica*, 7(1-2):67–81, 1972.

[515] M. Simonovits. Note on a hypergraph extremal problem. In *Hypergraph Seminar*, pages 147–151. Springer, 1974.

[516] H. S. Snevily. A sharp bound for the number of sets that pairwise intersect at $k$ positive values. *Combinatorica*, 23(3):527–533, 2003.

[517] V. T. Sós. Remarks on the connection of graph theory, finite geometry and block designs. *Colloquio Internazionale sulle Teorie Combinatorie (Roma, 1973)*, 2:223–233, 1976.

[518] V. T. Sós, P. Erdős, and W. Brown. On the existence of triangulated spheres in 3-graphs, and related problems. *Periodica Mathematica Hungarica*, 3(3-4):221–228, 1973.

[519] J. Spencer. Minimal completely separating systems. *Journal of Combinatorial Theory*, 8(4):446–447, 1970.

[520] E. Sperner. Ein Satz über Untermengen einer endlichen Menge. *Math Z*, 27(1):585–592, 1928.

[521] T. Szabó and V. H. Vu. Exact $k$-wise intersection theorems. *Graphs and Combinatorics*, 21(2):247–261, 2005.

[522] E. Szemerédi. On sets of integers containing $k$ elements in arithmetic progression. *Acta Arithmetica*, 27(1):199–245, 1975.

[523] J. Talbot. Chromatic Turán problems and a new upper bound for the Turán density of $K_4^-$. *European Journal of Combinatorics*, 28(8):2125–2142, 2007.

[524] T. S. Tan. Traces without maximal chains. *Electronic Journal of Combinatorics*, 17(1):N16, 2010.

[525] T. Tao. A variant of the hypergraph removal lemma. *Journal of Combinatorial Theory, Series A*, 113(7):1257–1280, 2006.

[526] T. Tao. A correspondence principle between (hyper)graph theory and probability theory, and the (hyper)graph removal lemma. *Journal d'Analyse Mathématique*, 103(1):1–45, 2007.

[527] T. Tao. A symmetric formulation of the Croot-Lev-Pach-Ellenberg-Gijswijt capset bound. *https://terrytao.wordpress.com/2016/05/18/a-symmetric-formulation-of-the-croot-lev-pach-ellenberg-gijswijt-capset-bound/*, 2016.

[528] H. T. Thanh. An extremal problem with excluded subposet in the Boolean lattice. *Order*, 15(1):51–57, 1998.

[529] C. Timmons. On $r$-uniform linear hypergraphs with no Berge-$K_{2,t}$. *Electronic Journal of Combinatorics*, 24:P4.34, 2017.

[530] N. Tokushige. On cross $t$-intersecting families of sets. *Journal of Combinatorial Theory, Series A*, 117(8):1167–1177, 2010.

[531] N. Tokushige. The eigenvalue method for cross $t$-intersecting families. *Journal of Algebraic Combinatorics*, 38(3):653–662, 2013.

[532] L. M. Tolhuizen. New rate pairs in the zero-error capacity region of the binary multiplying channel without feedback. *IEEE Transactions on Information Theory*, 46(3):1043–1046, 2000.

[533] I. Tomon. Decompositions of the boolean lattice into rank-symmetric chains. *Electronic Journal of Combinatorics*, 23(2):P2–53, 2016.

[534] I. Tomon. Almost tiling of the boolean lattice with copies of a poset. *Electronic Journal of Combinatorics*, 25(1):1–38, 2018.

[535] P. Turán. On an extremal problem in graph theory (in Hungarian). *Matematikai s Fizikai Lapok*, 48:0436–452, 1941.

[536] Z. Tuza. Critical hypergraphs and intersecting set-pair systems. *Journal of Combinatorial Theory, Series B*, 39(2):134–145, 1985.

[537] Z. Tuza. A generalization of saturated graphs for finite languages. In *Proceedings of IMYCS 86', MTA SZTAKI Studies*, volume 185, pages 287–293. MTA Számitástechnikai és Automatatizálási KutatóIntézet, Budapest, 1986.

[538] Z. Tuza. Inequalities for two set systems with prescribed intersections. *Graphs and Combinatorics*, 3(1):75–80, 1987.

[539] Z. Tuza. Extremal problems on saturated graphs and hypergraphs. *Ars Combinatorica*, 25:105–113, 1988.

[540] S. Ulam. *Adventures of a Mathematician.* Charles Scribner's Sons, New York, 1976.

[541] V. N. Vapnik and A. Y. Chervonenkis. On uniform convergence of the frequencies of events to their probabilities. *Teoriya Veroyatnostei i ee Primeneniya*, 16(2):264–279, 1971.

[542] J. Wang and H. Zhang. Cross-intersecting families and primitivity of symmetric systems. *Journal of Combinatorial Theory, Series A*, 118(2):455–462, 2011.

[543] X. Wang, H. Wei, and G. Ge. A strengthened inequality of Alon-Babai-Suzuki's conjecture on set systems with restricted intersections modulo $p$. *Discrete Mathematics*, 341(1):109–118, 2018.

[544] M. Ward. Note on the order of the free distributive lattice. *Bulletin of the American Mathematical Society*, 52(5):423–423, 1946.

[545] M. Watanabe. Arrow relations on families of finite sets. *Discrete Mathematics*, 94(1):53–64, 1991.

[546] M. Watanabe. Some best possible bounds concerning the traces of finite sets II. *Graphs and Combinatorics*, 11(3):293–303, 1995.

[547] M. Watanabe and P. Frankl. Some best possible bounds concerning the traces of finite sets. *Graphs and Combinatorics*, 10(2-4):283–292, 1994.

[548] A. B. Watts, S. Norin, and L. Yepremyan. A Turán theorem for extensions via an Erdős-Ko-Rado theorem for Lagrangians. *arXiv preprint arXiv:1707.01533*, 2017.

[549] G. Wiener. Rounds in combinatorial search. *Algorithmica*, 67(3):315–323, 2013.

[550] R. M. Wilson. The exact bound in the Erdős-Ko-Rado theorem. *Combinatorica*, 4(2-3):247–257, 1984.

[551] P. Wójcik. Union-closed families of sets. *Discrete Mathematics*, 199(1-3):173–182, 1999.

[552] B. Wu and Y. Peng. The connection between polynomial optimization, maximum cliques and Turán densities. *Discrete Applied Mathematics*, 225:114–121, 2017.

[553] B. Wu, Y. Peng, and P. Chen. On a conjecture of Hefetz and Keevash on Lagrangians of intersecting hypergraphs and Turán numbers. *arXiv preprint arXiv:1701.06126*, 2017.

[554] K. Yamamoto. Logarithmic order of free distributive lattice. *Journal of the Mathematical Society of Japan*, 6(3-4):343–353, 1954.

[555] L. Yepremyan. *Local Stability Method for Hypergraph Turán problems.* PhD thesis, McGill University Libraries, 2016.

[556] A. A. Zykov. On some properties of linear complexes. *Matematicheskii sbornik*, 66(2):163–188, 1949.

# Index